Topics in
Current Physics

47

Topics in Current Physics   Founded by Helmut K. V. Lotsch

Photoacoustic, Photothermal and Photochemical Processes

at Surfaces and in Thin Films

Edited by P. Hess

With Contributions by
A. C. Boccara   G. Busse   H. Coufal   R. W. Dreyfus
D. Fournier   P. Hess   L. Konstantinov   A. Mandelis
E. Matthias   R. M. Miller   U. Netzelmann
A. Neubrand   J. Pelzl   A. C. Tam   F. Träger

With 234 Figures

Springer-Verlag Berlin Heidelberg New York
London Paris Tokyo Hong Kong

Professor Peter Hess

Physikalisch-Chemisches Institut der Universität, Im Neuenheimer Feld 253,
D-6900 Heidelberg 1, Fed. Rep. of Germany

ISBN-13:978-3-642-83947-4   e-ISBN-13:978-3-642-83945-0
DOI: 10.1007/978-3-642-83945-0

2154/3150-543210 – Printed on acid-free paper

Preface

This volume comprises review articles based on talks presented at the 49th WE-Heraeus seminar, which was held at the Physikzentrum, Bad Honnef, November 21–23, 1988. This seminar was truly international and was attended by leading scientists in the field. It was the generous support of the WE-Heraeus Foundation that made possible this meeting and also this book resulting from the seminar. Therefore, I would like to thank the WE-Heraeus Foundation and their representatives involved in this seminar, Werner Buckel and Dieter Röss, directory; Volker Schäfer, manager; and Albrecht Bischoff, organizer, for their help and assistance.

The reviews published in this second volume cover photoacoustic, photothermal and photochemical processes occurring at surfaces and in thin films. This includes the interaction of high-power laser pulses with surfaces, leading to desorption, ablation and surface damage. Besides pulsed irradiation and analysis in the time domain, the modulation of cw radiation with a single frequency, analysis in the frequency domain, and multifrequency modulation are also treated. The instrumentation used in this field of research is reviewed, with emphasis on mass spectrometric, optical and acoustic detection methods. The topics considered in detail are the mechanisms involved in laser-induced desorption, ablation and surface damage, the analysis of thin films and interfaces using thermal and acoustic waves, depth profiling, nondestructive testing, frequency-modulated time delay domain spectrometry, heat diffusion in random media, surface acoustic waves and magnetic resonance in ferromagnetic films.

As the editor of this volume, I am indebted to Dr. Lotsch of Springer-Verlag, Heidelberg, for his valuable suggestions and assistance during the planning stage. I would also like to express my sincere appreciation to members of the physics editorial department for assistance during the final processing of the manuscripts.

Heidelberg, April 1989   *Peter Hess*

Contents

List of Contributors

A.C. Boccara
Laboratoire d'Optique Physique, ER 5 du CNRS, ESPCI, 10 rue Vauquelin,
F-75231 Paris Cedex 05, France

G. Busse
Institut für Kunststoffprüfung und Kunststoffkunde der Universität Stuttgart,
Pfaffenwaldring 32, D-7000 Stuttgart 80, Fed. Rep. of Germany

H. Coufal
IBM Research Division, Almaden Research Center, 650 Harry Road, San Jose,
CA 95120-6099, USA

R.W. Dreyfus
IBM Thomas J. Watson Research Center, P.O. Box 218, Yorktown Heights,
NY 10598, USA

D. Fournier
Laboratoire d'Optique Physique, ER 5 du CNRS, ESPCI, 10 rue Vauquelin,
F-75231 Paris Cedex 05, France

P. Hess
Physikalisch-Chemisches Institut der Universität Heidelberg,
Im Neuenheimer Feld 253, D-6900 Heidelberg 1, Fed. Rep. of Germany

L. Konstantinov
Institute of Applied Mineralogy, Bulgarian Academy of Sciences,
92 Radovski Blvd., 1000 Sofia, Bulgaria

A. Mandelis
Department of Mechanical Engineering, University of Toronto
5 King's College Road, Toronto, Ontario, M5S 1A4, Canada

E. Matthias
Fachbereich Physik, Freie Universität Berlin, Arminallee 14,
D-1000 Berlin 33, Germany

R.M. Miller
Unilever Research, Port Sunlight Laboratory, Quarry Road East,
Bebington, Wirral, Merseyside, L63 3JW, UK

U. Netzelmann
Institut für Experimentalphysik AG VI, Ruhr-Universität Bochum,
Universitätsstr. 150, D-4630 Bochum 1, Fed. Rep. of Germany

A. Neubrand

Physikalisch-Chemishes Institut der Universität Heidelberg,
Im Neuenheimer Feld 253, D-6900 Heidelberg 1, Fed. Rep. of Germany

J. Pelzl

Institut für Experimentalphysik AG VI, Ruhr-Universität Bochum,
Universitätsstr. 150, D-4630 Bochum 1, Fed. Rep. of Germany

A.C. Tam

IBM Research Division, Almaden Research Center, 650 Harry Road,
San Jose, CA 95120-6099, USA

F. Träger

Physikalisches Institut der Universität Heidelberg, Philosophenweg 12,
D-6900 Heidelberg 1, Fed. Rep. of Germany

1. Introduction

Peter Hess

With 4 Figures

This chapter provides an elementary introduction to the characteristic features of photoacoustic, photothermal and photochemical processes occurring in heterogeneous systems and their detection and analysis. An overview is presented of the recent advances and developments described in the different chapters of the book. This includes the development of theoretical models in the time and frequency domains, and the fundamental mechanisms involved in photoacoustic, photothermal and photochemical processes. The analysis of these phenomena by different experimental techniques and their application in spectroscopy, the field of transport processes, and nondestructive evaluation, etc., is discussed.

1.1 Laser Excitation and Induced Processes

1.1.1 Laser Excitation

Lasers are finding a wide variety of applications in the investigation of photoacoustic, photothermal and photochemical processes at interfaces. With their specific qualities, the ability to deliver light of extremely high power, extremely high spectral purity and/or extremely short duration, they are the most important radiation sources in the field today.

One of the most important properties is that of high spectral purity. The laser wavelength determines the nature of the excitation process. In Chap. 2. electronic excitation in adsorbate systems is considered by *Träger*, including the excitation of surface plasma resonances in small metal particles adsorbed on a substrate. The resonant excitation of molecular multilayer systems from the IR to the UV spectral region is treated by *Hess* in Chap. 3. Vibrational excitation in the IR and electronic excitation in the VIS and UV regions are the processes investigated in the majority of the studies.

Another peculiarity of laser radiation is the high power density and photon flux which can be achieved with these light sources. This property allows the realization of extremely high heating rates at surfaces in the 10^{10} K/s region for nanosecond laser pulses and in the 10^{15} K/s range for fs pulses. This is of great interest in the study of the dynamics of surface processes such as desorption and ablation, as discussed in Chaps. 2 and 3. In addition, nonlinear processes such as multiphoton excitation are accessible in the high power range. In Chap. 4, *Matthias* and *Dreyfus* present a survey on laser–surface interactions in the MW/cm^2 to GW/cm^2 region considering metallic and wide band gap ionic materials. The processes discussed in more detail

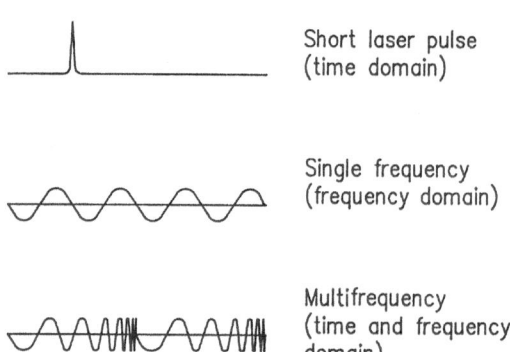

Short laser pulse
(time domain)

Single frequency
(frequency domain)

Multifrequency
(time and frequency
domain)

Fig. 1.1. Temporal behavior of laser radiation for time domain, frequency domain and combined time and frequency domain analysis

range from individual particle emission to ablation and surface damage originating from single and multiphoton absorption.

The high directionality and beam quality of many laser sources provides the basis not only for localized excitation and heating, but also for localized optical detection of density gradients, geometrical changes of shapes, etc. This will be discussed in more detail in the section considering experimental techniques.

The temporal properties of the laser radiation employed determine the principles of detection and analysis. Short laser pulses are used for detection in the time domain as shown in Fig. 1.1. Even microsecond to nanosecond laser pulses enable time-resolved detection of many processes such as transport phenomena. In the near future, ultrashort laser pulses will become increasingly available commercially and this will have an important impact on future developments. Of course, there is a trade off which relates the duration of light pulses to their spectral purity by the uncertainty principle. Thus, a picosecond pulse covers a frequency range of $5\,cm^{-1}$ and a femtosecond pulse, $5 \times 10^3\,cm^{-1}$. This spread over a larger frequency range, however, does not create problems in many photothermal and photoacoustic experiments. As shown schematically in Fig. 1.1, cw laser radidation is also employed for photothermal and photoacoustic analysis. The cw radiation is either modulated with a single frequency, to perform analysis in the frequency domain, or the modulation frequency is varied as indicated in Fig. 1.1. The latter technique is relatively new and possesses features intermediate between the frequency and time domains. The nature of this method, including instrumentation, recent developments and applications is discussed by *Mandelis* in Chap. 8.

1.1.2 Laser-Induced Processes

With laser radiation a variety of processes can be induced at surfaces and in thin films. Figure 1.2 gives an overview, illustrating schematically a large number of effects that may be stimulated by laser radiation. The absorption of one or several photons can lead to direct quantum effects, e.g., the dissociation of the absorbing molecule and the desorption of the fragments. It has been proved that direct photochemical decomposition of an adsorbed molecule by a laser photon is possible and that multiphoton photoemission of electrons occurs at metal surfaces, see Chaps. 2–4.

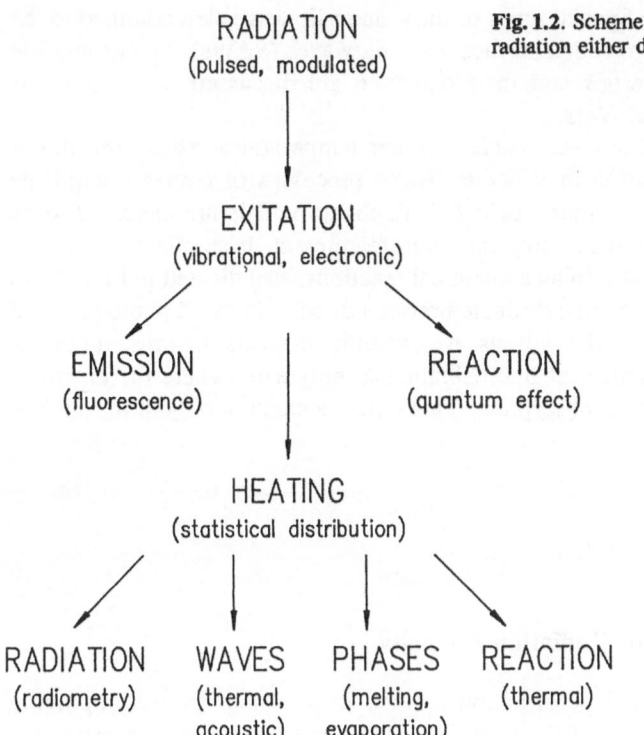

RADIATION
(pulsed, modulated)

Fig. 1.2. Scheme of the processes induced by laser radiation either directly or indirectly by heating

EXITATION
(vibrational, electronic)

EMISSION
(fluorescence)

REACTION
(quantum effect)

HEATING
(statistical distribution)

RADIATION
(radiometry)

WAVES
(thermal,
acoustic)

PHASES
(melting,
evaporation)

REACTION
(thermal)

Recent experimental results indicate a limited efficiency of these quantum processes. Direct photodecomposition and high-order multiphoton photoemission seem to be confined to the surface region. This is also true for the direct emission of photons following the excitation process and it is therefore difficult to observe IR and UV fluorescence from the surface.

The energy exchange between the excited state and other states and degrees of freedom often occurs on the picosecond time scale in condensed phases. Thus, laser irradiation normally leads to transient heating in the irradiated zone. This transient and localized heating process causes a series of effects as indicated in Fig. 1.2.

Transient heating results in variations of the infrared thermal radiation emitted from the irradiated sample region. This effect is used in photothermal radiometry for remote sensing and nondestructive testing, as discussed in Chap. 6 by *Tam* and in Chap. 10 by *Busse*. Another collective effect connected with transient or modulated heating is the creation of thermal waves. The special properties of these thermal waves and their application in thin film analysis, depth profiling and nondestructive evaluation are reviewed in Chap. 5 by *Coufal*, in Chap. 7 by *Miller* and in Chap. 9 by *Busse*, respectively. The disadvantage of thermal waves is their efficient attenuation. The detector has to probe the temperature within one thermal diffusion length of the excited area. Therefore, it is often advantageous to detect the acoustic waves generated by the transient temperature profile. The latter causes thermal expansion and a corresponding stress profile leading to the generation of sound waves. These sound waves are essentially unattenuated and thus propagate over long distances, as discussed in Chaps. 5 and 6. Special types of these acoustic waves are confined to the surface and they are therefore called surface acoustic waves (SAWs). The generation of these waves, their properties and their detection are discussed in Chap. 10 by *Konstantinov, Neubrand* and *Hess*.

If the transient heating process reaches higher temperatures, phase transitions such as melting or evaporation may occur. These processes of course complicate the analysis and are therefore often neglected. Realistic models are expected to be developed in the near future describing the main features of these effects.

High temperatures may also induce chemical reactions, as indicated in Fig. 1.2. In some cases it can be difficult to discriminate between directly induced photochemical processes and thermally induced reactions. An example of this is the present controversy concerning the mechanism of photoablation in polymers, where the chemical network is destroyed by photon absorption. This effect occurs not only at the surface but also in the bulk.

1.2 Detection Schemes

1.2.1 Temporal Variation of Radiation Intensity

The temporal behavior of the perturbation applied to a system by the impinging radiation determines the nature of the resulting effects and the appropriate detection scheme. Pulse methods use an impulse perturbation and the response of the system is monitored in the time domain (Fig. 1.1). In the quantitative description, the input waveform is modified by the impulse response function, giving the output waveform by a convolution integral. The physical basis of these pulsed methods is the fact that a narrow pulse contains many frequencies probing the sample simultaneously. The high peak powers of pulsed laser sources can be used to produce intense thermal transient, and time-gated detection allows effective discrimination against spurious signals. On the other hand, pulse measurements possess a low duty cycle of excitation.

Frequency domain measurements are realized by applying a single modulation frequency to a continuous laser beam (Fig. 1.1). The effect of the system consists in a modification of the amplitude and phase. This is described quantitatively by the system transfer function. Multiplication of the input spectrum by this transfer function directly yields the output spectrum. Thus, analysis is simpler in the frequency domain. The much more efficient duty cycle of this method permits a reduction of the amount of energy absorbed by the sample per unit time, and thus, its application to fragile materials. A more detailed discussion is given by *Miller* in Chap. 7.

As mentioned previously, multifrequency modulation can be performed to realize a technique intermediate between the frequency and time domains. This technique uses a variable modulation frequency sweeping linearly over the same range in a short frequency chirp, as shown in Fig. 1.1. An important advantage of this method is its ability to yield information in both domains simultaneously. The theory of this so-called frequency-modulated spectrometry is presented by *Mandelis* in Chap. 8. He also discusses the instrumentation and first applications of this technique.

1.2.2 Detection Methods

An advantage of the pulse methods is the high optical power density that can be achieved. This enables the experimenter to study nonlinear processes such as multiphoton excitation. The intense thermal transients may also induce phase changes, desorption and ablation of material, etc. Figure 1.3a shows schematically the emission of different particles, including electrons, ions, molecules and clusters, following irradiation with a narrow high power pulse. One of the most important methods of detecting these processes is time-of-flight mass spectrometry, which presents information on the chemical nature, yield and kinetic energy of emitted species. This technique also elucidates the mechanism involved, e.g., the photochemical or photothermal features. Examples are discussed in Chaps. 2–4. A number of detection methods utilize a second low-power probe laser beam to observe photothermal effects induced by the excitation laser. The probe laser beam either travels parallel to the surface and monitors the refractive index gradient caused by a temperature gradient or acoustic wave front, or the probe beam hits the surface and is deflected by the buckling surface as indicated in Fig. 1.3b. Examples of these pump–probe detection schemes are presented in several of the chapters.

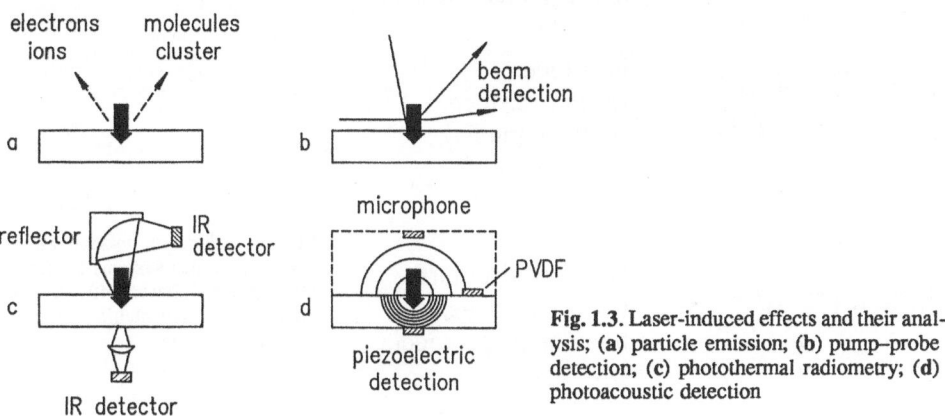

Fig. 1.3. Laser-induced effects and their analysis; (a) particle emission; (b) pump–probe detection; (c) photothermal radiometry; (d) photoacoustic detection

Another optical detection method is that of photothermal radiometry based on blackbody radiation emitted from the heated spot. As shown in Fig. 1.3c, the IR detector may be positioned at the side of the excitation source or on the other side of a thin sample. A scanning technique suitable for nondestructive evaluation of larger samples is described in Chap. 9 by *Busse*. He also gives several examples of remote inspection of materials.

Figure 1.3d presents three different configurations for the detection of acoustic waves with a pressure sensor. If the irradiated sample is in contact with a gas atmosphere, a cheap electret microphone can be used for detection. An interesting example of this detection scheme is given by *Pelzl* and *Netzelmann* in Chap. 12. Surface acoustic waves can be detected with a piezopolymer (PVDF) in contact with the irradiated surface, as discussed in Chap. 10. In a similar way, bulk sound waves can be monitored with piezoelectric transducers in contact with any surface of the sample.

1.3 Interface Systems

1.3.1 Homogeneous Phases with Ideal Boundaries

A simple model for treating photothermal and photoacoustic phenomena in interface systems is to assume two semi-infinite homogeneous phases, e.g., a gas and a solid, separated by an ideal surface as shown in Fig. 1.4. This means that spatially homogeneous optical and thermal properties are assumed for each phase. To describe heat diffusion, for example, the classical Fourier equation is used in the isotropic media and interfaces are taken into account by appropriate boundary conditions.

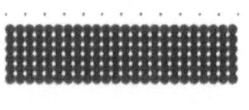

Interface
(ideal surface,
homogeneous media)

Interface
(rough surface,
random media)

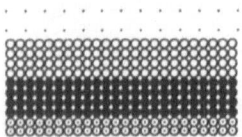

Layered structure
(ideal surfaces,
homogeneous media)

Fig. 1.4. Schematic presentation of several interface systems including homogeneous phases with an ideal interface, random media and a multilayer system

Energy deposition is governed by the optical properties and the energy absorbed by the sample contributes to signal formation. Only the heat generated within one thermal diffusion length of the sample will be able to reach the surface and influence the signal detected there. The thermal diffusion length, however, is a function not only of the material properties but also of the modulation frequency, and it decreases with increasing modulation frequency. Thus, a smaller depth is probed in this case. This is the basis of the depth profiling capability of this method. Several aspects of theory and spectroscopic depth profiling with thermal waves are discussed by *Miller* in Chap. 7.

1.3.2 Random Media

The photothermal or photoacoustic signal depends on the energy absorbed by the sample, as mentioned previously. Therefore, materials with low optical quality such as rough samples or opaque substrates can be investigated. Disordered structures such as amorphous or sintered materials, as shown schematically in Fig. 1.4, play an increasing role in materials science. Thus, there is considerable interest in describing the properties of such random media. New concepts in the theoretical treatment of diffusion in random media are introduced by *Fournier* and *Boccara* in Chap. 11. They apply the concept of fractal geometry to the description of rough surfaces and to their thermal behavior when acting as a heat source. To analyze heat diffusion in random structures, the dimensionality of the diffusion process is introduced and is discussed in the framework of percolation theory. It is shown that geometrical parameters such as the fractal dimensions of the surface and network can be obtained from simple photothermal experiments. Experimental data is obtained by observation of the average surface temperature evolution at an opaque sample surface with an IR detector, after irradiation with short laser pulses.

1.3.3 Films and Layered Structures

An important application of the photothermal and photoacoustic techniques is in the analysis of thin films. The observation of thermal and acoustic wave phenomena allows the determination of thermophysical film properties, while optical properties may be obtained by varying the laser wavelength. This topic is treated in detail by *Coufal* in Chap. 5. The influence of the thermal diffusion length on the signal generation process makes depth profiling and the inspection of subsurface features possible, but limits the photothermal analysis to a thin layer near the sample surface. These techniques are thus especially suited to thin film analysis.

In many cases, more complicated layered materials with two or more layers have to be analyzed, as indicated schematically in Fig. 1.4. One well-studied example is the characterization of ferromagnetic films and layers, as described in detail by *Pelzl* and *Netzelmann* in Chap. 12. The great interest in these films is due to the different magnetic properties of the films compared to the bulk material. To optimize specific properties, for example, sandwich layers with alternating magnetic properties can be produced. Several thermal wave detection techniques were applied to these ferromagnetic films for materials characterization, nondestructive evaluation of their magnetic properties and the investigation of the fundamental problems of magnetism. Chapter 12 deals, in particular, with the lateral and depth resolution of these techniques.

Systems with complicated layered or distributed structures are often found in biological systems. Here, depth profiling by a nondestructive method is of great importance. Examples are given by *Miller* in Chap. 7.

Improved characterization of surfaces and interfaces is of special interest, especially with regard to buried interfaces, because the information available on these transition regions is rather limited. The application of photothermal and photoacoustic methods to this problem is considered by *Tam* in Chap. 6.

1.4 Applications

1.4.1 Spectroscopy

Energy deposition in the sample is determined by the wavelength dependence of the optical absorption coefficient. Due to radiationless deactivation, part of the absorbed energy is released as heat and creates the signal. As previously stated, only the heat released within a diffusion length will be detected at the surface. Thus, thermal and also acoustic properties of the sample will affect the observed spectra and complicate their analysis. In most cases, a quantitative interpretation of the complicated signal generation and detection processes will not be possible. Therefore, the main advantage of these methods is not the quantitative signal analysis, yielding, for example, optical absorption coefficients, but the investigation of materials and layered structures with low optical quality, where conventional transmission spectroscopy cannot be applied, see Chap. 5. This is especially true for biological samples, where a depth profiling of the chromophore distribution may only be achievable using photothermal techniques, see Chap. 7. In several chapters, a variety of examples are given of the determination of specific spectroscopic information by means of photothermal or photoacoustic analysis.

1.4.2 Distribution of Energy

If the absorbed photon initiates a quantum process, e.g. quantum desorption, a simple quantitative analysis is possible using the law of energy conservation. Measurement of the kinetic energy of the desorbed particles using a time-of-flight technique, for example, allows the determination of the binding energy in the case of quantum desorption.

Efficient energy exchange normally leads to a thermal distribution of at least part of the energy in the irradiated region and a quantitative analysis is much more difficult. Normally, it is not possible to measure the transient temperature generated by light absorption. As a consequence, access not only to spectroscopic but also to thermophysical sample properties is limited. The situation is even more complicated if a certain part of the absorbed photon energy induces chemical reactions, while the other part contributes to the heating effect, due to effective collisional deactivation. In principle, the amount of energy channeled into quantum processes and into energy dissipation can be determined from this kind of experiment.

1.4.3 Transport Processes

The dissipation of heat from the irradiation region can be described quantitatively and yields accurate values for the thermal diffusivity. Examples are given by *Mandelis* in Chap. 8, where thermal diffusivities were determined for several materials employing the multifrequency modulation technique.

However, heat diffusion can also be studied in disordered and highly inhomogeneous materials. These measurements yield the experimental data needed for the development of new models for the theoretical description of inhomogeneous materials. The new ansatz in this direction, presented by *Fournier* and *Boccara* in Chap. 11, is based on fractal geometry. To compare the concept with the experiment, the geometrical aspects of heat diffusion through random structures were selected. The advantage of the heat diffusion process in this respect is the fact that this transport process can be studied with a relatively simple setup in a nondestructive and noncontact experiment.

1.4.4 Nondestructive Evaluation

One of the most important and well-developed applications is that of non-destructive evaluation (NDE). This method is based on the fact that the interaction of thermal waves with faults and changes in morphology lead to signal changes, which can be used as qualitative criteria for inhomogeneities in the material. In this case, depth profiling is not only performed at a single spot, but a scan technique is employed, where either the optical beam is moved across the sample or the position of the sample is changed. The application of this method to metals, semiconductors, ceramics and polymers, including coatings, is discussed in Chap. 9 by *Busse*.

1.5 Discussion of the Literature

Since the advent of laser radiation sources, the number of papers published in the field of photoacoustic and photothermal phenomena at surfaces and in thin films has increased drastically. There are already several books available on the subject in the form of review articles written by experts or as monographs [1.1–5]. Since the Third International Topical Meeting on Photoacoustic and Photothermal Phenomena in 1983, the invited and contributed talks have been published [1.6–8]. These proceedings volumes give an up-to-date view of the current activities and progress made since the previous conference.

References

1.1 Yoh-Han Pao (ed.): *Optoacoustic Spectroscopy and Detection* (Academic, New York 1977)
1.2 A. Rosencwaig: *Photoacoustics and Photoacoustic Spectroscopy* (Wiley, New York 1980)
1.3 D.S. Kliger (ed.): *Ultrasensitive Laser Spectroscopy* (Academic, New York 1983)
1.4 E. Lüscher, R. Korpiun, H. Coufal, R. Tilgner (eds.): *Photoacoustic Effect: Principles and Applications* (Vieweg, Braunschweig 1984)
1.5 V.P. Zharov, V.S. Letokhov: *Laser Optoacoustic Spectroscopy*, Springer Ser. Opt. Sci., Vol. 37 (Springer, Berlin, Heidelberg 1986)
1.6 J. Badoz, D. Fournier (eds.): *Photoacoustic and Photothermal Spectroscopy*, J. de Phys., Colloq. C6 (Les Editions de Physique, Les Ulis 1983)
1.7 L. Bertrand, P. Cielo, R. Leblanc, J.P. Monchalin, B. Mongeau (eds.): Proc. 4th Int. Topical Meeting of Photoacoustics, Thermal and Related Sciences, Can. J. Phys. **64**, 1023–1344 (1985)
1.8 P. Hess, J. Pelzl (eds.): *Photoacoustic and Photothermal Phenomena*, Springer Ser. Opt. Sci., Vol. 58 (Springer, Berlin, Heidelberg 1988)

2. Desorption Stimulated by Electronic Excitation with Laser Light

F. Träger

With 20 Figures

This chapter gives an overview of laser desorption promoted by electronic excitation. The primary motivation for such experiments is to investigate the underlying reaction mechanisms and elucidate the involved kinetics. Besides this scientific interest, a deeper understanding of laser-induced desorption is of considerable importance for applied areas like material ablation, etching or laser-induced damage of optically transparent materials. Examples of electronic excitation of the substrate or an adsorbate on the surface are described and the mechanisms outlined. Special emphasis is placed on experiments in which a collective electron oscillation precedes the desorption. This novel effect is not only interesting for the understanding of photodesorption phenomena in general, but also offers important applications for the generation of metal particles on surfaces with a very uniform size distribution. Future developments of the field include investigations of more complex systems, a more complete characterization of the desorption products, and the use of subpicosecond laser pulses.

2.1 Stimulated Desorption - An Overview

Why and how does a chemical bond between atoms or molecules break? This question can be asked from different points of view: If molecules are considered, the bond breaking is called dissociation or fragmentation; if an adsorbate comes off from a surface, the process is called desorption; and if a bulk material decomposes, the phenomenon is referred to as ablation or evaporation. This chapter addresses one of these topics, the field of stimulated desorption, and particularly focuses on *laser-induced* processes where an *electronic* excitation precedes the rupture of the surface chemical bond. Desorption is defined here as a reaction in which the rate of detached atoms, molecules or radicals is relatively small and the surface does not undergo macroscopic damage. In the following, a general overview of stimulated desorption will be given in order to place the electronically induced processes in the proper context.

Desorption can be stimulated by a variety of methods. Thermal heating [2.1,2], bombardment with electrons or photons [2.3-23], with ions [2.24,25], even strong transient electric perturbations [2.26,27] or acoustic waves [2.28-32] can promote desorption. If atoms or molecules are detached from a surface by electromagnetic radiation, very different photon energies can be involved, ranging from well below 1 eV for infrared light to hundreds of eV for synchrotron radiation. Depending on the magnitude of the energy, vibrational or electronic excitation is possible and it turns out that both can precede desorption. The field of electronically stimulated desorption is commonly referred to as DIET (Desorption Induced by Electronic Transitions) [2.5,6,14,16,17,21,23]. DIET is possible by either photon or electron bombardment. Desorption stimulated with ions has also been studied extensively in the low energy regime (sputtering) and for highly energetic particles. In the latter category, very heavy ions such as ^{252}Cf with energies of up to several MeV have been used. Such experiments are known as plasma desorption mass spectrometry (PDMS) [2.25].

A significant difference of desorption phenomena stimulated by electromagnetic radiation as compared to the classical method of sputtering by particle bombardment is that the photons carry essentially no momentum. Consequently, the kinetic energy of the desorption products ejected by photon absorption cannot be due to direct momentum transfer but must come from a different process involving some kind of repulsive interaction. This raises the question of how electronic excitation energy is converted into atomic motion. Of course, the necessary momentum for desorption can be derived from the lattice of the bulk. This is conceivable since the electronic relaxation rates are usually fast, making the coupling to the phonon bath of the substrate very efficient, so that the primary channel for electronic deexcitation is heating. Nevertheless, many cases have been established where thermal effects are absent or play a minor role, and desorption occurs as a *direct* result of electronic excitation. Such processes have been discussed extensively in the literature and a general picture for the desorption was proposed by *Menzel, Gomer* [2.33] and *Redhead* [2.34] (MGR model). It is

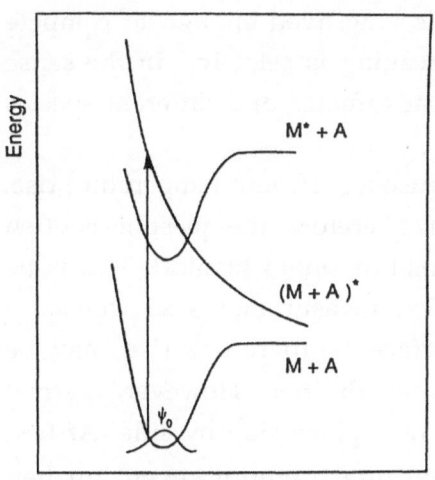

Energy

M* + A

(M + A)*

M + A

ψ_0

Distance from the surface R

Fig. 2.1. Electronic energy levels of an adsorbate-substrate system. Desorption can take place if a repulsive state is populated. The rupture of the bond has to compete with transitions from the excited state to the ground state by which the excitation energy is transferred to the substrate

also of relevance for the laser-induced desorption processes discussed in detail in Sect. 2.2.2 and can be explained by considering the energy level diagram depicted in Fig. 2.1. Similar to the well-known description of dissociation in molecular physics, the MGR model starts with a sudden Franck-Condon transition to a repulsive potential energy surface. As a result, the adsorbate acquires kinetic energy by moving away from the surface on the antibonding excited state potential curve. Bond breaking occurs beyond a critical distance from the surface where the particle has gained sufficient energy to escape. As a competing process, quenching of the excitation by energy transfer to the substrate can take place and the adsorbate is recaptured. Therefore, desorption reflects a delicate interplay of adsorbate exitation causing an increasing distance of the particle from the surface and conversion of electronic energy into substrate excitation. The MGR picture does not specify the exact nature of the electronic excitation, nor does it assume details of the quenching process. In fact, the electronic excitation can be very different, not only with regard to the populated state but also, in the case of excitation with electromagnetic radiation, to the required photon energy and the number of absorbed photons. Single-photon or multiphoton transitions can precede the desorption and single-particle or collective excitations can be involved. In practice, a detailed interpretation of desorption studies is often hampered by incomplete knowledge of the nature of the electronic transition. In any case, it seems that localization of the electronic

13

excitation is important in a state that is long-lived enough to compete with quenching. Such "direct" bond breaking is selective, in the sense that the excited particles come off but adsorbates of a different species remain on the surface.

If quenching and heat generation cause a sufficient temperature rise, desorption occurs as a thermal reaction. Therefore, the question is often raised whether a given process is thermal or photochemical. This issue can be difficult to resolve and detailed measurements as well as a determination or estimate of the surface temperature rise may be necessary in order to distinguish between the two. However, thermal and photochemical processes can also take place side by side. At first glance this seems to complicate the interpretation even further. However, the observation of thermal, in addition to nonthermal desorption can help considerably to identify photochemical reactions unambiguously.

The variety of possible excitation channels and bond breaking mechanisms is paralleled by the large number of different desorption products, such as neutrals on the one hand and ions on the other, being detected as atoms, molecules, fragments or clusters. Even desorption of large organic molecules without fragmentation has been observed under high-energy, heavy ion bombardment. However, if the desorption rates are large, the products are often not easily identified, since additional effects, such as plasma formation in front of the surface and collisions, severely hamper the detection of the species desorbed originally and make reliable measurement of the kinetic energy difficult.

The motivation for experiments on stimulated desorption is twofold: First, it is essential to investigate the reaction mechanisms and to elucidate the involved kinetics. Besides this scientific interest, a deeper understanding of stimulated desorption is of considerable relevance to several applied areas. In order to illustrate the great variety of different experiments and the technical importance of the field, a few examples will be given in the following. Others can be found in the literature.

- Desorption of gases by synchrotron radiation [2.35] and X-rays [2.36] from the walls of vacuum chambers was investigated in order to ensure that ultrahigh vacuum conditions can be maintained in accelerators and storage rings such as LEP (Large Electron Positron storage ring) at CERN so that the desired beam lifetimes of hours are achieved.

- Laser-induced desorption, sometimes denoted as "photon sputtering" (for reviews and summaries see e.g. [2.3,4,7-13,18-20]) is of great practical significance, e.g., for etching, ablation [2.37] or chemical vapor deposition and related developments for the preparation of circuitry in microelectronics fabrication. Laser desorption and ablation of highly refractive materials like tungsten, molybdenum or carbon has also been used in combination with adiabatic expansions to generate metal clusters of these materials [2.38].

- The study of damage mechanisms involving desorption of optically transparent materials is of major importance for the construction and application of high intensity lasers. Laser-induced desorption of such materials as BaF_2 [2.39-41] has been attributed to the excitation of surface states by multiphoton processes. Also, the ablation of metals can be understood by multiphoton absorption as the initial step [2.42] (see also Chap. 4, this volume).

These examples illustrate that stimulated desorption is a large field with many different facets. Desorption can occur wherever photons or particles are incident on a surface. In the following, laser-induced desorption and the underlying mechanisms will be treated. The available energy of the laser photons usually does not exceed about 6 eV. Therefore, valence excitation, ionization or fragmentation are possible, but processes where inner shell excitation would be involved are out of reach.

2.2 Desorption Induced by Laser Light

2.2.1 General Considerations

Figure 2.2 shows the principle of a laser desorption experiment. To ensure well-defined conditions, measurements are often performed under ultrahigh vacuum conditions. The sample is usually attached to a manipulator. It can be heated for cleaning and annealing. The substrate can also be cooled to cryogenic temperatures to condense adsorbates on the surface. The adsorbate-substrate system is illuminated with laser light. Desorbing particles are detected with a quadrupole or time-of-flight mass spectrometer.

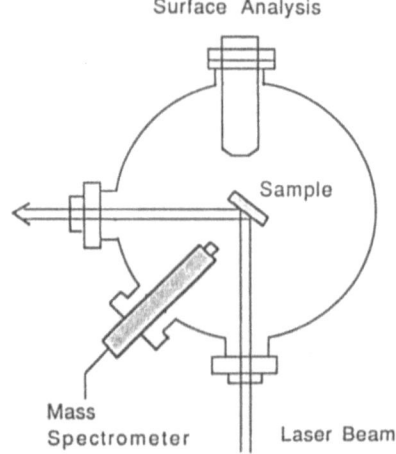

Surface Analysis

Sample

Mass Spectrometer

Laser Beam

Fig. 2.2. Schematic picture of a laser desorption experiment. The adsorbate/substrate combination is illuminated with laser light and the desorption products are detected with a quadrupole or time-of-flight mass spectrometer

Laser-induced desorption can take place by different mechanisms. Often it is simply promoted by thermal heating, see e.g. [2.20,43-58]. Light is absorbed at the substrate surface, the surface temperature rises and the particles desorb. This is quite similar to the classical method of temperature programmed desorption (TPD), which is a standard tool in surface science for the investigation of binding energies and reaction dynamics. In TPD, the heating rate is typically restricted to a few kelvin per second. In laser-induced desorption, however, the rates can be many orders of magnitude larger and amount to as much as 10^{10} K/s. An important characteristic of (laser-induced) thermal desorption is that a threshold is observed. This means that a minimum temperature rise of the surface and therefore a minimum laser intensity is required to initiate desorption. Even for the seemingly simple case of laser-induced thermal desorption, problems are encountered, and the question whether the kinetics are identical for low and high heating rates still seems to be controversial.

In addition to thermal bond breaking after absorption of light at the surface, desorption can be stimulated by vibrational excitation of the adsorbate molecules with infrared laser light [2.4,12,59-62]. One possible mechanism involves transfer of energy from the intramolecular potential into the surface potential so that the molecule comes off. To rupture the bond, absorption of many infrared photons is necessary to compensate for the binding energy. The dominating channel for energy dissipation, however, is quenching of the vibrational energy by coupling to the phonon bath of the substrate. The surface temperature rises, causing desorption. This means that the energy is absorbed by a *resonant* process initially but the desorption is *thermal*, a phenomenon denoted as "resonant heating". It is responsible for the lack of molecular or isotopic selectivity in such experiments. Resonant desorption is also observed if the adsorbate or the adsorbate-substrate system is electronically excited. This will be discussed in detail in the next section.

The following points deserve special attention to ensure that laser-induced desorption measurements do not suffer from artifacts. First, the laser beam which is reflected from the sample surface should not illuminate the inner walls of the vacuum system. If the light interacts with surfaces other than the sample it can stimulate additional desorption signals which are hardly interpretable. Therefore, the laser beam should leave the apparatus through a window. Second, the desorption rate should be kept low to prevent collisions of the desorbed particles with each other in front of the surface. Otherwise, the time-of-flight distributions are perturbed and the original translational energies are not determined correctly. At high light intensities the analysis of such experiments is further complicated by the formation of a plasma in front of the surface. Ionization of the products by the light field, collisions and dissociation can make the identification of the species originally desorbed from the surface virtually impossible. Third, a correction for the ion drift time in the quadrupole filter has to be applied, see [2.63]. This is of special importance in experiments where the signal-to-noise ratio necessitates a small distance between the sample and the ionizer. Finally, the laser beam should have a well-characterized profile that does not contain "hot spots" for example. The profile should preferably be Gaussian or have a homogeneous intensity distribution.

2.2.2 Desorption Stimulated by Laser-Induced Electronic Excitation

Laser-induced photodissociation and photodesorption associated with electronic excitation is a new field and the number of experiments is small compared to that of processes stimulated by electron bombardment for example. Still, different types of excitation have been observed and different mechanisms established. Electronic excitation may either be accomplished in the adsorbate, in the substrate or in the adsorbate-substrate complex. The majority of experiments have been performed with high-intensity pulsed lasers. Desorption of ions as well as neutrals has been observed. In the following, examples of the different types of experiments will be given and the mechanisms described. Further details can also be found in other review articles on the same topic [2.4,8,13,18,23]. Even though the nature of the electronic excitation has not always been identified, and a classification of experimental work is therefore difficult, the following overview has been subdivided into experiments where an electronic excitation in the adsorbate or in the substrate was induced with laser light.

(a) Adsorbate Excitation

Polanyi and co-workers have investigated the photodissociation and photodesorption of CH_3Br at submonolayer coverages from a LiF(100) single crystal surface with pulsed ultraviolet laser light [2.64]. In subsequent experiments, H_2S and other molecules were also studied [2.65,66]. The dielectric substrate was prepared by cleaving in air, mounting in a UHV chamber and subsequent heat treatment. Auger spectroscopy indicated that the fresh surface was clean. The crystal was held at a temperature of 115 K and was dosed by means of a stainless steel capillary. The CH_3Br coverage Θ was on the order of $\Theta \sim 0.01$ monolayer. An excimer laser was used to illuminate the surface at an angle of 85° to the normal. It was operated at a wavelength of either $\lambda = 222$ nm or $\lambda = 308$ nm, powers ranged from 1 to 6 mJ/pulse, focused to an area of ~ 0.3 cm^2 at the substrate surface. This corresponds to an intensity of 0.3 - 2.0 MW/cm^2. The molecules and fragments desorbed from the surface were detected at an angle of 5° to the surface normal. For this purpose they were ionized, mass-separated in a quadrupole filter and counted by an ion multiplier. During the ionization with 100 eV electrons, the CH_3Br molecules dissociate, resulting in a cracking pattern with >90% CH_3^+ ions. These radicals were used to monitor the desorption of CH_3Br in time-of-flight measurements. The spectra show a fast and a slow component, see Fig. 2.3. The yield of the slow component is approximately a factor of 100 higher than that of the fast one. The observation of two classes of particles in the time-of-flight spectrum implies that two different processes occur simultaneously if the surface is illuminated with laser light, i.e., photodissociation and photodesorption. The fast products are attributed to photodissociation of CH_3Br:

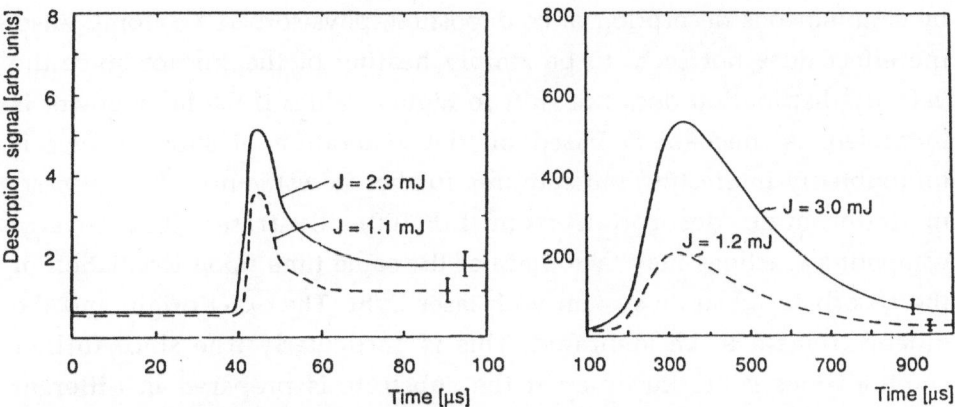

Fig. 2.3. Time-of-flight spectra for CH_3^+. The left hand side displays the "fast" component, which is attributed to photolysis of CH_3Br on LiF(100), the right hand side shows the "slow" component, i.e. the distribution measured for photodesorption of CH_3Br with subsequent cracking in the ionizer of the mass spectrometer. (After [2.64])

$$CH_3Br(ad) \rightarrow CH_3(g) + Br(ad/g)$$

resulting in the release of CH_3. The abbreviation "ad" stands for adsorbed and "g" for gas phase species. The slow component in the time-of-flight spectrum is assigned to the desorption of CH_3Br with subsequent cracking in the ionizer of the mass spectrometer. Photodissociation is a consequence of the absorption of light with $\lambda = 222$ nm, a wavelength close to the peak of the CH_3Br gas phase absorption spectrum. This interpretation is confirmed by irradiation with light of $\lambda = 308$ nm, which is not absorbed by the molecules and where no detectable flux of high-energy CH_3 radicals can be observed. Obviously the process transfers the released energy to the methyl rather than to the substrate. Therefore, quenching of the electronically excited state by the surface is minimized and dissociation after population of a repulsive electronic state takes place *directly* in less than 1 ps, i.e. on a time scale shorter than one vibrational period. In contrast, photodesorption of intact CH_3Br, resulting in the slower CH_3 component of the time-of-flight spectrum, seems to originate from absorption of light in the LiF substrate crystal. The mechanism of this desorption process is not entirely clear. It is speculated that the sudden deposition of energy, presumably by absorption of light in color centers of the LiF, causes an acoustic shock wave which kicks off the methylbromide molecules. This is corroborated by simultaneous desorption of co-deposited, physisorbed Xe atoms. Also, the effect does not seem to be simply heating of the surface since the velocity distribution does not shift to higher values if the laser power is increased. A mechanism based on the generation of shock waves is undoubtedly interesting but requires further investigation. In any case, methylbromide desorption from LiF(100) illustrates that several competing reactions may take place at the same time upon irradiation of the absorbate-substrate system with laser light. The exact origin and the kinetics remain to be identified. This is particularly true since further complications come into play if the substrate is prepared in different ways. It has been demonstrated [2.65], for example, that the resulting methyl translational energy distribution for a well-annealed LiF(100) surface is much narrower than for an unannealed crystal. Also, the translational energy is substantially higher for a well-annealed surface and peaks at $E_T = 1.7 \pm 0.1$ eV with a width of 0.54 eV. In addition, measurements of the angular distribution of the CH_3 photofragments give different results for annealed and unannealed substrate surfaces. Furthermore, temperature programmed thermal desorption studies carried out by *Tabares* et al. [2.67] provide evidence that the CH_3Br molecules do not wet the LiF substrate but form clusters. This is in agreement with experiments on small metal particles on LiF. For such systems, fractional-order thermal desorption is observed [2.68].

Laser desorption of CH_3Br from LiF(100) was also investigated by *Tabares* et al. [2.67]. Their results seem to disagree with those outlined above. However, since the conditions of the experiment were rather different, the results and the conclusions cannot be readily compared. *Tabares* and associates invoke a model based on collisions within the surface layer to explain the occurrence of slow fragments. It is argued that impulsive collisions of fragments could directly eject adsorbed CH_3Br molecules. It is concluded that a small fraction of molecules desorbs via a thermal process.

Other experiments under well-characterized conditions, i.e. in ultrahigh vacuum, were reported by *Chuang* and coworkers [2.69-73]. They have studied, for example, the photodissociation and desorption of CH_2I_2 from sapphire crystals and silver films. The adsorbate/substrate system was characterized by X-ray photoelectron spectroscopy, Auger spectroscopy and thermal desorption. The CH_2I_2 molecules were pumped by light of a XeCl laser at $\lambda = 308$ nm into the first excited and antibonding state. The molecules dissociate quickly into CH_2I radicals and iodine atoms with a high quantum yield. The desorption products were detected by a quadrupole mass spectrometer operating in time-of-flight mode. The photodissociation and photodesorption were investigated as a function of the laser fluence and the surface coverage. The authors point out the following important experimental observations:

- At surface coverages of one monolayer or less the dominant desorbed neutral species are CH_2I and I. No molecular CH_2I_2 or I_2 or any ionic species are detected in a large range of laser fluences. This is in contrast to CH_3Br on LiF (see above) where the majority of the desorbed particles are the parent molecule CH_3Br. Also, the mass distribution obtained under laser irradiation is substantially different from that produced in a thermal desorption experiment, where CH_2I_2 is the main species. It is detected as CH_2I^+ and I^+ after cracking of the CH_2I_2 in the ionizer of the mass spectrometer.

- For a given laser fluence, the desorption yield increases almost linearly with coverage up to about $\Theta = 0.8$ (Fig. 2.4). Similarly, the yield increases linearly with fluence for a fixed coverage. No threshold for photodissociation or photodesorption is found.

- The kinetic energy distribution of the desorbed CH_2I is independent of the coverage and of the laser fluence up to at least 500 mJ/cm^2.

- The quantum yield for photodissociation lies between 0.01 and 0.1, a value substantially lower than the corresponding yield for the molecule in the gas phase.

Of particular interest in these studies is the demonstration of characteristic electronic excitation effects. Since desorption of the parent molecule is not observed, one can anticipate molecular selectivity in photodesorption associated with fragmentation. Indeed, an experiment in which NH_3 was coadsorbed in a 1:1 mixture with CH_2I_2 on Al_2O_3 at 90 K showed only desorption of CH_2I and I. Desorption of ammonia, which is not electronically excited with laser light of $\lambda = 308$ nm, was not detected. In contrast to CH_3Br on LiF, these results also suggest that thermal or shock wave contributions play a negligible role. Another important feature of the experiments carried out by *Chuang* and *Domen* is the observation of explosive desorption effects at high surface coverage

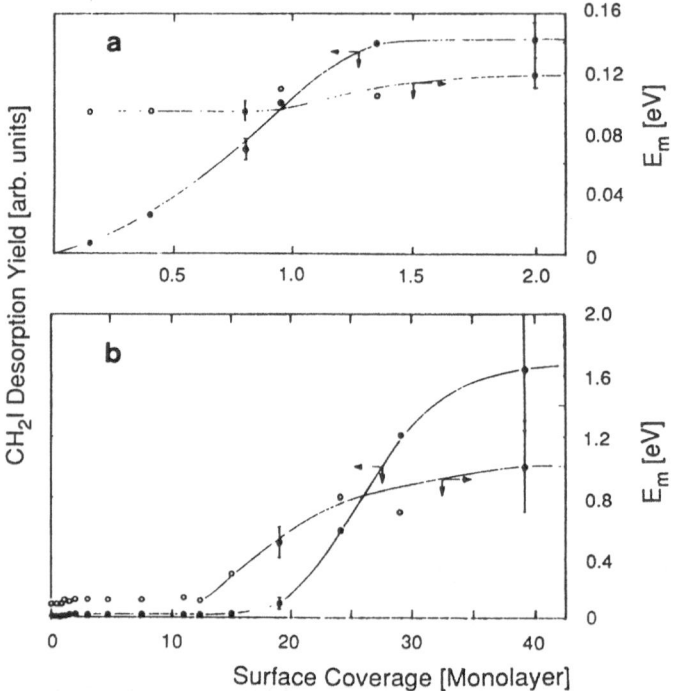

Fig. 2.4. Desorption yield and kinetic energies of CH_2I as a function of **(a)** low and **(b)** high surface coverage for CH_2I_2 on Al_2O_3 at 90 K. The laser fluence was 350 mJ/cm^2. (After [2.69])

($\Theta \geq 15$). Explosive desorption manifests itself in a very high desorption yield, that can be two or three orders of magnitude larger than for $\Theta = 1$, and in a dramatic rise of the kinetic energy E_m of the products (Fig. 2.4). It increases by more than a factor of ten and can exceed 1 eV. In this regime, desorption of CH_2I_2 is also detected and the molecular selectivity becomes less pronounced. It is pointed out that the phenomenon results from the deposition of energy in a layer, the large thickness of which prevents dissociation of excited molecules and causes deexcitation by converting the electronic excitation into translational energy of the involved molecules. Even if the molecules dissociate, they can quickly recombine because of the many other surrounding particles, and release the absorbed photon energy as kinetic energy. This causes local heating leading to thermal expansion and explosion. Such a process seems to be quite similar to photoablation of organic polymers with UV laser light [2.37]. The explosion is only observed above a certain threshold value for the laser fluence that depends on the coverage.

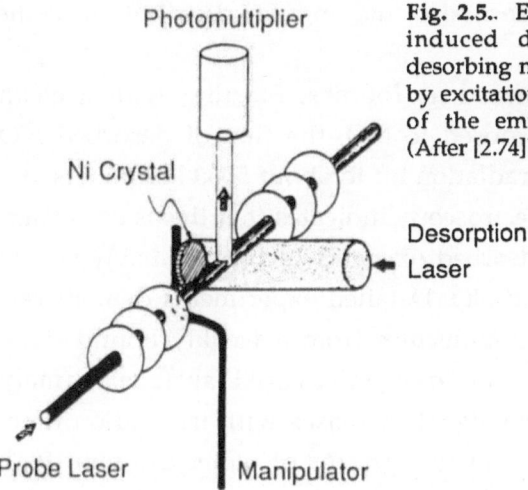

Photomultiplier

Ni Crystal

Desorption
Laser

Probe Laser   Manipulator

Fig. 2.5.. Experimental arrangement for laser-induced desorption studies in which the desorbing molecules are detected state selectively by excitation with a tunable laser and observation of the emitted resonance fluorescence light. (After [2.74])

The experiments described above have in common that the main source of information on the desorption products and kinetics of the reaction is mass spectrometry combined with time-of-flight measurements. Recently, *Weide* et al. [2.74] and *Budde* et al. [2.75] have also determined the internal state population of the detached molecules to characterize the desorption process more completely. The experimental arrangement is shown schematically in Fig. 2.5. Nitric oxide, NO, adsorbed on an initially clean Ni(100) surface at 140 K was irradiated with UV excimer laser light at $\lambda = 193$ nm. In place of mass spectrometry, the desorbed molecules were detected by laser induced fluorescence. For this purpose, frequency doubled light from a tunable excimer laser pumped dye laser with a wavelength of around $\lambda = 226$ nm passed parallel to the surface at a distance of about 25 mm and excited the desorbed NO molecules electronically. The emitted fluorescence was monitored by a photomultiplier. The technique has the advantage that the light is in resonance only with the desorbed particles, so that the detection is molecularly selective. In addition, the excitation with a narrow band laser permits the selective population of certain vibrational-rotational energy levels and the measurement of occupation numbers for each particular quantum state of the NO molecules. The experiments thus provide a fully state-resolved analysis of molecules photodesorbed from a surface. The translational energy was derived from time-of-flight measurements performed by variation of the delay between the desorption and probe laser pulses. Also, the angular distribution of the NO molecules was measured.

The results can be summarized as follows. Starting with a clean Ni(100) surface continuously exposed to NO, the flux of desorbed NO increases with the duration of irradiation up to about 5000 laser shots at a power of $\cong 2 \, mJ/cm^2$. Auger spectroscopy indicates that this is correlated with an increasing amount of adsorbed oxygen which presumably results from the dissociation of adsorbed NO. Detailed experiments demonstrate that the laser light only desorbs molecules from a weakly bound state, which is identified as NO on an oxygen-covered surface, namely Ni(100)-O. This explains why the signal increases with irradiation time. The time-of-flight spectra show that two groups of molecules with high and low translational energies desorb from the surface. The angular distribution of the "slow" molecules was found to be cosine. This and other results suggest that they originate from thermal desorption caused by light absorption in the substrate. The "fast" NO molecules, on the other hand, exhibit an angular distribution that peaks in the direction of the surface normal. The kinetic energy of the molecules is systematically larger for the $^2\Pi_{3/2}$ configuration and increases with internal energy for the $^2\Pi_{1/2}$ as well as for the $^2\Pi_{3/2}$ branch up to values of almost 3000 K for $E_{trans}/2k_B$. Thermal desorption, on the other hand, would already take place at a temperature of 240 K. Also, the surface temperature rise upon laser illumination is relatively small. For these reasons, the "fast" molecules are attributed to a nonthermal desorption process. The authors point out that a remarkable result of their study is the observation of a pronounced spin-orbit selectivity at low rotational quantum numbers for the molecules desorbed via the nonthermal exit channel, see Fig. 2.6. This phenomenon is tentatively interpreted on the basis of an elementary scattering event, in which desorption can be considered as a "half collision", so that a qualitative comparison with molecular scattering processes seems to be justified. Theoretical treatments for scattering of open shell 2Π molecules from flat surfaces indicate that two sets of interaction potentials have to be taken into account and quantum mechanical interference finally causes a trend of the relative populations like the one observed experimentally (Fig. 2.6). Similar experiments on NO desorbing from a Pt surface [2.76-78] and NO desorbing from Ag(111) [2.79] have also been reported. In the latter

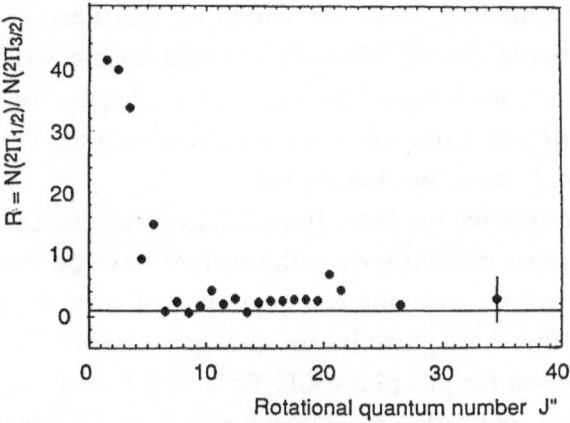

Fig. 2.6. Ratio of the populations of the $^2\Pi_{1/2}$ and $^2\Pi_{3/2}$ states as a function of the rotational quantum number J" for NO photodesorbed from Ni(100)-O. (After [2.75])

experiment the internal energy distribution of the desorbed NO molecules was determined by resonance enhanced multiphoton ionization. Again, two maxima in the time-of-flight spectra are found and attributed to thermal and nonthermal desorption channels.

Whereas the velocity distributions of the desorbed species in the experiments described above were either thermal or had a fast non-thermal component, other measurements have revealed the presence of very slow, "subthermal" desorption products. For example, *Higashi* [2.80] has investigated excimer laser induced desorption at λ = 193 nm of trimethylaluminum from Al_2O_3, SiO_2 and SiO/Si and found trans-lational energies of only 0.025 eV. Maxwell-Boltzmann fits to the velocity distribution show that this corresponds to a temperature of ~ 150 K. The surface temperature, on the other hand, was 300 K. The photon energy at λ = 193 nm being 6.4 eV, direct excitation of the Al-C bond would result in about 3.5 eV of excess energy. Direct electronic coupling of the energy into the substrate is thought to be negligible since the band gap of the solid is larger than 6.4 eV. It therefore appears that almost complete vibrational accommodation of the excess energy occurs *prior* to methyl desorption. This, however, requires a rather long-lived excited state in which the methyl can reside on the surface for a number of vibrational periods. Desorption with subthermal velocity distributions has been predicted theoretically [2.81] and has been previously observed experimentally [2.82-84]. It is argued [2.80,81] that the phenomenon occurs for desorption from shallow potential wells for which no exit barriers exist. The molecules or fragments desorb "without communicating with higher energy states", which causes a nonstatistical energy distribution. This idea is substantially different from other mechanisms (see above) where the electronic transition is directly dissociative.

In connection with experiments on laser-induced desorption, the following question is raised: what influence does the surface have on the cross section for light absorption and photodissociation of adsorbed molecules as compared to the free particles in the gas phase? For instance, the electronic absorption for gas phase CH_3Br is very broad and it is not surprising that CH_3Br molecules weakly adsorbed on LiF(100) also absorb light of a similar wavelength (see above, [2.64,65]). On the other hand, the presence of the surface can quench the electronic excitation, making the cross section for photon stimulated processes smaller than that for the gas phase molecules. The influence of different substrates has been demonstrated, for example, for the decomposition of CH_2I_2 and subsequent desorption of the products [2.69,72] from sapphire and Ag films. It was found that desorption from sapphire can be readily detected whereas desorption from the silver surface at submonolayer coverage is not observed at all, due to rapid electronic relaxation. The interpretation of such experiments with molecules on metal surfaces, however, can be complicated by simultaneous thermal desorption, since the metal substrate absorbs light rather strongly as compared to a transparent sapphire crystal for example. In another experiment with CH_3Br on a brominated Ni(111) surface [2.85,86], it was also found that the photodynamics is strongly perturbed. Nonetheless, complete quenching does not take place and photolysis of CH_3Br is observed. For a semiconducting substrate the situation seems to be similar to that for insulating materials like sapphire. *Creighton* has investigated the photochemistry of $Mo(CO)_6$ adsorbed on Si(100) [2.87] and concludes that energy relaxation does not compete strongly with the photodecomposition reaction. The quantum yield for decomposition is near unity and is not significantly modified by the presence of the semiconductor Si(100) surface. The cross section σ_r for photodecomposition of $Mo(CO)_6$ on Si(100) is surprisingly large [2.87]. The result of $\sigma_r = 5 \pm 3 \times 10^{-17}$ cm^2 falls in the range of values determined for the absorption cross section of gas phase $Mo(CO)_6$. In general, an influence of the surface on the cross sections for light absorption and photodecomposition is reflected in most experiments. As expected, coupling to substrate excitation is very pronounced for metal surfaces. However, even in this case there is not necessarily a complete quenching of the electronic excitation on a time scale in which photodesorption can occur. This is demonstrated by desorption of metal atoms from the surface of small metal particles via a nonthermal process (see below) [2.88]. Another example reported recently [2.89] is the UV-induced photochemical decomposition of $Fe(CO)_5$ and the desorption of CO from a Ag(110) surface. Obviously, the photodecomposition of $Fe(CO)_5$ is so fast that it is not quenched even by direct contact with the silver surface.

In other experiments on laser-induced desorption stimulated by electronic excitation of adsorbates *Creighton* (see above) [2.87] has studied the photodecomposition and photodesorption of $Mo(CO)_6$ adsorbed on Si(100). *Bartosch* et al. have investigated the same molecule on a Si(111) surface [2.90]. They particularly address the problem of how one can distinguish unambiguously between thermal and photoelectronic processes and present measurements of the surface temperature along with a model calculation for the temperature rise during laser irradiation. In a subsequent paper, the same authors have also studied metal carbonyls such as $W(CO)_6$ and $Fe(CO)_5$ on Si(111)7×7 [2.91]. Another interesting example of laser-induced desorption is the work by *Pflaum* et al. [2.92-94]. They have demonstrated that multiphoton stimulated desorption or fragmentation of sodium halide clusters can serve as a tool for structural investigations of these particles and for an interpretation of the "magic numbers" observed in the ionic mass spectra. It is suggested that direct electronic dissociation takes place without an intermediate state of vibrationally hot clusters. Desorption stimulated by electronic excitation has also been investigated in connection with laser etching experiments [2.95,96].

(b) Substrate Excitation

This type of experiment involves excitation of insulating, semiconducting and metallic substrates to stimulate desorption. Included in this section are studies where adsorbate molecules are desorbed from a substrate as well as experiments in which molecules are detached from each other in thick layers, i.e., from molecular crystals rather than from a substrate surface. Early work of the latter kind was carried out for instance by *Antonov* et al. [2.97], who have investigated desorption of polyatomic molecular ions such as adenine and anthracene induced by UV laser radiation. It was found that ions with kinetic energies of up to 3 eV were ejected and that the velocity distribution was nonthermal. Details of the underlying mechanism could not be derived. However, the importance of electronically excited states to stimulate the desorption was emphasized. In a similar experiment [2.98], rhodamine 6G molecules were desorbed with picosecond laser pulses of $\lambda = 532$ nm. The surprising result was that molecular ions of the dye can be detached without significant fragmentation. This is quite similar to the photodetachment, photodissociation and photochemistry work carried out by *Nishi* and collaborators [2.99]. They have condensed NH_3 and H_2O as well as mixtures of the two species on a quartz plate and illuminated the resulting ice film, which was up to 200 µm thick, by light from an excimer laser operating at wavelengths of $\lambda = 193$ or 248 nm. Neutral particles released from the solid surface were ionized by electron impact, the ions were extracted by ion optics, mass selected in a quadrupole filter and detected with a channeltron. Time-of-flight spectra for different molecular ions were recorded in order to clarify the detachment mechanisms involving both one- and two-photon excitation processes. The results of this work can be summarized as follows: In most cases, the translational energy of the ejected molecules does not depend on the laser fluence. The energy is of the order of the intermolecular interaction of the constituent surrounding molecules and ranges from 0.03 to about 0.2 eV. Furthermore, photodesorption is less effective if the ices are annealed, i.e., the surface roughness affects the efficiency significantly but does not alter the translational energy of the desorbed molecules. The measured rate of desorbed parent molecules did not change appreciably even after as many as 10^6 laser pulses. If the laser illumination was interrupted for several minutes, certain large photoproduct molecules of a mixture of NH_3 and H_2O such as N_2H_4 or NH_2OH showed relatively strong signals, indicating a rearrangement in the solid film during the dark period. Finally, the ejection of water molecules from pure H_2O ice was detected upon irradiation with laser light of $\lambda = 248$ nm, exhibiting a quadratic laser energy dependence. Clearly, these results eliminate the possibility of photodesorption by heating with the UV laser light. This is especially supported by the independence of the translational energy of the products from the laser fluence. The authors also rule out any influence of charged particles or plasma effects that might be induced by the laser field. It is suggested that H_2O molecules are detached as a result of the population of an electronically excited state, very likely by an exciton transition that has an energy of about 8.5 eV and that is attributed to the $1b_1 \rightarrow 3s$ Rydberg transition. Similarly, solid ammonia shows a strong absorption for ArF laser light with a sharp spike at $\lambda = 194$ nm, which is attributed to a Wannier exciton state of the crystal. The exciton propagates to the surface, finally approaching an edge molecule weakly bound to the crystal. A surface exciton in a Wannier state is thought to be energetically very stable. In addition, the energy transfer from the bulk to the surface is estimated to occur in 10^{-13} to 10^{-14} s, which is much shorter than the photodissociation time of free NH_3 molecules. Therefore, photodesorption should be able to compete with other channels of energy relaxation. The force acting on a surface molecule and causing desorption is ascribed to "electronic exchange repulsion between an excited molecule and a ground state molecule". The authors also present a model for molecular desorption from solids composed of different small molecules and discuss a mechanism for ion detachment in accordance with the work of *Antonov* et al. [2.97]. It is pointed out that edge molecules play a special role since they have particularly low coordination numbers and lower binding energies than other surface molecules. They desorb preferably and neighboring particles become the new edge molecules. It is expected that the total photodecomposition rate is proportional to the number of edge molecules. Consequently, rough surfaces should exhibit efficient detachment. As mentioned above, this is in agreement with the experimental observations.

In semiconductors, electron-hole pairs can be generated by absorption of visible or UV laser light. These electron-hole pairs are rather long-lived and can migrate over relatively large distances. Therefore, electron-hole pairs, even if created below the surface, have a chance to interact with the adsorbed molecules, causing rupture of the chemical bond. Such experiments involving photogenerated carriers were among the first observed light-induced desorption processes (see e.g. [2.100-102] and references therein). In the early studies, high-pressure mercury lamps or Xe arc lamps were used. This illustrates that photogenerated carriers can induce desorption so efficiently that high-intensity lasers may be unnecessary. An essential difference of these processes associated with electron-hole pairs as compared to direct electronic excitation of the adsorbate (see above) is that the molecules can desorb *without* fragmentation.

Recent experiments along these lines were performed by *Ying* and *Ho* [2.103]. They have studied the photoreactions of NO on Si(111)7×7 at 90 K over a wavelength range from the visible to the near ultraviolet. This permits direct observation of the effects of band-gap transitions in Si and of the role of photogenerated carriers. The measurements were performed under UHV conditions. The Si(111) samples were exposed to NO and the adsorbate/substrate system was characterized by HREELS (high-resolution electron-energy-loss spectroscopy) and temperature programmed thermal desorption. The HREEL spectra show that NO adsorbs both molecularly and dissociatively on Si(111)7×7. The sample was illuminated with 514 nm radiation of an Ar^+ laser or frequency doubled light of $\lambda = 257$ nm. Alternatively, a Xe arc lamp with appropriate glass filters was used. As a result of irradiation with light, significant changes of the vibrational spectra can occur. They indicate desorption of molecular NO. This is confirmed by measurements with a mass spectrometer. The light-induced desorption effect is nonthermal since the signal rises almost instantaneously if the laser is incident on the sample, whereas the time constant for the surface temperature rise is much larger. The wavelength dependence of the desorption signal was studied with the Xe arc lamp over the range 325 - 440 nm. The result is shown in Fig. 2.7 [2.103]. The desorption signal is enhanced if the photon energy is above the first band-gap transition of silicon at 3.37 eV, which corresponds to a wavelength of $\lambda \cong 370$ nm. If the wavelength decreases further, the desorption rate rises, follows the dependence of the absorption coefficient of Si in a certain wavelength interval and reaches a saturation value. Since the lowest electronic excitation energy and the dissociation energy of NO in the gas phase are above 5 eV but desorption can be stimulated with photon energies as low as 2.4 eV, it is concluded that the observed effect is not due to direct photoexcitation of the adsorbed NO. Instead, a mechanism based on electron-hole pairs is held to be responsible for the desorption. As a first step, carriers are generated by absorption of laser light. Part of the electron-hole pairs migrate to the

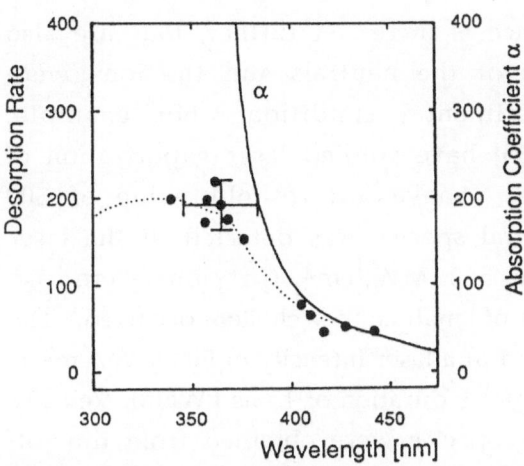

Fig. 2.7. Rate of NO photodesorption from Si(111)7×7 at 90 K versus wavelength of Xe-lamp radiation. The dotted line is a model fit (for details see text), the solid line is the absorption coefficient of Si in units of 10^3 cm^{-1}. (After [2.103])

surface and interact with the NO molecules. As a consequence, the bonds break and the molecules desorb. Since an energy barrier is present, reactions are induced only by hot carriers and not by those that have undergone appreciable energy loss through collisions with phonons. Clearly, only carriers created within a distance from the surface on the order of the mean free path can react with the NO. With this model, the observed wavelength dependence of the desorption rate can be explained (see dotted line in Fig. 2.7). Also, the power dependence and the time profile of the signals can be understood by extending an earlier model calculation [2.104] to the conditions of NO desorption from Si(111). It turns out, however, that the mechanism described here does not necessarily pertain to other molecules. A similar experiment in which ethanol on Si(111)7×7 was irradiated with Ar+ ion laser light of λ = 514 and 257 nm or with broad spectral radiation in the range ~250 to ~900 nm showed no photolytic reactions but only thermal desorption due to irradiative heating of the sample [2.105]. Other experiments on laser sputtering of semiconducting materials have been described in [2.8,13,102,106].

In addition to these experiments, desorption studies on clean metal surfaces have been performed. For a given substrate, a threshold for desorption of neutrals is found. This indicates that the laser-induced electronic excitation is followed by substrate heating and thermal desorption. If the photon fluence is increased further, ions are also detected. The kinetic energies of the neutrals and the ions were measured for different experimental conditions. For example, *Viswanathan* and *Hussla* [2.52,53] have studied laser vaporization of polycrystalline copper surfaces. Above a threshold value of 300 MW/cm^2, desorption of neutral species was detected. If the laser intensity was increased above 500 MW/cm^2, Cu$^+$ ions were also observed. In addition, desorption of small copper clusters occurred. The ratio Cu : Cu$_2$: Cu$_3$ was 100 : 12 : 1 at a laser intensity of 500 MW/cm^2, a wavelength of λ = 248 nm and a pulse duration of 15 ns FWHM. Velocity distributions of the evaporating species were obtained from time-of-flight spectra. They were fitted to Maxwell-Boltzmann distributions for both neutral and ionized species. It turns out that the translational energies are very high: they amount to 2.3 eV for the neutrals and 7 eV for the ions. The authors derive temperatures from the Maxwell-Boltzmann fits and quote values as large as 27 000 K and 81 000 K for the neutrals and ions, respectively. Even if it is assumed that the desorption process is explosive (see above) or that superheating of the substrate takes place, it is hard to believe that a thermal mechanism (which would justify the Maxwell-Boltzmann fits) can accelerate particles to such extraordinary velocities. Thus, the mechanism of the process and the origin of the fast particles remained unclear. One possible explanation may be the formation of a plasma in front of the surface. Its presence can strongly modify the velocities of the neutrals and the ions so that the time-of-flight spectra and the translational energies do not reflect the sample temperature during the rupture of the surface bond. Recent work [2.107,108], carried out with a combination of laser-induced fluorescence and Langmuir probe measurements following photoablation of Cu with wavelengths of λ = 193, 248 and 351 nm, gives new insight into the atom and ion behavior. It is emphasized that plasma potentials are present even at submonolayer removal rates so that the ions are accelerated. In addition, it is proposed that the high ion energies can be transferred to the neutral atoms by resonant charge exchange collisions of the type Cu$^+$ + Cu$^0 \rightarrow$ Cu^0 + Cu$^+$ where Cu$^+$ and Cu^0 denote fast ionic and neutral species, respectively.

The translational energy of ions photoablated from crystalline silver has recently been investigated by *Helvajian* and *Welle* [2.109,110]. In single laser shot experiments, they measured the nascent velocity distributions of Ag^+ and Ag_2^+ under ultrahigh vacuum conditions at fluences near the threshold for ion production. The results reveal that the ions are ejected with constant energy. The kinetic energy distribution is centered at 9 ± 1 eV and has a width of 3 eV. The mean energy is independent of the ion mass. Also, there is no dependence on the laser wavelength. There is, however, a wavelength dependence of the relative population ratio of the monomer and dimer ions. Obviously, the velocity distribution cannot be explained by a thermal process. In earlier experiments, the high kinetic energy of laser-desorbed ions was ascribed to excitation involving inverse Bremsstrahlung absorption in a plasma with a high density of electrons and ions above the surface. The experiments of *Helvajian* and *Welle*, however, cannot be explained along these lines since energetic ions are already observed for threshold ablation where plasma effects are thought to be negligible. Several other mechanisms are ruled out as well. An explanation consistent with the experimental results is that the ions are generated on, or very near, the surface and are accelerated by a transient light-induced electric field or via a collective charge excitation in the substrate. Possible mechanisms for this are discussed but none is found to be fully consistent with the data.

Similar to other experiments described above, the desorption probability for metals is strongly affected by the roughness of the surface under study. Therefore, significant differences of the threshold for detection of neutrals and ions are observed if a single crystal or a poly-crystalline metal sample of the same material is bombarded with laser light [2.111]. The energy "window" in which desorption of neutrals occurs exclusively and no ions are detected varies with surface morphology and depends on the substrate material. Multiphoton absorption as the initial step has been shown to be associated with metal ablation [2.42]. Further work is certainly needed for a detailed under-standing of laser-induced desorption of neutrals and ions from metal surfaces. Other work on the evaporation of solid materials can be found in [2.47] and [2.48].

2.3 Laser-Induced Desorption Stimulated by Surface Plasmon Excitation

2.3.1 Method and Experiment

In our recent experiments [2.88,112], a novel desorption mechanism has been discovered. Figure 2.8 shows the principle of the experiment: metal particles on a LiF(100) single crystal substrate are illuminated with laser light and photodesorbed individual atoms are detected with a quadrupole mass spectrometer. The experiments have several important features that make them different from related work carried out earlier. First, desorption was stimulated with continuous-wave laser light of low intensity and desorption could be observed with light intensities down to as little as a few milliwatts. Second, visible laser light was used. Third, a collective rather than a single electron excitation was stimulated and preceded the desorption. Due to the resonant character of the excited surface plasmon oscillation, the desorption rate exhibits a resonance as a function of laser wavelength. The position of this resonance depends on the particle size. Consequently, for a given laser frequency, only metal particles in a certain size range interact with the light. As will be described in Sect. 2.3.4, this opens up new possibilities of manipulating the size distribution of metal particles with laser irradiation.

The metal particles were made by deposition of sodium atoms from a thermal atomic beam. The atoms impinge on the surface and are weakly adsorbed. Surface diffusion commences and lasts for the mean residence time of the atoms on the surface, which is given by the Frenkel equation and determined by the binding energy and the surface temperature. The

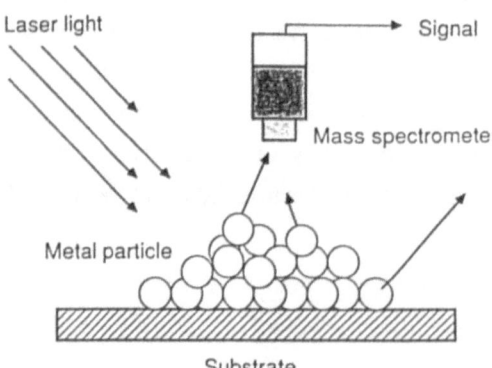

Fig. 2.8. Schematic view of atom desorption from the surface of small metal particles by excitation with visible laser light

atoms can redesorb into the gas phase, a process which resembles inelastic scattering. If an atom, however, meets either a defect, another metal atom or a particle, its binding energy is strongly increased and it cannot escape from the surface potential. This gives rise to the growth of metal particles on the substrate. They are bound to surface defects so that their density approximately equals the defect density. Therefore, the average particle size can be controlled simply by varying the sodium coverage on the surface. Absolute values of the particle size were determined by direct use of the deposition process [2.68,113]. For this purpose, the atomic beam with known constant flux was directed onto the surface and the rate of inelastically scattered sodium atoms was detected as a function of time. During the cluster growth, each atom can meet a metal particle in an increasingly shorter period of time. It is bound more strongly, thus preventing desorption into the gas phase. Consequently, the rate of inelastically scattered atoms decreases as a function of time and directly reflects the growth of the metal particles. Figure 2.9 shows such a scattering signal as a function of time. A quantitative theory has been developed in order to describe the time dependence of the scattering rate [2.113]. As can be seen from Fig. 2.9, a fit of the theory describes the experimental result very well. From this fit the average particle size and density were obtained. For the integral sodium coverages typically deposited in our experiments, the particles had radii ranging from 100 to 700 Å. The particle density was about

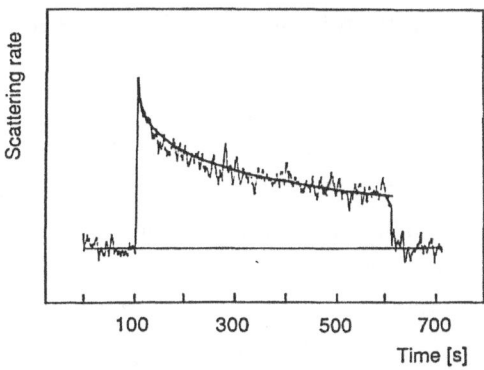

Fig. 2.9. Rate of inelastically scattered Na atoms as a function of time measured during the deposition of sodium on LiF(100) from an atomic beam of constant flux. The decrease of the signal reflects the growth of small metal clusters on the substrate and can be used to determine the average particle size

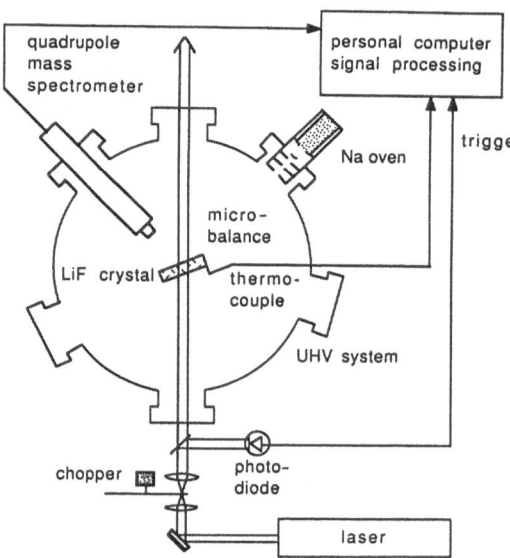

Fig. 2.10. Experimental arrangement for desorption studies of metal atoms from the surface of small metal particles by irradiation with Ar^+ or Kr^+ laser light

$5 \times 10^8 / cm^2$. This means that the distance between clusters was always large compared to the size.

Figure 2.10 shows the experimental arrangement. It consists of an ultrahigh vacuum system with a base pressure of 10^{-10} mbar. The LiF single crystal substrate can be cleaned by heat treatment, annealed and cooled to liquid nitrogen temperature. Its temperature is measured with a thermocouple. A thermal atomic beam of sodium is generated in the UHV chamber. The atoms impinge on the LiF(100) single crystal surface, where they form small metal particles as outlined above. Alternatively, the atoms can be collected on a quartz crystal microbalance for measurements of the Na coverage. An argon or krypton ion laser beam with an unfocused Gaussian profile is directed onto the sample at an angle of incidence of 30° to the surface normal, passes through the transparent LiF crystal and leaves the apparatus through a window. A laser power meter monitors the optical transmission. This enables us to measure the optical extinction and determine the quantum yield of the desorption process. Detached Na atoms are detected with a quadrupole mass spectrometer operating in single ion counting mode. The counting rate is stored in a computer together with the surface temperature. Before starting an experiment, the sample is baked at 700 K for about two hours. This heat treatment removes contamination and active sites for adsorption. Subsequently, the sample is cooled to 90 K. It is then exposed to the sodium atomic beam for a chosen period of time ranging from ten to several hundred seconds.

In order to determine the kinetic energy of the desorbing atoms, time-of-flight measurements were performed. For this purpose, the cw laser beam was chopped at a 100 Hz repetition rate. This gave light pulses of 5 μs duration. In addition, measurements were carried out with a high power frequency doubled Nd:YAG laser with a pulse duration of 3 ns. The time interval between each laser pulse and the arrival of the sodium ions at the detector behind the quadrupole mass spectrometer was measured with a time-to-pulse-height converter and a multichannel analyzer. In order to determine the true velocity with which the atoms desorb from the surface, the experimental velocity distribution was deconvoluted from the drift time distribution of the ions in the mass spectrometer.

2.3.2 Results

A signal displaying the desorption rate as a function of time is shown in Fig. 2.11. The laser beam was switched on and off at 10 s intervals in order to discriminate between signal and background. The signal immediately rises to its maximum value if the laser beam is incident on

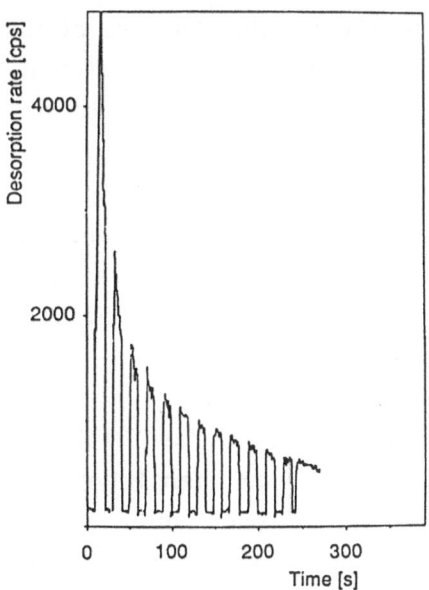

Fig. 2.11. Laser-induced desorption signal for Na particles of 50 nm radius. The detected rate of Na atoms is shown as a function of time for several laser-on/laser-off periods. The excitation wavelength is $\lambda = 514$ nm, the intensity of the cw laser beam $I = 113/cm^2$

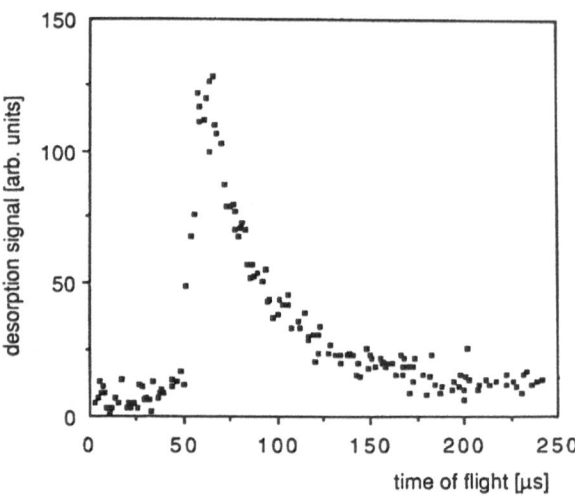

Fig. 2.12. Time-of-flight spectrum of Na atoms desorbing from the surface of small sodium particles. The measurement was made with laser light pulses of 5 μs duration, 514 nm wavelength and 113 W/cm^2 intensity. The most probable atom velocity of $v = 1.7 \times 10^5$ cm/s has been obtained from the displayed spectrum by smoothing and deconvolution from the drift time distribution of the ions in the mass spectrometer

the sample and drops off very rapidly if the laser beam is blocked. This constitutes the first indication of a nonthermal process. The signal also decays as a function of time. This is quite similar to observations in earlier experiments on vibrationally stimulated desorption and is attributed to the depletion of certain preferred adsorption sites on the surface. They are characterized by relatively low coordination numbers and low binding energies. A second reason for the decrease of the desorption signal will be discussed in detail below. The intensity dependence of the desorption rate was also investigated. In order to investigate the kinetics of the process, time-of-flight measurements have been performed. The result of such an experiment is shown in Fig. 2.12. The desorption rate as a function of the laser excitation wavelength is displayed in Fig. 2.13. Furthermore, the quantum yield was determined and the dependence of the signal both on the laser intensity (Fig. 2.14) and on the particle size (Fig. 2.15) was measured.

The experimental results can be summarized as follows:

- Desorption of sodium atoms can be readily detected when the light is incident on the sample. The amplitude of the signal typically amounts to several thousand counts per second. Even with a cw light intensity as low as 40 mW/cm^2, a signal can still be observed.

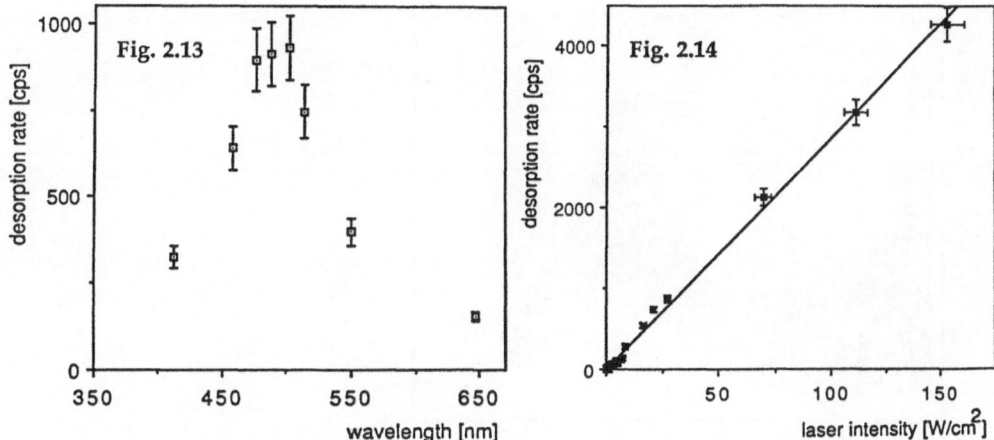

Fig. 2.13. Dependence of the laser-induced desorption signal at t=0 (see Fig. 2.11) on the laser wavelength. The mean cluster radius was 50 nm and the data were taken at a laser intensity of $I = 22\,\mathrm{W/cm^2}$

Fig. 2.14. Rate of desorbing Na atoms as a function of laser intensity. The dependence is linear down to a continuous laser power of several milliwatts. No threshold is observed

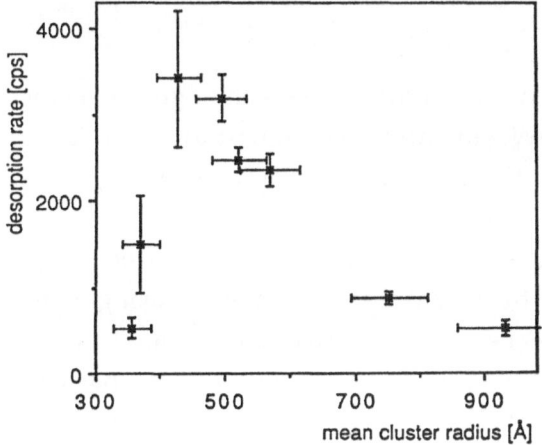

Fig. 2.15. Laser-induced desorption rate as a function of the mean particle radius. The excitation wavelength is $\lambda = 514$ nm

- Desorption starts immediately after the laser is turned on and stops promptly at the beginning of a dark period. The desorption rate decreases with time (see Fig. 2.11 and left hand side of Fig. 2.16). For moderate light intensities, the signal does not recover, even after many minutes *without* illumination.

- If the sample is illuminated with light intensities above 70 W/cm^2 for several minutes, a second desorption signal superimposed on the first one is observed (see right hand side of Fig. 2.16). It obviously results from thermal heating.

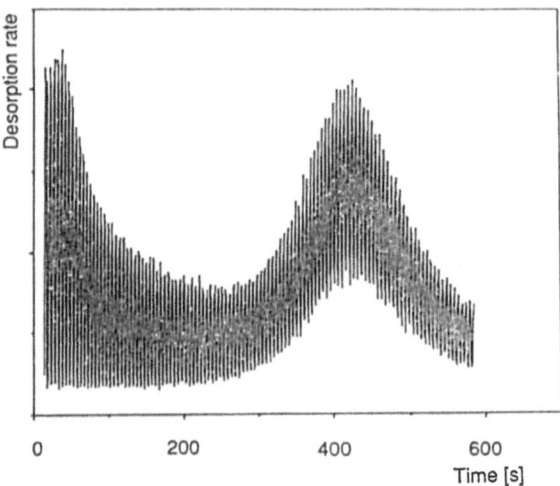

Fig. 2.16. Laser-induced desorption signal for Na particles of 50 nm radius. The excitation wavelength is $\lambda = 514$ nm. A relatively large laser intensity of 120 W/cm² was used. The laser was turned on and off at 5 s intervals. The signal resulting from nonthermal desorption decreases as a function of time (left hand side) until the temperature rise becomes large enough to promote thermal desorption after about 400 s (right hand side) in addition to the nonthermal effect. This second maximum is not observed for low laser intensities

- The desorption rate depends linearly (Fig. 2.14) on the light intensity in the range from about 40 mW/cm² up to the highest available laser intensity of 160 W/cm². No threshold for the desorption signal is found.

- The photodesorption yield strongly depends on the laser frequency. The frequency dependence of the signal was examined by using eight Ar⁺ ion and Kr⁺ ion laser lines between $\lambda = 410$ and 647 nm. The data were taken at a laser intensity of 22 W/cm² and with a mean particle radius of 50 nm. As can be seen from Figure 2.13, which depicts the desorption yield as a function of the laser wavelength, major desorption occurs for the blue and green Ar ion laser lines at $\lambda = 488$ and 514 nm. If laser wavelengths further in the red or ultraviolet are applied, the signal gradually disappears. The spectrum has a center frequency of $\lambda = 490\pm5$ nm and a full width at half maximum of $\Delta\lambda = 90$ nm. This corresponds to photon energies of 2.54 eV and 0.46 eV, respectively.

- The signal depends on the particle size (Fig. 2.15). If desorption is stimulated with laser light of $\lambda = 514$ nm, the detected rate is negligibly small for clusters with radii below 10 nm, reaches a maximum around R = 45 nm and drops off for larger sizes. Similarly, the peak position of the desorption spectrum varies with particle size, i.e., is red shifted for larger clusters.

- The time-of-flight spectrum exhibits a single peak (Fig. 2.12) the position of which does not depend on the laser intensity. This has been verified by measurements with 5 μs Ar⁺ laser pulses as well as with 3 ns YAG laser pulses of different intensity. The position of the maximum of the spectrum corresponds to a kinetic energy of 0.4 eV of the desorbed atoms. Time-of-flight measurements have also been performed with λ = 488 nm and with different distances between the sample and the ionizer. The results are compatible with those obtained for λ = 514 nm.

- A fraction, typically 20% of the total coverage, can be desorbed with the laser at a given light intensity. The exact percentage depends on the particle size, on the excitation wavelength and on the light intensity. A comparison of the optical extinction (measured with a laser power meter) with the number of photodesorbed atoms gives a quantum efficiency of about 10^5 photons per desorbed atom. This number, however, constitutes a lower limit for the efficiency of the process since a considerable fraction of the measured extinction is due not to absorption but to (elastic) light scattering.

2.3.3 Interpretation

Since the LiF crystal is transparent to light and the laser is operated at relatively low power, desorption due to direct substrate heating can be excluded. This is supported by estimates and by measurements of the temperature rise. Also, the desorption yield displays a resonant dependence on the laser frequency, which cannot be explained by the absorption spectrum of the substrate. Obviously, the metal particles are responsible for the absorption of light. A theoretical treatment of the interaction of small metal particles with electromagnetic waves was proposed by *Mie* as early as 1908 [2.114]. He also introduced the first multipole expansions. The Mie resonances are now mostly denoted as surface plasma resonances, i.e., collective oscillations of the conduction electrons at the metal surface (see also [2.115,116]).

In the past, a number of groups have studied the spectra of small sodium particles in a beam [2.117-119], in nobel gas matrices [2.120] or alkali halide crystals such as NaCl [2.121]. Also, such spectra have been calculated theoretically [2.122]. A comparison of the literature data with the results obtained in our experiments shows that the absorption is in the same spectral range and has a similar width. The spectra are also redshifted with increasing particle size and become broader. Therefore, desorption in our experiment is attributed to excitation of a surface plasma resonance.

To elucidate the desorption mechanism, one has to establish whether the bond breaking is a *direct* result of surface plasmon excitation without strong coupling to the phonon bath or whether the excitation energy is predominantly converted into heat so that the particle temperature rises and the atoms desorb thermally. The latter possibility is known as "resonant heating" and has been observed for vibrationally stimulated desorption, see Sect. 2.2.1. Evidence of nonthermal desorption in the present experiment comes from a number of experimental findings:

- Thermal desorption requires a certain threshold for the light intensity. In our experiment no such threshold is observed.
- One can easily estimate that absorption of light only leads to a very moderate temperature rise of the metal particles, typically much less than the $\Delta T \approx 200$ K required for thermal desorption [2.68,113].
- The kinetic energy of the ejected atoms is very high and does not depend on the laser intensity.
- Thermal desorption is observed in addition to the effect considered here (Fig. 2.16).

The linear dependence of the rate on the light intensity suggests that a single photon process is responsible for stimulating the desorption, i.e., an atom is ejected as a result of a single surface plasmon absorption/deexcitation cycle. Part of the excitation energy is required to break the bond of a surface atom (0.65 eV [2.113]). One deexcitation channel of the collective electron oscillation is the conversion of energy into a localized single electron excitation [2.123]. If this state is antibonding, desorption occurs. The quantum yield of 10^{-5}, however, indicates that photo-desorption has to compete with other relaxation channels, particularly with energy transfer to the substrate.

The desorption of atoms from small particles is very similar to the nuclear photoeffect which is observed when an atomic nucleus absorbs gamma rays and a collective oscillation of the protons against the neutrons is excited [2.124]. As a consequence, neutrons may be ejected from the nucleus. In view of this analogy, there is hope that the advanced theoretical methods developed in nuclear physics may help to explain in more detail which mechanism is responsible for the ejection of atoms from metal particles.

Another interesting point is that a collective dipole oscillation can also be induced in sodium particles of totally different size. *Knight* and coworkers have discovered in a gas phase experiment that a plasma resonance like this can be excited in clusters as small as the pentamer [2.119]. This means that even for sodium particles that contain as few as five atoms, the electrons are at least partially delocalized.

2.3.4 Applications - Towards Monodisperse Particles on Supports

Clusters, see [2.125-130], are particles intermediate between atoms and molecules on the one hand and condensed matter on the other. Therefore, their properties are size dependent. In order to fully utilize these interesting characteristics and synthesize new materials with adjustable properties, e.g. for catalysis, photography or other applications, it is highly desirable to make the cluster size distribution as narrow as possible. One of the key issues in cluster physics and chemistry is thus to generate particles that are monodisperse or have a very narrow distribution. Attempts have been made to generate such systems on surfaces by the deposition of particles on a substrate from a monodisperse beam. However, an unsolved problem is that of how and whether fragmentation can be avoided when the particles hit the surface. An alternative way might be to change the given distribution on a surface in a well-defined way and make it as narrow as possible. The desorption mechanism described above offers new ways of manipulating the size distribution along these lines for two reasons [2.131,132]. First, the surface plasmon excitation exhibits a resonance as a function of laser frequency. Second, the transition frequency of this resonance depends on the particle size. Similarly, a resonance as a function of particle size is observed for a fixed wavelength. Consequently, the ablation rate depends resonantly on the particle size and is different for different laser wavelengths. The result of this kind of measurement is shown in Fig. 2.17. Particles with mean radii of $R_0 = 65$ nm, for example, interact preferably with red laser light of $\lambda = 647$ nm whereas particles with $R_0 =$

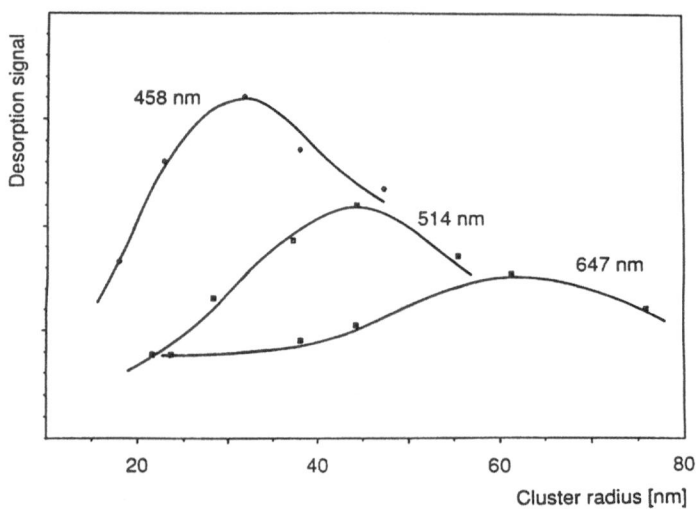

Fig. 2.17. Dependence of the laser-induced desorption signal on the mean particle radius for fixed laser wavelengths of λ = 458, 514 and 647 nm

30 nm resonantly absorb blue light of λ = 458 nm. Therefore, at a given laser wavelength, a change of the size distribution takes place. It is correlated to a variation of the ablation rate as a function of illumination time. This variation is the interplay of two effects. First, the number of atoms which can desorb decreases as a function of time due to the overall decrease of the surface coverage. Second, the ablation can result in a shift in or out of resonance depending on the position of the laser wavelength with respect to the (size dependent) absorption profile. The combination of these two effects determines the change of the ablation rate and of the size distribution. The process naturally comes to an end if the size has decreased so much that the particles are out of resonance. As will be shown below, this results in a considerable narrowing of the size distribution.

A model has been developed to describe this process quantitatively [2.131,132]. For a fixed laser wavelength λ, the absorption resonance curve as a function of radius is approximated by a Gaussian centered at R_λ. The assumption of a symmetrical absorption profile is supported by experimental and theoretical work on plasma resonances of particles in matrices [2.121]. Cluster size distributions f(R), on the other hand, are known to be asymmetric. Finally, the number of desorption sites on a cluster of radius R is given by aR, where a depends on the shape of the cluster [2.68]. The ablation rate dN/dt originating from particles in the size interval [R,R+ΔR] at time t is then given by

$$\frac{dN}{dt} = \sigma_0 I a N_c Q f(R) R \exp\left(-\frac{(R-R_\lambda)^2}{2\sigma^2}\right) \Delta R,$$

where I is the number of incident photons/cm^2·s, σ_0 the cross section for absorption of light at $R = R_\lambda$, Q the quantum efficiency, and N_c the number of illuminated clusters on the surface. The change of the cluster size distribution as a function of illumination time t is directly related to the rate dN/dt. To obtain the time dependence of f(R), the change of each cluster size during the ablation process was calculated, followed by an integration over the total size distribution.

The computation has been done for a number of cases with different particle sizes, excitation frequencies and laser intensities. Two examples will be described here. For both, the wavelength of the incident light was $\lambda = 488$ nm. An absorption resonance curve which peaks at 36 nm and has a full width at half maximum (FWHM) of 15 nm was used. Mean cluster radii of $R_0 = 33$ nm and 57 nm were chosen. The FWHM of the distributions were 11.7 and 20.3 nm, respectively. Figure 2.18 displays the obtained ablation rates as a function of time for the small (33 nm) and large (57 nm) particles. Since the laser is turned on and off during the experiment at 10 s intervals in order to distinguish between signal and background, the same procedure has been included in the calculation. Figure 2.19 shows the change of the cluster size distribution corresponding to the ablation rates of Fig. 2.18. The different traces (dashed lines) refer to the distribution assumed initially on the surface (0) and obtained 60 s (1), 120 s (2), 180 s (3) and 240 s (4) after starting the illumination with laser light, respectively. The assumed absorption profile (solid line) for light of $\lambda = 488$ nm, centered at $R_0 = 36$ nm, is also included.

The ablation rate exhibits different dependences as a function of illumination time, see Fig. 2.18. For the smaller clusters, a monotonic decrease is found, see Fig. 2.18a. For the larger ones we see an initial rise followed by a plateau and a decrease, see Fig. 2.18b. Both the position of the maximum of the rate as shown in Fig. 2.18b and the time dependence of the signals in general sensitively depend on the light intensity. Small

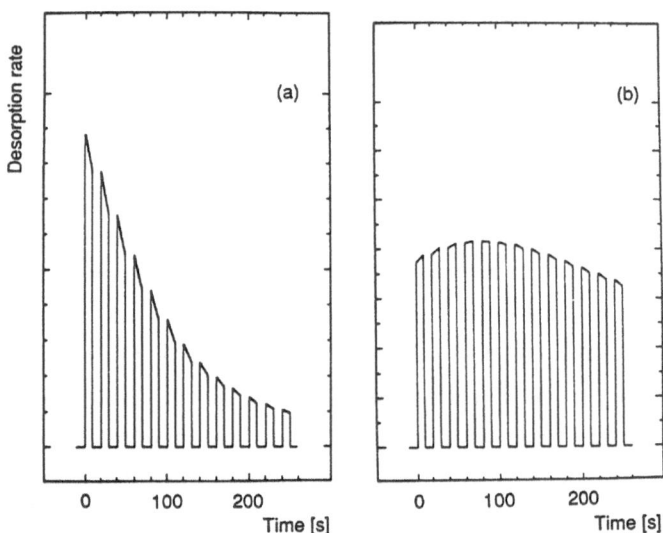

Fig. 2.18a,b. Calculated change of the ablation rate during laser illumination with $\lambda =$ 488 nm. The initial average cluster size was $R_0 = 33$ nm **(a)** and $R_0 = 57$ nm **(b)** with a FWHM of 11.7 and 20.3 nm, respectively

intensities give rise to a rather slow increase or decrease in the desorption rate.

As can be seen from Fig. 2.19, the change in the distribution is very different for the two size ranges. For $R_0 = 33$ nm, the larger particles in the distribution interact more strongly with the light than the smaller ones. They shrink and move out of resonance. Since the ablation comes to an end at about the same position of the absorption profile for all particles, a considerable narrowing of the distribution results. Figure 2.19a shows that this narrowing is as large as a factor of 6 after 240 s. It proceeds further if the illumination is continued and typically amounts to a factor of 10. The size manipulation can be repeated for different wavelengths to make the distribution even narrower.

The situation is quite different for particles with an initial mean size of $R_0 = 57$ nm (Fig. 2.19b). In this case only the smaller clusters of the distribution can interact with the incident light. The result is a narrowing of the initial distribution on one side. In addition, those clusters that interact with the light are shifted *across* the absorption profile. Initially, they interact more and more efficiently with the light and the ablation rate increases, see Fig. 2.18b. Subsequently, they move

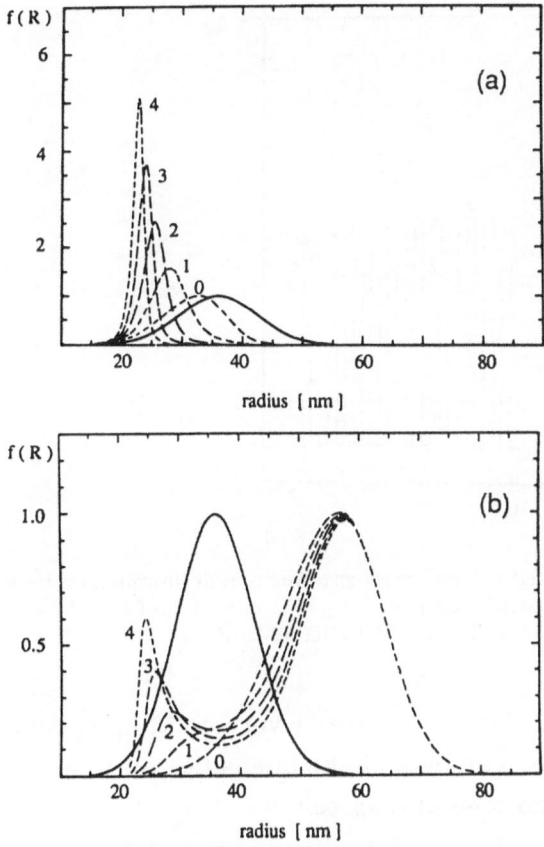

Fig. 2.19a,b. Calculated changes of the particle size distributions during laser illumination corresponding to the ablation rates of Fig. 2.18. The different traces (dashed lines) refer to the distributions assumed initially on the surface (0) and obtained 60 s (1), 120 s (2), 180 s (3) and 240 s (4) after starting the illumination, respectively. The absorption profile (solid line) for light of $\lambda = 488$ nm is also included

out of resonance as decribed above. When the process comes to an end, a second peak of the size distribution has grown.

In order to verify this model, the ablation rates for many different cases, particularly for the experimental conditions used in the above examples have been measured. The results are shown in Fig. 2.20. The average initial cluster size was $R_0 = 33$ nm in Fig. 2.20a and $R_0 = 57$ nm in Fig. 2.20b. The clusters were illuminated by laser light of $\lambda = 488$ nm with an intensity of $I = 100$ W/cm^2. For the small particles, a decrease in the rate is observed starting at the very beginning of the illumination. In contrast, the larger particles give rise to ablation rate changes where an initial increase followed by a broad maximum is observed. A comparison

47

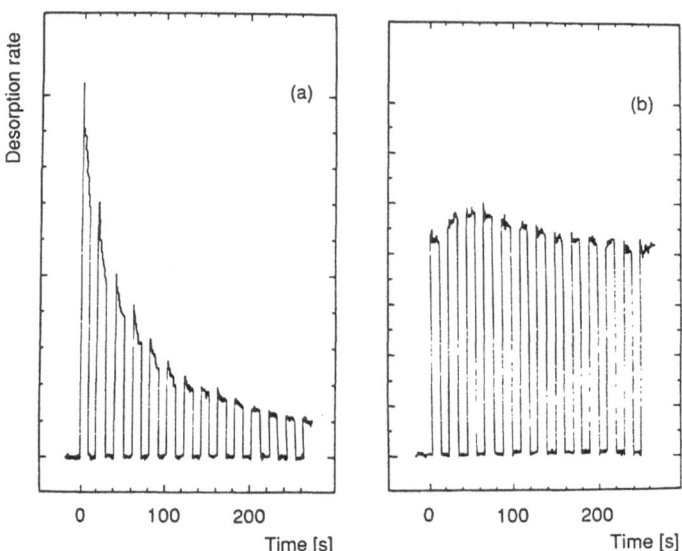

Fig. 2.20a,b. Experimentally determined ablation rates as a function of illumination time with laser light. The initial average particle sizes were $R_0 = 33$ nm **(a)** and $R_0 = 57$ nm **(b)**. Both measurements were made with $\lambda = 488$ nm and $I = 100$ W/cm^2

with Fig. 2.18 shows that the signal shapes are in very good agreement with the theoretical predictions. For other light intensities and cluster sizes the measured and calculated rates also agree.

The excellent agreement of the theoretical predictions with experiment for many different sets of parameters provides convincing evidence for the validity of the presented model and for the narrowing of the cluster size distribution under illumination with laser light. Further experiments with the specific aim of verifying the predicted manipulation of the size distributions also using in situ electron microscopy are in progress.

Another interesting application of plasmon stimulated desorption might be chemistry on the surface of metal clusters. It seems possible that the localized electron excitation, into which the plasma oscillation decays, might also populate a repulsive state in an adsorbed molecule by which it is detached from the surface. Such an effect could even be molecularly selective. In preliminary experiments it has already been demonstrated that as little as one third of a molecular monolayer of ammonia on the surface of the sodium clusters has a dramatic influence on the desorption signal. In the future the desorption of other metal clusters, particularly silver particles, will be investigated.

2.4 Conclusions and Outlook

The investigation of laser-induced desorption stimulated by electronic excitation is a very new field and it seems that many other exciting processes remain to be discovered. Major efforts, however, will be needed to achieve a detailed understanding of the mechanisms observed so far. For this purpose, the measurement of a larger variety of experimental parameters and their influence on the signals would be very helpful. For example, all the relevant parameters of the incident light field should be varied, such as the wavelength, the angle of incidence and the polarization. Of particular importance is the variation of the excitation wavelength to measure the spectral dependence of the desorption signal. Also, the desorption products have to be characterized more completely. In addition to a determination of their mass, kinetic energy and charge, the angular dependence of the desorption rate and the internal state distribution should be measured. Furthermore, a careful characterization of the substrate surface before and after desorption is essential for a complete analysis. In the future, experiments with ultrafast light pulses will certainly provide additional information, particularly on nonequilibrium processes. First results with subpicosecond laser pulses for NO desorbed from a Pd(111) surface have just become available [2.133].

Acknowledgements: The author gratefully acknowledges the continuous support of Prof. G. zu Putlitz and the fruitful and pleasant collaboration with W. Hoheisel, K. Jungmann, U. Schulte, M. Vollmer and R. Weidenauer on the laser desorption work involving small sodium particles.

References

2.1   D. Menzel: In *Chemistry and Physics of Solid Surfaces IV*, ed. by R. Vanselow, R. Howe, Springer Ser. Chem. Phys., Vol. 20 (Springer, Berlin, Heidelberg 1982) p. 389

2.2   J.T. Yates: In *Methods of Experimental Physics*, Vol. 22, ed. by R.L. Park, M.G. Lagally (Academic, Orlando, FL 1985) p. 425

2.3   P.D. Townsend: Surf. Sci. **90**, 256 (1979)

2.4   T.J. Chuang: Surf. Sci. Rep. **3**, 1 (1983)

2.5   N.H. Tolk, M.M. Traum, J.C. Tully, T.E. Madey (eds.): *Desorption Induced by Electronic Transitions, DIET I*, Springer Ser. Chem. Phys., Vol. 24 (Springer, Berlin, Heidelberg 1983)

2.6   M.L. Knotek: Rep. Prog. Phys. **47**, 1499 (1984)

2.7   D. Bäuerle (ed.): *Laser Processing and Diagnostics*, Springer, Ser. Chem. Phys., Vol. 39 (Springer, Berlin, Heidelberg 1984)

2.8   T. Nakayama, M. Okigawa, N. Itoh: Nucl. Instrum. Meth. **B1**, 301 (1984)

2.9   J.E. Rothenberg, R. Kelly: Nucl. Instrum. Meth. **B1**, 291 (1984)

2.10   R. Kelly, J.J. Cuomo, P.A. Leary, J.E. Rothenberg, B.E. Braren, C.F. Aliotta: Instrum. Meth. **B9**, 329 (1985)

2.11   R. Kelly, J.E. Rothenberg: Nucl. Instrum. Meth. **B7/8**, 755 (1985)

2.12   T.J. Chuang: J. Vac. Sci. Technol. **B3**, 1408 (1985)

2.13   T. Nakayama, N. Itoh: In [2.14], p. 237

2.14   W. Brenig, D. Menzel (eds.): *Desorption Induced by Electronic Transitions, DIET II*, Springer Ser. Surf. Sci., Vol. 4 (Springer, Berlin, Heidelberg 1985) p. 237

2.15   W. Drachsel, U. Weigmann, S. Jaenicke, J.H. Block: In [2.14], p. 245

2.16   D. Menzel: Nucl. Instrum. Meth. **B13**, 507 (1986)

2.17   R.F. Haglund, Jr., R.G. Albridge, D.W. Cherry, R.K. Cole, M.H. Mendenhall, W.C.B. Peatman, N.H. Tolk D. Niles, G. Margaritondo, N.G. Stoffel, E. Taglauer: Nucl. Instrum. Meth. **B13**, 525 (1986)

2.18   T.J. Chuang: Surf. Sci. **178**, 763 (1986)

2.19   D. Bäuerle: *Chemical Processing with Lasers*, Springer Ser. Mater. Sci., Vol. 1 (Springer, Berlin, Heidelberg 1986)

2.20   P.C. Stair, E. Weitz: J. Opt. Soc. Am. **B4**, 255 (1987)

2.21   R.H. Stulen, M.L. Knotek (eds.): *Desorption Induced by Electronic Transitions, DIET III*, Springer Ser. Surf. Sci., Vol. 13 (Springer Berlin, Heidelberg 1988)

2.22   S. Jaenicke, A. Ciszewski, J. Dösselmann, W. Drachsel, J.H. Block, D. Menzel: In [2.21], p. 236

2.23   P. Avouris, R.E. Walkup: to appear in Ann. Rev. Phys. Chem. **40** (1989)

2.24   A. Benninghoven, F.G. Rüdenauer, H.W. Werner (eds.): *Secondary Ion Mass Spectrometry*, Chemical Analysis, Vol. 86 (Wiley, New York 1987)

2.25   E.R. Hilf, H.-F. Kammer, K. Wien (eds.): *PDMS and Clusters*, Lecture Notes in Physics, Vol. 269 (Springer, Berlin, Heidelberg 1987)

2.26   P.K.D. Feigl, F.R. Krueger, B. Schueler: Organic Mass Spectrom. **18**, 442 (1983)

2.27   F. Mayer, F.R. Krueger, J. Kissel: In *Ion Formation from Organic Solids, IFOS III*, ed. by A. Benninghoven, Springer Proc. Phys., Vol. 9 (Springer, Berlin, Heidelberg 1986) p. 169

2.28   C. Krischer, D. Lichtman: Phys. Lett. **44A**, 99 (1973)

2.29   P. Taborek: Phys. Rev. Lett. **48**, 1737 (1982)

2.30   M. Sinvani, P. Taborek, D. Goodstein: Phys. Lett. **95A**, 59 (1983)

2.31   S.R.J. Brueck, T.F. Deutsch, D.E. Oates: Appl. Phys. Lett. **43**, 157 (1983)

2.32   P.M. Ferm, S.R. Kurtz, K.A. Pearlstine, G. McClelland: Phys. Rev. Lett. **58**, 2602 (1987)

2.33   D. Menzel, R. Gomer: J. Chem. Phys. **41**, 3311 (1964)

2.34   P.A. Readhead: Can. J. Phys. **42**, 886 (1964)

2.35   O. Gröbner, A.G. Mathewson, H. Störi, P. Strubin, R. Souchet: Vacuum **33**, 397 (1983)

2.36   E.M. Williams, F. Le Normand, N. Hilleret, G. Dominichini: Vacuum **35**, 141 (1985)

2.37   R. Srinivasan: Science **234**, 559 (1986)

2.38   J.B. Hopkins, P.R.R. Langridge-Smith, M.D. Morse, R.E. Smalley: J. Chem Phys. **78**, 1627 (1983)

2.39   J. Reif, H. Fallgren, H.B. Nielsen, E. Matthias: Appl. Phys. Lett. **49**, 930 (1986)

2.40   H.B. Nielsen, J. Reif, E. Matthias, W. Westin, A. Rosén: In [2.21], p. 266

2.41   E. Matthias, H.B. Nielsen, J. Reif, A. Rosén, W. Westin: J. Vac. Sci. Technol. **B5**, 1415 (1987)

2.42   E. Matthias, S. Petzoldt, A.P. Elg, P.J. West, J. Reif: In *Laser Induced Damage in Optical Materials*, NBS Special Publication (1987)

2.43   L.P. Levine, J.F. Ready, E. Bernal: J. Appl. Phys. **38**, 331 (1967)

2.44   L.P. Levine, J.F. Ready, E. Bernal: IEEE J. QE-4, 18 (1968)

2.45   G. Ertl, M. Neumann: Z. Naturforsch. A **27**, 1607 (1972)

2.46   K. Christmann, O. Schober, G. Ertl, M. Neumann: J. Chem. Phys. **60**, 4528 (1974)

2.47   R.A. Olstad, D.R. Olander: J. Appl. Phys. **46**, 1499 (1975)

2.48   R.A. Olstad, D.R. Olander: J. Appl. Phys. **46**, 1509 (1975)

2.49   J.P. Cowin, D.J. Auerbach, C. Becker, L. Wharton: Surf. Sci. **78**, 545 (1978)

2.50   T. Kawai, T. Sakata: Chem. Phys. Lett. **69**, 33 (1980)

2.51   G. Wedler, H. Ruhmann: Surf. Sci. **121**, 464 (1982)

2.52   I. Hussla, R. Viswanathan: Surf. Sci. **145**, L488 (1984)

2.53   R. Viswanathan, I. Hussla: In [2.7], p. 148

2.54   I. Hussla, H. Coufal, F. Träger, T.J. Chuang: Ber. Bunsenges. Phys. Chem. **90**, 240 (1986)

2.55   E.G. Seebauer, A.C.F. Kong, L.D. Schmidt: J. Vac. Sci. Technol. **A5**, 464 (1987)

2.56   R.B. Hall, A.M. DeSantolo: Surf. Sci. **137**, 421 (1984)

2.57   R.B. Hall, A.M. DeSantolo, S.J. Bares: Surf. Sci. **161**, L533 (1985)

2.58   A.A. Deckert, S.M. George: Surf. Sci. **182**, L215 (1987)

2.59   J. Heidberg, H. Stein, E. Riehl: Phys. Rev. Lett. **49**, 666 (1982)

2.60   B. Schäfer, P. Hess: Chem. Phys. Lett. **105**, 563 (1984)

2.61   T.J. Chuang, I. Hussla: Phys. Rev. Lett. **52**, 2045 (1984)

2.62   J. Heidberg, H. Stein, H. Weiss: Surf. Sci. **184**, L431 (1987)

2.63   M. Buck, P. Hess: J. Electron Spectrosc. Relat. Phenom. **45**, 237 (1987)

2.64   E.B.D. Bourdon, J.P. Cowin, I. Harrison, J.C. Polanyi, J. Segner, C.D. Stanners, P.A. Young: J. Phys. Chem. **88**, 6100 (1984)

2.65   E.B.D. Bourdon, P. Das, I. Harrison, J.C. Polanyi, J. Segner, C.D. Stanners, R.J. Williams, P.A. Young: Faraday Discuss. Chem. Soc. **82**, 343 (1986)

2.66   St.J. Dixon-Warren, I. Harrison, K. Leggett, M.S. Matyjaszczyk, J.C. Polanyi, P.A. Young: J. Chem. Phys. **88**, 4092 (1988)

2.67   F.L. Tabares, E.P. Marsh, G.A. Bach, J.P. Cowin: J. Chem. Phys. **86**, 738 (1987)

2.68   M. Vollmer, F. Träger: Surf. Sci. **187**, 445 (1987)

2.69   K. Domen, T.J. Chuang: Phys. Rev. Lett. **59**, 1484 (1987)

2.70   T.J. Chuang, K. Domen: J. Vac. Sci. Technol. **A5**, 473 (1987)

2.71   K. Domen, T.J. Chuang: J. Chem. Phys. **90**, 3318 (1989)

2.72   K. Domen, T.J. Chuang: J. Chem. Phys. **90**, 3332 (1989)

2.73   A. Mödl, K. Domen, T.J. Chuang: Chem. Phys. Lett. **154**, 187 (1989)

2.74   D. Weide, P. Andresen, H.-J. Freund: Chem. Phys. Lett. **136**, 106 (1987)

2.75   F. Budde, A.V. Hamza, P.M. Ferm, G. Ertl, D. Weide, P. Andresen, H.-J. Freund: Phys. Rev. Lett. **60**, 1518 (1988)

2.76   L.J. Richter, S.A. Buntin, R.R. Cavanagh, D.S. King: J. Chem. Phys. **89**, 5344 (1988)

2.77   S.A. Buntin, L.J. Richter, R.R. Cavanagh, D.S. King: Phys. Rev. Lett. **61**, 1321 (1988)

2.78   D. Burgess, Jr., R.R. Cavanagh, D.S. King: J. Chem. Phys. **88**, 6556 (1988)

2.79   W.C. Natzle, D. Padowitz, S.J. Sibener: J. Chem. Phys. **88**, 7975 (1988)

2.80   G.S. Higashi: J. Chem. Phys. **88**, 422 (1988)

2.81   J.C. Tully: Surf. Sci. **111**, 461 (1981)

2.82   J.R. Arthur, T.R. Brown: J. Vac. Sci. Technol. **12**, 200 (1975)

2.83   K.C. Janda, J.E. Hurst, J.P. Cowin, L. Wharton, D. J. Auerbach: Surf. Sci. **130**, 395 (1983)

2.84   C.J.S.M. Simpson, J.P. Hardy: Chem. Phys. Lett. **130**, 175 (1986)

2.85   E.P. Marsh, F.L. Tabares, M.R. Schneider, J.P. Cowin: J. Vac. Sci. Technol. **A5**, 519 (1987)

2.86   E.P. Marsh, M.R. Schneider, T.L. Gilton, F.L. Tabares, W. Meier, J.P. Cowin: Phys. Rev. Lett. **60**, 2551 (1988)

2.87   J.R. Creighton: J. Appl. Phys. **59**, 410 (1986)

2.88   W. Hoheisel, K. Jungmann, M. Vollmer, R. Weidenauer, F. Träger: Phys. Rev. Lett. **60**, 1649 (1988)

2.89   F.G. Celii, P.M. Whitmore, K.C. Janda: Chem. Phys. Lett. **138**, 257 (1987)

2.90   C.E. Bartosch, N.S. Gluck, W. Ho, Z. Ying: Phys. Rev. Lett. **57**, 1425 (1986)

2.91   N.S. Gluck, Z. Ying, C.E. Bartosch, W. Ho: J. Chem. Phys. **86**, 4957 (1987)

2.92   R. Pflaum, K. Sattler, E. Recknagel: Chem. Phys. Lett. **138**, 8 (1987)

2.93   R. Pflaum, K. Sattler, E. Recknagel: In [2.113], p. 103

2.94   R. Pflaum, E. Recknagel: Z. Phys. **D12**, 249 (1989)

2.95   W. Sesselmann, E.E. Marinero, T.J. Chuang: Surf. Sci. **178**, 787 (1986)

2.96   F.A. Houle: Phys. Rev. Lett. **61**, 1871 (1988)

2.97   V.S. Antonov, V.S. Letokhov, A.N. Shibanov: Appl. Phys. **25**, 71 (1981)

2.98   V.S. Letokhov, V.G. Movshev, S.V. Chekalin: Sov. Phys.-JETP **54**, 257 (1981)

2.99   N. Nishi, H. Shinohara, T. Okuyama: J. Chem. Phys. **80**, 3898 (1984)

2.100   D. Lichtman, Y. Shapira: CRC Crit. Rev. Solid State Mater. Sci. **8**, 93 (1978)

2.101   N.V. Hieu, D. Lichtman: J. Vac. Sci. Technol. **A1**, 1 (1983)

2.102   T. Nakayama: Surf. Sci. **133**, 101 (1983)

2.103   Z. Ying, W. Ho: Phys. Rev. Lett. **60**, 57 (1988)

2.104   Y. Shapira, R.B. McQuistan, D. Lichtman: Phys. Rev. **B 4**, 2163 (1977)

2.105   Z. Ying and W. Ho: Surf. Sci. **198**, 473 (1988)

2.106   T. Nakayama, H. Ichikawa, N. Itoh: Surf. Sci. **123**, L693 (1982)

2.107   R.J. von Gutfeld, R.W. Dreyfus: Appl. Phys. Lett. **54**, 1212 (1989)

2.108   R.W. Dreyfus, R.J. von Gutfeld: In Technical Digest, Conference on Lasers and Electro-Optics (Optical Society of America, Washington, DC 1989) paper THD3

2.109   R.P. Welle, H. Helvajian: AIAA Thermophysics, Plasmadynamics and Lasers Conference, San Antonio, Texas, June 27-29 (1988)

2.110   H. Helvajian, R.P. Welle: J. Chem. Phys., to be published

2.111   K. Höning, W. Hoheisel, M. Vollmer, F. Träger: to be published

2.112   W. Hoheisel, U. Schulte, M. Vollmer, R. Weidenauer, F. Träger: Appl. Surf. Sci. **36**, 664 (1989)

2.113   M. Vollmer, F. Träger: Z. Phys. D **3**, 291 (1986)

2.114   G. Mie: Ann. Phys. **25**, 377 (1908)

2.115   H. Raether: Springer Tracts Mod. Phys., Vol. 88 (1980)

2.116   J.A.A.J. Perenboom, P. Wyder, F. Meier: Phys. Rep. **78**, 171 (1981)

2.117   D.M. Mann, H.P. Broida: J. Appl. Phys. **44**, 4950 (1973)

2.118   J. Hecht: J. Appl. Phys. **50**, 7186 (1979)

2.119   W.A. de Heer, K. Selby, V. Kresin, J. Masui, M. Vollmer, A. Châtelain, W. D. Knight: Phys. Rev. Lett. **59**, 1805 (1987)

2.120   H. Abe, K.-P. Charlé, B. Tesche, W. Schulze: Chem. Phys. **68**, 137 (1982)

2.121   M.A. Smithard, M.Q. Tran: Helv. Phys. Acta **46**, 869 (1974)

2.122   W.T. Doyle, A. Agarwal: J. Opt. Soc. Am. **55**, 305 (1965)

2.123   T. Inagaki, K. Kagami, E.T. Arakawa: Appl. Opt. **21**, 949 (1982)

2.124   G.F. Bertsch, P.F. Bortignon, R.A. Broglia: Rev. Mod. Phys. **55**, 287 (1983)

2.125   W.A. de Heer, W. D. Knight, M.Y. Chou, M.L. Cohen: Solid State Physics 40, 93 (1987)

2.126   F. Träger, G. zu Putlitz (eds.): *Metal Clusters* (Springer, Berlin, Heidelberg 1986)

2.127   M. Moskovits (ed.): *Metal Clusters* (Wiley, New York 1986)

2.128   P. Jena, B.K. Rao, S.N. Khanna (eds.): *Physics and Chemistry of Small Clusters* (Plenum, New York 1987)

2.129   S. Sugano, Y. Nishina, S. Ohnishi (eds.): *Microclusters,* Springer Ser. Mater. Sci., Vol. 4 (Springer, Berlin, Heidelberg 1987)

2.130   G. Benedek, T.P. Martin, G. Pacchioni (eds.): *Elemental and Molecular Clusters*, Springer Ser. Mater. Sci., Vol. 6 (Springer, Berlin, Heidelberg 1987)
2.131   R. Weidenauer, M. Vollmer, W. Hoheisel, U. Schulte, F. Träger: J. Vac. Sci. Technol. **A7**, 1972 (1989)
2.132   M. Vollmer, R. Weidenauer, W. Hoheisel, U. Schulte, F. Träger: Phys. Rev. **B**, in press
2.133   J.A. Prybyla, T.F. Heinz, J.A. Misewich, M.M.T. Loy, J.H. Glownia: In Technical Digest, Conference on Quantum Electronics and Laser Science (Optical Society of America, Washington, DC 1989) paper THNN3

3. Time-of-Flight Analysis of IR and UV Laser-Induced Multilayer Desorption and Ablation

Peter Hess

With 22 Figures

This chapter reports recent results from investigations of multilayer desorption and ablation from molecular van der Waals films by high power IR and UV laser pulses. The information obtained by the time-of-flight technique about the mechanisms involved and about the characteristic features of the induced processes is discussed in detail. Photoablation in the IR and UV spectral regions is compared, to gain insight into wavelength-specific behavior.

3.1 Background

3.1.1 Lasers in Surface Science

In recent years lasers have been attracting increasing interest in surface science. Properties such as brightness (high photon flux), narrow bandwidth (high monochromaticity) and short pulse length (picosecond pulse width or shorter) allow new studies to be made of the complicated processes occurring at the gas-solid interfaces.

Important applications of lasers in surface science include the state-selective investigation of the dynamics of molecule-surface interactions, the study of surface properties using lasers as probes and the laser-induced variation of the state of the surface. The latter category of applications includes melting, desorption, ablation, etching and deposition.

One of the oldest fields in surface science is that of the investigation of desorption processes. This may be due to the fact that desorption is one of the simplest surface phenomena yielding insight into static and dynamic surface properties. A well-established technique is that of thermal programmed desorption, the source of most of the information available on the binding energies at surfaces.

Laser-induced desorption was first reported in 1967 by *Levine* et al. [3.1]. Since then, the field has been expanding rapidly and several excellent review articles have recently been published on the subject [3.2–4].

3.1.2 Laser-Induced Desorption and Ablation

Pulsed laser-induced desorption offers ways of studying the dynamics of surface processes. The monochromaticity of laser radiation allows a selective excitation of defined states and in ideal cases it may be possible to confine energy deposition to the substrate, adsorbate or multilayer film covering the substrate. In this review it is mainly the latter process, often called adsorbate-mediated desorption and ablation, that is discussed in detail.

The availability of short laser pulses allows the realisation of rapid heating rates not achievable by conventional methods. Using microsecond to nanosecond laser pulses, heating rates of 10^8–10^{12} K/s have been achieved, whereas for femtosecond pulses, heating rates as high as 10^{15} K/s have been reported [3.5]. These high laser heating rates should be compared with the traditional heating rates of about 10 K/s or less in thermal programmed desorption in order to appreciate the new opportunities offered by lasers in the study of surface processes.

Due to the high heating rates, laser-induced desorption is believed to be a transient temperature method. What may be observed essentially is isothermal desorption at the highest transient temperature achieved. Of course, the terms "temperature" and "heating" should only be used if the deviations from thermodynamic equilibrium are relatively small. Even in this case, the localized laser-induced temperature jump can lead to interesting new kinetic effects. The very high laser heating rate may successfully compete with other physical or chemical processes such as phase transitions or chemical decomposition. Thus, the probability of accessing an entropically favored process at the high temperatures achieved can be increased dramatically. For example, if the melting process cannot follow the rapid state change, overheating may take place and direct laser sublimation may be observed. The experimental finding that even large molecules can be desorbed intact with a pulsed laser, whereas slower heating rates lead to effective fragmentation, has been interpreted kinetically as a competition between the rate of desorption and the rate of reaction [3.6–8]. This explanation is based on efficient energy exchange between the surface bonds and the chemical bonds of the physisorbed molecules.

To explain the experimental fact that large biomolecules and other thermally labile and/or nonvolatile compounds can be desorbed intact at high heating rates, a different mechanism based on a nonstatistical concept has also been suggested [3.9]. Here, it is assumed that the energy flow between the weaker van der Waals type bonds and the ordinary chemical bonds can be inefficient enough to allow the desorption of internally lukewarm molecules at high heating rates, if the energy flows from the substrate to the adsorbate [3.10]. Figure 3.1 gives a graphical illustration of the underlying picture, namely, the interaction of an isolated molecule with a substrate and its desorption by a substrate-mediated excitation of the surface bond or by absorption of a photon.

The desorption of large molecules is one example of an active field of research, where a controversy exists concerning the mechanism involved. Another example is the photoablation of polymers with UV lasers, where photochemical and photothermal mechanisms have been suggested. Traditionally, the term desorption is used for systems where mono- or submonolayer amounts of molecules leave the substrate surface, whereas the term ablation was originally coined for polymer systems to describe the emission of many layers of the chemical network by a single UV laser pulse. In this review the term ablation is used more generally to indicate that many layers leave the condensed system irrespective of the forces holding the condensed phase together (van der Waals or exchange forces) and irrespective of the energy

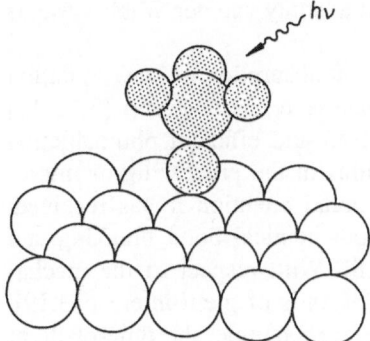

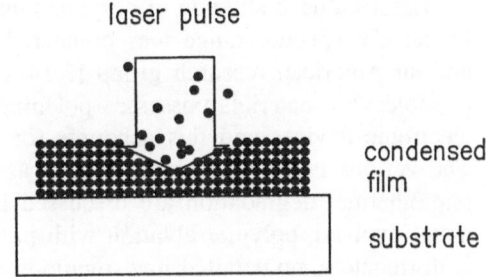

Fig. 3.1. Schematic representation of an isolated molecule on a surface, which may be desorbed by a substrate mediated process or absorption of a photon

Fig. 3.2. Schematic representation of localized multilayer ablation by a single laser pulse

of the photons of the laser pulse (IR or UV photons). The reason for using the term ablation in this more general sense is that new phenomena are observed in the case of multilayer ablation which show similar features irrespective of the forces involved and photon energies employed. Figure 3.2 illustrates schematically the detachment, by a single laser pulse, of several layers from a molecular film attached to a substrate. This phenomenon is the main topic of this review.

3.1.3 Scope of Review

This review deals mainly with the resonant excitation of molecular adsorbates or films employing IR and UV laser pulses. When a photon beam is incident on an optically transparent or weakly absorbing substrate covered with a molecular adsorbate, photons will be mainly absorbed by the adsorbed species in a suitable wavelength region. With pulsed IR lasers, e.g. a line-tunable CO_2 laser, an efficient excitation of internal vibrational modes of molecules in the submonolayer, monolayer and multilayer coverage region is possible. Studies involving electronic excitation of molecular adsorbates and films as well as molecular solids using UV and visible lasers will also be discussed in detail. For electronic excitation, excimer lasers are of increasing importance.

As molecular systems, both van der Waals films of polyatomic molecules condensed at low temperatures on a substrate and polymers consisting of a network of chemical bonds will be considered. These molecular systems possess strongly different binding energies, that the use of photons with varying energy (IR, UV) provides insight into the mechanism of laser-induced desorption and ablation. The main question of interest is which part of the absorbed photon energy is utilized for direct chemical bond breaking and which part relaxes and induces chemical processes and phase transformations essentially by heating.

Desorption due to resonant absorption of IR photons by adsorbed molecules in the monolayer and multilayer coverage region was first reported by *Heidberg* et al. 1980 [3.11]. This field developed rapidly as described in the two reviews published so far [3.2, 3], and just recently, UV lasers were used to study van der Waals systems (as discussed in more detail later).

Laser-induced ablation of polymers due to resonant absorption of laser radiation in the UV spectral range was pioneered by a Japanese research group [3.12, 13] and an American research group [3.14, 15]. The clean and efficient photoablation of molecular materials possesses potential applications in the processing of microelectronic devices and this accounts for the widespread attention it has received. The various mechanistic aspects such as direct photochemical bond breaking and photothermal degradation are discussed in [3.16–18]. With respect to the mechanism involved, polymer ablation with pulsed CO_2 lasers is of great interest [3.19]. Unfortunately, no detailed investigations employing, for example, the time-of-flight (TOF) technique are available for polymer ablation in the IR spectral region. The TOF technique does provide extensive and detailed information on desorption and ablation processes. It is based on a time-resolved detection of the molecules leaving the surface. The principles of TOF detection will be discussed, including problems encountered in signal analysis, and the application of this technique to resonant excitation of molecular multilayer systems will be reviewed.

3.2 Desorption and Ablation

3.2.1 Resonant IR and UV Excitation

Transitions between vibrational levels of isolated molecules are usually described by simple models such as the harmonic oscillator or the anharmonic oscillator. Quantum mechanics gives the selection rules for the allowed optical transitions, which require a change of the dipole moment during the transition. These models are sufficient to understand single-photon absorption processes. To induce IR chemistry, normally more than ten vibrational quanta are needed. It turns out that in fact isolated polyatomic molecules with four or more atoms can be brought to reaction if irradiated with IR radiation of appropriate frequency and sufficient intensity. To explain this behavior of polyatomic molecules under high power irradiation, the two-ladder, three-region model has been introduced, [3.20, 21]. This model assumes that the first few photons are absorbed in one vibrational mode (ladder 1) in a high-order multiphoton process ($n = 3$–7) defining region 1. At higher levels of excitation, rapid intramolecular energy exchange with other vibrational modes is assumed. Optical pumping in this second region, called the "quasicontinuum", is stepwise and incoherent. At even higher energies in region three, the possibility of dissociation exists.

In region 1, a mode-selective laser excitation is possible, however, this mode selectivity is destroyed by efficient intramolecular energy transfer in the quasicontinuum. Therefore, IR multiphoton dissociation is molecule selective, but not bond selective under collision-free conditions.

In a condensed phase of polyatomic molecules or in a molecular film, efficient intermolecular energy exchange between molecules comes into play. This intermolecular coupling makes it extremely difficult to accumulate 20 or 30 IR photons in one molecule and to achieve multiphoton dissociation. In fact, there is no clear, generally accepted example of an isotope separation process stimulated at the surface by selective IR laser-induced desorption. This indicates that molecule selectivity is difficult to obtain in the IR spectral region for condensed phases due to extremely fast intermolecular energy transfer. There is no evidence of a selective coupling between an excited vibrational mode and the adsorption potential, as indicated in Fig. 3.3, leading to nonstatistical desorption. On the other hand, it has been suggested that surface phonons may communicate well with the low-frequency mode of the adsorption potential but a bottleneck exists for the flow of energy into the chemical bonds of the adsorbed molecule [3.9]. This effect would allow a nonstatistical behavior in the case of energy flow from the substrate to the adsorbed molecules, see Fig. 3.3.

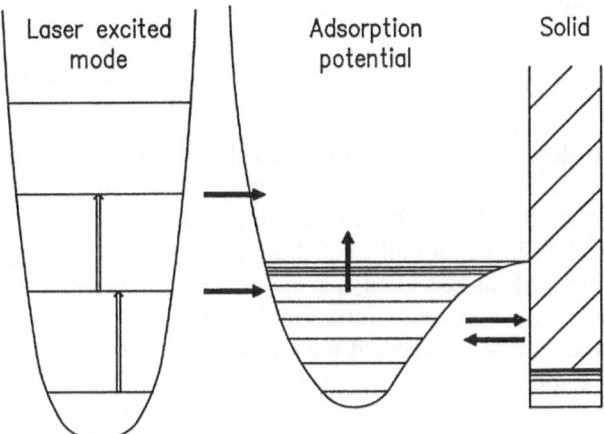

Fig. 3.3. Energy situation for IR excitation considering the vibrational levels of the excited vibrational mode, the adsorption potential and the phonons in the solid

For single photon electronic excitation, quantum theory predicts the strongest transitions between lower and upper states with essentially equal internuclear distances. Transitions to neighboring vibrational states will occur with reduced probability. Changes in the electronic distribution are always accompanied by a change in the dipole moment. Thus, electronic transitions are observed for all molecules. There are many possibilities concerning the nature of the excited state and the relative positions of the internuclear distances in the upper and lower states [3.22]. Figure 3.4 shows three different examples of laser-induced photo-dissociation processes. The first example is coupling with a repulsive upper state possessing no energy minimum, the second shows a transition to an upper bound state with slightly larger internuclear distance, leading to dissociation by internal conversion, and the third case illustrates the direct dissociation into electronically excited products by excitation of a bound state with energy in excess of the dissociation energy. The direct or indirect photodecomposition, for example, by absorption of a UV photon is possible not only for isolated molecules but also for adsorbed species and molecules located on top of a condensed phase. It is clear, however, that the higher density of molecules will lead to increasing competition between energy exchange processes and directly induced photochemical reactions.

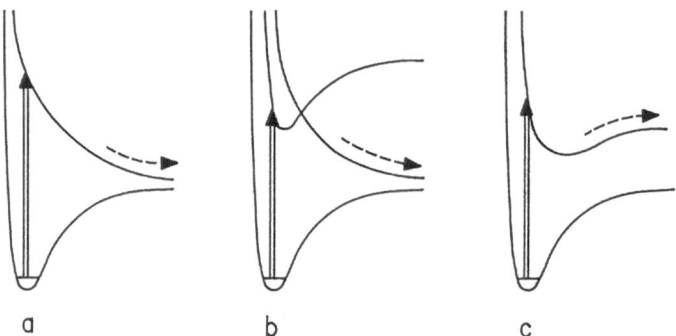

Fig. 3.4a–c. Energy situation for UV excitation considering (a) direct dissociation to ground state products by excitation of a repulsive state, (b) upper state excitation followed by internal conversion, (c) direct dissociation into electronically excited products

3.2.2 Mechanisms

One of the simplest models to describe the transition from the condensed to the gaseous state is that of desorption of a molecule by a single quantum event. This process may also be called quantum desorption. In this case, the absorption of a photon gives rise to a single desorbed molecule or molecular fragment if photochemistry is involved, and therefore detection of the desorbed species yields a direct probe of microscopic processes occurring at the surface, if further collisions are excluded.

The following characteristic behavior is expected for a quantum desorption process involving detachment of a molecule from the surface by interaction with a single photon:

1) For a given submonolayer coverage, the desorption yield increases linearly with laser fluence in the case of a one-quantum process, quadratically in the case of a two-photon process, etc.
2) The kinetic energy of the desorbed particles and the fragmentation pattern in the case of photodissociation is independent of laser fluence, but depends however on the binding energy and the laser wavelength.
3) The desorption process possesses molecular selectivity. It may be possible to desorb a specific isotope. Depending on the laser wavelength, a specific bond or group of bonds may be broken, generating specific photofragments.
4) No fluence threshold exists for desorption or dissociation.

There are only a few examples reported in the literature showing desorption behavior in accordance with these criteria. The disruption of a bond following the laser excitation process is usually called a photolytic reaction. Quantum desorption can be understood by assuming direct dissociation and/or desorption of the molecule due to absorption of a photon. The characteristic features are therefore comparable with those of the photoelectric effect in the case of electrons. The energetic situation is described by the law of energy conservation. As the process is localized in a single molecule, the photon energy is partly used to break a chemical bond and to supply the interaction energy with the surface. The rest of the photon energy is found as excitation energy of the internal degrees of freedom and/or kinetic energy of the fragments.

Photothermal desorption and ablation can be distinguished from quantum desorption because it possesses different properties. Mode selectivity and molecular selectivity are lost, due to rapid intramolecular and intermolecular energy exchange in the condensed phase. At very high heating rates, heating will be local in the irradiated region and even there deviations from complete thermodynamic equilibrium may occur. If we assume that desorption is mainly determined by the transient temperature achieved in the heated region due to effective energy transfer, the following characteristic behavior can be expected:

1) For a given coverage, the desorption or ablation yield varies in a strongly nonlinear way with the laser fluence.
2) The kinetic energy of the desorbed particles and the fragmentation pattern in the case of photodissociation depend on the laser fluence.
3) No molecular (isotope) selectivity can be expected.
4) A pronounced detection threshold is observed caused by the strong increase of the yield with the fluence.

There are many examples of this kind of behavior as will be discussed in detail later on. This is especially true for multilayer systems, where a small contribution due to quantum desorption may not be observable because photothermal desorption is extremely efficient.

A quantitative description of the photothermal effect is very difficult to achieve because the induced processes are not localized in the excited molecule. The deposited energy spreads over a macroscopic region including molecules not excited originally. The distribution of the energy into various degrees of freedom during the stimulated processes is very complicated in a molecular system and cannot be modelled on a molecular level.

A simple theory which is able to explain the basic features of TOF experiments, namely the fluence dependence of the ablation yield and the fluence dependence of the kinetic energy of the molecules leaving the surface, was developed in [3.23]. This model assumes fast equilibration of the excitation energy before desorption or ablation starts and is based on the following assumptions and simplifications:

1) The molecular film absorbs the laser radiation homogeneously. For an optical absorption coefficient of about 10^3 cm^{-1} and a film thickness of about $1\,\mu$m this is a good approximation. This leads to a nearly homogeneous heating process in the resonantly excited adsorbate or film. For short laser pulses, heat conduction to the surroundings may be negligible to a first approximation.

2) The final transient temperature reached by the laser-induced heating process can be calculated by integration, if the temperature dependence of the heat capacity is known. In the case of complete thermodynamic equilibration, the enthalpies needed for phase transitions, e.g. for melting, have to be included. On the other hand, it is possible to model overheating effects by assuming modified heat capacity functions, for example.

3) To determine the number of desorbed or ablated molecules and their kinetic energy, the statistical distribution function for a system with a variable number of molecules is used.

4) Only molecules with an enthalpy above the vaporization or sublimation enthalpy in the high energy part of the Boltzmann distribution can leave the surface. The number of these molecules in the tail states can be calculated by statistical concepts.

5) In statistical theory, the probability of emission of a molecule is independent of the depth in the multilayer film. A decreasing probability of emission with increasing distance from the top layer, due to cage effects, can be included by inserting a flexible weighting function.

6) The time-of-flight of the molecules to the detector can be estimated on the basis of a Maxwellian distribution or by considering a modified Maxwellian distribution including a stream velocity in the case of high yields.

3.2.3 Energy Considerations

It is interesting to compare the photon energy with the corresponding binding energies in the molecular system to see whether a single quantum process is energetically possible at the considered wavelength. In the molecular systems discussed here, weak van der Waals bonds and/or chemical bonds hold the condensed phase together. The van der Waals films considered mainly in this review consist of stable polyatomic molecules interacting with each other and the surface by van der Waals forces. According to statistical theory, the much weaker van der Waals bond should be broken first if the system is heated. Typical van der Waals binding energies are in the 0.3 eV range, and therefore even an IR photon may have enough energy for a single photon process (e.g. at 3 μm). In most experiments, powerful CO_2 lasers are employed and here 2–3 photons are typically needed to desorb a molecule, see Fig. 3.5. Thus quantum desorption from a van der Waals film by resonant excitation of internal vibrational modes with a CO_2 laser should show a quadratic or higher fluence de-

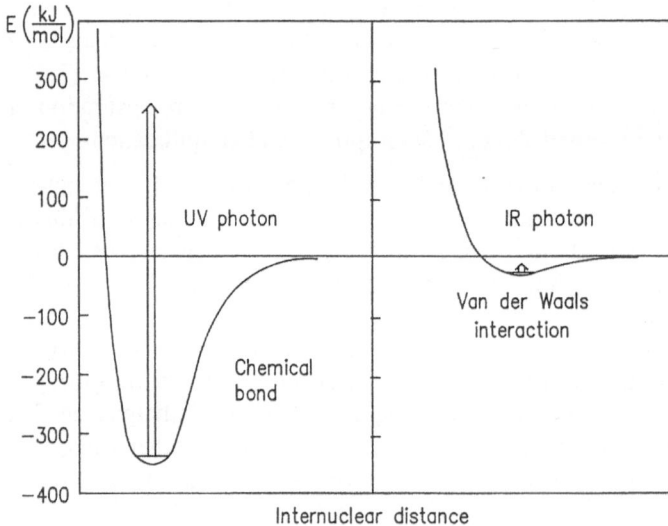

Fig. 3.5. Comparison of the energy for typical IR and UV photons with the interaction energy of a chemical bond and a van der Waals bond

pendence. With UV photons, on the other hand, not only van der Waals bonds but also chemical bonds can be broken from an energetic point of view (Fig. 3.5). In this case, the quantum yield tells us what fraction of the absorbed photons leads to photolysis and what fraction essentially heats the sample and finally breaks the van der Waals bonds. In this latter case, quantum yields greater than one may be observed, whereas in a chemically bonded network, the quantum yields are considerably lower.

Organic polymers are held together by a one-, two- or three-dimensional network of chemical bonds. To ablate material from such a network, chemical bonds must be destroyed. To overcome the chemical forces in a single photon process, UV photons are needed. As mentioned previously, pulsed excimer lasers can be used to induce photoablation. For the ArF, KrF and XeF lasers radiating at 193 nm, 248 nm and 351 nm, the photon energy is 6.4, 5.0 and 3.5 eV, respectively. This is larger than or comparable to the binding energy of a typical chemical bond e.g. 3.6 eV for C–C bonds as shown schematically in Fig. 3.5. This comparison shows that from the energetic point of view a direct photochemical decomposition of the chemical network is possible with excimer lasers. It has been suggested that the excitation of Rydberg states, in particular, should lead to direct bond dissociation.

We conclude from this discussion that the chemical interaction energy is comparable to the photon energy in the UV spectral region, whereas the van der Waals interaction energy is comparable to the energy of IR photons. Therefore, many IR photons are needed to dissociate a chemical bond. As mentioned before, multiphoton dissociation of an isolated polyatomic molecule follows a statistical pathway. Thus, there is little doubt that if polymer ablation is possible by the absorption of many IR photons this process is mainly due to heating, because IR multiphoton dissociation of a chromophore within the chemical network of a polymer may not be possible.

3.3 Time-of-Flight Technique

3.3.1 Experimental Determination of Time-of-Flight Distributions

The time-of-flight (TOF) technique has attracted increasing interest recently in the study of pulsed laser-induced desorption and ablation. Two different instruments are employed in TOF mass spectrometry, as shown in Fig. 3.6. The first type of instrument may consist of two lasers: one for desorption of neutrals and partial ionization and a second one for postionization in the gas phase to increase the number of ions. These ions are accelerated into and separated in an ion drift tube and detected with a channelplate detector. The advantage of this method lies in the registration of the whole mass spectrum of the different species desorbed by a single laser pulse. Therefore, this instrument is mainly used in laser mass spectrometry [3.24–27], e.g. to study the desorption of nonvolatile and thermally labile organic compounds.

The second TOF method is also based on pulsed laser desorption of neutrals. After a certain flight distance, the neutrals are ionized by electron bombardment in the ionizer of a quadrupole mass spectrometer. An ion produced with high abundance

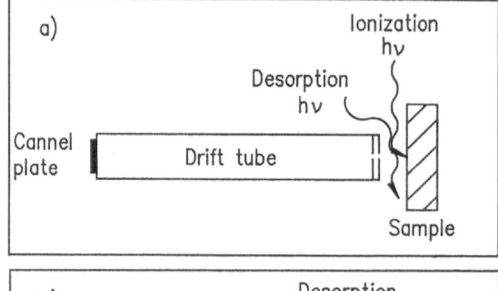

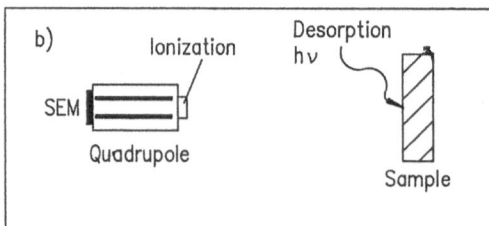

Fig. 3.6a,b. Schematic representation of the laser-based TOF techniques. (a) UV laser ionization and separation by a drift tube. (b) Electron impact ionization and separation by a quadrupole mass spectrometer

is selected for time-resolved detection in the quadrupole mass spectrometer. This technique allows the accurate registration of TOF distributions for selected masses. In the following, only the application of this latter method will be discussed in detail.

The TOF method using a quadrupole mass spectrometer enables accurate determination of the translational energies of molecules desorbed by IR or UV laser pulses. The neutral particles released from the surface are ionized with relatively low efficiency by electron impact. The ionized particles are extracted by ion optics into the quadrupole mass filter for mass selection and time-resolved detection. The drift time of the ions in the analyzer depends on the ionizer potential and is approximately proportional to the square root of the mass. This delay time can be measured as described in more detail later. After subtraction of the delay time, the TOF distribution can be transformed into a velocity distribution. The velocity and translational energy of a particle can be calculated from its mass, the flight time and the flight distance.

In the simplest case, the velocity distribution can be fitted to a Maxwellian distribution defining a translational temperature and an average translational energy. However, non-Maxwellian distributions are very often observed, due, for example, to collisional perturbations caused by high emission yields, in the case of multilayer ablation, responsible for a narrowing effect. Before information on the desorption process can be extracted from a measured TOF distribution an accurate correction of the ion drift time needs to be made. The accuracy needed for extraction of valid information depends on the flight distance employed. An estimation of the ion drift time from the parameters of the mass spectrometer may not be accurate enough and, therefore, an experimental determination should be preferred.

The ion drift time can be measured, for example, by generating ions in the ionization chamber with a pulsed high power laser. This experiment was performed with CO_2 laser pulses, which were focused onto the metal housing of the ionization chamber to generate Fe^+ ions [3.28]. The result is not a single drift time but a drift time distribution.

A second method of measuring the ion drift time is by switching the ionizer's extractor voltage e.g. by applying the accelerating potential in the form of a short voltage pulse [3.29]. The results were found to be in good agreement with the laser ionization method [3.28].

Information on the ion drift time can also be obtained from TOF distributions measured for fragment ions of different mass but this method does not seem as accurate as the former ones [3.28].

With an increase in flight distance, the influence of the drift time correction decreases. Therefore, it might be a reasonable procedure to take into account only a mean value of the ion drift time for large flight distances. For small flight distances of a few centimeters, however, the TOF distribution and the ion drift time distribution may possess a comparable halfwidth and thus, to obtain meaningful results, the measured drift time distribution has to be taken into account. Usually it is assumed that the desorption process occurs so fast that it will not influence the TOF analysis. In fact, if desorption occurs in less than 10^{-7} s and the time of flight is longer than 10^{-4} s the desorption process may be regarded as instantaneous.

From the corrected TOF distributions, the following information about the desorption process and the species desorbed can be obtained in principle:

1) translational temperature, 2) stream velocity, 3) emission yield, 4) identification of species.

To determine these properties, it is necessary to fit the measured TOF distributions to theoretical distributions derived from a desorption model. In the following section, these models will be discussed in more detail.

3.3.2 Theory of Time-of-Flight Distributions

Laser-induced desorption and ablation from a condensed phase at a very high heating rate may be a nonequilibrium process. However, the relaxation time yielding an equilibrium distribution of the translational energy is very short. Therefore, it is reasonable to expect that the velocity of desorbed species can be described by equilibrium concepts. Of course, this may not be true for other degrees of freedom of the system.

As a simple model, we may assume that the temperature versus time relationship of the transient heating process is not a significant variable. The particles are desorbed in a sharp pulse, essentially at the highest temperature achieved in a time that is short compared with the transit time of the pulse to the detector. A quantitative analysis is possible if we assume a Maxwellian distribution of velocities, as is characteristic, for example, of an effusive source [3.30–32]. In this case, the number of molecules desorbing from the area A_s into the solid angle $d\omega$ per unit time with a velocity between v and $v + dv$ is given by [3.32]

$$\frac{dN}{dt} d\omega = C A_s \left\{ \frac{m}{2\pi k_B T} \right\}^{3/2} v^3 \exp\left(-mv^2/2k_B T\right) dv\, d\omega \quad , \tag{3.3.1}$$

where C is a constant with the dimensions of a number density. The solid angle $d\omega = A_d/l^2$ depends on the area A_d of the detector and the distance l between emitting surface and detector.

The mass spectrometer measures the density of particles n and not the flux of particles. These two properties are related by

$$dn = \frac{1}{A_d} \frac{1}{v} \frac{dN}{dt} d\omega \quad . \tag{3.3.2}$$

This yields the following expression for the number density:

$$dn = C \frac{A_s}{l^2} \left(\frac{m}{2\pi k_B T} \right)^{3/2} v^2 \exp\left(-\frac{mv^2}{2k_B T} \right) dv \quad . \tag{3.3.3}$$

The velocity distribution must be converted into a time distribution because the experiment gives the density of particles as a function of time at the ionizer, which is at a distance l from the surface.

With the time of emission t' and the time of arrival t, the time of flight is obtained from $\tau = t - t'$ and the velocity from the target to the ionizer is given by

$$v = \frac{l}{t - t'} = \frac{l}{\tau} \quad . \tag{3.3.4}$$

The variable is transformed to τ using

$$dv = \left(-\frac{l}{\tau^2} \right) d\tau \quad . \tag{3.3.5}$$

This yields

$$dn = C' \frac{A_s}{l^2} \left(\frac{m}{2\pi k_B T} \right)^{3/2} \frac{l^3}{\tau^4} \exp\left[-\frac{m}{2k_B T} \left(\frac{l}{\tau} \right)^2 \right] d\tau \quad . \tag{3.3.6}$$

The density of particles $n(t)$ is obtained by integrating over all desorption times. As long as the desorption can be considered to be instantaneous, we can determine the value of the integral

$$\int_0^\infty C' \, dt = \Delta n \, \Delta t = a \tag{3.3.7}$$

from the total number of molecules desorbed.

Introducing the expression for the most probable velocity

$$v_0 = \frac{l}{t_0} = \sqrt{\frac{2k_B T}{m}} \quad , \tag{3.3.8}$$

we obtain the final result

$$n(t) = a \frac{A_s}{l^2 \pi^{3/2}} \frac{t_0^3}{t^4} \exp\left(-\frac{t_0^2}{t^2} \right) \quad . \tag{3.3.9}$$

Equations of this type have been reported in a series of papers [3.30–33].

The translational temperature of the desorbing molecules can be determined from the time t_m where the TOF distribution (3.3.9) reaches its maximum:

$$t_m = t_0 / \sqrt{2} \tag{3.3.10}$$

this yields the expression

$$T = \frac{m}{4k_\text{B}} \frac{l^2}{t_m^2} \; .$$ (3.3.11)

Equation (3.3.11) allows one to determine the TOF temperature from the position of the maximum in the measured TOF distribution.

In the following, a modification of the basic TOF distribution is considered, describing additional effects observed in the case of efficient desorption, e.g. for multilayer ablation. In [3.30] it was already suggested that the velocity distribution may be superimposed on a center-of-mass motion if the number of desorbed molecules is high. A more detailed model can be found in [3.34]. These authors assume a vaporization of particles in a small region of the surface. This hot gas bubble begins to expand against the surface and imparts momentum to it. Thereby, the center of mass of the particles gains a stream velocity in the laboratory coordinate frame. An observer moving with the center of mass would observe an expanding gas bubble. As the number density decreases further and further, the velocity distribution corresponding to a certain equilibrium temperature finally becomes frozen. This equilibrium temperature is a kinetic temperature which may not be in equilibrium with other degrees of freedom and differs from the surface temperature.

The Maxwellian velocity distribution of an effusive beam is therefore modified in this case by introducing the stream velocity u, which leads to a narrowing of the half-width of the TOF distribution. The corresponding velocity distribution is usually written in the following form for a density detector such as an electron bombardment analyzer [3.34–36]:

$$f(v) = Nv^2 \exp\left[-\frac{(v-u)^2}{v_0^2}\right] \; .$$ (3.3.12)

This distribution is not only used to describe pulsed laser-induced desorption and ablation, but also to study effusive molecular beams and supersonic nozzle beams [3.36, 37]. In the thermodynamic description, the stagnation enthalpy of the particles is partially converted into a kinetic energy of directed mass flow, expressed by the stream velocity. The rest enthalpy is the thermal energy characterized by the most probable random velocity defined in the center of mass coordinate frame. The degree of energy conversion into directed mass flow is measured by the Mach number M:

$$M = \sqrt{\frac{2}{\gamma}} \frac{u}{v_0} \; ,$$ (3.3.13)

where γ is the specific heat ratio. The number M^2 gives a measure of the ratio of the kinetic energy of directed mass flow to thermal energy. In a supersonic beam, very high Mach numbers above 1000 can be achieved, whereas in a pulsed laser ablation experiment the value may be below 2 in most cases.

3.3.3 Analysis of Time-of-Flight Distributions

The correct recording and analysis of TOF distributions is a difficult problem. First of all, the shape of the distribution can be changed by varying the instrumental parameters of the mass spectrometer. Therefore, only with great care can experimental artifacts be avoided. Second, the corrections applied to the TOF profile by the ion drift time in the quadrupole and by the stream velocity are often relatively small, but still produce strong effects. This means that only an accurate analysis yields the correct results.

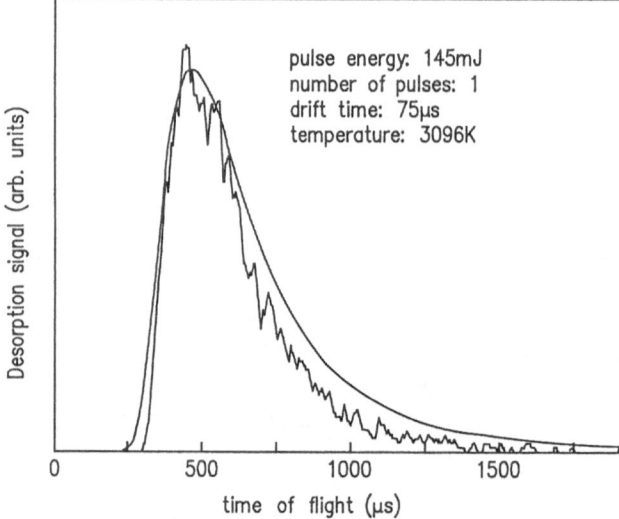

pulse energy: 145mJ
number of pulses: 1
drift time: 75µs
temperature: 3096K

Fig. 3.7. Single pulse TOF distribution measured for ablation of $C_6H_{10}C_{12}$ from a multilayer film [3.28]. The calculated distribution is a Maxwell distribution for $T = 3096\,K$ and an ion drift time correction of 75 μs

Figure 3.7 shows an example of the narrowing effect observed in the case of strong IR laser ablation from a van der Waals film of $C_6H_{10}Cl_2$ [3.28]. The measured distribution is compared with the corresponding pure Maxwellian distribution possessing the same maximum. This measured TOF distribution, however, can also be fitted to a pure Maxwellian by changing the ion drift time correction by a factor of two and the TOF temperature as displayed in Fig. 3.8. This example clearly illustrates the important role played by the ion drift time correction, which should be determined with the highest accuracy obtainable. Figure 3.9 shows a fit of the modified Maxwell distribution to the same measured distribution. This yields the lowest temperature and a stream velocity of 393 m/s.

3.4 IR Laser-Induced Desorption and Ablation

3.4.1 Overview of Systems

IR laser-induced desorption and ablation by resonant vibrational excitation of an adsorbate has been studied in recent years in systems ranging from monolayer coverage to condensed multilayers of polyatomic molecules. The most frequently used source of radiation was a line-tunable pulsed CO_2 laser in the 10 μm spectral region.

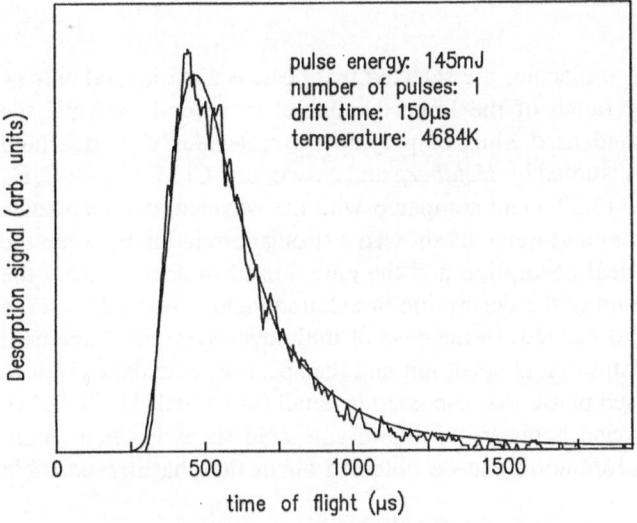

pulse energy: 145mJ
number of pulses: 1
drift time: 150μs
temperature: 4684K

Fig. 3.8. TOF distribution of Fig. 3.7 fit to a Maxwell distribution for $T = 4684\,K$ with an ion drift time correction of $150\,\mu s$

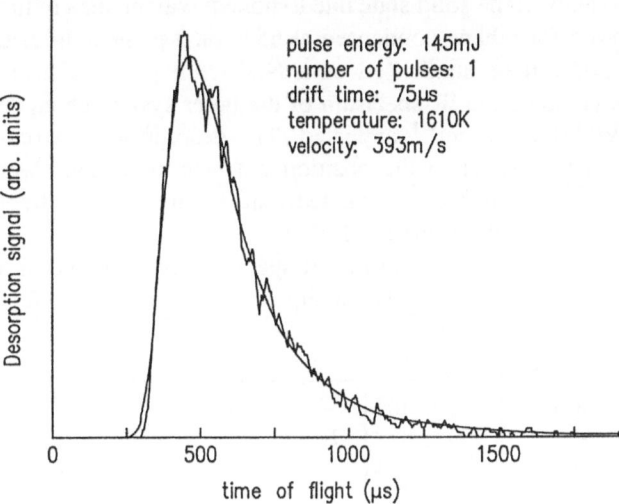

pulse energy: 145mJ
number of pulses: 1
drift time: 75μs
temperature: 1610K
velocity: 393m/s

Fig. 3.9. TOF distribution of Fig. 3.7 fitted to the modified distribution. The calculated distribution belongs to $T = 1610\,K$ and a stream velocity of $u = 393\,m/s$. The ion drift time correction is $75\,\mu s$

Only a few molecular systems have been studied employing other IR lasers, among them the desorption of CO and NO molecules in the $5\,\mu m$ region and ammonia near $3.4\,\mu m$.

Depending on coverage, the resonant desorption and ablation effect is influenced to a greater or lesser extent by the optical and thermal properties of the whole system, including the substrate. Therefore, several molecules were studied on different substrates such as dielectrics, semiconductors and metals to obtain information about substrate heating effects, heat conduction, etc.

Detailed studies were performed for the following systems: CH_3F–NaCl [3.11, 38-40]; C_5H_5N–KCl, Ag [3.41, 42]; CCl_4–Ge [3.43]; SF_6–NaCl [3.44]; C_6H_5CHO–Ge [3.45]; CH_2ClCH_2Cl–Ge [3.46]; $C_6H_{10}Cl_2$–Ge, Au [3.47, 48]; NH_3,ND_3–Cu, NaCl, Ag [3.49, 50]; C_2H_4–NaCl [3.51]; $C_6H_5CF_3$–C [3.52]; NO–C [3.52]; C_6H_6–Au, NaCl [3.53].

3.4.2 Wavelength Effects

To study wavelength effects, molecules are selected that possess a strong and narrow infrared band in the tuning range of the lasers available. In several systems, the optical absorption of the condensed films of polyatomic molecules was measured (see for example the systems studied by *Heidberg* and co-workers CH_3F [3.38]; C_2H_4 [3.51]; $C_6H_5CF_3$ [3.52]; NO [3.52]) and compared with the wavelength dependence of the desorption yield. These measurements showed a strong correlation between the position and strength of optical absorption and the emission of molecules from the surface. Usually, the maximum of the desorption or ablation yield coincided with the center of the vibrational band excited. In the case of multilayer systems, agreement is expected between the ablation yield spectrum and the spectroscopic data given in the literature for the condensed phase. As discussed in detail for CH_2ClCH_2Cl [3.46], the shift of the vibrational band between the liquid and solid states is often small, and therefore no detailed information can be obtained about the phase responsible for energy deposition.

From the spectroscopic point of view, interesting systems are those where the band splits into several components in the solid state due to nonequivalent sites in the crystal. This effect was observed for ethene condensed at 65 K on a plane substrate [3.54] and below 30 K for solid ethene condensed on a NaCl film [3.51]. During annealing at temperatures above 35 K, the IR spectrum of the latter system changed and a single band with a FWHM of 35 cm^{-1} appeared. The desorption spectrum agreed with this broad band with respect to the position but was much narrower with a FWHM of about 8 cm^{-1}. The authors speculated that this narrowing effect may be caused by a multiphoton excitation process [3.51].

Another molecule with a distinct difference in the IR spectrum of the liquid and solid states due to splitting is benzene, as shown in Fig. 3.10. The ablation yield

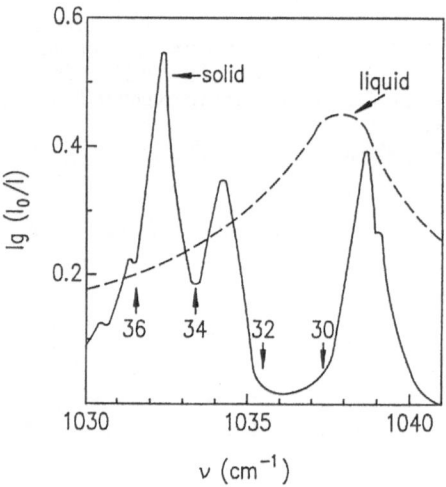

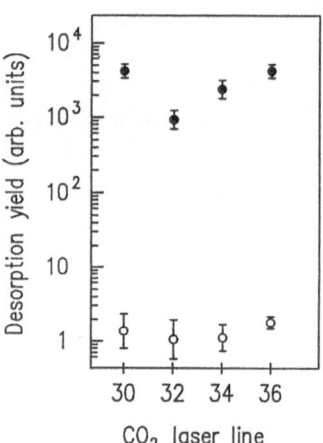

Fig. 3.10. Spectrum of the ν_{18} vibrational band of solid and liquid benzene. The arrows indicate the CO_2 laser lines employed for studying the wavelength dependence

Fig. 3.11. Spectral dependence of the desorption yield for two films of benzene: (●) 500 monolayers; (○) 75 monolayers, laser fluence 1.3 J/cm^2

spectrum (Fig. 3.11) and also the wavelength dependence of the kinetic energy of the species desorbed follow the solid-state spectrum in the case of thicker films and are not consistent with the spectrum of the liquid phase [3.53]. This is surprising because it indicates that energy deposition occurs in the solid state despite the relatively high translational temperatures of the ablated molecules. This effect could be easily understood if, at the high heating rates employed, the laser-induced process is sublimation caused by overheating of the solid phase. Direct laser sublimation without melting is possible at relatively small deviations from thermodynamic equilibrium, which cannot be excluded at heating rates of about 10^{10} K/s. To confirm this interpretation, the optical absorption of the solid benzene film needs to be measured in situ for the deposited film.

For a series of molecules with similar 10 μm bands in the liquid and solid state, condensed films with a thickness of several micrometers were studied to determine TOF distributions as a function of wavelength. The mass spectrometer signal decreased drastically from the center of the infrared band to the wing, and therefore many laser pulses had to be averaged in the wing region. From these TOF distributions, the relative ablation yield could be estimated by integrating the distributions, and in addition the translational temperatures were determined as a function of wavelength. Such an analysis was performed for CCl_4 [3.43], CH_2Cl CH_2Cl [3.46], C_6H_5CHO [3.45] and $C_6H_{10}Cl_2$ [3.47]. Typically, the TOF temperatures changed by no more than a factor of two, whereas the ablation yield varied by about two orders of magnitude in the investigated wavelength region. Furthermore, an Arrhenius relationship between the ablation yields and the reciprocal TOF temperatures was found, giving an activation energy in the range of the interaction energy of the condensed phase. This strongly suggests a thermally activated process due to resonant heating of the condensed phase with a wavelength dependence determined by the wavelength dependence of the optical absorption coefficient.

This interpretation is valid despite the fact that in these experiments the ion drift time was not measured but estimated from the parameters of the mass spectrometer and was possibly wrong. In view of our recent results described in [3.28], we believe that the correct drift time correction would have resulted in TOF distributions narrower than Maxwellian distributions at the relatively high ablation yields. The consequence of a fit of the measured distributions to modified Maxwell distributions would have been the inclusion of a stream velocity and lower translational temperatures without changing the main conclusions.

As an example of this kind of wavelength effect (which can only be observed in the infrared spectral region with narrow absorption bands) the results reported in [3.47] for *trans*-1,2-dichlorocyclohexane are presented. Figure 3.12 illustrates the IR spectrum of this molecule with the excited band in the 10 μm region. Figure 3.13 shows the wavelength dependence of the ablation yield and the TOF temperatures. Note the logarithmic scale in the yields plot. This strong variation of the number of emitted molecules makes TOF experiments near the wings of the IR band very difficult. Figure 3.14 shows the Arrhenius plot. The activation energy of 33 kJ/mol obtained from the slope is somewhat smaller than the estimated heat of vaporization of about 40–50 kJ/mol.

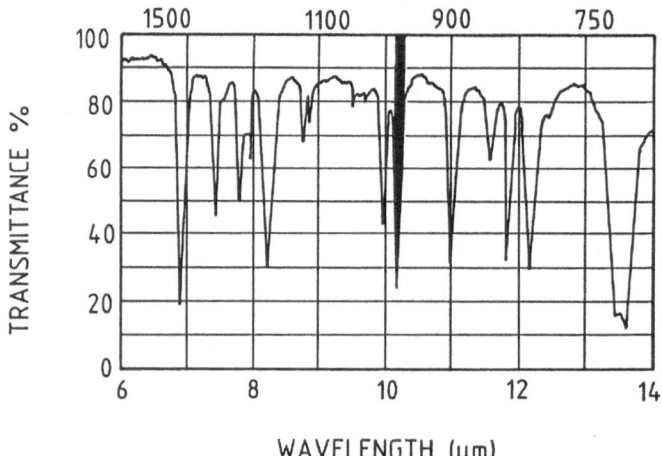

Fig. 3.12. IR absorption spectrum of liquid *trans*-1,2 dichlorocyclohexane. The bar indicates schematically the spectral region investigated with the CO$_2$ laser

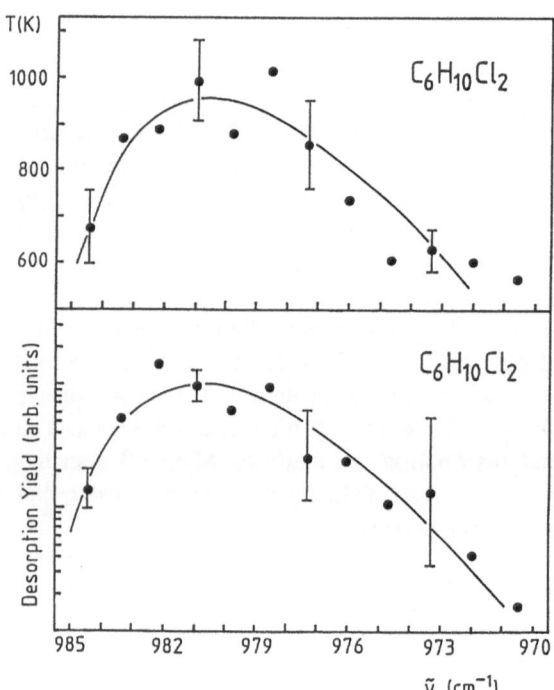

Fig. 3.13. Spectral dependence of the translational temperature (*upper part*) and desorption yield (*lower part*) determined for *trans*-1,2-dichlorocyclohexane

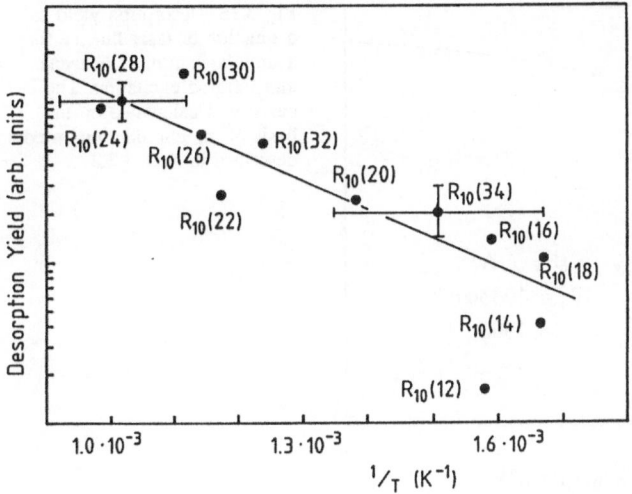

Fig. 3.14. Semilog plot of the desorption yield versus reciprocal translational temperature for resonant desorption of *trans*-1,2-dichlorocyclohexane

3.4.3 Fluence Dependence

As discussed previously, the fluence or intensity dependence of the desorption or ablation yield is different for quantum and thermal processes. Therefore, this dependence is an important criterion for mechanistic studies. Unfortunately, it is very difficult to determine the yield versus fluence dependence in a large dynamic range without changing other parameters. For example, the contribution of substrate heating may increase with increasing fluence, and therefore change the behavior. It is often also very difficult to vary the laser fluence by a substantial amount without changing the beam quality. For these and other reasons, most results obtained until now have not been very conclusive.

Most of the fluence dependences reported in the literature so far have been nonlinear and consistent with a thermal process. In a few cases, the plots were interpreted in terms of two-photon or multiphoton processes [3.38, 52]. As already mentioned, the absorption of two, three or more photons would be necessary in most systems to initiate a genuine quantum effect, i.e. desorption of the excited molecule without thermal equilibration. To study the strong nonlinear behavior characteristic of a thermal process, a change of the desorption and ablation yield over 5–6 orders of magnitude needs to be investigated. This is not the case for most of the systems studied so far. As an example of this type of data, Fig. 3.15 shows the ablation yield of benzene. The points are experimental results [3.53] and the dependence represented by the full line was calculated on the basis of the photothermal model [3.23]. The kinetic energy of the desorbed species as a function of fluence is another sensitive criterion for obtaining information on the mechanism involved. This criterion can easily be applied to TOF experiments, if the position of the TOF maximum is considered as a rough measure of the kinetic energy of the species desorbed. A quantum desorption process would imply that the position of the TOF maximum is not shifted with increasing laser fluence. This kind of behavior has not been observed in IR photodesorption experiments. In several molecular systems it has been verified

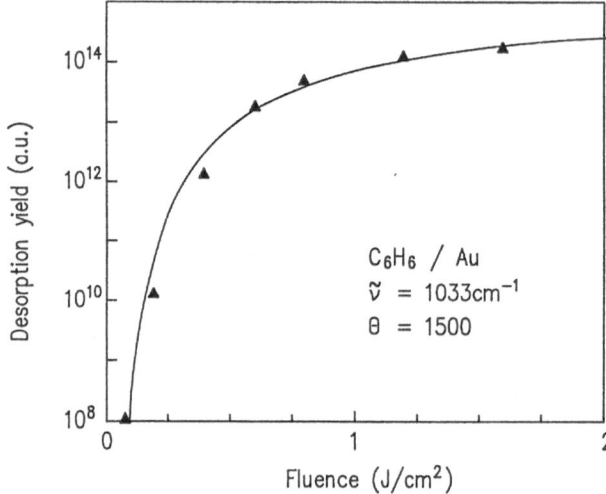

Fig. 3.15. Desorption yield as a function of laser fluence for a multilayer film of benzene and infrared excitation. The curve was calculated on the basis of the photothermal model described in Sect. 3.2.2

that the TOF maximum was shifted to shorter flight times by increasing the laser fluence. This is consistent with a higher translational temperature of the desorbing species at higher fluences and with a thermal desorption mechanism.

A quantitative analysis, taking into account additional effects such as collisional narrowing, substrate heating, etc. had not hitherto been performed for a desorption system. Figure 3.16 shows the shift of the TOF maximum as a function of fluence measured for benzene films [3.53]. The full line gives the result of a model calculation assuming photothermal desorption and ablation [3.23]. There are several reasons why the same excellent agreement cannot be expected for the kinetic energy as is obtained for the yield.

3.4.4 Selectivity and Mechanisms

To the best of our knowledge, there is no generally accepted example of a bond or mode selective process stimulated by vibrational excitation in the IR region for an

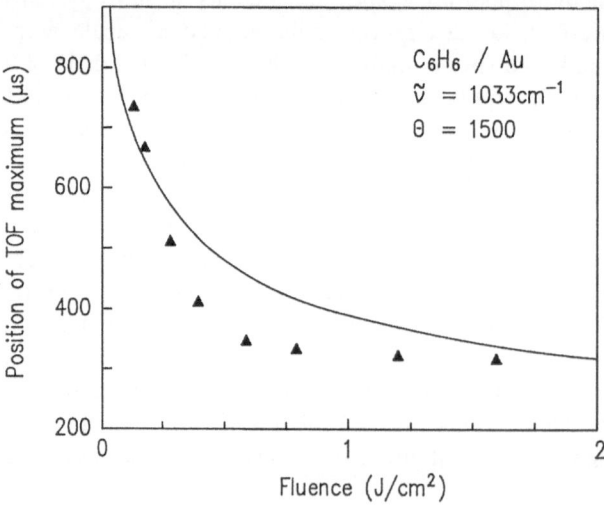

Fig. 3.16. Shift of the position of the TOF maximum as a function of fluence for a multilayer film of benzene and infrared excitation. The curve was calculated on the basis of the photothermal model (Sect. 3.2.2)

isolated molecule. Therefore, there is no real reason to expect that this ultimate level of selectivity can be achieved for molecules interacting with other molecules in the adsorbed or condensed state.

On the other hand, molecule selective desorption following resonant vibrational excitation is a topic discussed in several papers. In an IR photodesorption experiment involving H_2, HD and D_2 coadsorbed on porous Vycor glass at 120 K, excitation of the presumably physisorbed HD molecules in the 2.73–2.88 μm spectral range resulted in a preferential desorption of HD and a corresponding deuterium enrichment in the gas phase [3.55]. These results were not confirmed when the experiment was repeated under more carefully controlled conditions [3.56].

Molecule-selective desorption of CH_3F coadsorbed with C_2H_6 on a NaCl film was reported in [3.57] when CH_3F was resonantly excited at 976 cm^{-1}. No C_2H_6 molecules were detected with the mass spectrometer, which did not absorb in this spectral region. It is not known whether this system was also studied by another group. In summary, we conclude that there is no confirmed and generally accepted example of molecule or isotope selectivity following resonant IR excitation of an internal vibrational mode of adsorbate molecules.

A strong nonlinear fluence dependence of the desorption and ablation yield was observed in several systems. This dependence can be described quantitatively by a photothermal model. However, a suggestion has also been made that this nonlinearity is due to multiphoton absorption. The main argument in favor of the photothermal model is that it also describes the threshold behavior, which depends on the thickness of adsorbate film. This cannot be understood on the basis of a multiphoton excitation process, moreover the measured threshold fluences are too low for such a nonlinear absorption process.

The observed dependence of the kinetic energy of emitted molecules on laser fluence is also consistent with a photothermal process, despite the fact that only a qualitative description of the functional dependence was achieved.

In summary, the published data puts forward convincing evidence for a more or less complete equilibration of the absorbed radiation energy before IR laser-induced desorption or ablation occurs in molecular systems such as van der Waals films.

3.5 UV Laser-Induced Desorption and Ablation

3.5.1 Overview of Systems

The utilization of intense laser pulses in the visible and ultraviolet spectral region to stimulate desorption of adsorbates and ablation from condensed films and polymers has recently been the focus of considerable attention. In several studies, the TOF technique was employed to obtain information on processes induced by laser radiation. With visible light, desorption in the submonolayer coverage range and ablation in the multilayer coverage range were investigated by substrate heating. Examples are the desorption and ablation of O_2 and CO from polycrystalline copper [3.58], where subthermal TOF distributions were observed corresponding to temperatures below that of the surface. Time-of-flight studies were performed in the infrared spectral region for the systems CO-Fe (110) [3.32] and CH_3F-NaCl [3.59] and in the ultraviolet (248 nm) for the system CO-Cu (100) [3.60], resulting also in subthermal translational temperatures. A similar effect was observed for the excimer laser photolysis (193 nm) of methyl-aluminum complexes (trimethylaluminum) chemisorbed on oxide surfaces [3.61]. Methyl molecules were found to desorb with reasonable efficiency and Maxwell-Boltzmann fits to the corresponding TOF distributions yielded temperatures around 150 K even though desorption occurred from surfaces at room temperature. Thus, low kinetic energies were observed not only for desorption of intact molecules, but also for laser-induced surface reactions.

Kinetic energies and populations of internal states were reported for laser-induced desorption of NO from a platinum foil [3.62, 63]. These experiments were performed by exposing the platinum surface to a saturation dose and maintaining a low background pressure of NO during the experiment. Two desorption channels were found in this system, one with internal state populations which were well fit by Boltzmann distributions that correspond to the peak surface temperature and a second one displaying non-Boltzmann population distributions. In this latter channel, the kinetic energies exceeded the peak surface temperature by a factor of about 5 and the vibrational populations by a factor of as much as 50.

Submonolayers of CH_3Br physisorbed on a LiF (001) surface at 115 K were examined by TOF mass spectrometry using pulsed UV radiation [3.64, 65]. Translational energy and angular distributions were measured for photofragments and photodesorbed molecules. Single photon adsorbate photolysis produced the photofragments CH_3 and Br at 222 nm, which were characterized by the TOF spectrum. In addition, photodesorption of CH_3Br was detected. The nonthermal desorption of these species was attributed to UV absorption by the LiF crystal. In similar experiments with H_2S as the adsorbate, not only photodissociation but also photoreactions at the surface were observed, when the H_2S coverage on LiF (001) approached one monolayer. Again, efficient desorption of the adsorbate was detected due to absorption of UV light by the LiF substrate. It is suggested that the excitation of color centres in the crystal generates a photoacoustic shock, which is responsible for desorption.

Ablation from condensed multilayer films of polyatomic molecules, including chemical effects, was investigated recently for a number of molecules by means of resonant electronic excitation. Photoablation and photodissociation of surface molecules was observed for thick condensed films (200 μm) of pure NH_3 and H_2O and mixtures of these molecules at 193 nm [3.67]. The translational energies of the ablated molecules were in the range of the intermolecular interaction energies but were not changed with increasing laser fluence in the studied range. Films containing NH_3 released atomic hydrogen with a translational energy comparable to gas phase photolysis of NH_3 molecules and also molecular hydrogen with a translational temperature close to the surface temperature. Photofragmentation and photoablation was reported for multilayer films of CH_3Br physisorbed on LiF at 30 K [3.68]. One photon fragmentation was found at 193 nm, leading to broad, non-Boltzmann distributions for CH_3 and Br, influenced by subsequent changes of the primary states by collision processes. Molecules were ejected collisionally but also by a thermal mechanism, which, however, was not very efficient.

For H_2S physisorbed on LiF(001), photodissociation into H and HS was observed in the submonolayer region at 193, 222 and 248 nm [3.69, 70]. The translational energy distributions of the H photofragment were structured due to vibrational excitations of the HS fragment. Photoreactions within the adsorbate layer occurred at H_2S coverages > 0.1 monolayer, where different translational energy distributions were observed for the product H_2, interpreted as being due to direct and indirect dynamics [3.70]. At all coverages, photodesorption of H_2S molecules with low translational energies (0–0.5 eV) was detected and ascribed to the substrate-mediated photoacoustic shock wave mechanism. In the multilayer coverage region, photoablation of H_2S molecules with translational energies up to several eV was observed. This effect was explained by intermolecular quenching of electronically excited H_2S [3.70].

A detailed study of electronically excited photodissociation, desorption and ablation was also performed for CH_2I_2 molecules adsorbed on Al_2O_3 and Ag surfaces [3.71–73]. Excitation of CH_2I_2 into an antibonding state at 308 nm produced CH_2I radicals and I atoms. For these photofragments, a linear yield dependence and an invariance of the kinetic energy as a function of laser fluence was observed which is characteristic of the direct detection of processes resulting from electronic excitation. In the multilayer coverage region, efficient photoablation of CH_2I_2 molecules occurred, characterized by very high yields and high kinetic energies of the ejected particles.

For UV laser-induced desorption and ablation of NO at 220-270 nm from condensed films, not only the translational but also the internal energy distributions were studied [3.74]. The only significant product observed by the TOF technique were NO monomers in the electronic ground state. There were two distinct TOF components leading to bimodal TOF distributions under conditions where both mechanisms were detected. A slow peak with translational energies up to 0.06 eV was Maxwellian, possessing a broad angular distribution. The yield of this component increased strongly with layer thickness and exponentially with laser pulse energy. The fast TOF peak was non-Maxwellian with an average energy of 0.22 eV, an angular distribution which peaked toward the surface normal and a yield increasing linearly with laser pulse energy. Resonance enhanced multiphoton ionization (REMPI) was employed to probe vibrational and rotational populations in the fast peak, suggesting incomplete equilibration.

Resonant desorption and ablation from benzene films condensed on quartz and copper substrates was investigated by resonant excitation at 248 nm [3.75]. At the fluence values of 70 mJ/cm^2, no chemical effects such as the formation of naphthalene were observed for the transparent substrate. However, in the case of transient heating of a copper substrate, chemical reactions *were* clearly detected. Figure 3.17 shows the mass spectrometer signal for benzene (78) and naphthalene (128) as a function of the number of laser pulses irradiating the same spot. For the first ten laser pulses, the naphthalene concentration is far above the impurity level. Broadened and structured TOF distributions were found for the first laser pulses irradiating a multilayer film of benzene, as exhibited in Fig. 3.18. This picture illustrates the shift of the position of the maximum in the TOF distributions. The TOF distributions obtained in the later stages of ablation, e.g. after about 50 pulses, could be fitted to a modified Maxwellian distribution, as shown in Fig. 3.19. Gas chromatic analysis

Fig. 3.17. Mass spectrometer signal as a function of UV laser ▶ pulses (248 nm) irradiating the same spot of a benzene film condensed on Suprasil I (benzene: $m/e = 78$; naphthalene: $m/e = 128$)

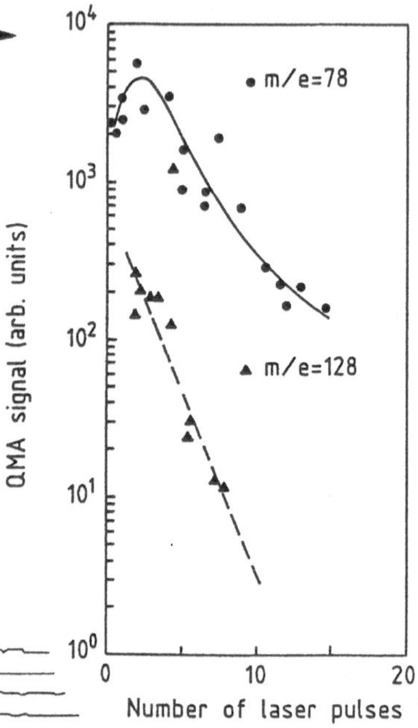

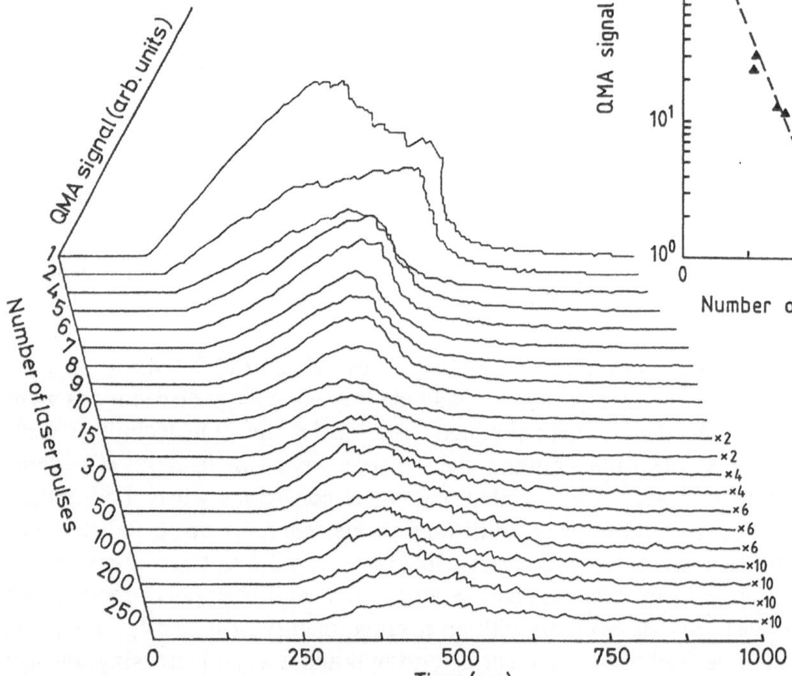

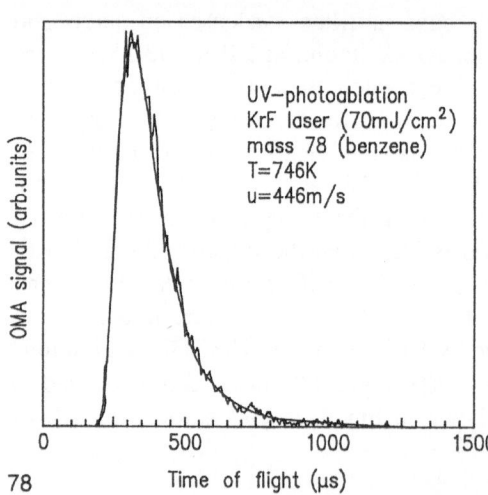

Fig. 3.18. TOF distributions as a function of the number of UV laser pulses irradiating the same spot of a benzene film condensed on a glass substrate

UV–photoablation
KrF laser (70mJ/cm^2)
mass 78 (benzene)
T=746K
u=446m/s

Fig. 3.19. Single shot TOF distribution measured after about 50 UV laser pulses had irradiated the same spot. A fit to the modified Maxwell distribution yields $T = 746\,\text{K}$ and $u = 446\,\text{m/s}$

of liquid benzene irradiated at 248 nm under ablative conditions showed many products at higher laser fluences [3.76]. Above $100 \, mJ/cm^2$, biphenyl and as many as six dimeric products were formed, including phenyl 1,3- and 1,4-cyclohexadienes. When the fluence exceeded $300 \, mJ/cm^2$, new products such as carbon and naphthalene were found in the irradiated mixture. Therefore, TOF experiments need to be performed at higher fluences in order to investigate these chemical effects in more detail.

TOF distributions were also measured for cryogenic Cl_2 films condensed on a transparent substrate at 25–110 K [3.77]. The predominant species ablated was molecular chlorine, with less than 7% atomic chlorine. Translationally fast molecules were observed at 193, 248, 351 and 355 nm, but no vaporization was detected for 532 and 1064 nm. The number of molecules ablated and their kinetic energy increased with increasing laser fluence and film thickness, consistent with a photothermal process. The measured TOF distributions were narrower than Maxwellian distributions, as expected for the relatively high ablation yield observed. The data could be well fit to the modified Maxwellian distribution with a superimposed flow velocity.

Photodissociation and photoablation from condensed thick layers of CH_3I (150 K) by UV laser pulses at 266 nm were reported in [3.78]. At low fluences, the signal was dominated by direct photodissociation products. In TOF measurements of CH_3, the two decay channels known from gas phase photodissociation were identified. At fluences above $1.5 \, mJ/cm^2$, large amounts of material were removed from the solid, including molecular clusters. The authors point out that in this region the process becomes similar to the UV photoablation of polymers.

Despite a rapidly increasing number of papers published recently on polymer ablation, only two have applied the TOF method to a more detailed study of the nature and properties of the ablated species. In the first paper, polymethylmethacrylate (PMMA) samples were irradiated at 193 nm [3.79]. The TOF distributions of the monomer unit methylmethacrylate (MMA) were measured as a function of laser fluence. In the fluence range below $120 \, mJ/m^2$, they corresponded to Maxwell distributions, yielding temperatures from 800 K at $60 \, mJ/cm^2$ to about 3000 K at $120 \, mJ/cm^2$. The same common temperature was also determined for other ablation products. At higher fluence values, e.g. $300 \, mJ/cm^2$, the distributions broadened considerably and could no longer be fitted to Maxwell distributions. In addition, a new peak showed up in the TOF spectrum caused by fast particles with high kinetic energy, e.g. more than 10 eV for MMA. These fast particles showed a high directionality.

In the second polymer paper using TOF analysis, the photoablation of polystyrene was investigated at 193 nm [3.80]. Near the threshold fluence of $15 \, mJ/cm^2$, the main neutral ablation product was the styrene monomer. The maximum of the kinetic energy distribution of these molecules was located near 0.7 eV. On the other hand, the mean internal energy of the monomers was less than 2 eV. This indicates large deviations from a thermal distribution of the energy, because a kinetic energy of 0.7 eV would correspond to an internal energy of more than 10 eV.

3.5.2 Photochemical Effects

The energy of UV photons is in the range of chemical binding energies, thus a one-photon dissociation of adsorbed molecules and their desorption is possible using excimer laser pulses. In fact, photofragmentation has been observed for many polyatomic molecules, as discussed in the preceding section. Photofragmentation can be observed at low fluence values because no threshold fluence exists for direct photodissociation. A low coverage is advantageous for observation purposes because the contribution of photodesorption of intact molecules occurring simultaneously is small under these conditions.

If the photofragments and molecules possess a different mean kinetic energy, a bimodal TOF distribution will be observed and the efficiency of the two channels can be studied as a function of coverage, fluence, etc. Normally, the contribution of the photothermal channel increases drastically with increasing coverage and fluence.

Single-photon photodissociation has been described in the literature for the following molecules: NH_3 [3.67] CH_3Br [3.64, 68], H_2S [3.70], CH_2I_2 [3.72] and $(NO)_2$ [3.73]. In some cases the energy of the fragments was similar to that observed in the photolysis of the free molecules because the interaction energy in the van der Waals film is small, e.g. for H from NH_3 [3.67] and CH_3 and Br from CH_3Br [3.68]. Considerably smaller energies are also found due, for example, to collisional effects, and even for H, structured translational energy distributions indicative of vibrational excitation of SH fragments were described [3.70]. The observed widths of the TOF distributions for CH_2I and I obtained by photofragmentation of CH_2I_2 were rather undisturbed, suggesting that the particles experienced few collisions and suffered little energy loss during the desorption process. The large differences between the available energies and the observed translational energies were most likely due to internal excitation of the CH_2I molecules in the primary photodissociation step. Only about 11% of the available energy appeared as translational energy of the CH_2I fragment [3.74].

An important criterion for the direct decomposition of molecules at the surface by a laser photon is the fluence dependence of the TOF distribution. For a photochemical process, the peak position should not vary with laser fluence, but only the area under the TOF distribution. This was clearly observed for the CH_2I_2-Al_2O_3 system, where the shapes of the TOF distributions and the corresponding translational energies were independent of fluence in a wide range [3.73]. Furthermore, it was shown that the peak positions and signal shapes remained unchanged with the number of laser pulses, as shown in Fig. 3.20, despite the fact that the signal decreased gradually. This should be compared with the photothermal behavior shown in Fig. 3.18. With increasing coverage, an increase in signal strength was found as expected [3.73]. This is shown in Fig. 3.21. These results provide sound evidence of a direct photochemical decomposition, despite a low quantum yield for photodissociation, probably in the 10^{-2}–10^{-3} range [3.73]. The independence of the peak position from the laser fluence should be observable in the low coverage region, where collisions do not disturb the kinetic energy. A constant mean translational energy was also detected in the NO dimer system, however, in this case the TOF distributions were broader than Maxwell distributions [3.74]. The criterion was also verified in the CH_3Br-LiF(001) system at low coverages [3.66].

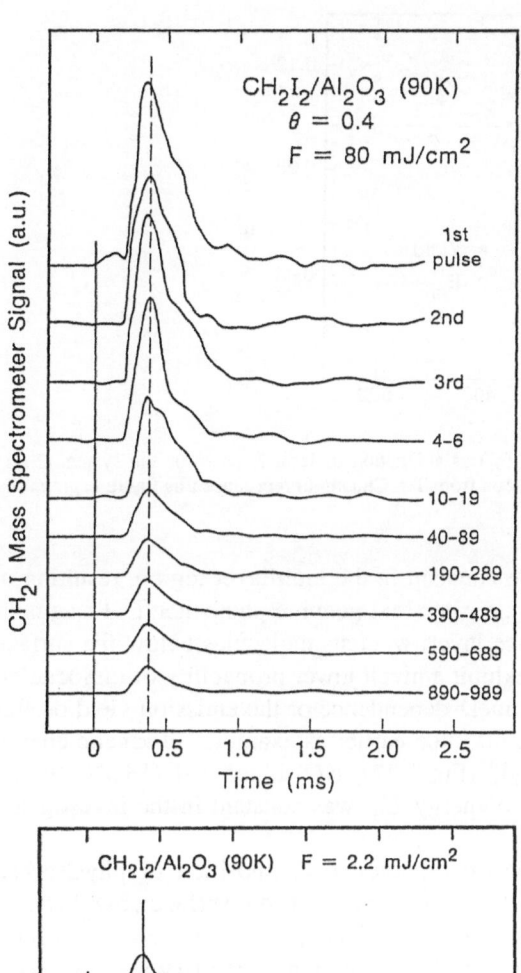

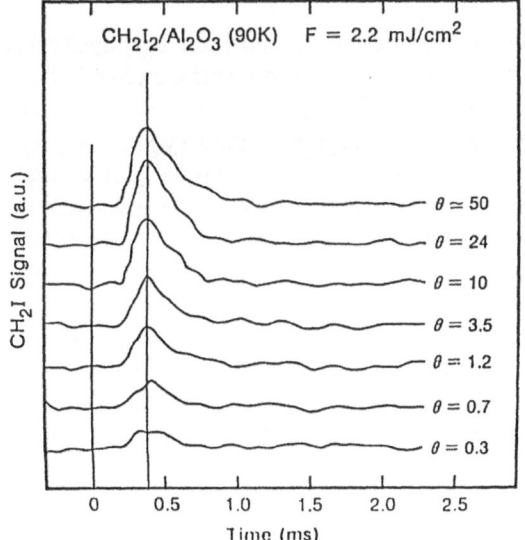

Fig. 3.20. TOF distributions of CH_2I as a function of laser pulses on CH_2I_2 sapphire at 0.4 monolayer coverage. Permission from T.J. Chuang to reproduce this unpublished figure is gratefully acknowledged

Fig. 3.21. TOF distribution of CH_2I at various coverages for the system CH_2I_2 sapphire. Permission from T.J. Chuang to reproduce this unpublished figure is gratefully acknowledged

A second important fluence criterion for the mechanism involved is the fluence dependence of the desorption yield. For a one-photon photodissociation process, a linear dependence of the desorption yield, at least in the submonolayer coverage range, is expected, as shown in Fig. 3.22 [3.73]. Depending on the morphology of

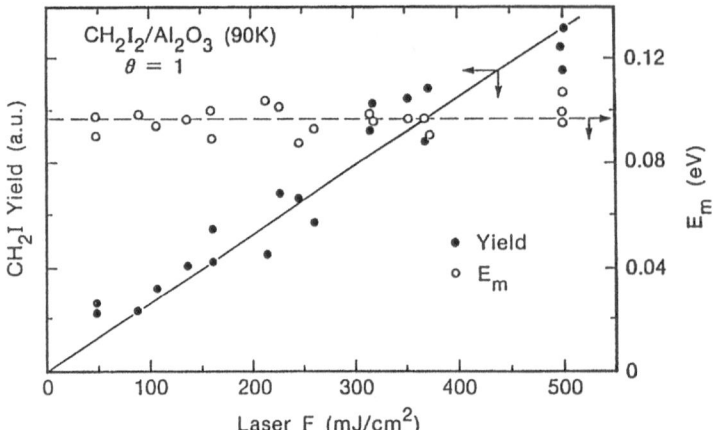

Fig. 3.22. CH_2I yields and kinetic energies (E_m) as a function of laser fluence for the system CH_2I_2 sapphire at monolayer coverage [3.73]. Permission from T.J. Chuang to reproduce this figure is gratefully acknowledged

the film, the surface area may become constant in the multilayer region, resulting in a saturation of the TOF signal. This indicates that genuine photochemical fragmentation occurs only from the top surface layer, whereas molecules below the surface layer may recombine, and therefore exhibit a much lower probability of dissociation due to the cage effect. The expected linear dependence of the emission yield on fluence could be detected in all the systems cited earlier as examples of kinetic energy independent of fluence, namely CH_2I_2 (Fig. 3.22), NO dimer and CH_3Br. Figure 3.22 shows that in addition the kinetic energy E_m was constant in the investigated fluence range [3.73].

This is sound evidence that direct one-photon decomposition of physisorbed molecules has been achieved with UV laser radiation. At low surface coverage and low laser fluence in particular, the photochemical effect can be seen in a nearly undisturbed form. In the monolayer and multilayer region, secondary reactions, collisional perturbations and the efficient emission of molecules from the surface make it more and more difficult to observe the primary photoproducts. At low laser fluences, however, it may be possible to detect the electronic effects in the multilayer regime as well, if there is no significant contribution from substrate heating. As a first approximation, it may be assumed that photochemical decomposition occurs mainly at the top surface layer. In deeper layers, the cage effect leads to recombination of the dissociation products, suppressing photolysis in the bulk. On the other hand, photothermal desorption and ablation increase with the thickness of the deposit for resonant excitation of the adsorbate and therefore become the dominant effect.

3.5.3 Photothermal Effects

The characteristic features of photothermal effects were clearly observed for resonant UV excitation of multilayer adsorbate systems. The influence of the substrate decreases with increasing thickness of the deposit and can be further reduced by choosing a transparent substrate material. The term "photothermal" will be used for those cases where at least some kind of local thermal equilibration of the excitation energy is achieved before the desorption or ablation process takes place. This implies fast energy relaxation between the excited level and other levels. Both intramolecular and intermolecular energy transfer must be fast. Thus, transient heating is not confined to molecules absorbing photons initially and molecular selectivity is lost. We wish to point out that this assumption does not necessarily mean that chemical equilibrium is achieved for the emitted species and that their internal and external degrees of freedom are in thermal equilibrium. In fact, we know of no system where such a complete thermodynamic equilibration has been proved for the emitted species, whereas many examples are known of incomplete equilibration.

The more or less complete conversion of the absorbed photon energy into heat generates a temperature profile in the irradiated region which is responsible for thermal desorption or ablation. The kinetic energy of the emitted molecules depends on this transient temperature and therefore the maximum of the TOF distribution will shift to smaller transit times with increasing laser fluence. The kinetic energy of the particles with the highest energy may be several eV. The majority of particles, of course, possess a much lower kinetic energy.

A temperature increase has a strong effect on the desorption and ablation yield. In partially and totally equilibrated systems, a strongly nonlinear increase of the yield with laser fluence can be expected. Photothermal ablation is not restricted to the very top of the deposit, but several layers may leave the adsorbate film simultaneously. Nevertheless, the ablation probability may not be equal for all layers in the multilayer adsorbate. The possibility of multilayer ablation with one pulse makes the photothermal effect orders of magnitude more efficient than the photochemical effect. The power density of absorbed radiation controls the photothermal effect. Due to the nonlinear yield dependence, a pronounced threshold behavior is observed in this case.

The effective threshold for photothermal ablation normally changes with the thickness of the deposit. This shows that a nonlinear absorption process, such as coherent or incoherent multiphoton absorption, is not responsible for the strong nonlinear yield dependence.

A convincing example of photothermal desorption and ablation was reported for the system NO on Ag(111) [3.74]. Condensed NO is an interesting system, because the solid is known to be completely dimerized with a dimer bond strength intermediate between that of typical van der Waals and chemical bonds. As already mentioned, essentially NO was detected with the mass spectrometer for UV excitation at 220–270 nm. The two components of the bimodal TOF distributions varied independently as a function of laser wavelength, pulse energy and film thickness. The slow peak was interpreted to be thermal in origin. The yield of this peak showed an exponential increase with laser fluence and a strong increase with NO layer thickness. The angular distribution was broad and approximately cosine. The corresponding peak in the TOF distribution could be fit to a Maxwell distribution. The latter result is somewhat surprising because, at least for the higher yields, a narrowing effect would have been expected.

Another well-characterized system where the features of photothermal behavior have been verified is CH_2I_2 on Al_2O_3 and Ag surfaces [3.71–73]. The desorption and ablation yields and the translational energy distributions were measured as a function of surface coverage, laser fluence and the number of laser pulses. In the multilayer coverage region at higher laser fluences, the nonlinear dependence of the yield and kinetic energy on the fluence was detected and the threshold behavior observed. The authors state that at high coverages, the yield per pulse may be 2–3 orders of magnitude higher than in the monolayer region and that the quantum yields may exceed unity for the top layers. They call this the regime of "explosive desorption", where a van der Waals system shows a similar behavior to that of the photoablation of organic polymers. Both parent molecules and the fragments CH_2I and I were observed in the explosive regime. The fragments can now originate from a photochemical effect as well as a photothermal effect, because thermal decomposition may occur at the higher transient temperatures involved. The TOF distributions consisted of a very fast component and a tail with a very broad translational distribution.

The TOF distributions obtained for photoablation of PMMA at 193 nm also point to photothermal effects [3.79]. For fluences below $120 \, mJ/cm^2$, the distributions measured for the monomer unit MMA and other fragments could be fitted to Maxwell distributions belonging to a common temperature. At higher fluences, directional non-Maxwellian distributions were observed which were ascribed to a jet-like expansion of the photoproducts. For polymer ablation, a more detailed study would be highly desirable also investigating the fluence dependence of the ablation yield and the kinetic energy for a number of photoproducts.

3.5.4 Real Systems

Considering isolated photochemical effects at the top adsorbate layer is an idealization. In reality, the substrate often plays an important role. This influence of course depends on the optical and thermal properties of the substrate and on the thickness of the overlayer. In the multilayer coverage region, the substrate effect on photochemical processes may be much smaller than on the photothermal processes [3.74]. Only in the limit of very thick adsorbate films will the behavior be independent of the substrate. In general, highly transparent materials with a large bandgap and a low thermal conductivity will have a much smaller influence than substrate material with a high optical absorption coefficient and a high thermal conductivity, such as metals. The substrate effects are numerous: They may change the optical properties of the first layer, act as a heat sink, and contribute to desorption by substrate-mediated heating or shock wave generation, etc.

Two main reasons are responsible for the fact that a realistic modelling of such a system is not possible. (1) The number of processes, including chemical reactions, energy exchange processes, transport processes, phase transitions, etc., is too high. (2) The system properties such as optical absorption coefficients, thermal diffusivities, heat capacities, relaxation times, etc., are either not accurately known or not known at all. This is a serious problem which limits the usefulness of model calculations.

3.6 Conclusions

The TOF method is a very powerful technique, providing us with detailed information on laser-induced desorption and ablation processes. This includes the identification of chemical species, their yield and their kinetic energy. Relative values for the last two properties are easy to obtain, whereas the determination of absolute values is a serious problem. The accurate analysis of the TOF profile in particular, e.g. the extraction of translational temperatures, is a problem that is not completely solved. Further theoretical developments are needed, including the study of collisional effects occurring in the case of photoablation of multilayers. With improvements in both theory and experimental technique, considerable progress can be expected within the next few years.

This review has discussed the application of the TOF method to photodesorption and photoablation. In the IR region, where pulsed IR lasers are employed for vibrational excitation, the results indicate a mainly photothermal behavior from submonolayer to multilayer coverage. This conclusion is supported by several observations, including the lack of selectivity, the wavelength and fluence dependence of the emission yield, and the fluence dependence of the kinetic energy of the particles leaving the surface.

The discussion presented above indicates the conditions necessary to achieve selectivity in laser-induced chemistry at surfaces. At low coverage and resonant adsorbate heating, a transparent substrate may serve as a heat sink, keeping thermal desorption effects relatively small. Under these circumstances, electronic effects may dominate the desorption behavior in the UV spectral region. It is becoming possible to detect genuine photochemical decomposition of molecules by UV laser photons. This process possesses bond or mode specificity, because the fragmentation pattern depends on the laser wavelength. This offers new scope for the study of the dynamics of energy exchange and bond-breaking processes at surfaces.

It is also clear that direct photodecomposition has molecular selectivity and may be used for isotope separation at surfaces. Selective laser chemistry may be confined essentially to the monolayer region. At higher coverage, photodissociation is, of course, also possible in the bulk region of the film; however, the cage effect causes recombination of the fragments. Thus, selective photochemical processes seem to be confined to the surface and may be considered as surface processes.

Photothermal desorption can be substrate-mediated, adsorbate-mediated or both. It is not confined to submonolayer and monolayer coverage, but becomes very effective in a multilayer adsorbate system for resonant excitation. This photoablation effect is observed for vibrational excitation in the IR and electronic excitation in the UV. The characteristic features proving that photoablation in condensed van der Waals systems is essentially a thermal effect have been verified by the TOF technique. This multilayer ablation effect, which is also called explosive desorption, is a real bulk effect, where the surface plays only a minor role. First TOF experiments performed with polymers also point to mainly thermal features in this case. In fact, polymer ablation possesses several properties that are similar to those observed for van der Waals films.

Acknowledgements: Financial support of this work by the German Ministry of Research and Technology (BMFT) under contract No. 13N5363 8 and the Fonds der Chemischen Industrie is gratefully acknowledged.

References

3.1 L.P. Levine, J.F. Ready, E.G. Bernal: J. Appl. Phys. **38**, 331 (1967)
3.2 T.J. Chuang: Surf. Sci. Rep. **3**, 1 (1983)
3.3 T.J. Chuang: Surf. Sci. **178**, 763 (1986)
3.4 R.B. Hall: J. Phys. Chem. **91**, 1007 (1987)
3.5 C.V. Shank, R. Yen, C. Hirlimann: Phys. Rev. Lett. **50**, 454 (1983)
3.6 R.B. Hall, A.M. DeSantolo: Surf. Sci. **137**, 421 (1984);
    R.B. Hall, A.M. DeSantolo, S.J. Bares: Surf. Sci. **161**, L533 (1985)
3.7 M.G. Sherman, J.R. Kingsley, D.A. Dahlgren, J.C. Hemminger, R.T. McIver: Surf. Sci. **149**, L25 (1985)
3.8 A.A. Deckert, S.M. George: Surf. Sci. **182**, L215 (1987)
3.9 R.N. Zare, R.D. Levine: Chem. Phys. Lett. **136**, 593 (1987)
3.10 R.N. Zare, J.H. Hahn, R. Zenobi: Bull. Chem. Soc. Jpn. **61**, 87 (1988)
3.11 J. Heidberg, H. Stein, E. Riehl, A. Nestmann: Z. Phys. Chem. (NF) **121**, 145 (1980)
3.12 Y. Kawamura, K. Toyoda, S. Namba: Laser Kenkyu **8**, 941 (1981)
3.13 Y. Kawamura, K. Toyoda, S. Namba: Appl. Phys. Lett. **40**, 374 (1982)
3.14 R. Srinivasan, V. Mayne-Banton: Appl. Phys. Lett. **41**, 576 (1982)
3.15 R. Srinivasan, W.J. Leigh: J. Am. Chem. Soc. **104**, 6784 (1982)
3.16 R. Srinivasan: In *Laser Processing and Diagnostics*, ed. by D. Bäuerle, Springer Ser. Chem. Phys., Vol.39 (Springer, Berlin, Heidelberg 1984) pp.343–354
3.17 J.T.C. Yeh: J. Vac. Sci. Technol. A **4**, 653 (1986)
3.18 R. Srinivasan: Science **234**, 559 (1986)
3.19 J.H. Brannon, J.R. Lankard: Appl. Phys. Lett. **48**, 1226 (1986)
3.20 A. Ben-Shaul, Y. Haas, K.L. Kompa, R.D. Levine: *Lasers and Chemical Change*, Springer Ser. Chem. Phys., Vol.10 (Springer, Berlin, Heidelberg, New York 1981)
3.21 V.S. Letokhov: *Nonlinear Laser Chemistry*, Springer Ser. Chem. Phys., Vol.22 (Springer, Berlin, Heidelberg, New York 1983)
3.22 G. Herzberg: *Molecular Spectra and Molecular Structure* (Van Nostrand Reinhold, New York 1950)
3.23 R. Braun: Diplomarbeit, University of Heidelberg, 1989;
     R. Braun, P. Hess: to be published
3.24 B. Spengler, U. Bahr, M. Karas, F. Hillenkamp: Anal. Instrum. **17**, 173 (1988)
3.25 L. Li, D.M. Lubman: Rev. Sci. Instrum. **59**, 557 (1988)
3.26 F. Engelke, J.H. Hahn, W. Henke, R.N. Zare: Anal. Chem. **59**, 909 (1987)
3.27 U. Boesl, J. Grotemeyer, K. Walter, E.W. Schlag: Anal. Instrum. **16**, 151 (1987)
3.28 M. Buck, P. Hess: J. Electron. Spectrosc. **45**, 237 (1987)
3.29 F.L. Tabares, E.P. Marsh, G.A. Bach, J.P. Cowin: J. Chem. Phys. **86**, 738 (1987)
3.30 H. Saltsburg: J. Chem. Phys. **42**, 1303 (1965)
3.31 R.A. Olstad, D.R. Olander: J. Appl. Phys. **46**, 1499 (1975)
3.32 G. Wedler, R. Ruhmann: Surf. Sci. **121**, 464 (1982)
3.33 B. Stritzker, A. Pospieszczyk, J.A. Tagle: Phys. Rev. Lett. **47**, 356 (1981)
3.34 N.G. Utterback, S.P. Tang, J.F. Friichtenicht: Phys. Fluids **19**, 900 (1976)
3.35 C. Peugnet: J. Appl. Phys. **48**, 3206 (1977)
3.36 H. Haberland, U. Buck, M. Tolle: Rev. Sci. Instrum. **56**, 1712 (1985)
3.37 D.M. Lubman, C.T. Rettner, R.N. Zare: J. Phys. Chem. **86**, 1129 (1982)
3.38 J. Heidberg, H. Stein, E. Riehl: Phys. Rev. Lett. **49**, 666 (1982)
3.39 J. Heidberg, H. Stein, E. Riehl: Surf. Sci. **126**, 183 (1983)
3.40 J. Heidberg, H. Stein, E. Riehl, Z. Szilágyi, H. Weiss: Surf. Sci. **158**, 553 (1985)
3.41 T.J. Chuang: J. Chem. Phys. **76**, 3828 (1982)
3.42 T.J. Chuang, H. Seki: Phys. Rev. Lett. **49**, 382 (1982)
3.43 B. Schäfer, P. Hess: Chem. Phys. Lett. **105**, 563 (1984)
3.44 J. Heidberg, H. Stein, A. Nestmann, E. Hoefs, I. Hussla: In *Laser-Solid Interactions and Laser Processing*, ed. by S.D. Ferris, H.J. Leamy, J.M. Poate (American Institute of Physics, New York 1979) p.49
3.45 B. Schäfer, P. Hess: Appl. Phys. B **37**, 197 (1985)

3.46 B. Schäfer, M. Buck, P. Hess: Infrared Phys. **25**, 245 (1985)
3.47 M. Buck, B. Schäfer, P. Hess: Surf. Sci. **161**, 245 (1985)
3.48 M. Buck, P. Hess: MRS Symp. Proc. **100**, 647 (1988)
3.49 T.J. Chuang, I. Hussla: Phys. Rev. Lett. **52**, 2045 (1984)
3.50 T.J. Chuang, H. Seki, I. Hussla: Surf. Sci. **158**, 525 (1985)
3.51 J. Heidberg, D. Hoge: J. Opt. Soc. Am. B **4**, 242 (1987)
3.52 J. Heidberg, C. Langoski, G. Neubauer, M. Folman: J. Electron Spectrosc. Relat. Phenom. **45**, 249 (1987)
3.53 M. Buck, P. Hess: Chem. Phys. Lett. **158**, 486 (1989)
3.54 D.H. Dows: J. Chem. Phys. **36**, 2833 (1962)
3.55 K.S. Suslik: US Patent No. 4010100 (1977)
3.56 D.B. McConnell: Ontario Hydro Research Division Report, No. 76-516-K (1976)
3.57 J. Heidberg, I. Hussla: J. Electron. Spectrosc. Relat. Phenom. **29**, 105 (1983)
3.58 C.J.S.M. Simpson, J.P. Hardy: Chem. Phys. Lett. **130**, 175 (1986)
3.59 J. Heidberg, H. Stein, E. Riehl, I. Hussla: In *Surface Studies with Lasers*, Springer Ser. Chem. Phys., Vol. 33 (Springer, Berlin, Heidelberg, New York 1983) pp. 226–229
3.60 D. Burgess, Jr., R. Viswanathan, I. Hussla, P.C. Stair, E. Weitz: J. Chem. Phys. **79**, 5200 (1983)
3.61 G.S. Higashi: J. Chem. Phys. **88**, 422 (1988)
3.62 D. Burgess, Jr., D.A. Mantell, R.R. Cavanagh, D.S. King: J. Chem. Phys. **85**, 3123 (1986)
3.63 D. Burgess, Jr., R.R. Cavanagh, D.S. King: J. Chem. Phys. **88**, 6556 (1988)
3.64 E.B.D. Bourdon, J.P. Cowin, I. Harrison, J.C. Polanyi, J. Segner, C.D. Stanners, P.A. Young: J. Phys. Chem. **88**, 6100 (1984)
3.65 I. Harrison, J.C. Polanyi, P.A. Young: J. Chem. Phys. **89**, 1475 (1988)
3.66 E.B.D. Bourdon, P. Das, I. Harrison, J.C. Polanyi, J. Segner, C.D. Stanners, R.J. Williams, P.A. Young: Faraday Discuss. Chem. Soc. **82**, 343 (1986)
3.67 N. Nishi, H. Shinohara, T. Okuyama: J. Chem. Phys. **80**, 3898 (1984)
3.68 F.L. Tabares, E.P. Marsh, G.A. Bach, J.P. Cowin: J. Chem. Phys. **86**, 738 (1987)
3.69 G.N.A. Van Veen, K.A. Mohamed, T. Baller, A.E. de Vries: Chem. Phys. **74**, 261 (1983)
3.70 I. Harrison, J.C. Polanyi, P.A. Young: J. Chem. Phys. **89**, 1498 (1988)
3.71 T.J. Chuang, K. Domen: J. Vac. Sci. Technol. A **5**, 473 (1987)
3.72 K. Domen, T.J. Chuang: Phys. Rev. Lett. **59**, 1484 (1987)
3.73 K. Domen, T.J. Chuang: J. Chem. Phys. **90**, 3318, 3332 (1989)
3.74 W.C. Natzle, D. Padowitz, S.J. Sibener: J. Chem. Phys. **88**, 7975 (1988)
3.75 M. Buck, P. Hess: to be published in Appl. Surf. Sci.
3.76 R. Srinivasan, A.P. Ghosh: Chem. Phys. Lett. **143**, 546 (1988)
3.77 L.M. Cousins, S.R. Leone: Chem. Phys. Lett. **155**, 162 (1989)
3.78 J. Kutzner, G. Lindeke, K.H. Welge, D. Feldmann: J. Chem. Phys. **90**, 548 (1989)
3.79 B. Danielzik, N. Fabricius, M. Röwekamp, D. von der Linde: Appl. Phys. Lett. **48**, 212 (1986)
3.80 D. Feldmann, J. Kutzner, J. Laukemper, S. MacRobert, K.H. Welge: Appl. Phys. B **44**, 81 (1987)

4. From Laser-Induced Desorption to Surface Damage

E. Matthias, R.W. Dreyfus

With 23 Figures

This chapter presents a brief survey of thermal and nonthermal effects in laser-surface interactions on metallic and wide band gap ionic materials, for nanosecond laser pulses in the power range MW/cm^2 to GW/cm^2. An attempt is made to develop a comprehensive view extending from individual particle emission to ablation and plasma formation, and emphasis is placed on the fact that thermal and nonthermal effects can occur simultaneously. The discussion will distinguish between the primary photon absorption and follow-up processes leading to particle emission. Examples are drawn from experiments on metal, sapphire, and fluoride surfaces.

4.1 Overview

Interest in the interaction between intense laser light and materials comes from different areas of basic research and technological developments, ranging from fusion to materials processing, surgery, lithography, etc. The different fields of interest are illustrated in Fig. 4.1 where a typical rate of material removal per laser pulse is plotted versus power density. It should be evident that the wide parameter space means that different physical processes can dominate in different regions. Also it is important to keep in mind that both the spatial and temporal shape of each laser pulse runs through several orders of intensity, as indicated in the insert of Fig. 4.1. This means that each laser pulse has the possibility of starting out with isolated atom removal, and with generation of surface defects, which in turn serve as absorption centers for the photons in the later part of the pulse. Likewise, in the spatial domain, the effects will be different in the center and on the circumference of the beam spot. We will return to this point later.

Above a certain intensity threshold that depends on the material, laser wavelength and pulsewidth, matter will be removed in quantities that cause macroscopic damage to the surface [4.1] and a plasma is generated, sometimes strong enough to serve as a source for soft X-rays and electrons [4.2]. Below such a "damage threshold", there is particle emission (electrons, neutrals, ions) and heating to a degree that depends linearly or nonlinearly on laser intensity. Hence, the definition of "damage threshold" only refers to macroscopic damage altering the morphology and the optical properties of the surface. On a microscopic scale there will be a defect or "local damage" as soon as an atom or ion is desorbed out of the surface layer. These isolated defects can play an important role in the development of macroscopic damage since they serve as absorption centers and therefore multiply rapidly within one laser pulse. For

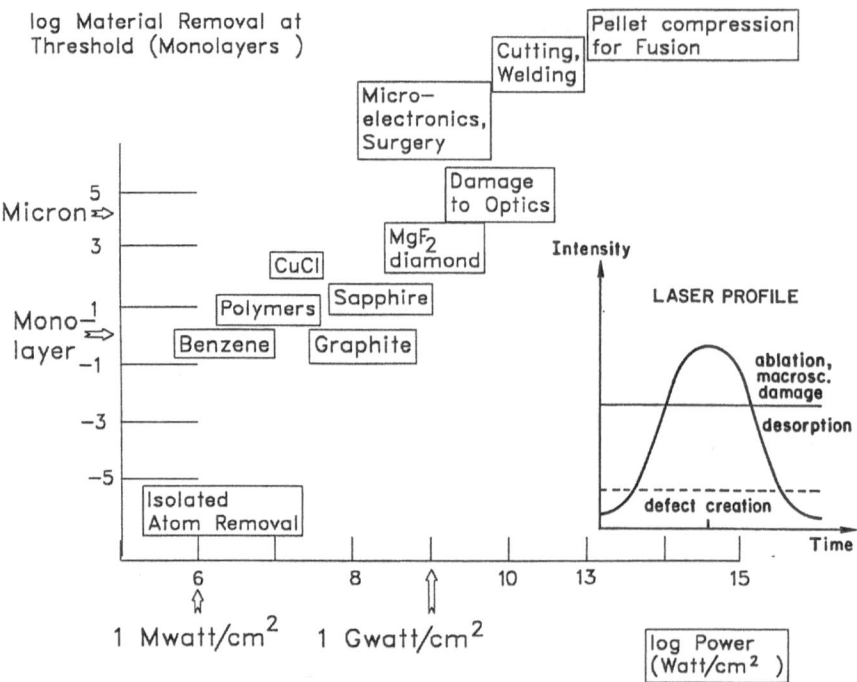

Fig. 4.1. Material removal in units of monolayers as a function of laser power density. Fields of application are indicated as well as a few substances of particular interest. The inset is to recall that a powerful laser pulse runs through the whole scenario

some experiments these initial defects are of primary concern; for other experiments they will be subservient to other reactions.

In this chapter we will restrict the discussion to metals and wide band gap ionic materials, and will not include laser etching of semiconductors [4.3], or the ablation of polymers or biological tissue [4.4, 5]. We will also exclude laser processing of metals or alloys with infrared lasers [4.6], although the basic thermal (but not plasma) effects are similar to the ones obtained by visible light. Neither will we include the field of laser-induced desorption of adsorbates on metals or semiconductors [4.7], nor the effects observed with subpicosecond lasers. Thus we limit the discussion to laser surface interactions with visible and UV light on a nanosecond time scale and with metals and ionic insulators.

Ever since high-power lasers became available, considerable efforts have been devoted to the understanding of laser damage of optical materials [4.8], and the 20th anniversary of the Boulder Damage Symposium in 1988 testified to the long-standing research in this area. Despite this, it is fair to say that there are many details of the physics of laser damage which are not understood, in particular those of the surface damage mechanism. The reason is the multiplicity of the processes for different materials on the one hand, and on the other the still limited quantitative information on an atom/surface level. With respect to the complexity of the process, it can be seen that laser ablation commonly means dealing with two highly nonlinear systems, i.e. the generation and avalanching of defects (Fig. 4.1 inset) and the nonlinear optical excitation ($nh\nu$) of a sequence of electronic transitions. Another problem is that ablation normally takes place in nanoseconds, but observations on the plume commonly require microseconds. Hence chronological details are lost. Although many systematic investigations have been made into the physics of laser damage [4.1, 8], more complementary data for one single system are needed to either support or dismiss detailed models about the damage mechanism. Unsolved problems are, for example, the absorption of light in surface layers or coatings of transparent materials, and the development of a plasma and its energy feedback to the surface via radiation, electron sputtering, thermal conduction, and shock wave impact. On the other hand, several facts, mostly concerning the primary energy absorption, are firmly established: Thermal effects in metals are normally the result of single-photon absorption [4.9, 10], although multiphoton absorption processes have occasionally been invoked for metals [4.11–14] and insulators [4.15–17]. The role of defects in the absorption of light [4.16, 18, 19], their influence on the damage threshold [4.20, 21] and the importance of thermal stress [4.22, 23] have been recognized. Furthermore, realistic models for the above-surface avalanche breakdown during laser irradiation have been developed [4.24–27].

Contributions towards a better understanding of laser-surface interactions in general have come from areas like DIET [desorption induced by (usually isolated) electronic transitions] as this minimizes interactions between defect sites [4.28–30], laser etching [4.3, 31–33], and photothermal research [4.34–36]. In all these fields, questions of energy absorption and dissipation in opaque and transparent materials as well as in absorbates have been studied. For thermal desorption, the rapid heating rate available with pulsed lasers was attractive [4.37]. Nonthermal desorption pathways were also explored, for example vibrational transitions in adsorbed molecules [4.38], desorption following plasmon excitation [4.39], photochemical processes in polymers [4.4, 40], and bond-breaking by excitation of electron-hole pairs [4.41]. Also, concepts developed for desorption triggered by inner-shell excitations [4.42] might become relevant, since, for follow-up processes, it makes no difference whether holes in the valence band are generated by Auger processes or multiphoton ionization.

Two principal properties of materials determine their response to laser radiation: (1) Linear or nonlinear *optical absorption* regulates the energy intake; and it is important to recognize that the degree of absorbency and its order usually change drastically even during a nanosecond laser pulse. (2) *Thermal properties* of the material, on the other hand, govern the sequential effects like thermal expansion and stress, particle desorption, damage threshold and plasma onset. As a consequence of this, metals, semiconductors, and insulators (soft or hard) often show a distinctly different ablation behavior, e.g. splashing and exfoliation. Further, we have to distinguish between two different *forces* driving the material removal from the surface. One is *thermal expansion*, which, according to its degree, is responsible for deformation, stress, evaporation, and shock wave generation. The *repulsion force* on the other hand leads to the emission of neutrals (via antibonding states), ions, explosive ablation, and again shock wave generation. Clearly, Coulomb repulsion, i.e. the absence of a prior attractive force, is inherently connected to electronic excitation, while thermal expansion is either the response of free charge carriers or the vibrational response of defects or molecules in the surface layer. In view of some polarized discussions in the past, we want to emphasize that *electronic excitation and heat always go together*. The question is only which one is dominant (but not exclusive), and that is in general determined by the material as well as the wavelength and pulsewidth of the laser radiation.

In Table 4.1, an attempt is made to list the effects of interest in this review, grouped according to intensity, material and order of photon absorption. Regarding laser–surface interactions, the list is by no means complete, for example, second harmonic generation in reflection on surfaces was not mentioned, although it is a powerful tool in surface physics [4.43]. The purpose of Table 4.1 is to remind the reader of the complexity of desorption and ablation processes and of the possibility of observing them using several different experimental techniques. In fact what is presently missing is some type of "*complete experiment*" in which relevant quantities extending from surface physics to an inventory of all plume species have been measured for *one system*. However, when comparing results obtained by unlike methods, one has to keep in mind that different techniques are sensitive to different processes and thereby might detect differing orders of photon absorption. For example, the photothermal deformation of surfaces represents a thermal process proportional to the adsorbed energy flux density, while particle or shock wave emission is often a sequence of multiphoton absorption and is therefore controlled by the power density. Another question is that of *surface sensitivity*; after all, light penetrates into the bulk, even for highly opaque materials. In the bulk certain defect states can exist that are not stable at the surface. Hence in each experiment one should carefully examine the depth sensitivity.

Table 4.1. Surface Processes and Detection Techniques for Metals and Ionic Materials

Intensity	Absorption	Material	Effect	Detection
All	Single photons	All	Heating	Surface Deformation
		Metals	Melting	Reflectance, Microscopy
		Metals	Evaporation	Mass spectroscopy
		All	Thermal desorption of neutrals	Kinetic energy, LIF
All	Single or multiphotons	All	Photoemission	Charge, Kinetic energy
			Fluorescence	Light emission
$\gtrsim 1\,\mathrm{MW/cm^2}$	Multiphoton	Insulators	Nonthermal desorption of neutrals and ions, incubation	Mass spectroscopy, LIF, Kinetic energy, Charge, Absorbance
$\gtrsim 10\,\mathrm{MW/cm^2}$	Single and multiphoton	Insulators	Ablation	Mass spectroscopy, LIF, Shock wave, Light emission, Reflectance, Microscopy, Particle scattering
$\gtrsim 100\,\mathrm{MW/cm^2}$	Single and multiphoton	Insulators	Explosive eruption and shock wave damage	Mass spectroscopy, Shock wave, Light emission, Microscopy, Particle scattering, Depth profiling
$\gtrsim \mathrm{GW/cm^2}$	Single and multiphoton	All	Plasma formation	Shock wave, Light emission, Plasma current, Interferometry

In the following, we will present selected examples of photothermal deformation, photoemission, desorption of neutrals, nonthermal desorption, ablation and shock wave damage. Some of the examples are chosen in order to emphasize the point that several effects can occur simultaneously. The discussion is separated for metals and insulators, owing to their significantly different optical and thermal properties.

4.2 Metals

Laser-surface interactions on metals have been studied intensely from three different points of view. One is the industrial processing of metals and alloys with lasers [4.1, 6]. The second is plasma generation for providing soft X-ray and electron sources [4.2]. The third area of interest is laser induced desorption [4.7, 30, 37, 38], both thermal and nonthermal, whereby metals play the role of well-controlled substrates that can easily be heated by light absorption. (Unfortunately, the optical absorption is often smaller than the reflection; hence when the reflection changes with temperature, wide uncertainties may be inflicted upon the absorption.) We will not be concerned here with the latter, instead we will try to follow the various steps that, with increasing intensity, eventually lead to plasma ignition. The sequence leading to surface damage of metals is generally thought to be [4.1, 44–46].

thermal expansion $\rightarrow$ melting $\rightarrow$ boiling $\rightarrow$ plasma.

In general, thermalization of the absorbed energy occurs in the nanosecond range (including the possibility of time constants attributable to Franck-Condon emission of phonons) [4.47], and the normal melting point is reached during pulse lengths of some tens of picoseconds [4.48, 49]. It has been suggested that for pico- and subpicosecond pulses, superheating occurs, and processes far from thermal equilibrium will dominate [4.50]. In addition, both evaporation and plasma formation require a finite time [4.51, 52] and therefore one would expect minimal plasma for subpicosecond pulses.

In Fig. 4.2, an instructive sketch is shown of the various thermal effects resulting in different surface morphologies during one laser pulse of 10 ns length on polished copper surfaces [4.44]. The picture conveys the variety of effects taking place *simultaneously* for a single laser shot of sufficiently high fluence. Slip lines arise from thermal stress [4.53]. Ripple patterns result from constructive interference between the incident light and scattered surface waves near the melting point [4.54]. These features will not be discussed here and the reader is referred to the pertinent literature. The question of interest here is particle emission in the various intensity ranges. Although we are not aware of any systematic studies along these lines, it is evident that photoemission will occur throughout all regimes indicated in Fig. 4.2. One would also expect that the spatial photoemission yield would correlate with the ridges of the ripple pattern. Charged particle emission has been observed in connection with surface cleaning [4.55], but in air it is likely to originate from an ill-defined adsorbate system. More in line with our interest is the evaporation of neutrals from the melt and the ignition of a plasma, which was extensively discussed in [4.1].

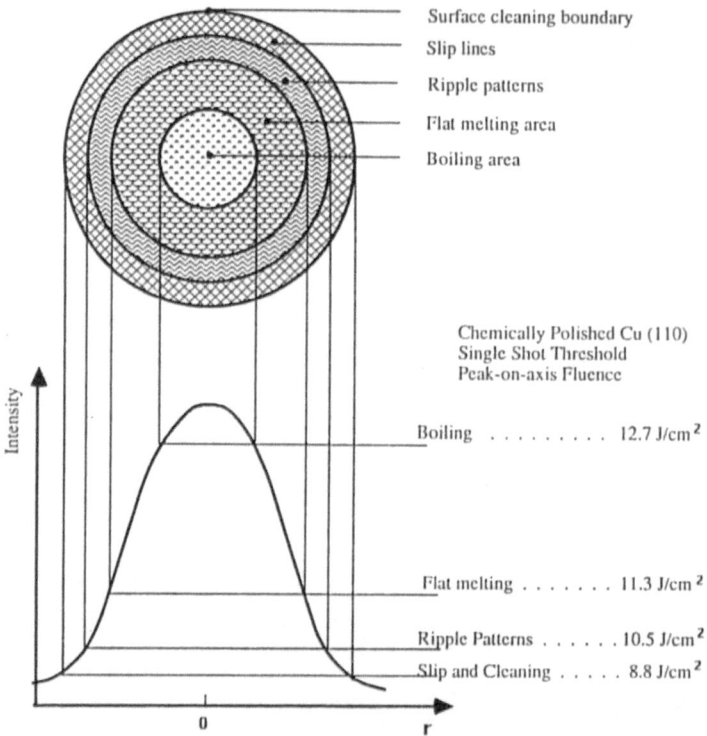

Surface cleaning boundary

Slip lines

Ripple patterns

Flat melting area

Boiling area

Chemically Polished Cu (110)
Single Shot Threshold
Peak-on-axis Fluence

Boiling . . . . 12.7 J/cm²

Flat melting . . . . 11.3 J/cm²

Ripple Patterns . . . . . . 10.5 J/cm²

Slip and Cleaning . . . . . 8.8 J/cm²

Fig. 4.2. Typical damage morphologies of metal surfaces generated by a single laser pulse. The onset fluences for 1064 nm radiation (pulse length 10 ns) on chemically polished Cu(110) in air are shown. (From [4.44], courtesy of M.F. Becker)

There are basically three different contributions causing the ionization of the vapor and the formation of a plasma: (1) In thermal equilibrium, a certain fraction of the gas is ionized, as described by the Saha equation [Ref. 4.1, eq. (5.14)] although it is not evident that this is a large amount. (2) Single-photon or multiphoton photoionization of neutral and excited atoms by the incoming light [4.45]. (3) Ionization of neutrals by inelastic collisions with photoelectrons that originate both from the surface layer and from photoionization in the gas. Once a sufficiently large degree of ionization is attained, inverse bremsstrahlung becomes the dominant absorption process for the laser light [4.1], which, in the case of infrared, no longer reaches the surface. Instead, further heating of the surface occurs by black body radiation and electron flux from the plasma. Again one should bear in mind that these are temporal sequences within one laser pulse taking place at the center of the spatial profile in Fig. 4.2. Perhaps Fig. 4.2 can also be considered a reason for aperturing the laser beam so as to provide a reasonably uniformly illuminated area.

In the next three sections, we will discuss the initial photon absorption mechanism for metals. Emphasis is placed on the fact that different techniques monitor different absorption modes, such as linear and nonlinear ones. Although multiphoton absorption usually first becomes important at higher intensities, there are scenarios where single and multiphoton absorption can compete, e.g. when multiphoton processes are resonantly enhanced, or when absorption takes place in a ladder-like multiple-photon absorption sequence. In this way one could easily imagine the coupling of intermediate states to the phonon bath to be a competing loss process in the multiple-photon ionization, via transference of heat to the lattice.

4.2.1 Photothermal Deformation

The temperature rise of a laser irradiated surface can be monitored in a noncontact manner by infrared detection [4.56] or by photothermal displacement spectroscopy [4.57]. The latter technique is more sensitive and allows temperature changes as low as 1 mK to be recorded for cw lasers. We are more interested here in transient heating by pulsed lasers and wish to present an experiment carried out by *Karner* et al. [4.34] on metals with fluences in the mJ/cm^2 range. The principle of the experimental arrangement is displayed in Fig. 4.3. As indicated, the technique is based on the change of the probe beam reflection angle when the surface bulges as a result of heating. In their experiment, *Karner* et al. used 20-ns pulses with energies of less than 1 mJ/pulse to observe the relaxation of the surface deformation.

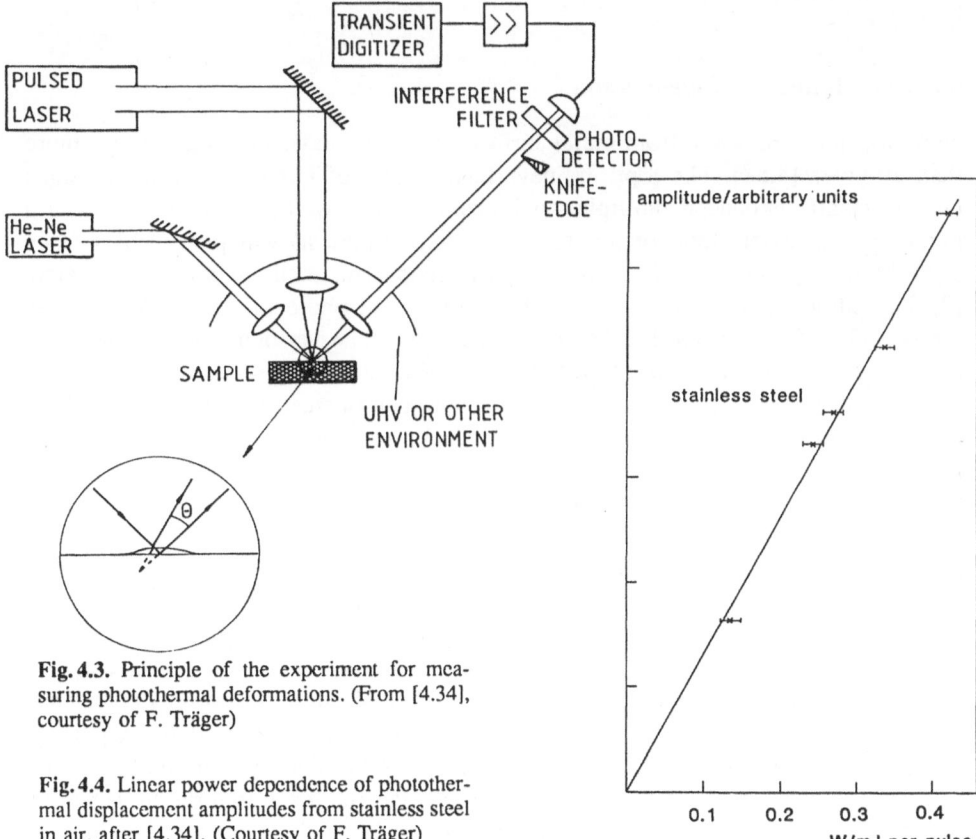

Fig. 4.3. Principle of the experiment for measuring photothermal deformations. (From [4.34], courtesy of F. Träger)

Fig. 4.4. Linear power dependence of photothermal displacement amplitudes from stainless steel in air, after [4.34]. (Courtesy of F. Träger)

This is determined by the thermal conductivity of the sample, and one result of the experiment was that the $1/e$ decay time of the bump multiplied by the thermal diffusivity was approximately constant, i.e. the larger the heat conductivity, the faster the decay of the deformation. The second result was that the displacement signal is proportional to the laser power, as shown in Fig. 4.4. This proves that heating is achieved by single-photon absorption. It must be noted, however, that the technique is not sensitive to any other absorption process. The wavelength used by *Karner* et al. was too large to cause single-photon ionization. Looking at their power densities (0.09–$0.9\,\text{MW/cm}^2$) it appears likely, however, that multiphoton ionization did occur simultaneously, particularly in an air environment. To emphasize this point in the next section we will show a result by *Strupp* et al. [4.13], who have observed three-photon photoemission from Cu(100) surfaces for power densities as low as $0.01\,\text{MW/cm}^2$.

One other comment is in order. In all experiments involving laser radiation impinging on surfaces, the question of surface temperature is of primary importance, particularly in view of thermally activated chemical processes [4.3]. In principle, the photothermal displacement technique is an ideal way for in situ determination of temperatures. However, although absorbed energy changes can be measured with this technique quite readily, the correct determination of absolute temperatures still presents a calibration problem if one is not certain about the thermal diffusivity.

4.2.2 Multiphoton Photoemission

Nonlinear photoemission from metal surfaces has been under investigation for more than 20 years [4.58]. The topic is, nevertheless, still of interest for three reasons: (1) Resonantly enhanced multiphoton ionization from well-prepared surfaces is a unique spectroscopic tool for investigating image states, as was proved by *Giesen* et al. [4.59], and more of this type of spectroscopic applications is to be expected. (2) The influence of temperature on the multiphoton photoemission is understood theoretically [4.11, 12] but has been avoided rather than verified, e.g., for nanosecond pulses. (3) The question of what fraction of photoelectrons originates from the bulk or, alternatively, from the surface has not been pursued systematically. In the following we shall elaborate on these points and present examples. One should bear in mind, however, that in the context of this chapter, multiphoton photoemission primarily plays the role of a source for holes and free electrons.

One example of multiphoton photoemission from well-defined *surfaces* in ultrahigh vacuum (UHV) was reported by *Strupp* et al. [4.13]. These authors observed 2- and 3-photon photoemission from a Cu(100) single crystal in UHV, using two different wavelengths. Their results are displayed in Fig. 4.5. They also made an attempt to measure the kinetic energy of the photoelectrons and found consistency for the 3-photon ionization with 590 nm radiation. For 308 nm, the power dependence (Fig. 4.5) and the kinetic energy departed from what was expected for a pure 2-photon process. These authors suggest that this behavior hints at a thermally assisted 2-photon mechanism. What was most surprising was the result that the higher-order process is much more efficient, as can be seen in Fig. 4.5. This was explained by

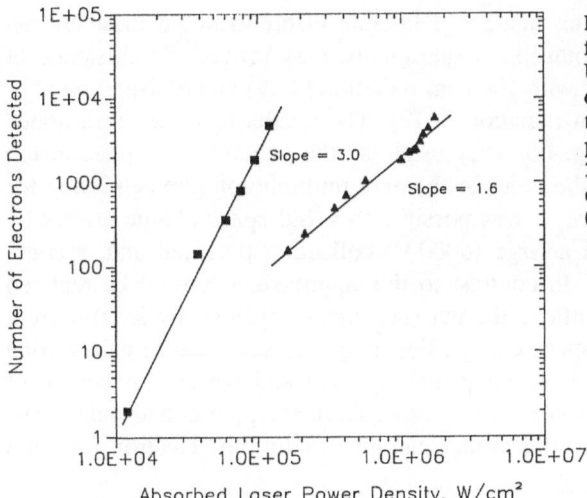

Fig. 4.5. Multiphoton photoemission from a clean Cu(100) surface under UHV conditions measured by *Strupp* et al. [4.13]. Squares denote data taken with 590 nm radiation with 1 μs pulse length, triangles are results for 308 nm wavelength and 20 ns pulse duration. The irradiated area was 0.11 cm². (Courtesy of P.G. Strupp)

a final state resonance enhancement. For us it is of interest to see that multiphoton photoemission can be observed well below 1 MW/cm² and, in the case of resonance enhancement, down to 0.01 MW/cm². To be specific, since the 2-photon process is supposed to be thermally assisted, while for the 3-photon process any thermal influence seems to be negligible, it would be most instructive to measure the thermal deformation and the multiphoton photoemission for Cu(100) simultaneously.

Bechtel et al. [4.11, 12] extended an earlier theory on the temperature dependence of photoemission and developed a theoretical expression describing a *thermally assisted multiphoton process*. They write the total emitted current density as a sum over all contributions up to the nth order, where n denotes n-photon photoemission:

$$J(r, t) = \sum_n J_n(r, t) \quad . \tag{4.2.1}$$

The current density for the n-photon process is given by

$$|J_n(r, t)| = a_n \left(\frac{e}{h\nu}\right)^n I^n (1 - R)^n A T^2 f\left(\frac{nh\nu - \phi}{k_B T}\right) \quad . \tag{4.2.2}$$

Here, I is the laser intensity, R the reflectivity, T the temperature, ϕ the surface work function, A the theoretical Richardson constant, and a_n an empirical constant. The temperature depends on space and time corresponding to the spatial and temporal shape of the laser pulse. The function f is called the Fowler function [4.12] and is tabulated in [4.60]. For $n = 0$, (4.2.2) reduces to the familiar Richardson equation describing thermionic emission, i.e. $f(n = 0) = \exp(-\phi/k_B T)$. *Bechtel* et al. [4.11, 12] pointed out that J_n contributions with n smaller than the maximum n required to exceed the work function represent photoemission from the high-energy tail of the Fermi-Dirac distribution at elevated temperatures. If present, they will change the slope of the electron yield curve which for pure n-photon photoemission should be proportional to I^n. Although they developed the theory including temperature effects, *Bechtel* et al. made a deliberate effort to avoid these by using 30-ps pulses. In carefully controlled experiments, they proved the existence of 4-photon photoemission from W with 1064 nm radiation [4.11] and of 2-photon photoemission from Ta with 532 nm radiation [4.12]. The results of these "milestone" experiments are presented in Fig. 4.6. They establish the unambiguous relationship between the work function and the order of the pure multiphoton photoemission for a given metal. In the above work, it was possible to avoid space charge effects by two features: first, they utilized a large (6000 V) collecting potential and, second, they utilized picosecond pulses. In contrast to this approach, it should be realized that under most laboratory conditions the net (i.e. uncompensated by ions) current is limited to $\sim 10^{-9}$ C/cm^2 by space charge. For longer pulses, such as nanosecond ones, sizeable thermal contributions are to be expected and we are not aware of any systematic studies of this problem up to now. Even for picosecond pulses, the temperature rise of "whiskers" on the surface can give thermionic electron emission for long periods.

Concerning bulk versus surface photoemission, these two contributions can be separated by measuring the polarization dependence of the photoemission yield. Employing this technique and using 30-ps pulses at 1064 nm (1.17 eV), *Lompre* et al.

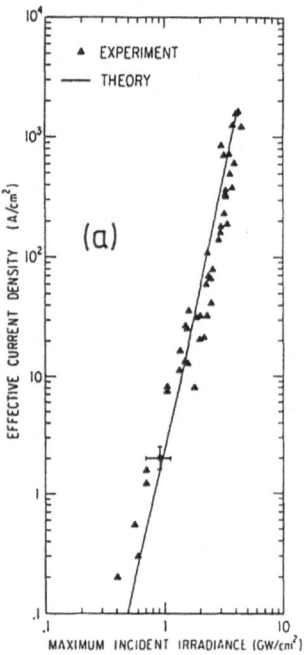
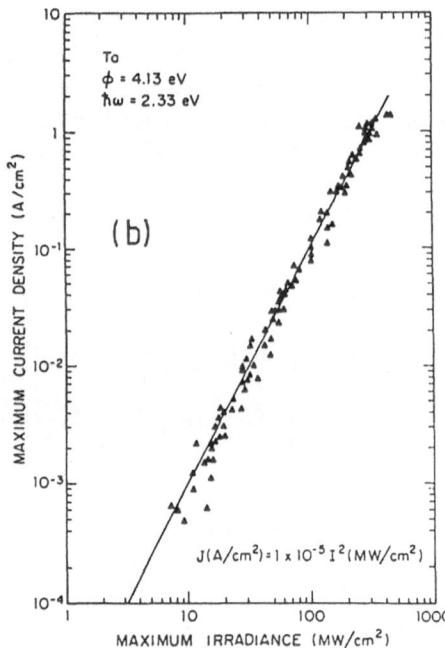

Fig. 4.6a, b. Multiphoton photoemission from W and Ta reported by *Bechtel* et al. [4.11, 12]. The data for W in (a) were measured with 30 ps pulses of 1064 nm with the target in 1×10^{-7} torr vacuum [4.11]. The effective current density is the total charge per pulse divided by laser area (0.95 mm^2) and pulse duration. The solid line indicates the theoretical expectation for a 4-photon process according to (4.2.2). In part (b) the two-photon photoemission from a Ta target in 10^{-8} torr vacuum irradiated with 21 ps, 532 nm laser pulses is shown [4.12]. The solid line represents the relation given in the figure. (Courtesy of N. Bloembergen)

[4.61] proved that 5-photon photoemission from gold originated solely from the surface. *Bechtel* et al. [4.12], on the other hand, observed that 2-photon photoemission from Ta, induced by 532 nm (2.33 eV) radiation, came partly from the bulk and partly from the surface. *R. Haight* [4.62] has some recent results on this point for semiconductor surfaces, which however lies outside of our stated area of interest. It is presently an open question whether or not this different behavior of Au and Ta is specific for the two materials or whether it is caused inherently by the order of the multiphoton photoemission. Certainly, this topic is of importance and needs further investigation.

4.2.3 Damage Threshold of Metals

In view of the large power densities needed to obtain the data in Fig. 4.6, one tends to be curious as to whether there was simultaneous emission of neutral metal atoms from these Ta and W surfaces. In other words, does photoemission play a role, even an indirect role, parallel to boiling, for the removal of material from metal surfaces? If so, then one would expect, for a given intensity, the ablation rate to depend on the work function and the electron density close to the Fermi edge. The question is, which technique is suitable for measuring such a dependence? Recently, *Petzoldt* et al. [4.63] have shown that the energy of the shock wave generated by a high power laser pulse impinging on a surface is correlated to the volume of the removed material. Hence, measuring the shock wave energy would be one way of monitoring a possible dependence of the ablation rate on multiphoton photoemission. In the following we will describe this kind of experiment and discuss first results obtained for metals [4.64].

The energy of shock waves generated by laser pulses and expanding away from the surface into the ambient atmosphere can conveniently be measured by the "mirage effect" [4.35, 63–67]. The shock pulse causes a transient change in refractive index of the gas close to the surface, which can be detected by the deflection of a NeHe probe laser beam running parallel to, and at a distance z (a few millimeter) above the surface. The principle of the experiment is illustrated in Fig. 4.7. A detailed analysis [4.67] shows that the deflection angle ϕ is given by the relation

$$\tan \phi = \frac{l_0}{2\epsilon_0 n_0} \frac{c_v}{c_p} \frac{\alpha}{k_B T} \frac{\partial p}{\partial z}\bigg|_{z=0} \quad , \tag{4.2.3}$$

where l_0 is the extension of the pressure pulse along the probe beam, T the temperature, $p = p_0 + \Delta p(t)$ the pressure, α the molecular polarizability, and n_0 the refractive index of the ambient gas. In the experiment, a sensitivity could be reached to detect a deflection angle of $\phi \sim 4 \times 10^{-7}$ rad, corresponding to a pressure gradient of 5×10^{-6} bar/mm. As indicated in Fig. 4.7, a Nd:YAG-pumped dye laser was used to produce the damage. Its intensity was varied, without changing the beam geometry, by suitable polarization optics. The HeNe laser beam probing the shock wave had a waist of typically 120 μm. Any deflection moved the intensity profile of the probing beam across a 40 μm slit (of a monochromator), resulting in a corresponding intensity change at a photomultiplier. Such a deflection signal S is nearly identical for

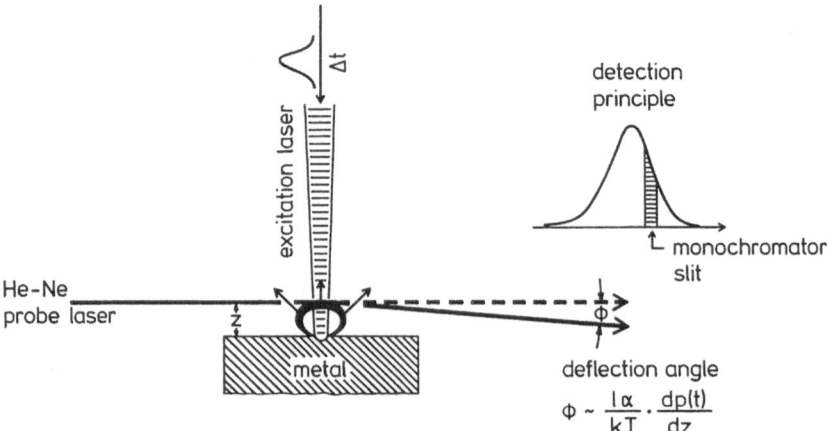

Fig. 4.7. Principle of the scheme for measuring shock wave energies by deflecting a probe laser beam by the transient change in refractive index

all materials, and proportional to dp/dt. Consequently, the time-integrated signal is proportional to the amplitude $\Delta p(t)$ of the pressure pulse, and its energy E can be derived from the deflection signal by proper integration [4.67]:

$$E \propto \int (\Delta p)^2 \, dt = \int dt \left(\int_0^t S \, dt' \right)^2 . \tag{4.2.4}$$

The proportionality factor relating the measured photocurrent to the energy carried away by the pressure pulse must be determined for each material. As was proved in [Ref. 4.63, Fig. 4], this energy E is, in turn, proportional to about the third power of the spot diameter. Hence, once these proportionalities have been properly calibrated, it is expected that the probe beam deflection technique will be capable of measuring the picograms of material removed by a single laser shot of a certain energy.

When applying the probe beam deflection technique to Al, Cu, and Ag, the intensity dependence displayed in Fig. 4.8 was obtained for the photoacoustic energy, as defined in (4.2.4). During the measurement, the intensity was varied at random, and each point in Fig. 4.8 represents an average of about ten individual points, each falling within the intensity interval used for averaging. A deflection signal is detectable first above a certain threshold value, which is approximately $5\,\text{MW/cm}^2$ for Al, $20\,\text{MW/cm}^2$ for Cu, and $80\,\text{MW/cm}^2$ for Ag. These threshold intensities should not be taken too seriously where the purity of the metal, its surface quality, and its reflectivity are not further specified. More physical information comes from the slope with which the photoacoustic energy rises above threshold, up to a point beyond which the data bend away from the straight line. This behavior can be seen in Fig. 4.8 in all three cases. It was suggested [4.64, 67] that as long as the data follow a straight line (in a log-log plot) this is evidence of *multiphoton ionization* playing a major role in the surface damage of metals, while the bending at higher intensity signals the onset of plasma formation.

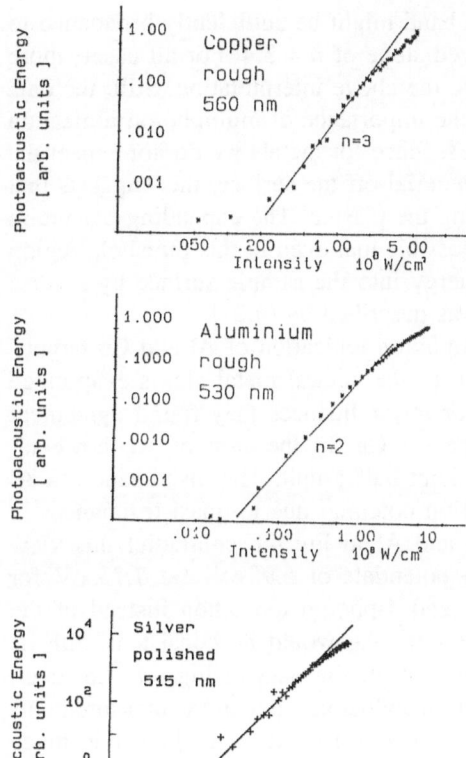

Fig. 4.8. Photoacoustic pulse energies according to (4.2.4) as a function of intensity for Cu, Al, and Ag [4.67]. The data were taken in air, with dye laser pulses of about 3 ns width and different wavelengths for each metal. Measurements on Cu and Al were done in multishot (N-on-1) mode, while the Ag data represent single shot results (1-on-1). The solid lines indicate the slope for a limited intensity range immediately above threshold

Notice that different wavelengths were used for the three metals in Fig. 4.8, in order to identify the correlation between the order of the multiphoton process and the work function. For *copper*, three photons of 560 nm (2.21 eV) are needed to overcome a work function of 4.65 eV; the data closely follow a line of slope $n = 3$ for about one order of magnitude in intensity. For *aluminium*, two photons of 530 nm (2.34 eV) suffice to match the work function of 4.28 eV. Again the data follow a slope $n = 2$ for more than an order of magnitude. For *silver* too, the data define a slope in an intensity range between 70 and 700 MW/cm^2, but in this case a slope of two was expected for 515 nm (2.41 eV) in view of a work function of 4.64 eV [polished Ag(100)]. However, a slope of $n = 3.4$ was observed.

Looking at the band structure, we notice that Al is a p-band metal with a high density of states up to the Fermi energy. This provides for efficient ionization by only two photons of total energy 4.68 eV. In Cu and Ag, on the other hand, the highest densities of states are in the $3d$ and $4d$ bands, respectively. For Cu, three photons with a total energy of 6.63 eV can ionize $3d$ electrons, leading to the slope of $n = 3$ in Fig. 4.8. In Ag, however, the $4d$ band has a binding energy of more than 8 eV, and even three photons of 2.41 eV each would not suffice to ionize $4d$ electrons. Nevertheless, at

101

temperatures near the melting point the $4d$ band might be sufficiently broadened to allow $4d$ ionization, resulting in the observed slope of $n = 3.4$. For all cases, more systematic results are needed to substantiate the above interpretation. Still, the data in Fig. 4.8 furnish the first evidence about the importance of multiphoton ionization for shock wave generation on metal surfaces. Since for metals we do not expect the holes to play a significant role in driving material off the surface, this suggests that the photoelectrons play a key role in igniting the plasma. The expanding plasma in turn generates the shock wave. It appears possible that even in this threshold region the photoelectrons couple the etch laser energy into the sample surface by inverse bremsstrahlung $\rightarrow$ shock wave $\rightarrow$ surface, as described by [4.27].

Trying to interpret their results on multiphoton ionization of Al and Cu targets, *Park* et al. [4.14] assumed direct ionization of the neutral metal atoms evaporated from the surface by the incoming light. For lower fluences they found agreement between their view and the observed slope for Cu. In the case of Al, however, the measured slope came out too low by about half a unit. The discrepancy in Al was explained by a lowering of the ionization potential due to quasi resonances in the multiphoton process. Our data for Cu and Al in Fig. 4.8 contradict this view of direct ionization of neutrals. Ionization potentials of 5.98 eV and 7.73 eV for Al and Cu, respectively, would require 3- and 4-photon ionization instead of the observed 2 and 3 photons. The slope of 3.4 for Ag would be consistent with its ionization potential of 7.57 eV. On the basis of all the data in Fig. 4.8, however, we favor the model of ionization by electron collisions, but more measurements are needed to settle this issue. Comparison of experiments with [4.14] is made significantly more difficult by two other factors: first, the authors imply that the supply of neutral atoms is constant over their fluence; however, because of these materials' high heat of vaporization, the slope of atom density versus fluence is even higher than the implied multiphoton effects; and second, the fluences quoted are well into the plasma region [4.27]. While the recent results of *Helvajian* and *Welle* [4.68] were obtained at atom removal fluxes that are microscopic compared to the majority of those of the present work, nevertheless, some of their implications about multiphoton effects in solids may be applicable to the present range of interactions.

The disagreement between our view and that of *Park* et al. [4.14] raises the question about the validity of conclusions drawn from slopes of shock wave energy or, in the case of *Park* et al., of positive ion yield measured in a poor vacuum (10^{-3} Torr) and with charge densities that imply a Debye length that is small compared to the dimensions of the collector. Apart from the fact that one always needs to examine the comparability of data from different measurements, the accuracy of the observed slopes is also an issue. Although the slopes in Fig. 4.8 carry a large uncertainty, it is far smaller than 1 unit. Corrections due to surface temperature, as implied by (4.2.2), are also unlikely to amount to as much as 1 unit. Hence the slopes are believed to characterize the dominant multiphoton process reasonably well. A more critical point is the physical state of the surface during a laser pulse of nanosecond duration (Fig. 4.2). Numerous experimental results [4.44, 47–49, 69] suggest that the surface is molten when the shock wave develops. In principle, that should not change the work function significantly compared to that of the poly-crystalline metal, but it would rule out any surface orientation dependence when

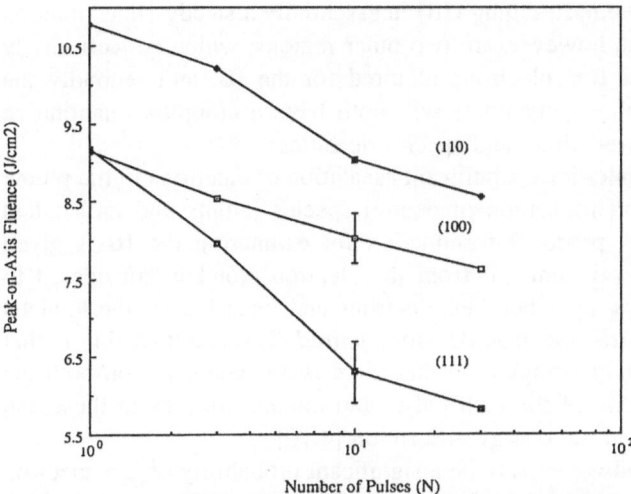

Fig. 4.9. Decrease of damage thresholds with increasing number of laser pulses on oriented chemically polished Cu surfaces in air. (From [4.44], courtesy of M.F. Becker)

using single crystals. Neither would one expect any difference when the data are taken in the multishot (N-on-1) or the single-shot (1-on-1) mode, as was done in Fig. 4.8 for Cu and Al (N-on-1) and for Ag (1-on-1). It comes as some surprise, therefore, that exactly those dependences have been observed by *Jee* et al. [4.44] for single-crystal metal surfaces. Their result is displayed in Fig. 4.9. It shows that the damage fluence differs for the (110) and (100) or (111) orientation even in the single-shot mode. Furthermore, the damage fluence decreases with the number of incident pulses for all surface orientations, whereby the threshold difference between the various orientations remains. This result provides strong evidence that the primary photon absorption is a most important factor in the development of damage. Surface defects not only change the absorption properties of the surface [4.20], they also reduce the local melting threshold. Multiple-shot irradiation, *Jee* et al. [4.44] argue, causes plastic slip deformation, which in turn increases the surface roughness and thereby enhances the light absorption and decreases the fluence threshold for melting. The lesson from this experiment is that the surface preparation, crystal orientation, ambient atmosphere (causing oxidation), and the irradiation mode all play a significant part in the primary energy absorption and hence in the resulting damage. Consequently, the observed slopes of the shock wave energy in Fig. 4.8 provide no concrete information about the order of the multiphoton process as long as the surface is not more clearly defined. They do, however, point the way to further research.

4.2.4 Plasma Formation

The transition of metal ablation on going from a thermal to the plasma region is primarily a question of the generation and recombination of ions in the plume. During an intense etch pulse, the general concept is that the laser energy is coupled into the plasma by inverse bremsstrahlung (IB) in essentially a steady-state situation [4.27]. Adjoining this region, however, are two other regions, which concern firstly the *generation* of the initial free electrons required for the IB, and secondly the *decay* of the plasma. The above quantities will both have a complex quantitative dependence on the laser power, time, and species densities.

One of the effects that underlies a significant escalation of electrons in the plume is multiphoton excitation and ionization of neutral species (atoms and molecules) in the etch plume [4.70]. A practical formulation for estimating the IB is given in [4.71]. Note that the energy removal from the electron cloud is considered to be due to Coulomb scattering by other free electrons and ions; hence, the total IB energy transfer goes as the free electron density *squared*. The result of this is that the energy transfer is relatively complex, in that there is a dependence on both the temporal and geometric profile of the etch pulse, and nonuniformities in these can cause significant alterations in the energy absorption [4.70].

In the etch plume, there also seems to be a significant probability of free electron recapture; this may occur by either three-body recombination (TBR) or dissociative attachment. As a first approximation, one can use TBR rates [4.72] to find that, for the first few millimeters of expansion, TBR is quite rapid if the electrons have less than 3 eV of energy. The reason for the energy limitation is that the TBR rate has an energy dependence varying as $E^{-4.5}$, i.e. a very strong dependence. Furthermore, inelastic electron scattering will only bring electrons down into the above-mentioned energy range at a rate dependent on $E^{-1.5}$, again there is no effect on higher-energy electrons. Overall, one now sees how a fluence threshold is built into the plasma formation: one has the loss of electrons, varying as energy to -4.5 power in the electron temperature, being subtracted from IB, which raises the electron temperature which, in turn is linearly dependent on electron density. In addition, once the tail of the electron distribution runs away to energies much greater than 3 eV, additional collisional ionization will occur and additional IB will set in. Two overall observations about the above are: the detailed free-electron behavior should be amenable to being modelled, and second, interesting laser wavelength dependences should be observable as these will distinctly alter both the electron temperature (depending on $nh\nu$ minus ionization energy) and the inverse bremsstrahlung loss rate (as λ^2).

4.3 Wide Band Gap Ionic Materials

The endeavor to understand the basic mechanisms of optical damage [4.8] led to a great number of investigations, focusing both on surface damage [4.20, 21, 73–76], and on damage in the bulk [4.17, 19, 23, 77–81]. Stimulating ideas also came from DIET studies of alkali halides [4.28–30]. In this chapter we will limit the discussion mostly to *sapphire* and to *fluoride crystals* for two reasons: (1) A large body of results is available for these crystals, and (2) the structure of the fluorides is comparatively simple so that they can serve as model substances.

Both the optical and the thermal properties of these materials are so distinct from those of metals that the primary photon absorption and the response of the system are drastically different. For example, light absorption does not in general lead to melting of the material, although free carrier heating in the bulk can raise the temperature close to the melting point [4.17, 79–82]. Experiments further show that, at least for laser pulses longer than tens of picoseconds, the material removal from crystal surfaces takes place in a violent explosive way [4.67, 73, 76]. Hence the notion of a neutral vapor cloud that becomes ionized after leaving the surface, which appeared to be a reasonable model for metals, does not apply to transparent dielectrics. Instead, other models have been proposed, in particular the avalanche breakdown model [4.25, 83–85], whereby one has to carefully distinguish between extrinsic and intrinsic laser-induced breakdown, as emphasized by *Braunlich* et al. [4.82]. Generalizing somewhat, the sequence with increasing intensity leading to surface damage of wide band gap materials can be visualized as

thermal expansion/thermal desorption
⇓
nonthermal desorption
⇓
explosive ablation (shock wave)
⇓
plasma formation

Actually, steps 1 and 2 may be inverted as there appear to be cases where photochemistry is the initial step in the ablation process and residual thermal energy continues the process as the original host crystal no longer retains its original chemical and structural inertness [4.10]. The various reactions can all be present in one single pulse (Fig. 4.1), or, alternatively, can be reached by properly limiting the laser intensity, wavelength, and pulse length. In the following sections we will present examples for each step.

One of the foremost questions for wide band gap materials is how light is absorbed; after all, these materials should be transparent for photon energies smaller than the band gap. The answer must distinguish between surface and bulk absorption. For the latter, *Braunlich* and his collaborators have shown in a set of carefully designed experiments [4.17, 79–82] that for intrinsic, i.e., defect-free crystals, the dominant prebreakdown energy deposition is by multiphoton absorption. This group employed both photoacoustic techniques [4.17, 79] and self-trapped exciton recombination luminescence [4.80, 81] to identify 4-photon excitation of valence electrons to the conduction band and subsequent free carrier heating for ultrapure crystals of NaCl and KBr. Of course, a higher concentration of free electrons in the lattice, either present initially or generated by the ionization of defects, would lead to avalanche breakdown. Therefore, it was pointed out early and by many workers in the field that lattice defects play a decisive role in photon absorption in transparent optical solids [4.1, 8, 19–21, 73–75], in particular at surfaces.

The situation regarding surfaces is still more complex, compared to that of the bulk. Any kind of surface treatment leaves extrinsic defect states, and little has been done with well-prepared and analyzed surfaces in the UHV. But even for stoichiometric surfaces, the precise band structure and the degree of surface reconstruction is known only in a few cases [4.86]. In addition, some of the conventional analytical techniques (LEED, Auger, EELS) can be applied to surfaces of dielectrics only with great difficulty, if at all, because of charge-up problems. On the other hand, surfaces do offer some advantages; the most outstanding of which is the larger variety of applicable analytical tools, e.g. particle emission, thermal deformation, reflectance, and second harmonic generation (SHG). Another advantage is the likelihood of accumulating F-centers at the surface, i.e. an unpaired valence electron of the metal atom at the site of a missing neutral halogen. It has been shown by cluster calculations that such a F-center at the (111) surface of a CaF_2 structure leads to occupied and unoccupied states in the upper half of the (bulk) band gap [4.16, 87]. We hypothesize that these surface F-centers are the dominant light absorption centers at clean surfaces of transparent alkali and alkaline earth halides. Their origin may be sought in external surface treatment, like polishing and cleaving, but may also be the result of defect production following multiphoton absorption. Extrinsic defects, determined by surface quality and environment, will probably be most important. Gross structural defects, adsorbates and impurities [4.20, 21, 73–75] all lead to an increased density of states in the band gap. Therefore, one can generally picture the light absorption at surfaces of "transparent" crystals to be inherently tied to defect sites. This is illustrated in Fig. 4.10.

Provided there are occupied states in the band gap, *single-photon absorption* will take place at low intensities and cause local heating, resulting in a photothermal deformation of the surface [4.10, 36, 88]. This effect will be discussed in Sect. 4.3.1. In connection with photothermal deformation we would like to call attention to the model of *Manenkov* et al. [4.22] suggesting that, in the bulk, multishot laser pulses of power levels far below the single-shot damage threshold build up stress due to the volume increase accompanying the production of defects. The accumulated

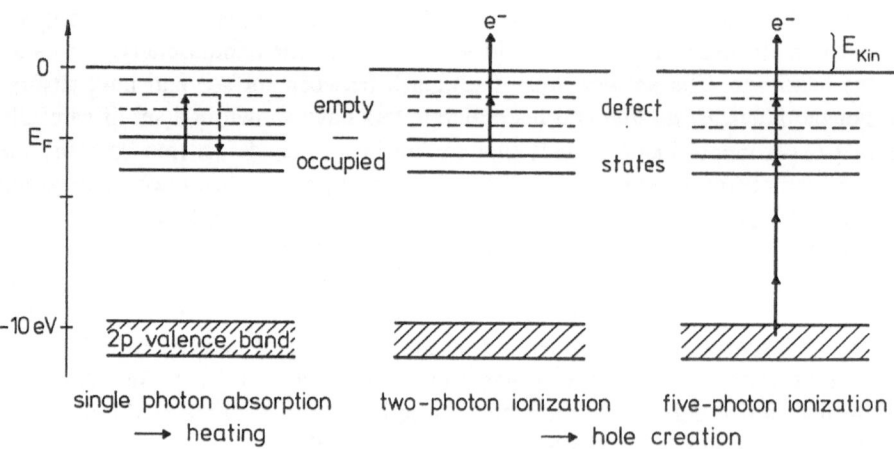

Fig. 4.10. Possible single- and multiple-photon absorption processes on surfaces of wide band gap crystals. Solid bars represent occupied and dashed bars unoccupied surface defect states in the band gap. An example would be a surface F-center, which, in the fluorides, is equivalent to a missing neutral fluorine [4.87]

stress could conceivably reduce the damage threshold, similar to what was observed in Fig. 4.9 for copper. Experimental evidence supporting this model was recently presented by *Casper* et al. [4.23] for multishot intrinsic bulk damage in KBr. The thermal deformation of surfaces might cause a similar reduction of the multishot damage threshold, however, to our knowledge this has not yet been investigated.

In this connection, the microscopic origin of thermal deformation is of interest. Generally, one would assume that the vibrational motion of the ions leads, with increasing temperature, to an increased mean separation between them. The temperature rise may be caused by excitation and radiationless decay of occupied defect states in the band gaps, as sketched in the left part of Fig. 4.10. Such a mechanism would suggest that surfaces with a high concentration of, for example, *F*-centers would bulge much more readily compared to more perfect surfaces, since the *F*-centers would ensure an increased energy deposition. This could be experimentally checked. The fact that defect accumulation causes a volume increase suggests that there may be still another contribution to the surface deformation, namely the accumulation of *F*-centers at the surface. If separated *F*-*H* pairs are generated in the bulk following multiphoton absorption [4.19] then there is a certain probability that some *F*-centers will diffuse to the surface layer and cause an expansion. Since rapid diffusion of *F*-centers requires elevated temperatures [4.89], this effect should not be significant at room temperature but should be looked for with heated samples.

As indicated in Fig. 4.10, at higher intensities *multiphoton ionization* becomes a likely event once the density of states in the band gap is sufficiently large to make the quasi-resonant states a dominant pathway. Photon absorption of low order (e.g. two photons) leads to ionization of occupied surface defect states (Fig. 4.10). This process is not bond-breaking, but it will change the electric properties of the surface. At still higher intensities, a multiphoton process bridging the band gap can occur, provided that it is supported by some intermediate states. Photoemission results supporting this view [4.16] will be presented in Sect. 4.3.3. It would probably be more correct to call such a process a multiple-photon sequential absorption, particularly since the coupling of the intermediate states to the phonon bath is a competing loss channel leading to heating. On the other hand, we expect that at high intensities there is still a well-defined phase relation between the initial and final states, justifying the term "multiphoton" ionization. In any case, a multiphoton transition across the bulk band gap would produce a hole in the valence band and would therefore be bond-breaking. As discussed above, *Braunlich* and collaborators have identified multiphoton transitions in the bulk of pure alkali halide crystals and shown that it leads to defect production and free carrier heating [4.17, 19, 79–82]. Here we take the view that at least in the surface layer multiphoton transitions across the band gap proceed resonantly enhanced. The intensities required for nonresonant multiphoton transitions in wide band gap materials would inevitably lead to dielectric breakdown in front of the surface, as will be discussed in Sects. 4.3.3 and 4.3.4. Hence the role of defects produced following electron-hole pair excitation is that they increase the density of states in the band gap, which in turn further promotes resonantly enhanced multiphoton absorption. This applies to transient defects within one laser pulse (cf. Fig. 4.1) as well as to longer-lived ones for consecutive pulses. Therefore, one would generally expect the rate of defect generation to be proportional to the number of defects already present, leading to an exponential increase of the defect density during irradiation.

Another difference between bulk and surfaces is that electrons excited to the conduction band near the surface can escape rather than serving for free carrier heating. The remaining holes in the valence band will charge the surface layer accompanied by lattice strain due to screening charges. Once the density of holes in the valence band exceeds a certain crucial value, the charges will repel, which could be the reason for the occurrence of explosive ablation off these materials [4.76]. In Sect. 4.3.4 we will discuss the results of some shock wave measurements on fluoride crystals [4.63]. These experiments actually provide the first evidence that there exist two types of surface damage mechanisms; one initiated by multiphoton ionization from the valence band and the other by dielectric breakdown.

4.3.1 Photothermal Deformation of Sapphire

Because of the absorption depth it is often difficult for metals to separate surface and bulk absorption effects. In contrast, for pure transparent materials there is virtually no bulk absorption and therefore the photothermal deformation can be used to monitor the surface absorption, which, in turn, is tied to defect states, as indicated in Fig. 4.10. To our knowledge, this has only been performed with Al_2O_3, for transparent materials [4.10, 36, 88]. In this section the main results of that experiment will be reviewed. Its motivation arose from the incongruity that sapphire appears to readily etch photochemically, yet there was no obvious mechanism for the absorption of 193 nm photons. The evidence for photochemical etching was based on only one type of laser induced fluorescence (LIF) result [4.90] as detailed later in this section, and it called for additional evidence. In addition, photothermal deformation yields information about the energy deposited in the surface and the degree of thermal equilibrium attained (in microseconds). The value can be compared to the melting (2040°C) and vaporization temperatures of sapphire in order to decide about the nature of the etching mechanism. The outcome was that, e.g. for 193 nm excimer laser radiation, a very large surface absorption, 3%, was observed, which however could only raise the surface temperature to about 430 K if the ablative energy were purely thermal and hence diffused into the bulk. Note that because of the nature of the photothermal deformation experiment, one actually has microseconds to establish the equilibrium and obtain a value for the temperature. This 430 K is far too low to account for ablation by thermal vaporization, which proves that the laser etching energy of sapphire must remain as an electronic excitation on the surface for at least nanoseconds and hence act as a photochemical dissociating agent. This observation explains the low damage threshold of sapphire.

The photothermal deformation of sapphire [4.10] was measured with an experimental setup quite similar to the one used by *Karner* et al. [4.34], shown in Fig. 4.3. Pulsed excimer laser radiation was focused to spots with a diameter between 0.4 and 1.0 mm, and the surface bump was probed by a HeNe laser of about 0.2 mm waist under oblique incidence off center. Sapphire samples of 100 μm and 250 μm thickness were used. The absorbed energy was determined by calibrating the angular deflection of the probe beam against the deflection obtained from the surface deformation of a Si wafer of the same thickness [4.88]. Such calibration rests on theoretical model calculations [4.36]. Hence the final surface temperature was derived from a comparison between the theoretically predicted and the measured slope of the bump. In this way it was also shown that unacceptably large bulk absorption would have been required to dominate over the surface absorption. A more direct verification of this was the fact that the whole sapphire plate bent because of the surface deformation. Since, for phase reasons, the average electric field is higher at the exit side, the energy absorption was 1.56 times as large at the exit surface as at the entrance surface. This caused a deflection of opposite sign compared to Si. Furthermore, the deformation was about three times larger for the thinner 100 μm sapphire plate.

The experimental results for the absorbed energy versus laser fluence are displayed in Fig. 4.11 for three different wavelengths. From the slope $n = 1$ in all three cases we conclude that single-photon absorption dominates and increases with increasing photon energy. It was further shown [4.10] that the surface absorption depends sensitively on the treatment before irradiation. For example, chemical etching reduced the amount of absorbed energy by about one order of magnitude at a given fluence at 193 nm. The reduction by chemical etching for 248 nm was about half of that. This observation not only proves again the light absorption in the upper surface layer, it also suggests that defects, probably associated with surface dislocation networks, are the absorbing centers. These defects may be in the form of dangling bonds or dangling bonds that have subsequently reacted with impurities. Although virtually nothing is known about the exact nature of these defects, it is conceivable that they give rise to an electronic density of states in the band gap, in the same

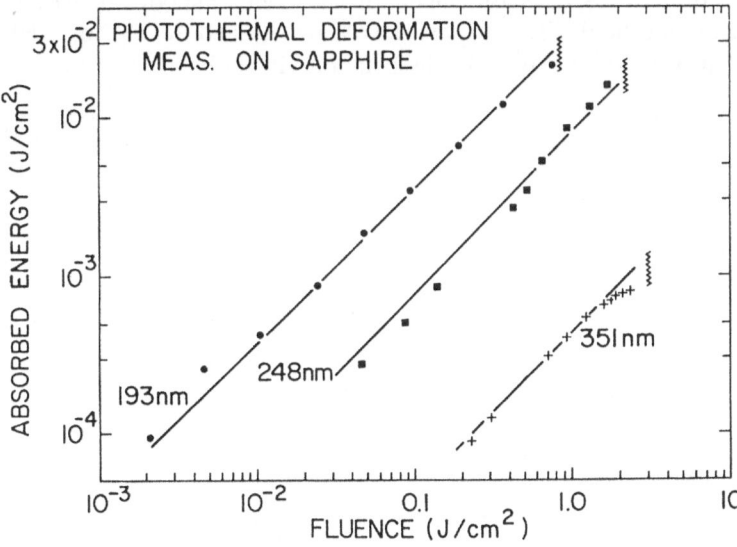

Fig. 4.11. Energy absorbed in the surface layer of sapphire versus fluence, measured by monitoring the photothermal deformation of 250 μm thick plates [4.10]. Etching thresholds are indicated by serrated marks. Note the identical increase in absorbed energy with increasing fluence for all three wavelengths

way as sketched in Fig. 4.10. Excitation of electrons may lead to heating, causing the surface to expand and in turn generate additional dislocations. In this connection it would be instructive to check for photoelectrons and measure their kinetic energy.

An independent cross-check of the surface temperature had been made earlier by measuring the internal temperatures (rotational and vibrational energies) of AlO leaving from the surface of sapphire near the ablation threshold [4.90–92]. In general, this method has yielded significant amounts of information about surface temperatures because gas phase collisions have a small probability of altering the internal energy of diatomics. This distribution can be easily derived from LIF employing the $(v')B\,^2\Sigma^+ \rightarrow (v'')X\,^2\Sigma^+$ electronic transition in AlO. Excitation and fluorescence were separated by observing different vibrational transitions in fluorescence than were used for excitation; for example, excitation: $v'' = 0 \rightarrow v' = 2$; fluorescence: $v' = 2 \rightarrow v'' = 2$. A typical LIF spectrum is presented in Fig. 4.12. The calculated spectrum in the insert shows by comparison that the rotational distribution corresponds to approximately 500 K. The low population of higher vibrations $v'' \geq 2$ in the electronic ground state was evidence that the AlO originated from the surface rather than being formed in the gas phase by recombination. Therefore the temperatures derived from the rotational and vibrational population are representative of the surface. In Fig. 4.13, the values obtained from LIF measurements on AlO near threshold are compared to the temperatures determined by photothermal deformation (calibrated against Si). Despite the scatter (primarily due to the subtraction involved in determining ΔT), the LIF data are consistent with the temperatures extrapolated from the surface deformation, which also gives confidence in the calibration procedure of the latter. Regarding the etching mechanism, these surface temperatures are over ten times too low to have produced the etching thermally, which would require temperatures around 4800 K or higher. Further evidence for photochemical etching comes from time-of-flight measurements of the kinetic energies of Al and AlO (Fig. 4.14). Near threshold one observes 3.8 eV for Al and 1.2 eV for AlO, which are about what one would expect for dissociation via antibonding states, but

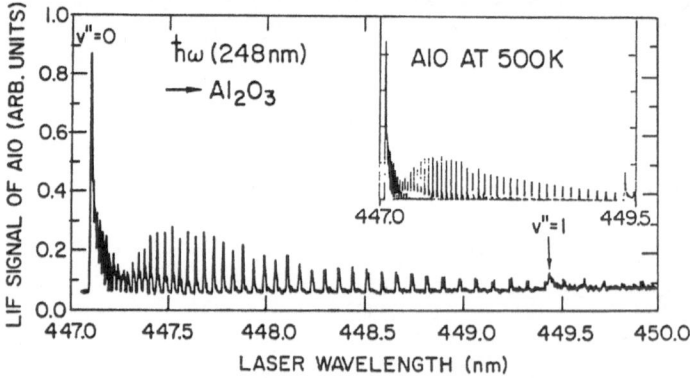

Fig. 4.12. Laser-induced fluorescence spectrum of AlO etched off Al$_2$O$_3$ near threshold, when each laser pulse removes ~ 0.4 nm [4.90, 92]. The rotational distribution calculated for $T = 500$ K is shown in the inset

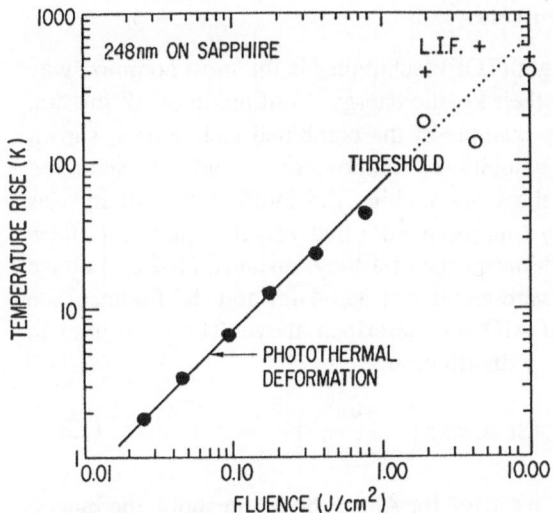

Fig. 4.13. Comparison of internal temperatures of AlO diatomics, etched near threshold, with temperatures obtained from photothermal deformation measurements below threshold. Open circles represent vibrational temperatures and crosses rotational temperatures derived from LIF spectra like the one in Fig. 4.12

they far exceed the critical temperatures of both Al and Al_2O_3 [4.92]. Presently there seem to be no conclusive results indicating the precise nature of the repulsive states, although proposals involving initial ejection of an oxygen cation appear attractive. Also, there is no detailed information about the subsequent thermal dissociation of the now nonstoichiometric lattice.

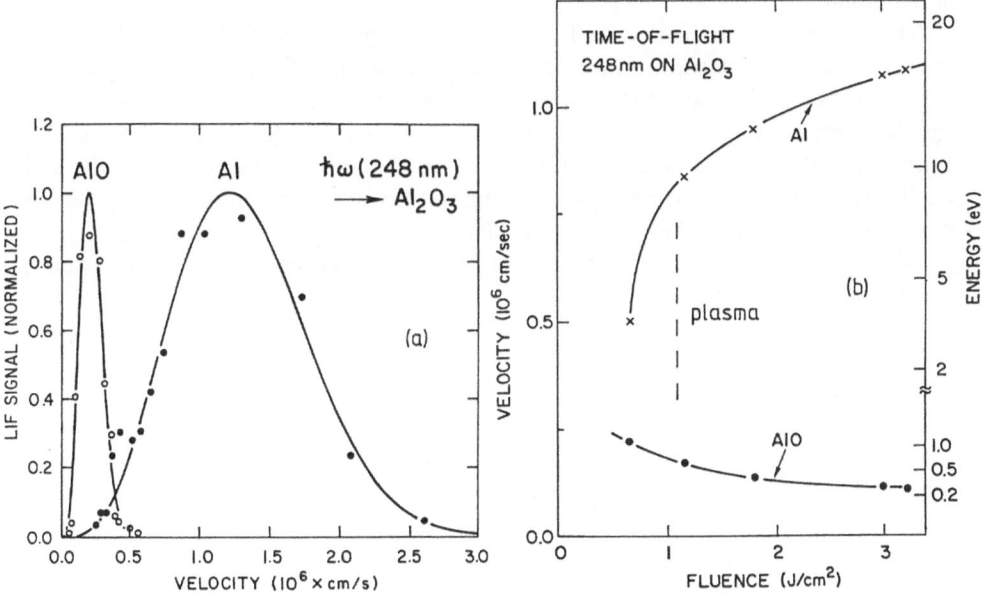

Fig. 4.14a, b. Time-of-flight results for Al and AlO departing from sapphire surfaces etched with 248 nm pulses of 20 ns duration. (**a**) Velocity distributions for a fluence around $2 \, J/cm^2$ [4.91]. Solid lines represent Boltzmann distributions (4.3.1) fitted to the data. (**b**) Velocity of the particles versus fluence [4.92]. The rapid increase for Al reflects the influence of plasma forces, and the onset of plasma formation is indicated by the dashed line. The etching threshold is $\sim 0.6 \, J/cm^2$

4.3.2 Nonthermal Desorption of Neutrals

A combination of LIF and time-of-flight (TOF) techniques is the most common way of identifying neutrals and measuring their kinetic energy. Continuing the discussion at the end of the previous section, an example of the combined technique is shown in Fig. 4.14a [4.90–92]. It shows the velocity distributions of Al and AlO sputtered with 248 nm light at a fluence well above the etching threshold. The velocity was determined by varying the time delay (microseconds) between the sputtering pulse and the dye laser pulse exciting the fluorescence at a fixed distance (1–2 cm) above the surface. Ground state Al atoms were excited at 394.4 nm and the fluorescence was observed at 396.1 nm; the LIF of AlO was described above. The solid lines in Fig. 4.14a represent a fit to the velocity distribution:

$$\text{LIF signal} \propto \text{particle density} = \text{const.} \, v_z^4 \exp\left(-\frac{mv_z^2}{2k_{\mathrm{B}}T}\right) \quad , \tag{4.3.1}$$

yielding $k_{\mathrm{B}}T = 13\,\text{eV}$ for Al and $k_{\mathrm{B}}T = 0.6\,\text{eV}$ for AlO. Above threshold, the kinetic energy varies with fluence as illustrated in Fig. 4.14b. One can see that the laser-sputtered Al atoms quickly gain energy with increasing fluence by interactions with the laser-produced plasma, involving acceleration of Al^+ ions and subsequent charge exchange collisions and/or electron-ion recombination [4.91]. In contrast, the AlO molecules lose energy with increasing fluence. Also, above $1\,\text{J/cm}^2$ the intensity of the LIF signal decreases rapidly. Both observations are explained by dissociation of those AlO molecules that have velocities overlapping with the plasma centered on the Al cloud.

This powerful combination of TOF techniques with LIF detection has often been used in the areas of desorption (of adsorbates), scattering of molecules on surfaces, and desorption following inner-shell excitations. The importance of the technique was the reason for presenting the instructive examples in Figs. 4.12 and 4.14. The contrasting situation is to detect only the ions, because of the convenience and sensitivity of mass spectrometers. In doing so, however, one may actually fail to observe the majority of the emitted particles.

We now turn to the underlying mechanism for the emission of neutral species, referring among others to the work of *Itoh* and collaborators [4.18, 31, 41, 93]. The emission of ground state and excited neutrals from wide band gap materials has been one of the key issues in the DIET field for years [4.28]. Here, however, instead of the low energy of laser photons, high-energy excitation was used, produced by either electron impact or high-energy photons (synchrotron radiation). The directional emission of neutral halogens from alkali halide crystals brought about by ruby laser radiation was demonstrated as early as 1975 in a "milestone"-type experiment by *Schmid* et al. [4.94]. It demonstrated the decay of H-centers at the surface and proved the formation of separated F-H pairs following four-photon absorption. The local atomic defects, produced by the departure of individual lattice constituents, can in turn serve as centers for further heating or resonantly enhanced multiphoton absorption and therefore play an important role in the development of damage. Interest in this problem has been rekindled by the phenomenon called incubation, i.e., the observation that the multishot damage threshold is generally lower than the one for single shots (compare the results for Cu crystals in Fig. 4.9 [4.44]). *Küper* and *Stuke* [4.52] proposed a phenomenological description of this effect by introducing an effective absorption coefficient, depending on the degree of incubation and on the laser intensity. The underlying physical effect may very well be the accumulated stress, as suggested by *Manenkov* et al. [4.22], mentioned above. Alternately, in the case of metals, surface morphology (roughening) on a micrometer-sized scale is suspected. For nonelemental solids, nonstoichiometry on or near the surface is also a possibility. Besides differential ablation, one should also consider the possibility of differential redeposition of vapor phase debris, which could in turn produce a nonstoichiometric surface. Separate, but related to this, is the fact that there exists no precise definition of the difference between nonstoichiometry and (optically absorbing) defects. The two perturbations actually blend smoothly one into the other.

It is of great interest in this connection to find out whether particle emission is observable below the macroscopic damage threshold and, if so, whether it can be correlated to the reduction of the damage threshold in the case of multishot irradiation. *Chase* and collaborators [4.95–97] have taken up this question and investigated the predamage emission in a number of crystals. One informative example [4.95] is the emission yield of neutral Na from NaF as a function of fluence for excitation with 5–8 ns pulses at 266 nm, shown in Fig. 4.15. The data were taken for consecutive shots on the same spot, with the sample in UHV. Obviously there is a threshold for the emission of Na atoms close to 2 J/cm^2, above which the yield rises almost linearly. This threshold lies at about 1/4 of that for optical damage. Above the damage threshold there is copious emission of neutral and charged particles, which is not of interest in this connection. From TOF velocity distributions, the authors derive that emission below the damage threshold is a cold process. For example, at a fluence of 3.8 J/cm^2 the velocity points to a particle temperature of 300 K, simi-

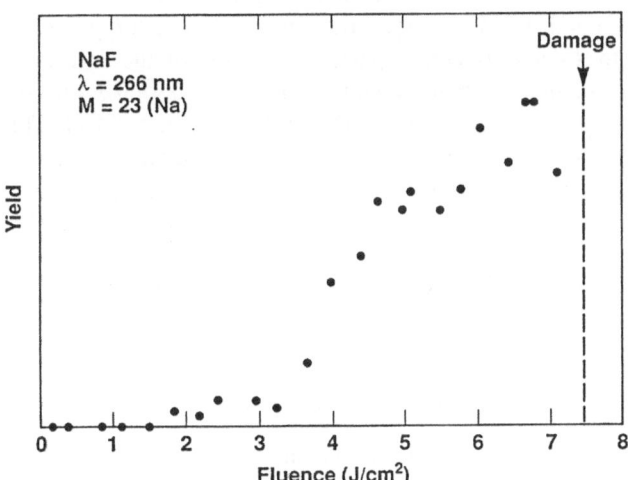

Fig. 4.15. Laser-induced emission yield of neutral Na from NaF for consecutive shots of increasing fluence on the same spot. The crystal was mounted in UHV and Na atoms were recorded by a quadrupole mass spectrometer; the laser pulsewidth was 5–8 ns. The macroscopic damage threshold is indicated by the dashed line. (From [4.95], courtesy of L.L. Chase)

lar to the sample temperature. Hence, the temperatures in Fig. 4.15 range between 300 K and 800 K for NaF, which is far too low for evaporation or sublimation, ruling thermal emission of Na$^{(0)}$ unlikely. This assumes, however, that kinetic energies obtained from TOF measurements correctly reflect surface temperatures; however, until now, the perturbing effects generally appear to increase, not decrease, the observed kinetic energies. The authors also observe an increase of the threshold for neutral emission with increasing wavelength and a correlation with the optical damage threshold [4.95]. Although no conclusive explanation was offered, *Chase* and *Smith* suggested that the emission might be caused by electronic excitation of defect states, similar to what we assumed in Fig. 4.10 for the photon absorption. Nevertheless, it is uncontested that the emission of neutrals is a precursor to optical damage and understanding its origin is imperative for a better comprehension of the damage mechanism.

As mentioned above, *Schmid* et al. [4.94] discovered the emission of neutral halogen atoms along the $\langle 110 \rangle$ and $\langle 211 \rangle$ directions of KCl, KBr, and NaCl, at intensities below the breakdown thresholds of these materials. The experiment was carried out under high vacuum conditions, using 30 ns pulses of a ruby laser (1.78 eV). By monitoring the halogen emission along the $\langle 110 \rangle$ direction as a function of laser intensity, it was established that for KCl a 4-photon absorption process was involved [4.94]. The directional dependence of the halogen emission was measured at room temperature with a photoplate detector [4.82], mounted in front of the crystal parallel to the (100) plane, equipped with a central hole permitting the laser beam to pass. An example of the conspicuous spatial emission pattern obtained by *Braunlich* et al. [4.82] for KBr and KCl is shown in Fig. 4.16, in the upper picture for bromine atoms emitted from KBr, and in the lower one for chlorine atoms coming off KCl. The anisotropic emission is thought to be the result of a halogen replacement collision chain along the $\langle 110 \rangle$ direction. The driving energy comes from the formation of halogen molecular ions (V_k centers) by four-photon absorption. By attracting an electron from the conduction band, the V_k-centers transform into self-trapped excitons, which in turn convert to F- and H-center pairs. A fraction of the H-centers move to the surface by a replacement collision chain where they dissociate and the neutral halogen atom is emitted in the direction of the molecular axis [4.82]. This experiment had no time resolution, therefore no information is available about the time interval elapsed between the generation of a defect and the final emission of halogen atoms from the surface. In other words, the depth from which these collision chains originate is unknown at present. Time-resolved experiments with directional sensitivity could decide whether the defects are formed near the surface or deep in the bulk, and would be worthwhile doing.

4.3.3 Multiphoton-Stimulated Emission of Charged Particles

Many experimental results of electron and ion emission from wide band gap materials indicate that the initial light absorption is a multiphoton process. From the beginning, the issue was the origin of the particles, was it emission from intrinsic dielectric surfaces or rather from surface contamination? In one of the early papers on the subject, *Rousseau* et al. [4.98] reported experiments with a Q-switch ruby laser

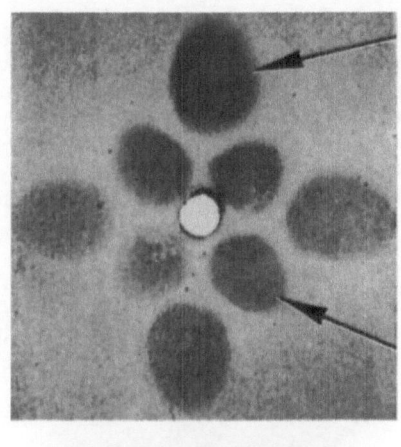

⟨110⟩

KBr

⟨211⟩

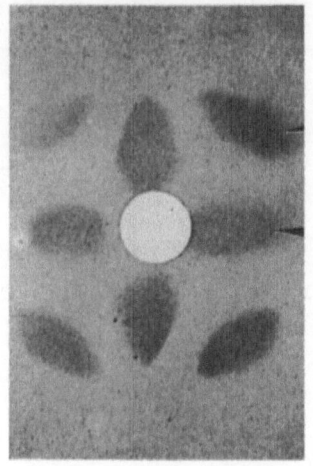

⟨110⟩

⟨211⟩

KCl

Fig. 4.16. Directional emission of neutral halogen atoms from the (100) planes of KBr (*upper part*) and KCl (*lower part*), measured by *Braunlich* et al. [4.82], using a Q-switched ruby laser at power levels below the damage threshold. The sample was mounted in 10^{-6} torr vacuum and the emission pattern was recorded by a silver nitrate coating on a glass plate. (Courtesy of P. Braunlich)

on optical materials under 10^{-5} torr vacuum conditions. At intensities below the macroscopic damage threshold, the authors observed both positive and negative charged particles. For negative particle emission two signals distinctly different in time were found and were recognized as photoelectrons ejected from the surface by plasma radiation (early signals $\sim 40\,\mathrm{ns}$), and electrons resulting from the plasma breakup (late signals $\sim 4\,\mu\mathrm{s}$). For both types of electrons, the emission yield was highly nonlinear with laser power (slopes 2.5 and 8 for the fast and slow electrons, respectively). From their observations, the authors concluded that "laser heating of surface contaminants" was the likely cause of the emission of electrons and ions.

More recent results, again obtained with intensities below the damage threshold, with a better vacuum (10^{-9}–10^{-10} torr), and laser cleaning before taking data, were reported by *Siekhaus* et al. [4.15] on wide band gap semiconductors and insulators. These authors measured the electron emission yield with a charge collector cup, with proper precautions to avoid surface charging. Their data for the electron yield as a function of fluence, taken with an incident photon energy of 1.16 eV for a few pulse lengths and three different materials, are displayed in Fig. 4.17. The observed slopes n are compatible with the optical band gaps (E_g) in all three cases (CdTe:

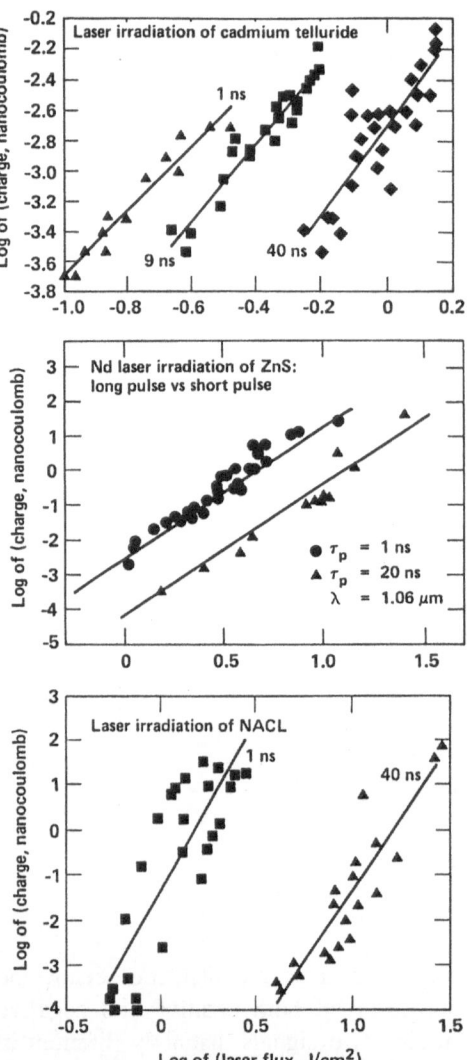

Fig. 4.17. Multiphoton photoemission from CdTe, ZnS, and NaCl, measured by *Siekhaus* et al. [4.15] using 1064 nm pulses of different length and fluence. The samples were mounted in UHV and the emitted charges were detected by a collector cup. The solid lines represent approximate slopes of the data. (Courtesy of L.L. Chase)

E_g = 1.5 eV, n = 2.1; ZnS: E_g = 3.88 eV, n = 4.4; NaCl: E_g = 8.75 eV, n = 7.2). The total charge emission was, however, much larger than what one would expect for multiphoton ionization of such an order, and the authors [4.15] suggest the possibility of multistep excitation processes supported by defect or surface states. Perhaps of even more importance is the observation that the dependence of the charge emission on pulse length is not consistent with that expected for pure multiphoton ionization. Instead, the emitted charge increases more rapidly with pulse length, indicating that additional processes are operative in enhancing the photoemission yield. Another result was that no sudden increase in the yield occurred near or even above the damage threshold, ruling out any contribution by avalanche ionization.

The suggestion that multistep or multiphoton absorption across the band gap is resonantly enhanced by defect states (Fig. 4.10) [4.15, 16] implies that the resulting particle emission should be wavelength dependent. Cluster calculations by *Rosén* and *Westin* [4.87] modelling *F*-centers (missing fluorine atoms) on (111) surfaces of CaF_2 and BaF_2 support these expectations. To check these predictions, *Reif* et al. [4.99, 100] carried out mass- and charge-selective desorption experiments on BaF_2(111), cleaved and mounted in UHV (2×10^{-10} torr), irradiated by pulsed tunable laser light at an intensity of 5×10^7 W/cm^2 in the range 410–450 nm. Four main features were observed: (1) The positive ions F^+, Ba^+, and $(BaF)^+$ were desorbed from the surface, of which F^+ is most surprising, since it requires double charge exchange. (2) A pronounced wavelength dependence was found, together with evidence that Ba^+ was emitted from two chemically different environments [4.100]. (3) There appeared to be an anticorrelation between the emission yield of Ba^+ and F^+, which corresponds to the Ba-F-F-Ba structural sequence along the ⟨111⟩ direction. (4) The positive ion emission ceased after a couple of thousand laser shots at an intensity of 5×10^7 W/cm^2. The reason for this last observation is presently not understood. Most likely it is a "cleaning" effect depleting the surface of defects and thereby diminishing the multiphoton absorption cross section. Such an interpretation would bear some resemblance to the observation of *Allen* et al. [4.21]. The physical interpretation of the findings (1)–(3), in particular the driving force for the emission of positive ions, is still subject to speculation. The observation of F^+, however, provides a strong hint that Coulomb repulsion is involved. We will return to this point in connection with Fig. 4.19.

Wavelength-dependent measurements of the charged particle emission from BaF_2(111) were further pursued in the green spectral range [4.16, 101]. Instead of the quadrupole mass spectrometer used for the experiments described above [4.95, 100], a charge sensitive detector (multiplier plus accelerator grid) with a much more favorable solid angle was utilized. In this way it was easy to register both the photoelectrons and the positive ion emission from the same sample. Again, the crystal was mounted and cleaved in UHV (2×10^{-10} torr) to avoid surface contamination. However, gross structural defects were produced by the cleaving action, which made itself felt by the fact that some spots were more active than others across the surface [4.101]. The measured wavelength dependence of both electron and positive ion yield at intensities of about 5 MW/cm^2 is shown in Fig. 4.18 for two mutually perpendicular polarization directions of the incident light. The left part of Fig. 4.18 (a and b) was measured for *p*-polarized light when the component of the electric field vector in the surface plane is along the ⟨$\bar{1}\bar{1}2$⟩ direction. For the right part (c and d) *s*-polarized light was used with the electric field vector along the ⟨$\bar{1}10$⟩ direction. The angle of incidence was in all cases 70°. The distinct wavelength and polarization dependence of both electron and ion yield in this spectral range is obvious. Also, electron and positive ion emission completely correspond to each other. We interpret this as evidence that the primary energy intake is by multiphoton *ionization*, which in turn triggers the subsequent ion emission. Although the conspicuous peaks in Fig. 4.18 remind us of the five components of the Ba 5*d* orbital [4.16], there is at present no explanation for the observed spectral dependence. Theoretical attempts [4.87] need further refinement before they will be able to predict spectra like those in Fig. 4.18.

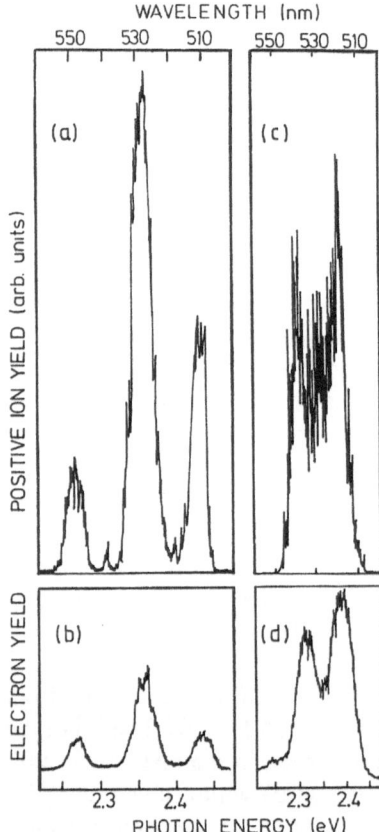

Fig. 4.18. Wavelength and polarization dependence of positive ion (a, c) and electron (b, d) emission from BaF$_2$(111) in the green spectral range at laser intensities of about 5 MW/cm^2. The measurements were carried out under UHV conditions [4.16]. The yield spectra to the left (a, b) are for p-polarized light, the ones to the right (c, d) for s-polarized light, provided the plane of incidence is along the ($\bar{1}\bar{1}2$) direction. When the plane of incidence lies in the ($\bar{1}10$) direction the polarization dependence will be reversed

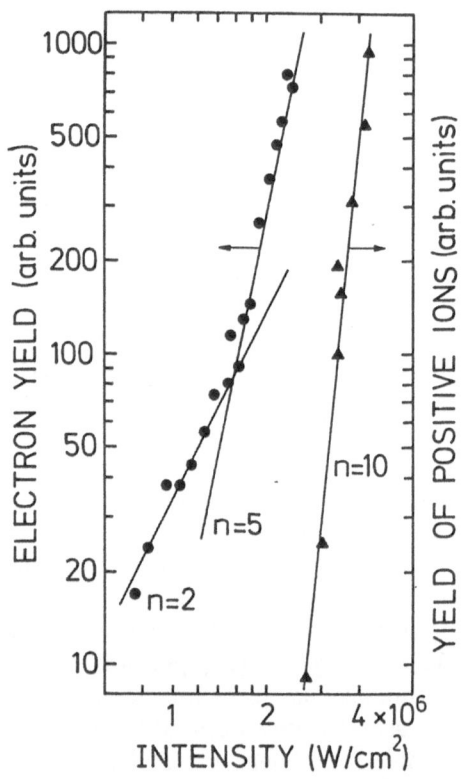

Fig. 4.19. Intensity dependence of electron (dots) and ion (triangles) yields from BaF$_2$(111), measured under UHV conditions with 518 nm radiation and a pulsewidth of about 8 ns [4.16]. The solid lines represent the slopes indicated in the figure

Note that the intensity used to obtain the data in Fig. 4.18 is roughly two orders of magnitude below the damage threshold. This makes it possible to measure the intensity dependence in order to identify the underlying multiphoton process. A necessary requirement for this is that the emission spectra keep their shape and remain unshifted. We found that this was indeed the case [4.16]. The measured intensity dependence of both electron and ion yield, taken at a wavelength of 518 nm, is presented in Fig. 4.19. The dots to the left pertain to electrons, the triangles on the right to positive ions. The electron data inform us that two types of multiphoton processes are occurring, a two-photon photoemission below 1.6 MW/cm^2, and for higher intensities a 5-photon photoemission. We identify these with the processes indicated in the middle and right parts of Fig. 4.10, respectively. In fact the occurrence of a two-photon ionization is proof of the existence of occupied surface defect states in the band gap. The onset of the 5-photon ionization at comparatively low intensities also indicates that this absorption process must be resonantly enhanced at the surface. A direct 5-photon excitation in the bulk, within a depth determined by the mean free path of the electrons, can be excluded because of the small cross section. The same argument rules out other bulk processes like the excitation of a mobile exciton dissociating at the surface [4.102] or the excitation of a conduction electron migrating to the surface.

The formalism for multiphoton transitions [4.103] allows an estimate of the increase in cross section for resonant versus nonresonant excitation. When the ratio of resonant (σ^{res}) to nonresonant (σ^{non}) cross sections is taken the generalized transition dipole moments [4.103] drop out and one finds for the two-photon absorption [4.104]

$$\frac{\sigma_2^{\mathrm{res}}}{\sigma_2^{\mathrm{non}}} = 4\left(\frac{\omega}{\Gamma}\right)^2 = 4 \times 10^4 \quad , \tag{4.3.2}$$

where Γ is the width of the intermediate state and ω the photon frequency. The ratio ω/Γ can be found from the width of the peaks in Fig. 4.18 to be approximately 10^2. Combining the ratio in (4.3.2) with a typical cross section for nonresonant two-photon ionization of 10^{-50} cm^{4}s [4.105, 106], one expects for the resonant two-photon absorption $\sigma_2^{\mathrm{res}} \approx 4 \times 10^{-46}$ cm^{4}s.

For a five-photon absorption which is resonant in its last two steps, *Reif* [4.104] obtained the ratio

$$\frac{\sigma_5^{\mathrm{res}}}{\sigma_5^{\mathrm{non}}} = 2304\left(\frac{\omega}{\Gamma}\right)^4 = 2.3 \times 10^{11} \quad , \tag{4.3.3}$$

assuming that the two intermediate levels have the same width, and again taking $\omega/\Gamma \approx 10^2$. Notice the huge enhancement of the five-photon absorption cross section even for only two resonant steps, which would increase the nonresonant cross section of $\sigma_5^{\mathrm{non}} \approx 10^{-140}$ cm^{10}s^4 reported in the literature [4.106, 107] to $\sigma_5^{\mathrm{res}} \approx 2 \times 10^{-129}$ cm^{10}s^4.

Knowing the cross section σ_n [cm^{2n}s^{n-1}], the number of electrons $N_{\mathrm{e}}^{(n)}$ emitted per laser pulse of duration τ by n-photon ionization can be calculated from the relation [4.104]

$$N_{\mathrm{e}}^{(n)} = \frac{N_0}{n-1} \frac{F^n}{\tau^{n-1}} \sigma_n \quad , \tag{4.3.4}$$

where N_0 is the number of absorbers and F [cm^{-2}] the photon flux per laser pulse and beam area. For a pulse length of $\tau = 6$ ns and a power density of 2 W/cm^2, using a cross section of $\sigma_5 = 2 \times 10^{-129}$ cm^{10}s^4, a ratio of

$$\frac{N_{\mathrm{e}}^{(5)}}{N_0} \approx 10^{-12} \tag{4.3.5}$$

would be expected for a two-photon resonant five-photon absorption. This ratio seems quite small, in particular in view of the electron emission results shown in Fig. 4.19. Making use of the fact that the emission yield of both two-photon and five-photon ionization is identical at 1.6 MW/cm^2, it is possible to derive via (4.3.4) the five-photon cross section from the more reliable two-photon cross section [4.104]. With $\sigma_2 = 4 \times 10^{-46}$ cm^{4}s and $\tau = 6$ ns one obtains $\sigma_5 \approx 2 \times 10^{-120}$ cm^{10}s^4, which would increase the ratio given by (4.3.5) by eight orders of magnitude. Certainly, more detailed investigations are needed on the resonance enhancement, and efforts should be made to vary the defect density at the surface, which is held responsible for the enhancement, in a well-controlled manner. Nevertheless, the above considerations together with the spectral dependence of the electron emission in Fig. 4.18 show that a resonantly enhanced five-photon ionization, as indicated by the slope in Fig. 4.19, is a defensible conclusion.

The increase of the positive ion yield with increasing intensity shows a I^{10} dependence as illustrated by the triangular data points in Fig. 4.19. A genuine 10-photon process appears unlikely for several reasons. First of all, a 10-photon absorption would amount to an energy of 24 eV, approximately 5 eV more than the binding energy of the Ba 2p band. Secondly, no electrons with this power dependence are observed, and yet the spectral yield of electrons and positive ions in Fig. 4.18 is very similar. Therefore it was suggested [4.16, 101] that the ten photons involved in the emission of one positive ion reflect the action of two holes in the valence band, each generated by 5-photon ionization, in order to repulse a positive ion. This notion would harmonize well with the observation of F$^+$ emission. It is analogous to the *Knotek-Feibelman* mechanism [4.42, 108] for the desorption of positive ions after innershell excitation. There, two adjacent holes are produced by interatomic Auger transitions, while in our case they would result from direct ionization. The two-hole mechanism has also been invoked by *Itoh* et al. [4.41, 93], and as a multihole concept by *Dreyfus* et al. [4.88], in order to explain the emission of neutrals from nonmetallic solids. *Itoh* and *Nakayama* [4.41] suggest that for oxides, as a result of coupling with phonons, two holes can be attracted to the oxygen ion site near the surface, leading to the desorption of neutral oxygen. In our case, Coulomb repulsion would have to be the driving force, and the measurement of the kinetic energy of the positive ions would be a crucial test of our hypothesis. The question is how the holes could meet. Although there is a certain probability that they are excited at adjacent fluorine ions [4.16, 104], it seems more likely that the holes find their way to the same site by diffusion and trapping at a defect [4.93]. But while one might hypothesize a two-hole localization [4.41], the problem of the long range repulsive interaction obstructing diffusion of two holes to one site still remains. More data, in particular the ratio of neutrals to ion yields, are needed to substantiate any speculation on this point.

4.3.4 Photoacoustic Determination of Damage Thresholds

The emission of neutral and charged particles from ionic crystals, as discussed in the previous sections, will cause localized defects on an atomic scale. As long as the desorption rate is small and the defect density stays below a certain limit, the optical properties are not affected. However, as discussed in connection with Fig. 4.10, each defect center might act either as an absorber, i.e. a local heat source, or assist multiphoton ionization from the valence band by means of its electronic density of states in the band gap. Hence, by gradually increasing the laser intensity, the defect density will be amplified exponentially, and eventually the macroscopic damage threshold will be reached. Such a mechanism presumes the laser light to be absorbed in the defect-rich surface layer of the crystal. At sufficiently high laser intensities, an alternative mechanism would be that of electrical breakdown [4.25] at or near the surface, triggered by some chance electron. As mentioned before, there had been a long-standing dispute about the question of whether avalanche breakdown or multiphoton absorption is responsible for bulk damage in transparent materials [4.77–82, 85]. Although the issue has been settled for very pure crystals [4.82], it was still open for surfaces until recently, when evidence was presented that both mechanisms can indeed occur at surfaces [4.63, 64]. It is now of great interest to examine in detail which of the two damage mechanisms will prevail and under what circumstances.

Since laser damage is inherently connected with shock wave generation, the probe beam deflection technique discussed in Sect. 4.2.3 and sketched in Fig. 4.7 is appropriate for studying the development of damage as a function of intensity. In Fig. 4.20, the energy of the acoustic pulses, defined in (4.2.4), is shown as a function of laser intensity at 530 nm for three different polished single crystals in air [4.63].

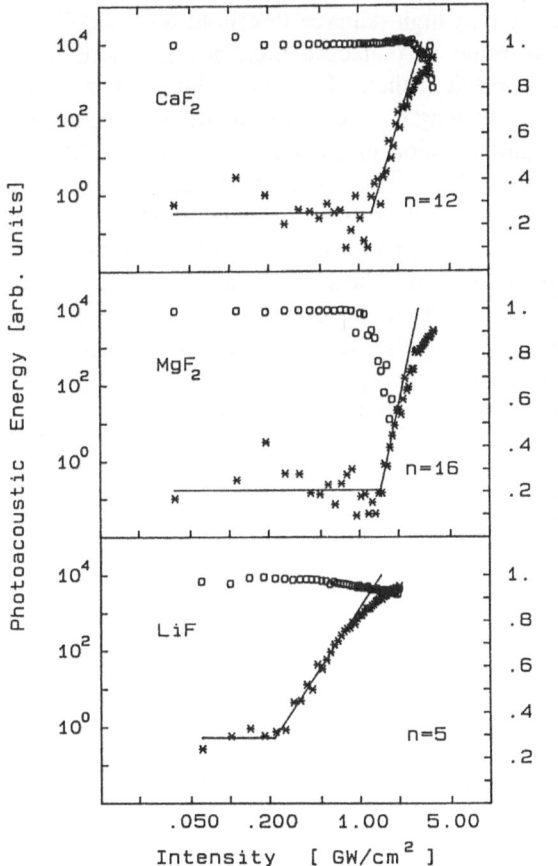

Fig. 4.20. Energy of the acoustic wave vs intensity resulting from the laser–surface interaction at a wavelength of 530 nm and 3 ns pulsewidth for three polished fluoride single crystals in air [4.63, 64, 67]. Stars denote single-shot acoustic beam deflection data, cf. Fig. 4.7, while circles indicate the results of simultaneous reflectivity measurements. Horizontal lines mark the background where no beam deflection signal is recorded. Sloping lines represent fits to the data in a selected intensity interval starting at threshold

The acoustic data (denoted by stars) were measured in a single-shot mode (1-on-1). The probing beam was at a distance of 5 mm above the surface to ensure complete thermalization of the shock wave [4.67]. Circles represent the reflected intensity of a second HeNe laser at 45° angle of incidence and serve as a cross-check for the acoustic data. The agreement between the increase of the acoustic pulse energy and the decrease of the reflectivity proves the consistency of both techniques [4.63]. The horizontal lines in Fig. 4.20 indicate the background level where no reproducible acoustic beam deflection can be obtained. The sloping lines define both the onset and the growth in energy of the pressure pulse. In other words, the intersection between horizontal and sloping lines is taken to represent the damage threshold. The slopes are thus obtained by fitting the data above threshold within a suitable intensity interval.

Two features in Fig. 4.20 are striking: one is the high damage threshold (above 1 GW/cm^2) of CaF_2 and MgF_2, as opposed to one a factor of 5 lower for LiF ($\sim 0.2 \text{ GW/cm}^2$). The other is the drastically different rise of the acoustic pulse energy for CaF_2 and MgF_2, on the one hand, and LiF on the other. These findings, together with additional data on fused silica and BaF_2 [4.67], support the view that there are two classes of materials with regard to their laser surface damage properties. High damage thresholds apparently correlate with steep slopes, low damage thresholds with considerably smaller slopes. A tentative interpretation of these observations is as sketched in Fig. 4.21. Surfaces with a high damage threshold are "inert" and the electric field can increase to the point of dielectric breakdown, where some "free" electrons multiply in an avalanche-like fashion. The avalanche rate is a highly nonlinear function of laser fluence, wavelength, ionization energies, and surface temperature [4.17, 25, 85]. An important question in this connection is whether the avalanche process occurs in the atmosphere in front of the surface, and is merely ignited by electrons occasionally available at the surface, or whether the surface is directly involved. Inspection of Fig. 4.20 shows that for CaF_2 and MgF_2 the threshold is independent of the material. On the other hand, these materials are so much alike that they are not suitable for ascertaining such precise details.

The remarkably different slope for LiF in Fig. 4.20 is taken as evidence that a completely different physical mechanism, namely multiphoton ionization, leads to surface damage. Both in this case and for BaF_2 [4.64, 67] the slopes agree

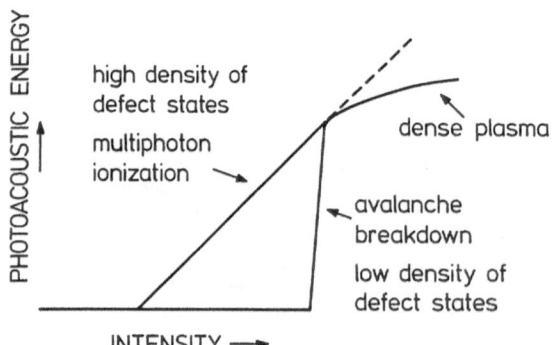

Fig. 4.21. Illustration of the two distinct mechanisms responsible for surface damage, as established by the experimental results shown in Figs. 4.20, 22, and 23

reasonably well with the number of photons required to bridge the band gap ($5 \times 2.34\,\text{eV} = 11.7\,\text{eV}$ compared to a band gap of $13.5\,\text{eV}$ for LiF [4.109]). As discussed in Sect. 4.3.3, a certain density of states is required to resonantly enhance the probability for a 5-photon process to such a degree that it occurs at intensities much lower than the ones required for avalanche breakdown, see Fig. 4.21. Apparently, a polished LiF (or BaF_2) surface in air carries a sufficient amount of adsorbates or defects to facilitate a 5-photon ionization. The nature of these adsorbates or defects is unknown at present. We suspect, however, that, in air, the fact that the solubility in water of LiF or BaF_2 is two orders of magnitude greater than that of either CaF_2 or MgF_2 points toward stronger dipolar forces and the adsorption of water on the former two crystals. Nonstoichiometry in the form of missing fluorine may be another possibility. For example, LiF forms bulk color centers very rapidly under ionizing (electron) irradiation compared to the other fluorides [4.110]. Gross structural defects can be duplicated, at least in a qualitative manner, by employing rough surfaces or damaging the spot by preceding laser shots. The drastic change in the damage threshold and slope between a polished and a rough CaF_2 surface is displayed in Fig. 4.22. The data were again taken with 530 nm, 3 ns laser pulses in a 1-on-1 mode [4.111]. The slope obtained for the rough surface is close to what one would expect from the band gap energy of CaF_2 ($12.1\,\text{eV}$ [4.112]). The result in Fig. 4.22 can be taken as evidence that not only the damage threshold but also the mechanism can be changed at will by altering the surface quality in a controlled manner, as suggested in Fig. 4.21. Obviously, experiments need to be carried out with surfaces that have been characterized as to their defect density.

It should be remembered that irradiation of surfaces in a multishot mode with intensities above the damage threshold prepares a roughened-up surface and results in a situation rather similar to the one shown in the lower part of Fig. 4.22. Therefore, apart from the first pulse, we would not expect to observe genuine avalanche

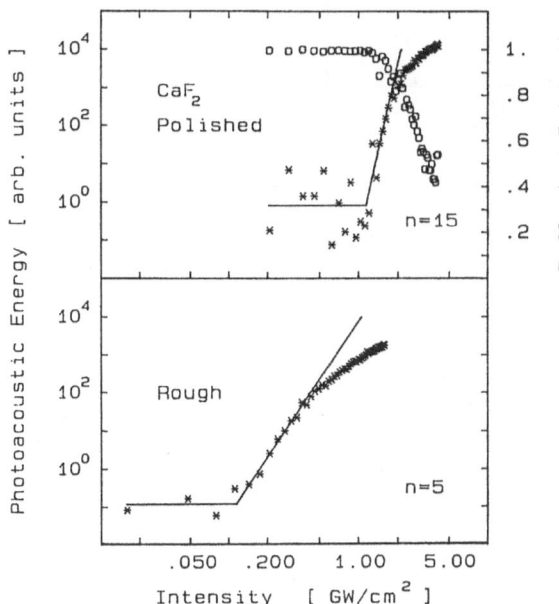

Fig. 4.22. Intensity dependence of the energy of the acoustic pulse resulting from the single-shot interaction of a 3 ns, 530 nm laser pulse with a polished and a rough CaF_2 surface in air [4.111]. The slopes of the data above threshold are indicated, while horizontal lines again mark the background. Circles in the upper part of the figure represent simultaneously recorded reflectivity data

breakdown phenomena for multishot irradiation at intensities below the limit of formation of a dense plasma. Or, phrasing it in another way, even if the first shot caused avalanche breakdown, the next pulses on the same spot would meet a considerably lowered threshold value and multiphoton ionization would set in (Fig. 4.21). A multishot experiment on BaF_2 using 530 nm, 3 ns laser pulses is shown in Fig. 4.23. Again, the slope resembles the band gap energy and the damage threshold is near the one found for rough CaF_2. The small scatter of the data in this case allows us to clearly distinguish between three intensity ranges, where completely different physical processes take place. Below the damage threshold ($0.2 \, GW/cm^2$) we have the range where desorption of individual particles (neutrals or ions) takes place, as discussed in Sect. 4.3.2 and 4.3.3. No shock wave is generated in this area. The second interval is between 0.2 and $\sim 1.0 \, GW/cm^2$. Here the laser light is absorbed by the surface and the shock wave energy indicates multiphoton ionization from the valence band. The order of the multiphoton absorption is material specific, and the damage threshold is determined by the band gap itself, the density of electronic states due to defects or adsorbates, and the photon energy.

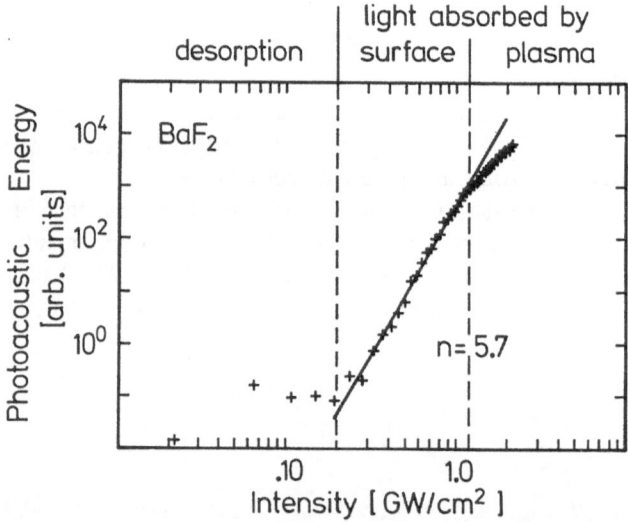

Fig. 4.23. Intensity dependence of the acoustic energy measured with 530 nm, 3 ns laser pulses in a multishot mode on a cleaved surface of BaF_2 in air [4.67]. Dashed lines mark the intensity range for which the data yield the fit represented by the solid line. For a discussion of the three different intensity ranges, see text

The origin of the shock wave in this intermediate intensity range can be understood as follows: If we assume that the shock wave is generated by plasma expansion, the plasma must still be faint and transparent to the incoming laser light. Remembering that the acoustic energy is proportional to the volume of the material removed [4.63, 67], we can view the slope in Fig. 4.23 as simply reflecting an increase of the amount of ablated material with I^n, where I is the laser intensity and n the order of the multiphoton absorption. With increasing intensity, the plume becomes gradually more dense until it finally forms a strongly absorbing plasma above $1.0 \, GW/cm^2$, which in turn decouples the laser light from the surface. We classify this as the third range, and take the deviation from a straight line in Fig. 4.23 as experimental evidence for it. In fact, the levelling off of the increase of the acoustic energy with increasing intensity can be observed for all shock wave experiments, independent of the material, see Figs. 4.8, 20, 22, and 23. Specifically for the fluorides, the formation of a dense plasma seems to occur generally above $1\,GW/cm^2$ and appears to be largely independent of whether the preceding mechanism was avalanche breakdown or multiphoton ionization. This observation is illustrated in Fig. 4.21 and appears quite reasonable, as avalanching energy is due to inverse bremsstrahlung, which depends only on the scattering frequency of free electrons. However, more data and complementary information from other experimental techniques are needed to further substantiate this model.

4.4 Concluding Remarks

An attempt was made in this review to scrutinize the present state of the art of laser–surface interactions with metals and transparent ionic materials. This was done by discussing selected experiments and what we thought were informative results. However, the discussion was mainly limited to the nanosecond time range, which implies restrictions for both conclusions and suggestions. Furthermore, most of the examples were chosen for visible or near-UV light, hence what was said might not be equally valid for the entire wavelength range from CO_2 to ArF lasers.

Primary photon absorption is well understood for metals and we are beginning to understand it for transparent materials. For the follow-up processes we have to differentiate between the various phenomena and materials. Thermal deformation, photoemission, and laser-induced thermal desorption are well understood, in particular for metals. Nonthermal desorption, damage thresholds and ablation mechanisms are still unsettled problems, even if some progress has been made both for metals and for transparent crystals. Shock wave generation near the damage threshold, where it still depends on material properties, is an open question. In general, however, the gradual transition from desorption of individual particles to damage by removal of macroscopic quantities of material is conceptually clear but needs to be confirmed in more detail. To achieve this, there is an urgent need for more "complete" experiments, where many techniques are utilized to investigate the various phenomena of laser–surface interactions for *one* single material rather than collecting a large body of incoherent results for many different samples.

Acknowledgements: This work was supported by a NATO Research Grant 5-2-05/RG No. 0020/88. Much of the work presented here was the result of the collaboration of the authors with many people. One of us (E.M.) would particularly like to acknowledge the fruitful collaboration with J. Reif, H.B. Nielsen, S. Petzoldt, A.P. Elg, M. Reichling, P. West, A. Rosén, and E. Westin, as well as clarifying discussions with T.A. Green. The following have also contributed to various concepts and clarifications: R. von Gutfeld, R. Srinivasan, and R. Walkup.

References

4.1. M. von Allmen: *Laser-Beam Interactions with Materials*, Springer Ser. Mater. Sci., Vol. 2 (Springer, Berlin, Heidelberg 1987)
4.2. G. Kühnle, F.P. Schäfer, S. Szatmari, G.D. Tsakiris: Appl. Phys. B **47**, 361 (1988)
4.3. B. Bäuerle: *Chemical Processing with Lasers*, Springer Ser. Mater. Sci., Vol. 1 (Springer, Berlin, Heidelberg 1986)
4.4. R. Srinivasan: Science **234**, 559 (1986)
4.5. See for instance the J. of Lasers in Medicine and Surgery
4.6. W. Witteman: *The CO_2 Laser*, Springer Ser. Opt. Sci., Vol. 53 (Springer, Berlin, Heidelberg 1987)
4.7. T.J. Chuang: J. Vac. Sci. Technol. B **3**, 1408 (1985); Surf. Sci. Rep. **3**, 1 (1983)
4.8. R.M. Wood: *Laser Damage in Optical Materials*, Adam Hilger Ser. Opt. Optoelectron. (Adam Hilger, Bristol 1986)
4.9. J.H. Bechtel: J. Appl. Phys. **46**, 1585 (1975)
4.10. R.W. Dreyfus, F.A. McDonald, R.J. von Gutfeld: J. Vac. Sci. Technol. B **5**, 1521 (1987)
4.11. J.H. Bechtel, W.L. Smith, N. Bloembergen: Opt. Commun. **13**, 56 (1975)
4.12. J.H. Bechtel, W.L. Smith, N. Bloembergen: Phys. Rev. B **15**, 4557 (1977)
4.13. P.G. Strupp, J.L. Grant, P.C. Stair, E. Weitz: J. Vac. Sci. Technol. A **6**, 839 (1988)
4.14. C.O. Park, H.W. Lee, T.D. Lee, J.K. Kim: Appl. Phys. Lett. **52**, 368 (1988)
4.15. W.J. Siekhaus, J.H. Kinney, D. Milam, L.L. Chase: Appl. Phys. A **39**, 163 (1986)
4.16. E. Matthias, H.B. Nielsen, J. Reif, A. Rosén, E. Westin: J. Vac. Sci. Technol. B **5**, 1415 (1987)
4.17. S.C. Jones, A.H. Fischer, P. Braunlich, P. Kelly: Phys. Rev. B **37**, 755 (1988)
4.18. N. Itoh: J. de Phys. **37**, C7–27 (1976)
4.19. P.F. Bräunlich, G. Brost, A. Schmid, P.J. Kelly: IEEE J. QE-**17**, 2034 (1981)
4.20. N. Bloembergen: Appl. Opt. **12**, 661 (1973)
4.21. S.D. Allen, J.O. Porteus, W.N. Faith, J.B. Franck: Appl. Phys. Lett. **45**, 997 (1984)
4.22. A.A. Manenkov, G.A. Matyushin, V.S. Nechitailo, A.M. Prokhorov, A.S. Tsaprilov: NBS Spec. Publ. **669**, 436 (1982)
4.23. R.T. Casper, S.C. Jones, P. Braunlich: Proc. Boulder Damage Symposium 1988, to be published
4.24. B. Steverding: J. Appl. Phys. **45**, 3507 (1974)
4.25. A.S. Epifanov: IEEE J. QE-**17**, 2018 (1981)
4.26. W.M. Manheimer, D.G. Colombant, J.H. Gardner: Phys. Fluids **25**, 1644 (1982)
4.27. C.R. Phipps, Jr., T.P. Turner, R.F. Harrison, G.W. York, W.Z. Osborne, G.K. Anderson, X.F. Corlis, L.C. Haynes, H.S. Steele, K.C. Spicochi, T.R. King: J. Appl. Phys. **64**, 1083 (1988)
4.28. N.H. Tolk, M.M. Traum, J.C. Tully, T.E. Madey (eds.): *Desorption Induced by Electronic Transitions, DIET I*, Springer Ser. Chem. Phys., Vol. 24 (Springer, Berlin, Heidelberg 1983);
W. Brenig, D. Menzel (eds.): *Desorption Induced by Electronic Transitions, DIET II*, Springer Ser. Surf. Sci., Vol. 4 (Springer, Berlin, Heidelberg 1985);
R.H. Stulen, M.L. Knotek (eds.): *Desorption Induced by Electronic Transitions, DIET III*, Springer Ser. Surf. Sci., Vol. 13 (Springer, Berlin, Heidelberg 1988)
4.29. Ph. Avouris, R. Kawai, N.D. Lang, D.M. Newns: J. Chem. Phys. **89**, 2388 (1988)
4.30. Ph. Avouris, F. Bozso, R.E. Walkup: Nucl. Instrum. Methods, B **27**, 136 (1987)
4.31. T. Nakayama, M. Okigawa, N. Itoh: Nucl. Instrum. Methods, B **1**, 301 (1984)
4.32. I.W. Boyd: *Laser Processing of Thin Films and Microstructures*, Springer Ser. Mater. Sci., Vol. 3 (Springer, Berlin, Heidelberg 1987)
4.33. F.A. Houle: Appl. Phys. A **41**, 315 (1986)
4.34. C. Karner, A. Mandel, F. Träger: Appl. Phys. A **38**, 19 (1985)
4.35. G.E. Jamieson, G.C. Wetsel, Jr.: In Proc. IEEE 1985 Ultrasonic Symposium, Vol. 1, p. 451
4.36. F.A. McDonald, R.W. Dreyfus, R.J. von Gutfeld: In Proc. IEEE 1987 Ultrasonic Symposium, p. 1179
4.37. R.B. Hall: J. Phys. Chem. **91**, 1007 (1987)
4.38. T.J. Chuang, I. Hussla: Phys. Rev. Lett. **52**, 2045 (1984)
4.39. W. Hoheisel, K. Jungmann, M. Vollmer, R. Weidenauer, F. Träger: Phys. Rev. Lett. **60**, 1649 (1988); see also Chap. 2 of this volume
4.40. E. Sutcliffe, R. Srinivasan: J. Appl. Phys. **60**, 3315 (1986)
4.41. N. Itoh, T. Nakayama: Phys. Lett. **92A**, 471 (1982)
4.42. M.L. Knotek: Rep. Prog. Phys. **47**, 1499 (1984)
4.43. J. Reif, P. Tepper, E. Matthias, E. Westin, A. Rosén: Appl. Phys. B **46**, 131 (1988)
4.44. Y. Jee, M.F. Becker, R.M. Walser: J. Opt. Soc. Am. B **5**, 648 (1988)

4.45. D.I. Rosen, J. Mitteldorf, G. Kothandaraman, A.N. Pirri, E.R. Pugh: J. Appl. Phys. **53**, 3190 (1982)
4.46. M. Newstein, N. Solimene: IEEE J. QE-17, 2085 (1981)
4.47. J.M. Hicks, L.E. Urbach, E.W. Plummer, H.-L. Dai: Phys. Rev. Lett. **61**, 2588 (1988)
4.48. J.M. Liu, H. Kurz, N. Bloembergen: Appl. Phys. Lett. **41**, 643 (1982)
4.49. D. von der Linde, N. Fabricius: Appl. Phys. Lett. **41**, 991 (1982)
4.50. R.W. Schoenlein, W.Z. Lin, J.G. Fujimoto, G.L. Eesley: Phys. Rev. Lett. **58**, 1680 (1987)
4.51. D. Kühlke, U. Herpers, D. von der Linde: Appl. Phys. Lett. **50**, 1785 (1987)
4.52. S. Küper, M. Stuke: Appl. Phys. B **44**, 199 (1987)
4.53. H.M. Musal, Jr.: NBS Spec. Publ. **568**, 159 (1980)
4.54. J.F. Young, J.S. Preston, H.M. van Driel, J.E. Sipe: Phys. Rev. B **27**, 1155 (1983)
4.55. F.E. Domann, M.F. Becker, A.H. Guenther, A.F. Stewart: Appl. Opt. **25**, 1371 (1986)
4.56. P. Cielo: J. Appl. Phys. **56**, 230 (1984)
4.57. M.A. Olmstead, N.M. Amer, S. Kohn, D. Fournier, A.C. Boccara: Appl. Phys. A **32**, 141 (1983)
4.58. M.C. Teich, G.J. Wolga: Phys. Rev. **171**, 809 (1968)
4.59. K. Giesen, F. Hage, F.J. Himpsel, H.J. Riess, W. Steinmann. Phys. Rev. Lett. **55**, 300 (1985)
4.60. L.A. DuBridge: Phys. Rev. **39**, 108 (1932)
4.61. L.A. Lompre, J. Thebault, G. Farkas: Appl. Phys. Lett. **27**, 110 (1975)
4.62. R. Haight, J.A. Silberman: Phys. Rev. Lett. **62**, 815 (1989)
4.63. S. Petzoldt, A.P. Elg, M. Reichling, J. Reif, E. Matthias: Appl. Phys. Lett. **53**, 2005 (1988)
4.64. E. Matthias, S. Petzoldt, A.P. Elg, P.J. West, J. Reif: NIST Spec. Publ. **756**, 217 (1988)
4.65. A.C. Boccara, D. Fournier, W. Jackson, N.M. Amer: Opt. Lett. **5**, 377 (1980)
4.66. G. Koren: Appl. Phys. Lett. **51**, 569 (1987)
4.67. S. Petzoldt: Diplomarbeit, Freie Universität Berlin (1988)
4.68. H. Helvajian, R. Welle: Private communication
4.69. R. Kelly, J. Rothenberg: Nucl. Instrum. Methods B 7/8, 755 (1985)
4.70. R.W. Dreyfus, R.J. von Gutfeld: CLEO'89, OSA 1989 Technical Digest Series, Vol. 11, THD3
4.71. T.W. Johnston, J.N. Dawson: Phys. Fluids **16**, 722 (1973)
4.72. Y.B. Zel'dovich, Y.P. Raizer: *Physics of Shock Waves and High Temperature Hydrodynamic Phenomena*, Vol. I (Academic, New York 1966)
4.73. N.L. Boling, M.D. Crisp, G. Dubé: Appl. Opt. **12**, 650 (1973)
4.74. W.H. Lowdermilk, D. Milam: IEEE J. QE-17, 1888 (1981)
4.75. J.O. Porteus, S.C. Seitel: Appl. Opt. **23**, 3796 (1984)
4.76. J.E. Rothenberg, R. Kelly: Nucl. Instrum. Methods B **1**, 291 (1984)
4.77. W.L. Smith, J.H. Bechtel, N. Bloembergen: Phys. Rev. B **15**, 4039 (1977)
4.78. T.W. Walker, A.H. Guenther, P. Nielsen: IEEE J. QE-17, 2053 (1981)
4.79. S.C. Jones, X.A. Shen, P.F. Braunlich, P. Kelly, A.S. Epifanov: Phys. Rev. B **35**, 894 (1987)
4.80. X.A. Shen, P. Braunlich, S.C. Jones, P. Kelly: Phys. Rev. Lett. **59**, 1605 (1987)
4.81. X.A. Shen, P. Braunlich, S.C. Jones, P. Kelly: Phys. Rev. B **38**, 3494 (1988)
4.82. P. Braunlich, S.C. Jones, X.A. Shen, R.T. Casper, P. Kelly: NIST Spec. Publ. **756**, 476 (1988)
4.83. B.S. Sharma, K.E. Rieckhoff: Can. J. Phys. **48**, 1178 (1970)
4.84. N. Bloembergen: IEEE J. QE-10, 375 (1974)
4.85. A. Vaidyanathan, T.W. Walker, A.H. Guenther: IEEE J. QE-16, 89 (1980)
4.86. U.O. Karlsson, F.J. Himpsel, J.F. Morar, F.R. McFeely, D. Rieger, Y. Yarmoft: Phys. Rev. Lett. **57**, 1247 (1986); Phys. Rev. B **34**, 7295 (1986)
4.87. A. Rosén, E. Westin, E. Matthias, H.B. Nielsen, J. Reif: Phys. Scr. T **23**, 184 (1988)
4.88. R.W. Dreyfus, F.A. McDonald, R.J. von Gutfeld: Appl. Phys. Lett. **50**, 1491 (1987)
4.89. T.A. Green, G.M. Loubriel, P.M. Richards, N.H. Tolk, R.F. Haglund, Jr.: Phys. Rev. B **35**, 781 (1987)
4.90. R.W. Dreyfus, R. Kelly, R.E. Walkup: Appl. Phys. Lett. **49**, 1478 (1986)
4.91. R.W. Dreyfus, R.E. Walkup, R. Kelly: Radiat. Eff. **99**, 683 (1986)
4.92. R.W. Dreyfus, R. Kelly, R.E. Walkup, R. Srinivasan: In *Excimer Lasers and Optics*, SPIE, Vol. 710 (1986) p. 46
4.93. N. Itoh: Nucl. Instrum. Methods B **27**, 155 (1987)
4.94. A. Schmid, P. Bräunlich, P.K. Rol: Phys. Rev. Lett. **35**, 1382 (1975)
4.95. L.L. Chase, L.K. Smith: NIST Spec. Publ. **756**, 165 (1988)
4.96. L.L. Chase, H.W.H. Lee: Proc. Boulder Damage Symposium 1988, to be published
4.97. H.F. Arlinghaus, W.F. Calaway, C.E. Young, M.J. Pellin, D.M. Gruen, L.L. Chase: J. Appl. Phys. **65**, 281 (1989)
4.98. D.L. Rousseau, G.E. Leroi, W.E. Falconer: J. Appl. Phys. **39**, 3328 (1968)
4.99. J. Reif, H. Fallgren, W.E. Cooke, E. Matthias: Appl. Phys. Lett. **49**, 770 (1986)

4.100. J. Reif, H. Fallgren, H.B. Nielsen, E. Matthias: Appl. Phys. Lett. **49**, 930 (1986)
4.101. H.B. Nielsen, J. Reif, E. Matthias, E. Westin, A. Rosén: In [4.28c], p.266
4.102. P.W. Levy, P.J. Herley: Mater. Sci. Res. **4**, 156 (1969)
4.103. H.B. Bepp, A. Gold: Phys. Rev. **143**, 1 (1966)
4.104. J. Reif: Optical Engineering **28**, 1122 (1989)
4.105. H.B. Bepp: Phys. Rev. **149**, 25 (1966)
4.106. J.M. Catalano, A. Cingolani, A.N. Minafra: Phys. Rev. B **5**, 1629 (1972)
4.107. P. Bräunlich, A. Schmid, P. Kelly: Appl. Phys. Lett. **26**, 150 (1975)
4.108. M.L. Knotek, P.J. Feibelman: Phys. Rev. Lett. **40**, 964 (1978)
4.109. C. Gout, F. Pradal: J. Phys. Chem. Solids **29**, 581 (1968)
4.110. D. Heath, P. Sacher: Appl. Opt. **5**, 937 (1966)
4.111. A.P. Elg: Laboratory Report, Freie Universität Berlin (1988);
        J. Reif, S. Petzoldt, A.P. Elg, E. Matthias: Appl. Phys. A **49**, 199 (1989)
4.112. G.W. Rubloff: Phys. Rev. B **5**, 662 (1972)

Note added in proof: Further literature on this topic can be found in a recently published special issue of Optical Engineering, Vol. 28, No. 10, 1989, edited by P.J. Kelly.

5. Photothermal Analysis of Thin Films

H. Coufal

With 17 Figures

Exposure of a sample to any form of radiation causes local heating via absorption and subsequent thermalization of energy. If the incident radiation is modulated or pulsed a transient heat source is generated in the sample. Thermal waves and, due to thermal expansion, acoustic waves are launched into the sample under study and interact with the sample before being detected by one of the numerous detection schemes available nowadays. By changing the incident wavelength, spectra of the sample can be recorded readily. Scanning of the excitation across the sample allows mapping of optical, thermal and acoustical properties of the sample under investigation. Any type of radiation can be utilized for the excitation of these phenomena. When using electromagnetic radiation, lasers are the preferred light source since they provide convenient access to an energy source of high spectral resolution, power, and, in the case of time resolved experiments, well-characterized pulse shape. In addition, lasers with their parallel beam facilitate the focusing of the energy into a small area. Other sources of energy that can be advantageous are particle beams, such as electron or ion beams.

5.1 Photothermal and Photoacoustic Effect in Thin Films

In spectroscopy, the sample is excited with a tunable light source. In conventional spectroscopy the incident, transmitted and reflected light intensity are recorded as a function of wavelength and from the numerical difference the light absorption in the sample is calculated. If the sample is a weakly absorbing thin film, then the incident and transmitted intensity have to be measured with high precision to determine the small absorbed fraction of the incident light. In photothermal detection schemes only energy absorbed in the sample contributes towards the signal, making these techniques the method of choice for spectroscopic studies of thin films. Due to thermal diffusion, only the heat deposited within a thin layer of the surface of the sample will be detectable at the surface, a feature allows spectroscopy of weakly absorbing thin films on an absorbing substrate and optically opaque films can be addressed by photothermal techniques. Spectroscopy of thin films is, therefore, an extremely important area of application for photothermal analysis.

Since thermal diffusion, sound propagation and possibly other energy transport processes, such as charge transport or excitonic processes in semiconductors, are involved in the generation of the signal, the associated material parameters can be studied.

Compared with more conventional thermal analysis methods, photothermal analysis has the advantage of noncontact generation of a well-defined heat source at the surface or in the volume of the sample of interest. With the high time resolution of some photothermal detection schemes, the transient temperatures at a sample surface or within the volume of the sample can now be determined in real time with very high time resolution. This allows us to study the thermal properties and the temperature dependence of other physical properties, of which phase transitions or charge distributions are just two examples. Due to the high sensitivity of many photothermal detection schemes, very low intensities suffice for excitation, hence, sample heating is negligible and temperature sensitive samples or sample properties can be studied.

Laser induced ultrasound allows the convenient noncontact generation of high frequency ultrasonic waves in a sample. Compared to conventional ultrasonic transducers, photothermal generation is always noncontact and free of transducer ringing. Coupling of the sound pulse from the transducer into the sample, subject to acoustic impedance matching etc., in the case of a transducer is no problem with laser generated ultrasound since the sound is generated in the sample, a feature which might prove convenient when studying bulk samples. For ultrasonic analysis of thin films, the wavelength of the probing pulse should be substantially less than the sample thickness. In this case, techniques based on conventional ultrasonic transducers are cumbersome, to say the least, due to the above-mentioned acoustical problems. With a sufficiently short and powerful laser pulse, absorbed in a thin sample, ultrahigh ultrasonic frequencies are easily generated, making the photoacoustic analysis of thin films relatively straightforward.

In imaging, a large number of techniques compete with photothermal imaging and microscopy. Optical microscopy is an extremely well developed, powerful and mature analytical technique. The signal-to-noise ratio in microscopy is excellent and parallel recording and processing of a frame is state of the art. Quite clearly, photothermal imaging with its relatively low signal-to-noise ratio and complicated signal generation and hence contrast mechanism will never be able to compete with light microscopy. Thermal wave and ultrasonic imaging of thin films are, however, two niches where photothermal imaging is becoming a powerful tool.

One of the key features of photothermal excitation is the fact that it does not require extensive sample preparation or physical contact with the sample. The same is true for a number of the detection schemes. In many applications, excitation and detection of the signal can be accomplished on the same side of the sample, enabling single sided, remote monitoring of the sample under study. This feature makes photothermal analysis a good candidate for industrial and even military applications.

A number of comprehensive books [5.1-4] have reviewed the field of photoacoustics and photothermal phenomena and there are also excellent review articles [5.5-9] which provide good coverage of the field. Up-to-date information on the status of the field can be found in recent special issues of journals that were dedicated to photoacoustics [5.10, 11] and the proceedings of international conferences on photoacoustics [5.12, 13]. The focus of this chapter will be to complement this substantial body of literature and to enable a prospective user of this technique to critically assess the potential of photothermal and photoacoustic methods in the analysis of thin films. References will be limited to the most recent developments in this rapidly growing field.

The physical principles of the signal generation process and selected detection schemes are briefly reviewed and various detection schemes are compared in detail. Typical applications in thermal analysis, ultrasonic testing and imaging of thin films are discussed and examples highlighting the advantages and drawbacks of photothermal and photoacoustic techniques are analyzed.

5.1.1 Signal Generation Process

In a prototypal photothermal experiment (Fig. 5.1) a homogeneous thin film with thickness T is excited with a modulated or pulsed light source. The light absorption in the sample can be characterized by a wavelength dependent optical absorption length $A\lambda$

$$A(\lambda) = \frac{1}{\beta(\lambda)}, \qquad (5.1.1)$$

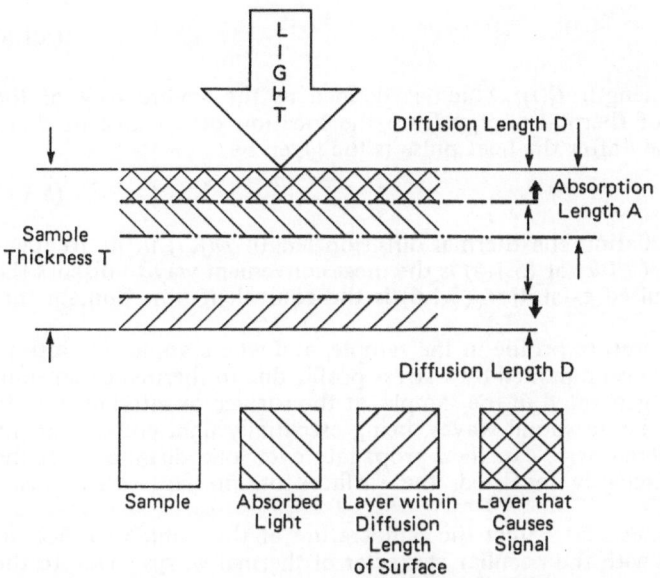

Fig. 5.1. Schematic of the photothermal signal generation process. For definitions of the absorption and diffusion length please see the text

with $\beta(\lambda)$ the commonly used optical absorption coefficient of the sample. Due to radiationless processes, part of the absorbed energy is released as heat. With the incident energy being either modulated or pulsed, the heat generation will show a corresponding time dependence. Via heat diffusion then, a temperature profile develops in the sample. For a heat source with the modulation frequency f, the heat diffusion can be described by the thermal diffusion length D

$$D(f) = \sqrt{\frac{k}{\pi \rho c f}} \,, \tag{5.1.2}$$

where k denotes the thermal conductivity, ρ the density and c the specific heat of the sample. With the thermal diffusivity

$$\alpha = \frac{k}{\rho c} \,, \tag{5.1.3}$$

the thermal diffusion length is a function of this material parameter and the modulation frequency

$$D(f) = \frac{\sqrt{2\alpha}}{2\pi f} \,. \tag{5.1.4}$$

The amplitude of a thermal wave is attenuated by a factor of $\exp(-2\pi)$, i.e., 2×10^{-3}, when diffusing one diffusion length [5.14]. Therefore, essentially only heat generated within one thermal diffusion length of a sample surface will be able to reach this surface. If the sample is in contact with another medium, thermal waves are reflected and transmitted at the interface. Heat diffusion extends into the second medium according to (5.1.2) or (5.1.4) with the material parameters of that medium instead of the sample properties. For pulsed excitation the thermal transit time τ is the relevant parameter. The temperature at a distance ℓ from a pulsed heat source reaches its maximum at a time τ after the excitation. For one-dimensional heat flow the thermal transit time is given by

$$\tau = \frac{\ell^2}{2\alpha} \ . \tag{5.1.5}$$

The thermal diffusion length $D(t)$, which is defined as the square root of the mean-square distance of thermal energy from the location of a transient point source of heat, at a time t after the heat pulse is the given by

$$D(t) = \sqrt{2\alpha t} \ . \tag{5.1.6}$$

For modulated CW radiation the thermal diffusion length $D(f, t)$ in its frequency-dependent definition (5.1.2) or (5.1.4) is the most convenient way to discuss the observed signals; for pulsed excitation (5.1.6) is the equivalent time domain formula.

The transient temperature profile in the sample, and where applicable also in the adjacent medium, is accompanied by a stress profile due to thermal expansion. Sound waves are hence generated in the sample, at the surface or interface and in the adjacent material. These sound waves, being essentially unattenuated in the frequency range considered here, can then propagate over long distances. If the heat generation is sufficiently localized, the surfaces or interfaces also buckle slightly.

All detection schemes that detect the temperature at the sample surface directly will have to deal with the peculiar character of thermal waves. Due to the fact that thermal waves are critically damped

- the detector has to probe the temperature within one thermal diffusion length of the excited area and
- only energy deposited within this distance is effective in signal generation.

The first restrictions can be overcome by indirect detection of the surface temperature change; sound waves generated by the photothermal phenomenon propagate essentially unattenuated over large distances. An acoustic detector can therefore be far from the illuminated area of the sample. For moderate light intensities, i.e., below the threshold for ablation of sample material, the amplitude of the observed acoustic signal will depend on the temperature distribution in the heated sample volume and the adjacent medium, both determined by the thermal diffusion length $D(f, t)$ and the optical absorption length $A(\lambda)$. The sound wave serves as a carrier of the thermal information. Due to this diffusive character of thermal waves, information obtained by the detection of radiation induced thermal and acoustic waves is completely different from conventional optical spectra. Some of the unique features and problems of photothermal and photoacoustic spectroscopy arise from this difference.

In an optically opaque sample with thickness $T > A(\lambda)$, such as shown in Fig. 5.1, the radiation induced heating can always be observed at the illuminated side of the sample. On the back side a temperature increase is, however, only noticeable when light is absorbed within one thermal diffusion length of the back side, i.e., when the condition $T - A(\lambda) < D(f, t)$ is met. This condition is a function of material constants of the sample such as $\beta(\lambda)$, k, ρ and c but also of an experimental variable, the modulation frequency f. As mentioned above, an acoustic transducer will be able to detect a signal on either side of the sample; on the back even if there is no sizeable temperature fluctuation reaching that side! Sample thickness is, therefore, only a minor concern for photoacoustic spectroscopy. It is, however, important to remember that the sound wave serves only as a carrier of the thermal information. The amplitude of the observed acoustic signal will depend on the temperature distribution in the heated sample volume and the adjacent medium, both determined again by the thermal diffusion length $D(f, t)$ and the optical absorption length $A(\lambda)$. For spectroscopy, therefore, the relation between these two parameters has to be analyzed.

Whenever the thermal diffusion length $D(f, t)$ is larger than the optical absorption length $A(\lambda)$, $D(f, t) > A(\lambda)$, and the sample thickness T is larger than the absorption length, $T > A(\lambda)$, all the incident energy is absorbed and contributes to the observed signal. A small variation of the absorption coefficient in this case does not change the observed signal at all. The signal is *saturated* and the observed signal is proportional to the incident energy and $(1 - R)$, R being the reflectivity of the sample. This condition can be avoided by using a thin sample or by increasing the modulation frequency until $D(f, t) < A(\lambda)$ is achieved.

A sample with an optical absorption length smaller than the sample thickness, $A(\lambda) < T$, absorbs all the incident light, hence no light is transmitted. Such an opaque sample is not accessible to conventional transmission spectroscopy. If the thermal diffusion length $D(f, t)$ is, however, smaller than the optical absorption length $A(\lambda)$, $D(f, t) < A(\lambda)$, the energy deposited within this thermal diffusion length is proportional to the absorption coefficient β of the sample. Because the thermal diffusion length decreases with increasing modulation frequency according to (5.1.2), this condition can, at least in principle, be fulfilled by modulating at a high enough frequency. Photothermal spectroscopy thus allows the recording of *spectra of opaque samples*.

At very high modulation frequencies, only light absorbed in the surface layer contributes to the signal. When using a lower modulation frequency the same surface layer and the adjacent layer generate the signal. By comparing spectra at various modulation frequencies, therefore, surface absorption can be readily distinguished from bulk absorption. In principle, the *depth profile* of a layered structure can thus be obtained in a nondestructive manner (Chap. 9).

In conventional optical transmission spectroscopy, the path length along which absorption occurs has to be known to allow a determination of the absorption coefficient, a task that can be cumbersome for rough samples, fibers or powders. In photothermal and photoacoustic spectroscopy, however, it is sufficient to ensure that the thermal diffusion length is smaller than the relevant sample dimension such as grain size or fiber diameter, to obtain a qualitative absorption spectrum. For many practical applications a qualitative analysis is sufficient, in which case *no sample preparation* is required. For quantitative analysis, however, and also in PT spectroscopy, extensive calibration procedures are required.

In spectroscopy, samples with a wavelength-variable absorption coefficient are, of course, of particular interest. Different parts of the spectrum might then have an absorption length $A(\lambda)$ that could vary from larger than the selected thermal diffusion length down to much shorter (Fig. 5.2). In this case, parts of the spectrum where $D(f, t) > A(\lambda)$, i.e., with a large absorption coefficient β, will be saturated. On the other hand, the thermal diffusion length $D(f, t)$ is a function of the modulation frequency and the thermal properties of the sample. According to the above considerations, spectra of the same sample will be dependent on the modulation frequency (Fig. 5.3). Furthermore, spectra of samples with identical optical but different thermal properties might be quite different. The same holds true for samples with identical light absorption but different fluorescence yields and therefore different heat generation. Photothermal and photoacoustic spectroscopy thus requires particular care in the selection of the modulation frequency, careful consideration of thermal sample parameters and thickness and an understanding of the signal generation process.

5.1.2 Detection Methods

Radiation induced transient effects, thermal as well as mechanical, have been studied for quite some time. Determining the temperature or the energy of a sample has traditionally been the domain of thermometry and calorimetry. Radiation induced mechanical effects were studied using interferometry or ultrasonic transducers. In the last few years interest in radiation induced thermal and acoustic processes has increased substantially, largely due to scientific and industrial applications of lasers. A number of classical detection schemes were adapted for this

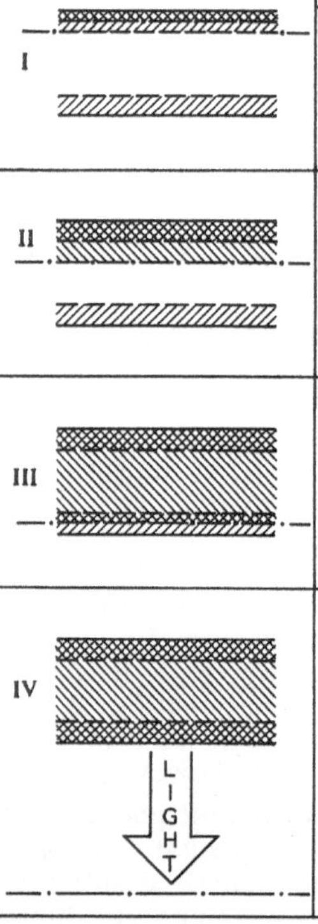

	Condition	Optically	Signal
I	A < D	Opaque	(1−R) None
II	D < A < T−D	Opaque	β(1−R) None
III	T−D < A < T	Opaque,	β(1−R) $\frac{1}{\beta}$(1−R)
IV	T < A	Transparent	β(1−R) β(1−R)

LIGHT

Fig. 5.2. Schematic of the photothermal signal generation process in a thermally thick sample as a function of absorption length A. The relationship between thermal diffusion length D and sample thickness T for an increasing $A(I-IV)$ is shown. For these cases, characterized by the relationship between the parameters, the variation of the photothermal signal with optical reflection coefficient R and absorption coefficient β of the sample is given for detection on the front and back sides, respectively. See Fig. 5.1 for the definition of the symbols and the text for more details

particular type of application and new detection methods were developed. Detectors tie into various stages of the signal generation chain described above. In photothermal detection schemes the radiation induced temperature increase in the sample or at the sample surface is monitored by measuring either the temperature directly or a temperature-dependent property of the sample or of the adjacent medium. Photo- and optoacoustic methods detect the acoustic waves caused by the radiation induced heating of the sample itself or a gas or a liquid that is in thermal and/or acoustical contact with the sample. For experiments with the emphasis on surface temperature, classical temperature sensors such as thermocouples are clearly the method of choice because they are easy to calibrate and convenient to use. For spectroscopic measurements, however, sensitivity and convenient coupling of the detector to the sample are the main concern.

The transient heating of a pyroelectric material can, for example, be detected via the induced electrical charge or current. As a matter of fact, many commercially available laser power meters function on this basis. A variation of this technique uses a *pyroelectric calorimeter* as the substrate for the sample of interest.

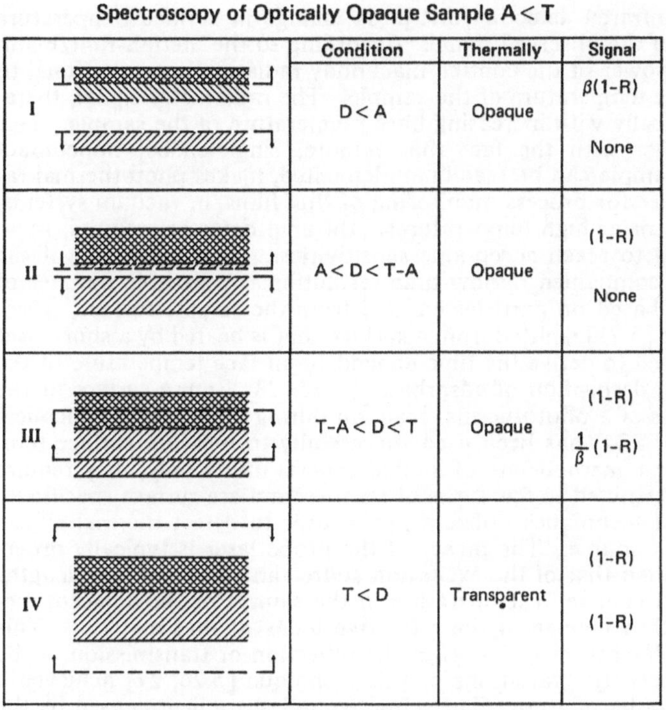

	Condition	Thermally	Signal
I	$D < A$	Opaque	$\beta(1-R)$ None
II	$A < D < T-A$	Opaque	$(1-R)$ None
III	$T-A < D < T$	Opaque	$(1-R)$ $\frac{1}{\beta}(1-R)$
IV	$T < D$	Transparent	$(1-R)$ $(1-R)$

Fig. 5.3. Schematic of the photothermal signal generation process in an optically opaque sample as a function of thermal diffusion length D. This shows the relationship between optical absorption length A and sample thickness T for an increasing $D (I - IV)$. For these cases, characterized by the relationship between the parameters, the variation of the photothermal signal with optical reflection coefficient R and absorption coefficient β of the sample is given for detection on the front and back, respectively. See Fig. 5.1 for the definition of the symbols and the text for more details

The thermal wave, generated at the front side of the sample, is detected on the back with the pyroelectric detector. The sample has to be thermally thin $T - A(\lambda) < D(f, t)$ to take advantage of this detection scheme. For a very thin sample, extremely short thermal transit times can be obtained according to (5.1.5). Utilizing a pyroelectric thin film calorimeter [5.15] a sensitivity of nanojoules combined with a time resolution of nanoseconds was achieved recently. This technique is, however, restricted to pyroelectric samples or thin films that can be directly deposited onto a pyroelectric substrate. Work is currently under way that will allow the deposition of a thin pyroelectric detector film on top of the substrate under study. *Thin film resistance thermometers* which are directly deposited onto the sample [5.16] compete with pyroelectric temperature sensors as far as time resolution is concerned. *Thin film thermocouples* are another promising temperature sensor for photothermal applications [5.17, 18]. These three sensors have the advantage that they are well understood and can be readily calibrated in absolute units. Their main drawback is that they require thermal contact with the sample and have to be deposited directly onto the sample of interest or the sample deposited onto the detector.

In contrast, *photothermal radiometry* can be utilized with bulk samples and is, moreover, a noncontact technique. Blackbody radiation from the sample is imaged onto a suitable infrared detector [5.19]. A change in surface temperature then effects a change of the observed signal. According to the Sterfan-Boltzmann law, the total radiant power of the emitted blackbody radiation is proportional to the fourth power of the temperature of the sample. The radiometry signal, therefore, increases dramatically with increasing base temperature of the sample. This unique feature, together with the fact that remote, single-ended, noncontact probing of a moving sample can be readily implemented, makes photothermal radiometry the ideal choice for process monitoring of thin films, in vacuum systems, in hostile atmospheres or at high temperatures. Infrared detectors require, however, cryogenic cooling to reach acceptable sensitivities. High sensitivity of the detector is typically accompanied by low time resolution. A sizeable number of detection schemes are based on particles emitted from the sample surface. Electrons [5.20] and atoms [5.21] emitted from a surface that is heated by a short laser pulse have been analyzed to derive the time-dependent surface temperature of the sample. Laser induced desorption of adsorbates [5.22, 23] from a surface or the analysis of the products of a photothermal reaction during laser-assisted chemical deposition of a metal [5.24] has been used successfully to measure surface temperatures. Most of these methods are of limited general utility only, they require UHV conditions, are restricted to few types of samples and are surface specific.

A large number of techniques utilize a *probe laser* to detect thermal effects caused by another light source. The power of the probe laser is typically orders of magnitude smaller than that of the excitation source and different wavelengths are commonly used. A change in temperature in the sample or the adjacent medium is associated with a change in the refractive index of that material. This change in refractive index causes a change in the reflection or transmission of the probe laser [5.25]. Recently, variations of this technique [5.26, 27] achieved a time resolution of the order of 10 ps. The refractive index gradient caused by the temperature gradient in the sample or the adjacent medium forms a transient thermal lens capable of deflecting a probe laser [5.28]. The same is true for the surface buckling due to localized heating [5.29]. In a similar fashion, acoustic wave fronts give rise to refractive index gradients, and these, along with surface displacement due to acoustic waves can be probed by lasers or other optical techniques [5.30]. All probe laser techniques have the advantage of optical excitation and probing and are therefore noncontact techniques. They do require, however, careful alignment of two lasers, and optically flat samples; criteria that are fairly easy to meet with thin films vacuum deposited on to flat substrates. A large number of publications in the semiconductor area report the use of photothermal deflection schemes to determine the properties of thin films on wafer substrates.

Piezoelectric transducers [5.31], attached to the sample under study convert the sound waves that are generated in the sample into an easily recorded electrical signal. Their main drawback is the requirement of good mechanical contact with the sample and problems inherent in acoustic detection, such as matching of the acoustic impedance of sample and transducer, susceptibility to acoustic noise and trade-offs between sensitivity and time resolution. Transducers using a change in capacitance or inductance [5.32] between sample surface and a reference electrode or inductor to monitor the surface displacement overcome many of the disadvantages of conventional piezoelectric ultrasonic transducers. The main area of application of ultrasonic detection of photothermal transient signals is in ultrasonic material testing; its use in spectroscopy is limited to weakly absorbing samples such as adsorbates or transparent materials. The same holds true for special transducers employed in the detection of surface acoustic waves [5.33]. If the sample can be in contact with a gas atmosphere, *microphones* are frequently used for detection (see, for example, [5.2]). Besides requiring a gas-filled cell to contain the sample, the frequency range of microphones is rather limited and the suppression of acoustic noise imposes restraints on such systems.

An evaluation of the pros and cons of the major photothermal and photoacoustic detection schemes shows that each of the detectors has its merits, making it the prime choice for certain applications or a particular type of sample. As should be evident from the above discussion, the signal generation and detection process can be rather complicated and may involve a large number of individual processes. A substantial loss in sensitivity and time resolution is associated with each diffusion or conversion process. These losses become particularly significant when detection occurs at the end of the signal generation chain. Microphone detection in particular has low signal generation efficiency and time resolution and requires the most elaborate theoretical models for the quantitative interpretation of the observed signals. However, even with these limitations, high signal-to-noise spectra can be obtained using a microphone. The spectra can be readily interpreted due to the long history and large body of literature associated with this approach. If highest time resolution is required for studies of extremely thin films, then pyro- and piezoelectric deflection schemes excel. Photothermal deflection schemes seem to be most popular in all applications dealing with semiconductor wafers and thin films on these wafers, i.e., samples of optical quality.

5.1.3 Instrumentation

A typical photothermal setup comprises a suitable light source, the detector and signal recovery electronics (Fig. 5.4). It is important to keep in mind that the signal generation process involves optical, thermal and possibly acoustic properties of the sample.

In spectroscopy, it is important, for example, to eliminate thermal and acoustic artifacts in the recorded spectra. To account for the wavelength dependence of the source, the thermal parameters of the sample and the characteristics of the detector, a reference sample is normally employed. Reference data are obtained in a single beam spectrometer before or after the sample spectrum or, in a dual beam arrangement, simultaneously with the sample spectrum. Due to the strong influence of the modulation frequency, it is imperative to record both data sets at the same modulation frequency with samples of identical or well-known thermal characteristics.

Similarly, for a thermal and acoustic analysis of the sample it is necessary to eliminate the influence of optical sample parameters and distinguish between the

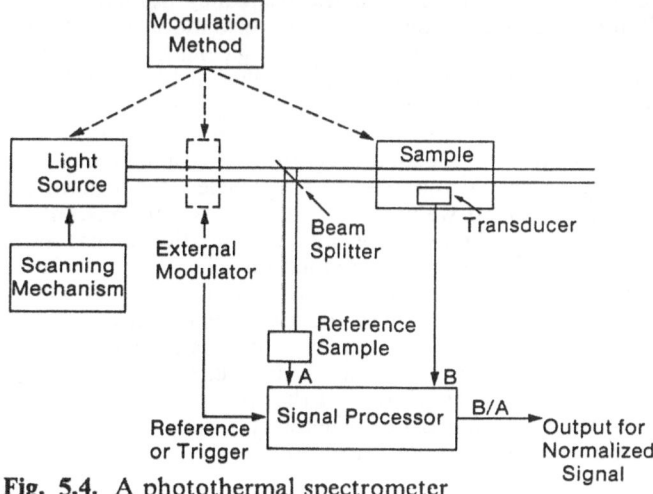

Fig. 5.4. A photothermal spectrometer

thermal and acoustical signal. The thermal diffusion process is quite different from sound propagation, thermal waves are critically damped within one wavelength to less than 1% of their initial amplitude; sound waves, however, propagate at the frequencies considered here over long distances. The thermal signal can therefore be readily suppressed by a simple delay line. If this is not feasible, the well-defined transit time of a sound pulse can then be utilized to discriminate electronically against the thermal signal by time gating techniques. Thermal diffusion across an interface depends on the thermal diffusivities of both materials; sound transmission is a function of their respective acoustic impedances. Couplers and filters can therefore be designed that transmit thermal but no acoustic energy. Another important difference between thermal and sound waves of the same frequency is the much longer wavelength of sound waves. For most thin film applications, the wavelength of the interrogating wave should be of the order of the film thickness. The ultrasonic analysis of a film requires, therefore, considerably higher modulation frequencies than the thermal analysis of the same film. The cost of high frequency electronics increases dramatically with increasing frequency, making photothermal analysis a less costly alternative and worth looking into.

The light source in a typical experiment will be intensity modulated by a suitable means, such as current modulation of a lamp or a mechanical chopper. Wavelength or polarization modulation are advantageous when small absorptions superimposed on a large background absorption are of interest. Where applicable, modulation of the absorption properties of the sample (Stark modulation) or other physical parameters (temperature) that allow modulation of the sample characteristics can be used.

Most commonly, sinusoidal or square wave modulation of the incident light intensity is employed [5.1-4]. In this case a lock-in amplifier is the appropriate tool for signal recovery. It should be emphasized that if a source of constant intensity is modulated in this way the energy per excitation cycle decreases with increasing modulation frequency. The signal at high modulation frequencies, desirable, for example, because of the above saturation considerations, will then be extremely weak. Excitation with a short light pulse has the advantage of generating a higher surface temperature than with the same energy in a longer pulse or a periodic excitation, due to the fact that heat loss during the short pulse can be neglected. This results in a superior signal/noise ratio but might damage the sample irreversibly or affect temperature-sensitive sample properties. With pulsed excitation, box-car integrators are frequently used to monitor the signal in a small time window [5.34]. Transient digitizers enable one to obtain the complete time domain signal and, if desired, transform this into the frequency domain [5.35]. Recently, another modulation scheme has emerged using noise modulation [5.36] of the light source and transient digitizers for data recording. Time domain type results can then be obtained from the data by cross correlation [5.37], and frequency domain data by Fourier analysis [5.35]. The advantage of this new technique is that many modulation frequencies are probed simultaneously and therefore depth profiles can be obtained in a small fraction of the time required for recording spectra subsequently at several modulation frequencies (for a more detailed discussion see Chaps. 8 and 9). If we assume a light source of constant intensity that is modulated by an external modulator, noise modulation makes more efficient use of the light intensity than any other modulation scheme.

Photoacoustic detection is frequently employed in FT-IR studies of thin films or powders. Here the intensity modulation due to the interference fringes can be utilized when a commercial rapid scanning FT-IR spectrometer is used as the light source. One should, however, be aware of the fact that this results in a much lower modulation frequency for the long wavelength side of the spectra as compared to the short wavelength part of the same spectrum. Therefore a spectrum might be partially distorted for high wave numbers due to saturation.

With conventional light sources, only a limited spectral, time and depth resolution can be obtained. Nevertheless, high sensitivity, instrumental simplicity and minimal sample preparation make photothermal detection an interesting alternative to more established techniques, such as diffuse or internal reflectance spectroscopy [5.38]. Photothermal detection, however, offers a unique combination of advantages that are not available with other techniques when combined with the high spectral and time resolution possible with laser excitation. This is evident for spectroscopy, but is also true for thermal and ultrasonic analysis of thin films when using photothermal effects to generate the probe that interrogates the sample under study. In the following section, examples will be presented underlining some of the above features.

5.2 Spectroscopy of Thin Films

Due to its peculiar signal generation and detection process, photothermal spectroscopy has several unique features and is quite different from conventional transmission spectroscopy. Photothermal spectra represent only heat released within the thermal diffusion length of the sample surface. As outlined above this makes photothermal spectroscopy the method of choice for thin film studies. Additional advantages of photothermal detection methods are (1) that they are not affected by scattered light and (2) the depth profiling capability of this technique. If these techniques are so powerful why are they not dominant in spectroscopy? Quite clearly, the fact that thermal and acoustic properties of the sample affect the observed spectra complicate the interpretation of the data considerably. Reference samples with well-defined optical, thermal and acoustical properties are required [5.39] and photothermal spectra of two samples with identical optical absorption are identical only if the samples have identical thermal and acoustic properties, the same depth profile, the same quantum efficiency for radiationless deexcitation and are recorded at the same modulation frequency. These complications let photothermal spectroscopy complement conventional techniques but not replace them [5.40].

5.2.1 Semiconducting Films

Semiconducting powders and wafers have been used for quite some time to demonstrate the power of photoacoustic detection, mainly using air-filled cells and microphones for detection. More recently, photothermal deflection schemes have been introduced into this field. With semiconductors being well defined and well understood samples, these experiments served at first to develop an understanding of the photoacoustic and later also of the photothermal signal generation process. Later, the advantages of these techniques over conventional spectroscopies were demonstrated and helped to establish credibility in this field. A large number of recent publications, reviewed for example in [5.4], show that the method is now well accepted in the semiconductor area and methods have been developed to derive the optical absorption coefficient from the data [5.41, 42]. Some of the key issues in semiconductor samples are associated with impurities and defects. Semiconductors are characterized by their band gap, and these impurities and defects cause additional states in the band gap that affect the electrical characteristics of semiconductor devices. Due to the dramatic change in optical density in the region close to the edge of the band, conventional optical spectroscopies with their rather limited dynamic range and very high background are not too suitable for this type of application. Many of the samples of interest are amorphous films, which typically have a rough surface. With conventional transmission spectroscopy, numerical corrections have to be introduced to account for the scattered light. Measurements of the diffuse reflectance are, however, by no means trivial, causing systematic errors in these corrections of the measured optical transmission. Photothermal methods excel here and enable one to determine the spatial distribution of the defects and to study the relaxation mechanisms of electrons from optically excited states to the ground state via nonradiative pathways, an issue of burning interest for photovoltaic and solid state laser devices.

The emphasis in the last few years has been on studies of amorphous semiconductor films. Due to interest in xerography and photovoltaics, a number of experiments focus on a-Si. Photothermal spectra complement a xerographic analysis of the material and [5.43], enable the effect of compensation on defects to be studied [5.44] and the derivation of the density of gap states in a-SiH to be ascertained [5.45-50]. Photoacoustic studies on other systems, many of them reported during the *International Conferences on Amorphous and Liquid Semiconductors*, demonstrate how photothermal deflection spectroscopy can be utilized in the study of amorphous Si-Ge alloys [5.51], the doping of a-Si films [5.52, 53], the optical properties of a-Ge films [5.54, 55] or materials such as a-SiN [5.56, 57] or CuInSe$_2$ [5.58]. Systems such as Cd$_{1-x}$Zn$_x$S have been studied with photothermal deflection spectroscopy and the results compared with those of conventional transmission spectroscopy [5.59]. Electroluminescent thin films, otherwise accessible only to sophisticated techniques such as time resolved fluorescence measurements, have been extensively studied with photoacoustic and photothermal methods [5.60-62].

As is evident from the large number of recent publications, photoacoustic and photothermal detection schemes are now at a point where they are able to make unique contributions towards the understanding of thin semiconducting films.

5.2.2 Dielectric and Metallic Films

Dielectric and metallic thin films are another area where photothermal detection methods can be advantageous. An example is the first direct measurement of the optical absorption of polyacetylene [5.63]. As a result of its complicated fibrillar morphology, polyacetylene is a very complex optical medium. Optical absorption spectra that can be compared with various theoretical models had to be inferred from transmission spectra of thin films. It was found that these films scatter several percent of the incident radiation diffusely. Transmission measurements cannot distinguish between scattering and absorption. Therefore absorption constants derived from transmission measurements, even when using specular or diffuse reflectance data for correction, can exhibit systematic errors that make a comparison between various theoretical models almost arbitrary. Photothermal deflection spectra provided the first reliable measurement of the absorption edge of polyacetylene and showed a Urbach-type behavior.

Electrode supported films are of considerable interest in electrochemical studies. One problem with this type of sample is that the sample is deposited onto a metal. Since the amplitude of the light intensity vanishes close to the metal, conventional spectroscopies suffer from a lack of sensitivity. This effect was clearly seen in FT-IR studies of Prussian Blue and cupric hexacyanoferrate on various electrodes [5.64]. In another experiment [5.65] the distance between a silver surface and the chromophore was varied using a varying number of transparent Langmuir-Blodgett spacer layers. Thus the amplitude of the electric field vector of the standing light wave in front of the silver was actually mapped.

Many of the problems in electrochemistry are usually addressed ex situ with surface scientific methods requiring ultrahigh vacuum. Ex situ transmission and reflection measurements complement these experiments. To ensure that the sample preparation is not introducing artifacts, in situ experiments are also desirable. Recently, a group was successful in recording photoacoustic spectra in situ not only of grown electrochemical films but also monitoring the spectra of growing films [5.66]!

5.2.3 Spectroscopy of Layered Films

Photographic color reversal films with their intricate but well-defined layered structure have served as a model system to demonstrate the depth-profiling capability of photoacoustic [5.67] and photothermal [5.68] detection schemes. As discussed in Sect. 5.1.1 and shown in Fig. 5.3, with decreasing modulation frequency a thicker layer of the sample contributes towards the observed signal. With a sample such as a color reversal film, the observed spectra depend strongly on the modulation frequency and would, at least in principle, allow the reconstruction of the depth-profile, i.e., optical density as a function of wavelength and depth, of the sample. Depth-profiling is discussed in more detail in Chap. 9.

5.2.4 Nonradiative Quantum Yield

Absorption spectroscopy is the prevalent application of photothermal and photoacoustic spectroscopy. Since radiationless deexcitation of the electronic excited state is a key step in the signal generation chain, the relevant quantum yield for nonradiative processes can be determined. To demonstrate the determination of nonradiative yields and to highlight the frequency dependence of spectra, let us consider a trilayered sample such as the one shown in Fig. 5.5, with a weakly absorbing layer 1 separated from another absorbing layer 3 by a transparent spacer layer 2. In the example discussed here, layer 1 contained Nd_2O_3 molecules whereas layer 3 was a silver film. The spectra shown in Figs. 5.6 and 5.7 were recorded at two different modulation frequencies with a thermal detector at the back of the sample [5.69]. At a low modulation frequency, an absorption-like spectrum is observed (Fig. 5.6). Light absorbed by the neodymium ions is converted into heat. This heat diffuses across the sample and is detected on the back side. At high modulation frequencies (Fig. 5.7), the much shorter thermal diffusion length does not allow heat generated by the neodymium ions to reach the detector. Only light that is transmitted by the Nd_2O_3 layer and then absorbed by the silver film will contribute to the observed signal. The observed spectrum is, therefore, a transmission spectrum of the neodymium film detected by a silver absorber with subsequent thermal detection. These two extreme cases correspond to Fig. 5.3, where a homogeneous sample was discussed. If the structure of the film or its thermal properties are known, additional information can be derived from these spectra. The thermal diffusivity or the thickness of the spacer layer can be derived from a comparison of the signals. A more important parameter is that of the quantum yield for radiationless deexcitation. In the transmission spectrum, Fig. 5.7, light that has been *absorbed* by the neodymium ions is missing. In the

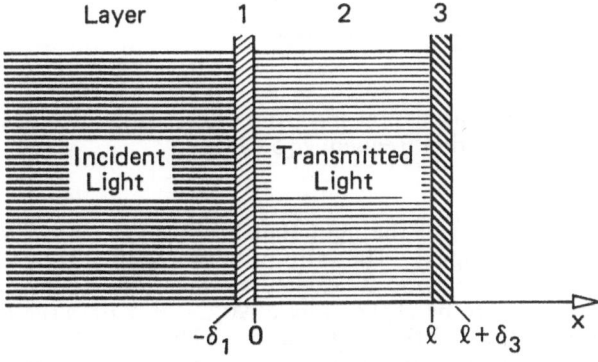

Fig. 5.5. A composite trilayered sample

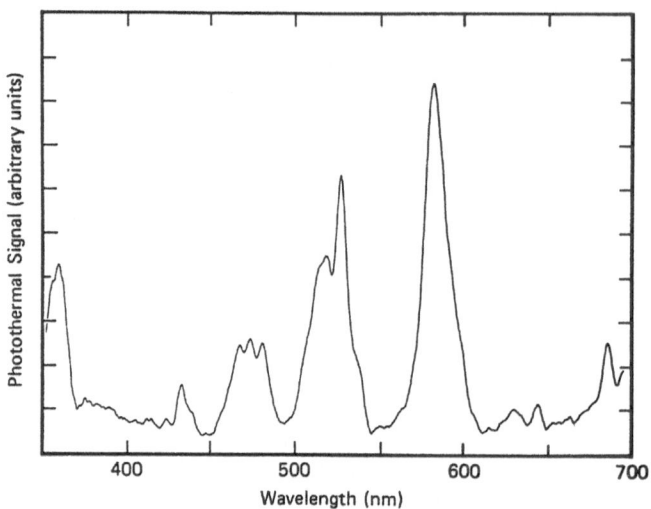

Fig. 5.6. Normalized photothermal absorption spectrum of 0.8×10^{15} Nd_2O_3 molecules in a 1 μm thick PMMA film coated on top of 0.1 mm undoped PMMA cast on a silver substrate. Recorded by out-of-phase detection at 2.2 Hz modulation frequency [5.69]

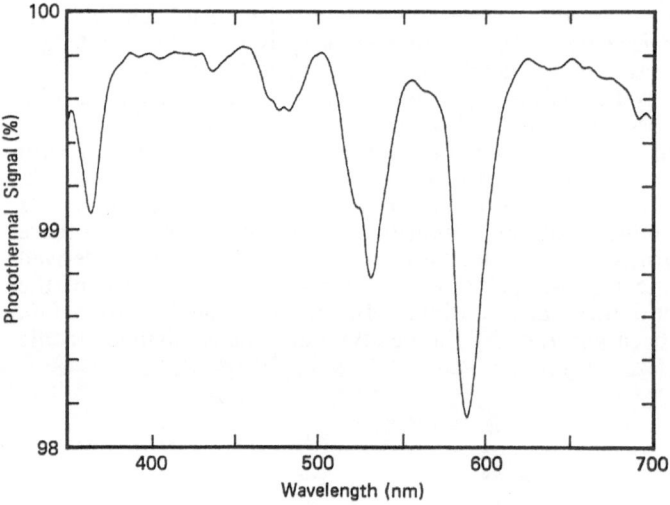

Fig. 5.7. Normalized photothermal transmission spectrum of the sample described in Fig. 5.6. Recorded at 88 Hz modulation frequency [5.69]

absorption spectrum, only that fraction of the light that has been *absorbed and converted into heat* contributes to the signal. The ratio of both signals is then the probability of nonradiative decay of the optically excited state. In the above example, this quantum yield turned out to be of the order of 90%. Since both signals are measured with the same detector and then ratioed, no calibration of the detector is required to determine the absolute quantum yields. This technique allows the measurement of quantum yields of thin films and even of monolayers [5.65].

5.3 Thermal Analysis of Thin Films

Photoacoustic and photothermal spectroscopy can be cumbersome due to the thermal diffusion step in the signal generation process. In the thermal analysis of thin films the same process is now advantageous. Photothermal excitation of a light source facilitates the implementation of a well-defined heat source for the probing of the thermal properties of thin films. The heat source requires no mechanical or thermal contact with the sample and can be tailored to the application by simply changing the temporal shape or wavelength of the light source and by changing the size and the location of the illuminated area. Combined with the high sensitivity and time resolution of many photoacoustic and photothermal detection schemes, the thermal analysis of temperature sensitive properties of extremely thin films becomes a viable proposition.

5.3.1 Thermal Diffusivity

The thermal diffusivity of a thin film can be determined by generating a transient heat source at the front of the sample and monitoring the subsequent temperature increase at the back. When using amplitude modulation of a light source to generate the heat source the amplitude or phase of the transmitted temperature wave can be analyzed [5.70]. For a film with thickness ℓ, an amplitude decrement A

$$A = \ln \frac{\Theta(0)}{\Theta(\ell)} \tag{5.3.1}$$

is observed, with $\Theta(0)$ and $\Theta(\ell)$ being the amplitude of the temperature oscillation on the front and back of the sample. For a thin film of diffusivity α, the frequency dependence of the amplitude decrement is given by

$$A = \frac{\ell}{\sqrt{\alpha}} \times \sqrt{\pi f} . \tag{5.3.2.}$$

In addition, a phase lag Δ, where

$$\Delta = \Phi(\ell) - \Phi(0) , \tag{5.3.3}$$

is observed between the temperature at the front and the back, with Φ being the phase angle between the modulated light source and the observed temperatures, with Δ being equal to A. Figure 5.8 shows a typical result for the modulation frequency response of the amplitude decrement of a thin plastic film. From the slope of the curve, the product of thermal diffusivity times ℓ^2 can be derived. If the film thickness is known, the thermal diffusivity is determined. The data in Fig. 5.8 also show the typical increase of the statistical error at higher modulation frequencies. Phase lag measurements (Fig. 5.9) have considerably better signal/noise ratio. For very thin films, however, their resolution is also insufficient. The solution to this problem is to measure at higher frequencies. Data for the same films as in Fig. 5.9 but recorded in the time domain with pulsed excitation are shown in Fig. 5.10. From these data the thermal diffusivities of thin films can be derived using numerical models. Experimental requirements for photoacoustic diffusivity measurements of the thermal diffusivity of thin films, first introduced by *Adams* and *Kirkbright* [5.71], are modest. Due to the limited bandwidth of microphone detection, photothermal methods are currently dominating the field. With the sample directly coated onto a pyroelectric sensor highest time resolution can be obtained and extremely thin films can be studied [5.72-76]. However, this technique involves special sample preparation. Other methods typically require optical access to the front and the back of the sample [5.75], a problem which can be overcome by measuring the lateral heat flow in the sample [5.74]. Still another

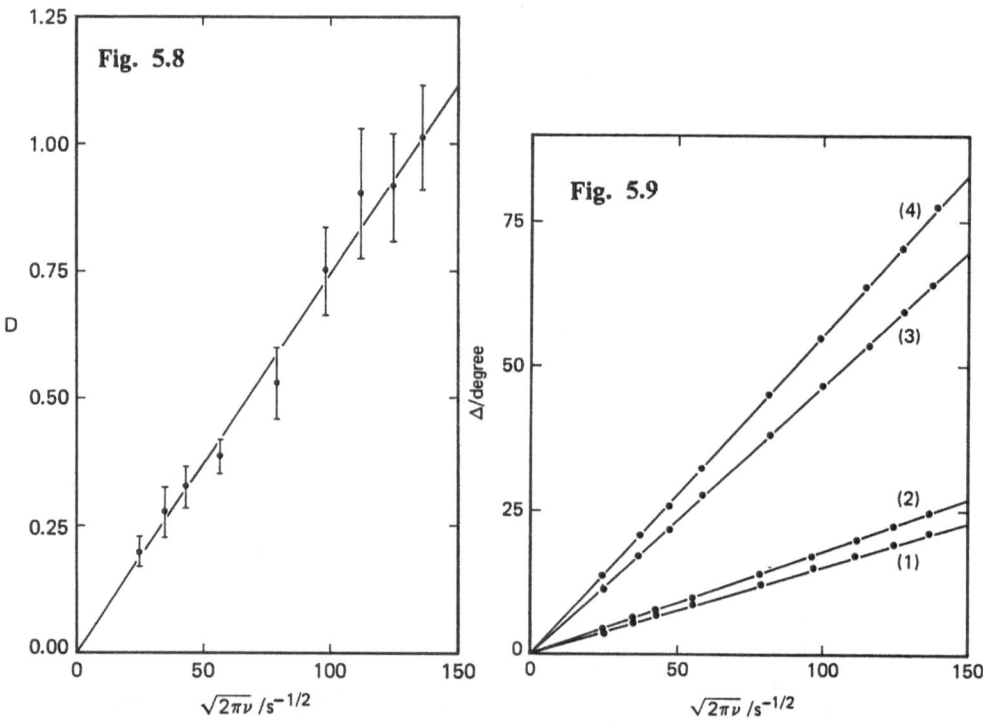

Fig. 5.8. Modulation frequency dependence of the amplitude decrement A of a 2.1 μm thick Novolac film [5.72]

Fig. 5.9. Modulation frequency dependence of the phase lag Δ for four Novolac films of thickness: (1) 0.65 μm, (2) 0.80 μm, (3) 2.1 μm and (4) 2.45 μm [5.72]

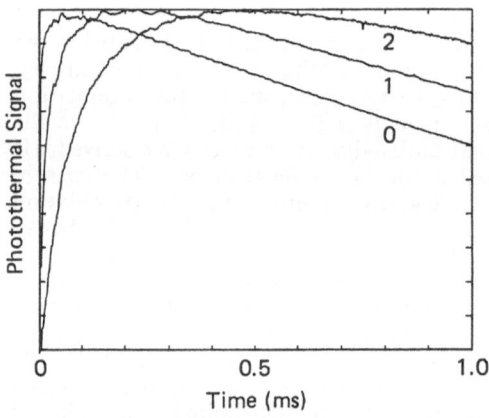

Fig. 5.10. Photothermal signal for a pyroelectric thin film calorimeter excited by a 10 μJ laser pulse with 15 ns pulsewidth: (0) bare calorimeter, (1) coated with a 0.65 μm, and (2) a 0.80 μm thick Novolac film [5.72]

way is to determine the surface temperature as a function of time while heat diffuses from the surface of the sample into the interior. Photothermal radiometry [5.76, 77] and photothermal deflection schemes [5.78, 79] have demonstrated their utility in this type of noncontact, single-ended measurement. Even photoacoustic schemes can be adapted for samples with inaccessible back sides [5.80] and corrections for the geometry of the cell improve accuracy considerably [5.81].

Laser excitation combined with one of the photothermal detection schemes facilitates measurements of thermal transport coefficients in thin films. Whereas frequency domain experiments are very straightforward in their interpretation, time domain experiments with pulsed light sources for excitation require extensive analytical or numerical efforts to derive transport coefficients from the observed data.

5.3.2 Film Thickness

The above discussion on thermal diffusivity assumed that the thickness of the sample can be derived by other methods. But quite clearly (5.3.2) can be utilized to determine the thickness of a sample with known thermal diffusivity. A large number of more conventional techniques, many of them mechanical or optical, are available to measure film thickness conveniently and with the desired precision. Most of the reported thermal thickness measurements demonstrate only the validity of models underlying the derivation of (5.3.2) and show that results determined with this technique are consistent with other measurements. Why bother with a technique whose accessible thickness range is severely limited by the thermal diffusion length (5.1.2) or (5.1.6)? One application might be the optical measurement of thin films on opaque substrates of low reflectivity such as SiO_2 on Si [5.82]. All of the optical measurements require an independent measurement of the refractive index of the material to derive the film thickness. By a combination of thermal and optical interferometric techniques, the thickness of the film can be determined in situ without an off-line measurement of the refractive index [5.83].

5.3.3 Phase Transitions

Heat deposited by a laser beam in a sample can increase the temperature of the sample, as discussed in Sect. 5.3.1 on thermal diffusivity or, if the sample undergoes a first-order phase transition, it can provide the latent heat of that transition without changing the temperature at all. This well-known phenomenon was first observed in the context of photoacoustics by *Florian* et al. [5.84]. Anomalies in the amplitude and the phase of the observed signal [5.85-87] were interpreted [5.88] as an oscillation of the liquid-solid phase boundary in the sample during the periodic illumination. Photoacoustic experiments have been complemented by photopyroelectric measurements of phase transitions [5.89]. These experiments on model systems established photothermal methods and the associated theory for the study of phase transitions. Most recently, these methods have been applied to the glass-crystal phase transition of semiconductors [5.90] and to real-time studies of laser annealing [5.91]. Figures 5.11 and 5.12 show representative data from this study. During the laser pulse, the sample, a thin tellurium film, is heated by the laser and subsequently cools down. With increasing laser fluence, the peak temperature of the film does not follow that increase due to the latent heat of the melting and boiling transition. At high time resolution (Fig. 5.11d) the sample melts during the first part of the laser pulse and is heated all the way up to the boiling point by the remainder of the pulse. During the cool down (Fig. 5.12), latent heat is released during crystallization, keeping the sample temperature constant for an extended period of time. The amount of latent heat is proportional to the amount of material left after boiling off part of the film. The crystallization is, therefore, finished earlier for samples of smaller mass, i.e., samples that remained at the boiling temperature longer due to larger laser fluence. Compared

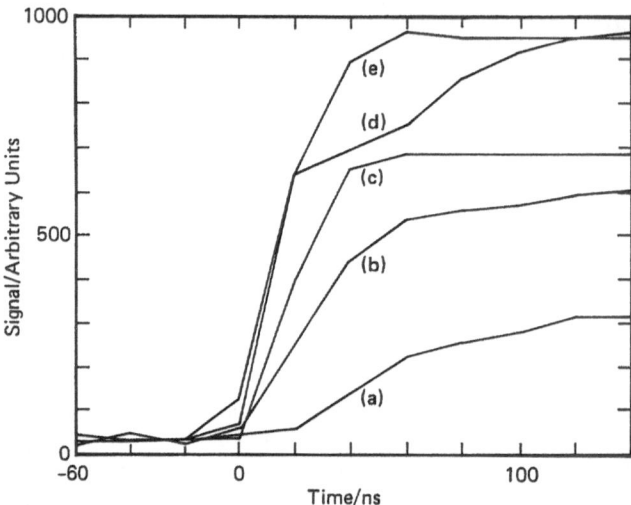

Fig. 5.11. Initial temperature during annealing of 30 nm thick Te films with one pulse from a XeCl excimer laser for five different fluences. The pulse energy is approximately tripled between pulses: (a) $0.7\,\mathrm{mJ/cm^2}$, (b) $2\,\mathrm{mJ/cm^2}$, (c) $7\,\mathrm{mJ/cm^2}$, (d) $24\,\mathrm{mJ/cm^2}$ and (e) $70\,\mathrm{mJ/cm^2}$ [5.91]

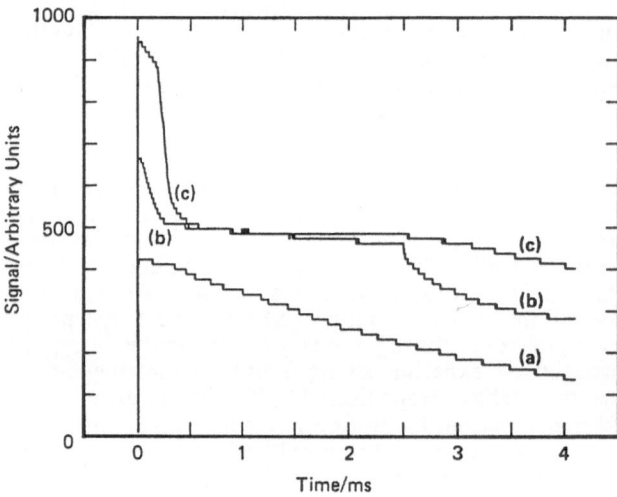

Fig. 5.12. Pyroelectric signal during annealing of 30 nm thick Te films with one pulse from a XeCl excimer for three different fluences. The pulse energy is increased by one order of magnitude between pulses: (a) $0.7\,\mathrm{mJ/cm^2}$, (b) $7\,\mathrm{mJ/cm^2}$ and (c) $70\,\mathrm{mJ/cm^2}$ [5.91]

with frequency domain experiments, time resolved data do not require sophisticated models for their interpretation. Frequency domain experiments fostered the understanding of the photoacoustic signal generation process but have only limited appeal to users of conventional thermal analysis equipment. Time domain studies might, however, be very attractive tools for studies of laser induced thermal processes, such as annealing and ablation. Their high time resolution combined with excellent sensitivity enables real-time studies of single transient events which are crucial in that type of research.

5.4 Ultrasonic Analysis of Thin Films

Pulsed lasers have been used by many authors as a convenient and flexible means for the generation of ultrasound in thin films or solids; for recent reviews on this subject see [5.92-94]. For the ultrasonic analysis of a thin film, the wavelength of the sound wave used to monitor the thickness of the sample and its acoustic properties should be less than the sample thickness. With a sound velocity of the order of 5×10^3 m/s, a film thickness of the order of 1 μm requires a pulse of less than 200 ps to meet this criterion. Conventional techniques are very cumbersome at the corresponding frequencies, whereas photothermal techniques are relatively straightforward.

As in the photothermal generation of a heat source (5.1.3), so the key advantage of photoacoustic generation of a sound source is the fact that the photoacoustic method requires no mechanical contact with the sample. The timing and location of the sound source can be selected by simple optical techniques and arrays of coherent sources can be readily implemented. A number of photothermal detection schemes are based on optical probing of the sample surface [5.30]. The detection of the ultrasonic pulse can, therefore, also be without mechanical contact with the sample. For semiconductor applications this means, amongst other things, that no contamination will be introduced by a coupling medium. In addition, the excitation source as well as the detector can be scanned readily across the sample. As long as the intensity of the excitation stays well below the thresholds for evaporation or ablation of the sample, no irreversible processes are anticipated and the method can be considered truly nondestructive.

Conventional piezoelectric ceramic and single crystal transducers are excellent mechanical resonators. When stimulating such a transducer with a short pulse, its mechanical resonance is excited, causing the transducer to ring for quite some time after the pulse. The time resolution of the transducer and hence the thickness of the sample that can be studied is limited by this effect. Numerical corrections or electronic filters can compensate partially for this transducer transfer function. Piezopolymers enable one to overcome this problem due to the large internal damping of these materials [5.95] associated, however, with a much lower conversion of mechanical energy into an electrical signal.

Coupling of the transducer to the sample of interest is another issue that can be eliminated in photothermal studies. Coupling of acoustic energy is a function of the acoustic impedance of the sample and the transducer. Impedance matching can be accomplished by selecting appropriate materials and geometries. Since the choice of piezoelectric materials is rather limited, it might not be possible to match the acoustic impedance of the sample of interest by this technique. Antireflection layers, the other alternative, unfortunately have a limited bandwidth and thus cause severe distortions in a pulsed mode. At the high frequencies desirable for thin film studies, attenuation in the coupling medium can become a serious problem. Light, however, can be readily coupled into virtually any material.

Lasers delivering nanosecond pulses have been available for quite some time at a reasonable cost. Generation of ultrasonic pulses with these lasers has, therefore, been dominating the literature, references [5.96-98] serve merely as examples of these applications. High power picosecond lasers are just becoming available commercially. The cost of picosecond lasers and the fast electronics required to take advantage of the ultrashort pulses for the analysis of thin films is, however, still rather prohibitive and maintenance of these lasers is labor intensive. At this time, therefore, only few researchers [5.99, 100] have been able to take advantage of this emerging diagnostic tool.

5.5 Nondestructive Evaluation of Thin Films

In a rigorous sense, spectroscopic, thermal and ultrasonic analysis of a thin film constitute nondestructive evaluation of the material, if excitation and detection of the probing thermal and ultrasonic waves are accomplished in a nondestructive way. Evaluation in a wider sense refers here to well-developed applications that are potentially of industrial interest. The unique feature of photothermal and photoacoustic techniques is that they allow nondestructive mapping of geometrical, optical, thermal and, where applicable, acoustic sample properties [5.101].

5.5.1 Depth Profiling

The depth profiling capability of photothermal and photoacoustic methods has been mentioned several times in this chapter and is covered in detail in Chap. 9. As far as thin films, semiconducting thin films in particular, are concerned, the key issue is to distinguish surface absorption from bulk and interface absorption. This problem can be addressed by recording photothermal or photoacoustic spectra of the sample of interest at two or more modulation frequencies (5.1.4) or at different times after pulsed excitation (5.1.6). Instead of recording these spectra, data can be obtained subsequently by modulating the light source with several frequencies at the same time and using Fourier transform techniques to retrieve amplitudes and phases of the individual frequency components [5.35]. With pulsed excitation, individual time domain signals can be recorded and analyzed to distinguish contributions due to the surface and bulk of the sample. From these data then, the location of the absorber can be inferred in analogy to the procedure used in Sect. 5.2.4, Figs. 5.6 and 5.7. These methods have been utilized to determine, for example, the surface passivation of amorphous silicon films [5.102], to separate volume from interfacial absorption in TiO_2/SiO_2 multilayered films [5.103] or to establish the influence of subsurface defects on plasma-sprayed coatings [5.104].

Thermal and acoustic waves can be utilized to measure or profile temperature- or pressure-dependent properties of the thin film. Pyroelectric [5.105] and piezoelectric properties of thin films can be determined readily with these techniques. Using time resolved measurements, the depth profile of charge or field distributions can also be determined. Techniques employing laser induced thermal [5.106] or pressure pulses [5.96] are the standard methods for measurements of charge or field distributions in dielectric films [5.106]. Here, the fact that only the excited layer of the sample induces an electrical signal and that the excitation sweeps across the sample is utilized to derive the depth profile.

Polymer electrets such as β-polyvinylidene fluoride (PVDF) are frequently used in pyro- and piezoelectric transducers for the detection of radiation induced transients. The polarization profile in these materials determines the ultimate time resolution of the transducer and serves as an example to highlight some of the differences between thermal and ultrasonic analysis of a sample. PVDF film from two manufacturers was used for the data shown in Figs. 5.13-16. Due to the completely different manufacturing processes the polarization profiles are quite different, the bulk properties, however, are almost identical. Using a 500 ps laser pulse, absorbed in a tungsten rod, a shock wave is generated. The piezoelectric

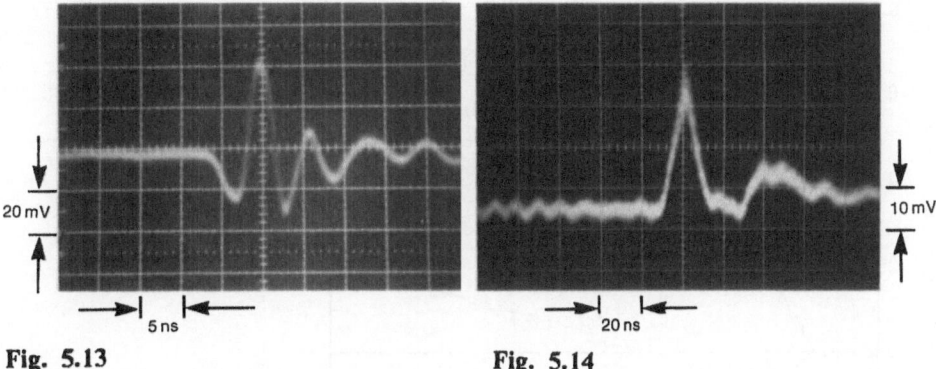

Fig. 5.13   **Fig. 5.14**

Fig. 5.13. Piezoelectric response of an ultrasonic transducer built from 9 μm thick β-PVDF from Kureha to an ultrasonic pulse generated by absorption of a 500 ps pulse with an energy of 1 mJ [5.105]

Fig. 5.14. Piezoelectric response of an ultrasonic transducer built from 24 μm thick SOLEF foil to the same pulse as used in Fig. 5.13 [5.105]

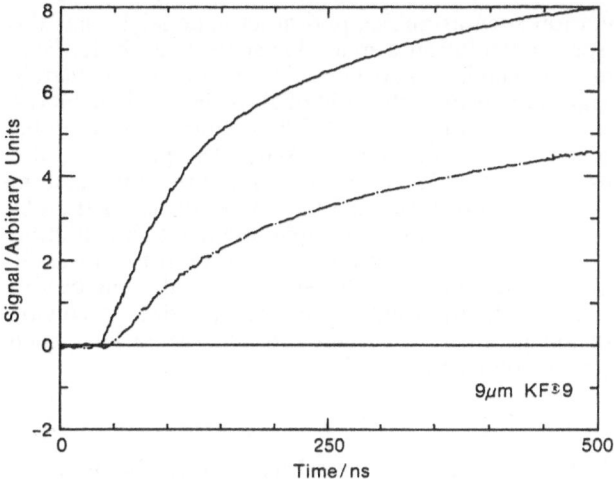

Fig. 5.15. Pyroelectric response of the sample of Fig. 5.13 excited directly with a 0.5 mJ pulse from a XeCl excimer laser with a duration of approximately 40 ns. The signals for excitation from both sides are shown [5.105]

response of the two foils to this shock wave is quite different. The pulse generates a response of approximately twice the halfwidth, but only half the amplitude in the second foil (Fig. 5.14). Photothermal analysis of the same foils and excitation with the same laser directly on the surface of the transducer show that the first foil is well poled. The risetime of the signal is fast and excitation from the front and back of the film results in almost identical signals (Fig. 5.15). This behavior would be expected for a homogeneously poled film. The photothermal response of the second film (Fig. 5.14) shows that one side of the transducer has a fast risetime, but the back side has a fairly thick unpoled layer and the polarization reaches the bulk value only far away from the surface.

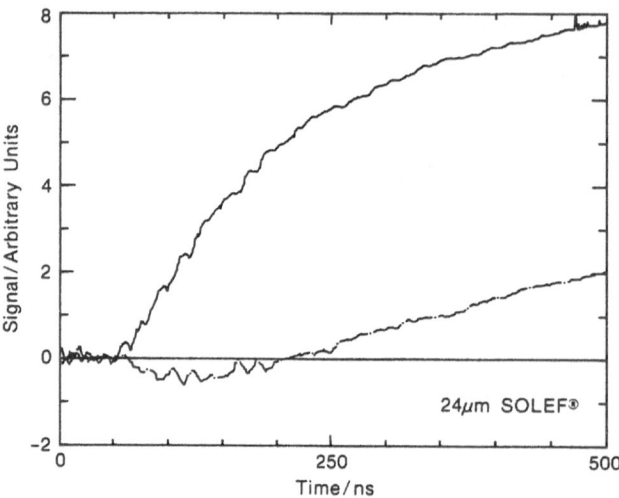

Fig. 5.16. Pyroelectric response of the sample of Fig. 5.14. The conditions of the experiment are otherwise identical to Fig. 5.15 [5.105]

For a qualitative interpretation such as above, photothermal analysis has the advantage that for the same spatial resolution a much lower time resolution suffices. With the bandwidth of commercial electronics limited to approximately 6 GHz, and that at very substantial expense, photothermal analysis might be the method of choice rather than photoacoustic analysis. The qualitative interpretation of acoustic signals is, however, straightforward. While the pressure pulse traverses the sample, it induces a signal that is a direct measure of the polarization. The depth of the sample is directly mapped into the time dependence of the observed signal. Since thermal pulses diffuse into the sample, the main contribution to the signal is always caused by the surface. While the pulse diffuses into the sample, successive layers of the sample add up to this surface signal. The deconvolution of the observed signal into a depth profile is, however, possible by solving the heat diffusion equation for this geometry. A number of algorithms have been developed to accomplish this deconvolution [5.107].

5.5.2 Imaging

Maps of the geometrical, optical, thermal and acoustical features can be obtained by scanning the excitation across the sample or by using masks to illuminate different combinations of pixels subsequently [5.108, 109]. With optical spectroscopy being a well-developed technique and with the parallel processing of all pixels, i.e., viewing or recording of one complete picture, now being state of the art, photothermal and photoacoustic techniques cannot compete in the area of geometry and optical imaging. Thermal and acoustical images do, however, bring some of their own unique properties to bear.

Due to the complicated signal generation process, interpretation of photothermal and photoacoustic images can be ambiguous [5.110, 111]. The capability of these microscopes to image subsurface features is, however, of considerable technological interest. Defect maps in thin dielectric films [5.112], maps of radiation induced defects [5.113] or thickness measurements of Si membranes [5.114] represent just a few examples of applications in thin films. The bulk of the application is, however, in the area of integrated circuits and thin film devices. Two machines have been marketed successfully using either an electron beam [5.115] or a laser for excitation [5.116]. Detection is accomplished with an ultrasonic

SEM   **SEAM**

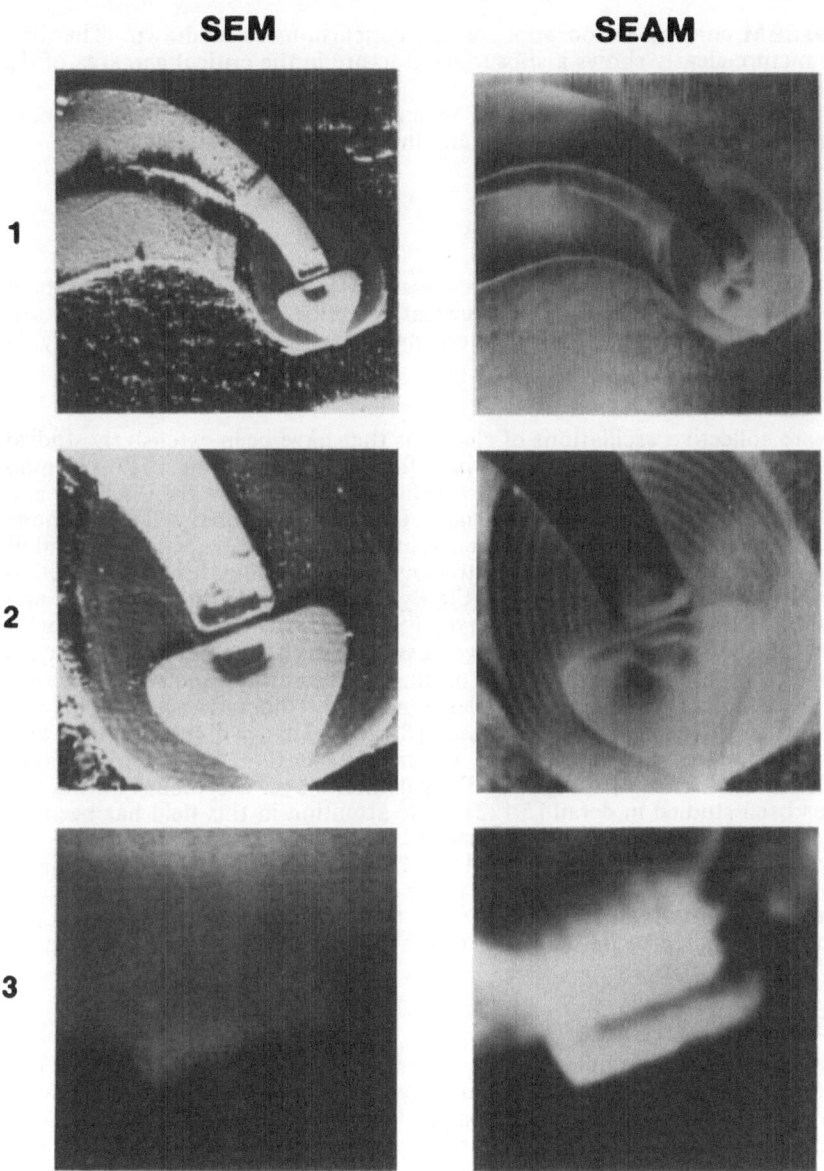

Fig. 5.17. Secondary electron microscopy (SEM) images of an inductive thin film head for magnetic recording compared with secondary electron induced acoustic images (SEAM) for increasing magnification

transducer inserted into a modified electron microscope or using laser beam deflection or changes in reflectivity to probe the temperature change.

The power of this imaging technique is evident in Fig. 5.17. A commercial thin film recoding head is imaged in an electron microscope. On the left side are shown conventional SEM secondary electron images; on the right side are the corresponding thermal wave images. Clearly visible in the thermal wave images are the copper contacts and the copper coil of the inductive head. At very high resolution, SEM contrast is too poor for any conclusions to be drawn. The thermal wave picture clearly shows a subsurface fracture in the critical gap area of the head.

The great potential of this imaging technique would justify considerable efforts being made to improve equipment and interpretation of the images.

5.6 Miscellaneous Thin Film Applications

Besides those applications discussed previously, a number of publications have dealt with the photoacoustic or photothermal detection of surface plasmons and the use of ferromagnetic resonance phenomena to excite the sample selectively.

5.6.1 Plasmon Detection

Plasmons are collective oscillations of electrons that have been extensively studied with conventional spectroscopic techniques for quite some time. The plasmon surface polariton is extremely sensitive to surface roughness and the thickness and dielectric constant of the adjacent materials. Using conventional cell-microphone techniques to detect the nonradiative decay, it was possible to demonstrate that the sensitivity of this method is superior to conventional attenuated total reflection measurements and that the influence of the dye monolayer on the amplitude and width of the resonance was readily observed [5.117]. With the same technique, it was shown that radiative and nonradiative deexcitation channels complement each other [5.118]. With diffraction gratings on the sample surface, instead of a prism coupler, the photothermal detection of plasmons was further enhanced [5.119] and even free standing Ag films were addressed [5.120]. The field enhancement at a silver surface due to resonant surface plasmon excitation has been observed [5.121] and the influence of Langmuir-Blodgett monolayer assemblies on the resonance has been studied in detail [5.122]. The attention in this field has been focused most recently on the properties of silver island films [5.123], on the influence of dye overlayers on these island films [5.124] and on the optical absorption of small, oblate silver spheroids [5.125]. After demonstrating that photoacoustic methods duplicate results obtained with other techniques, the field then developed into experiments that provide data that cannot be observed with any other spectroscopic tool.

5.6.2 Ferromagnetic Resonance

Ferromagnetic resonance, similar to conventional optical spectroscopy, allows one to tailor the excitation such that only one layer is excited in a photoacoustic or photothermal experiment. Optically opaque layers can be penetrated by microwaves and allow the excitation of a selected layer deep in the sample.

At first, microphones were employed to detect the ferromagnetic resonance with photoacoustic techniques [5.126]; photothermal deflection was explored next [5.127] and the effects of sinusoidal versus square wave modulation of the incident microwave power were compared [5.128]. Most recently, magnetic media were imaged with these techniques, demonstrating the power of this specialized tool for the three-dimensional mapping of defects in complex layered structures [5.129].

5.7 Conclusion

Using well-prepared and characterized test samples, photoacoustic and photothermal techniques have been able to establish credibility for the experimental methods and interpretation of the data. The thermal diffusion process makes these techniques unique in that it limits the access to a thin layer at the sample surface. This is a severe drawback with bulk samples. For thin films, however, thermal diffusion poses no major problems. It even allows the monitoring of subsurface features and the derivation of depth profiles. Due to the complex signal generation and detection process, the interpretation of photothermal and photoacoustic data is rarely unambiguous. Nevertheless, many of the applications are becoming established and in the foreseeable future photothermal and photoacoustic analysis might be able to complement conventional thin film analytical tools.

References

5.1   Y. H. Pao: *Optoacoustic Spectroscopy and Detection* (Academic, New York 1977)
5.2   A. Rosencwaig: *Photoacoustics and Photoacoustic Spectroscopy*, Chemical Analysis, Vol. 57 (Wiley, New York 1980)
5.3   V. P. Zharov, V. S. Letokhov: *Laser Optoacoustic Spectroscopy*, Springer Ser. Opt. Sci., Vol. 37 (Springer, Berlin, Heidelberg 1986)
5.4   A. Mandelis (ed.): *Photoacoustic and Thermal Wave Phenomena in Semiconductors* (Elsevier, Amsterdam 1987)
5.5   C. K. N. Patel, A. C. Tam: Rev. Mod. Phys. **53**, 517-550 (1981)
5.6   J. B. Kinney, R. H. Staley: Annu. Rev. Mater. Sci. **12**, 295-321 (1982)
5.7   G. A. West, J. J. Barrett, D. R. Siebert, K. V. Reddy: Rev. Sci. Instrum. **54**, 797-817 (1983)
5.8   A. C. Tam: Rev. Mod. Phys. **58**, 381-431 (1986)
5.9   H. Vargas, L. C. M. Miranda: Phys. Rep. **161**, 43-101 (1988)
5.10   Special Issue: IEEE Trans. UFFC-33 (5) (1986)
5.11   Special Issue: Appl. Phys. B **43**, 1 (1987)
5.12   Special Issue: Can. J. Phys. **64**, 9 (1986)
5.13   P. Hess, J. Pelzl (eds.): *Photoacoustic and Photothermal Phenomena*, Springer Ser. Opt. Sci., Vol. 58 (Springer, Berlin, Heidelberg 1987)
5.14   H. S. Carslaw, J. C. Jaeger: *Conduction of Heat in Solids* (Clarendon, Oxford 1960)
5.15   H. Coufal: In [5.10], pp. 507-512
5.16   R. Zenobi, J. H. Hahn, R. N. Zare: Chem. Phys. Lett. **150**, 361-365 (1988)
5.17   P. Baeri, S. U. Campisano, E. Rimini, J. P. Zhang: Appl. Phys. Lett. **45**, 398-400 (1984)
5.18   K. Tanaka, R. Satoh, A. Odajima: Jpn. J. Appl. Phys. **22**, 592-594 (1983)
5.19   P. E. Nordal, S. O. Kanstadt: Phys. Scr. **20**, 659-662 (1979)
5.20   R. T. Williams, M. N. Kabler, J. P. Long, J. C. Rife, T. R. Royt: In *Laser and Electron Beam Interactions with Solids*, ed. by B. R. Appleton, G. K. Keller (North-Holland, New York 1982) pp. 97-102.
5.21   B. Stritzker, A. Pospieszczyk, J. Tagle: Phys. Rev. Lett. **47**, 356-368 (1981)
5.22   I. Hussla, H. Coufal, F. Träger, T. Chuang: Ber. Bunsenges. Phys. Chem. **90**, 240-245 (1986)
5.23   M. Buck, B. Schafer, P. Hess: Surf. Sci. **161**, 245-254 (1985)
5.24   P. B. Comita, T. T. Kodas: Appl. Phys. Lett. **51**, 2059-2061 (1987)
5.25   A. Rosencwaig, J. Opsal, W. L. Smith, D. L. Willenborg: Appl. Phys. Lett. **46**, 1013-1015 (1985)
5.26   S. V. Bondarenko, E. V. Ivakin, A. S. Rubanov, V. I. Kabelka, A. V. Mikhailov: Opt. Commun. **61**, 155-158 (1987)
5.27   G. L. Eesley, B. M. Clemens, C. A. Paddock: Appl. Phys. Lett. **50**, 717-719 (1987)
5.28   D. Fournier, A. C. Boccara, N. M. Amer, R. Gerlach: Appl. Phys. Lett. **37**, 519-521 (1980)
5.29   M. A. Olmstead, N. M. Amer, S. Kohn: Appl. Phys. A **32**, 141-154 (1983)

5.30   For a review see for example J. P. Monchalin: In [5.10], pp. 485-499
5.31   A. C. Tam, C. K. N. Patel: Opt. Lett. **5**, 27-29 (198)
5.32   H. M. Frost: In *Physical Acoustics*, Vol. 14, ed. by W. P. Mason, R. N. Thurston (Academic, New York 1979) pp. 179-276
5.33   R. E. Lee, R. M. White: Appl. Phys. Lett. **12**, 12-14 (1968)
5.34   A. C. Tam, C. K. N. Patel: Appl. Opt. **18**, 3348-3358 (1979)
5.35   H. Coufal: J. Photoacoustics **1**, 417-428 (1984)
5.36   A. Mandelis: In [5.10], pp. 590-614
5.37   Y. Sugitani, A. Uejima, K. Kato: J. Photoacoustics **1**, 217-236 (1982)
5.38   P. R. Griffiths, J. A. de Haseth: *Fourier Transform Infrared Spectrometry* (Wiley, New York 1986)
5.39   H. Coufal: Appl. Opt. **21**, 104-109 (1982)
5.40   J. M. Bennett: Thin Solid Films **123**, 27-44 (1985)
5.41   K. Driss-Khodja, A. Gheorghiu, M.-L. Theye: Opt. Commun. **55**, 169-173 (1985)
5.42   T. Hata, T. Hatsuda, T. Komatsu: Jpn. J. Appl. Phys. **24**, 1463-1466 (1985)
5.43   G. P. Ceasar, M. Abkowitz, J.W.-P. Lin: Phys. Rev. B **29**, 2353-2355 (1984)
5.44   N. M. Amer, A. Skumanich, W. B. Jackson: J. Non-Cryst. Solids **59-60**, 409-412 (1983)
5.45   R. Mostefaoui, J. Chevallier, S. Meichenin, F. Auzel: J. Non-Cryst. Solids **77-78**, 307-310 (1985)
5.46   D. Jousse, E. Bustarret, F. Boulitrop: Solid State Commun. **55**, 435-438 (1985)
5.47   T. L. Chu, S. S. Chu, S. T. Ang, A. Duong, C. G. Hwang: J. Appl. Phys. **59**, 3122-3125 (1986)
5.48   M. Favre, H. Curtis, A. V. Shah: J. Non-Cryst. Solids **97-98**, 731-734 (1987)
5.49   K. Pierz, B. Hilgenberger, H, Mell, G. Weiser: J. Non-Cryst. Solids **97-98**, 63-66 (1987)
5.50   R. C. Van Oort, M. J. Geerts, J. C. Van Den Heuvel: J. Non-Cryst. Solids **97-98**, 1427-1430 (1987)
5.51   S. Guha, J. S. Payson, S. C. Agarwal, S. R. Ovshinsky: J. Non-Cryst. Solids **97-98**, 1455-1458 (1987)
5.52   A. H. Mahan, R. C. Kerns, G. Devaud: J. Electron. Mater. **12**, 1033-1049 (1983)
5.53   D. Jousse, J. C. Bruyere, E. Bustarret, A. Deneuville: Philos. Mag. Lett. **55**, 41-46 (1987)
5.54   J. S. Payson, R. C. Ross: J. Non-Cryst. Solids **77-78**, 579-582 (1985)
5.55   M. L. Theye, A. Gheorghiu, K. Driss-Khodja, C. Boccara: J. Non-Cryst. Solids **77-78**, 1293-1296 (1985)
5.56   C. H. Seager, J. A. Knapp: Appl. Phys. Lett. **45**, 1058-1059 (1984)
5.57   T. Hata, T. Hatsuda, T. Miyabo, S. Hasegawa: Jpn. J. Appl. Phys. Suppl. **25-1**, 226-228 (1986)
5.58   J. P. Roger, D. Fournier, A. C. Boccara: Thin Solid Films **128**, 11-20 (1985)
5.59   M. Fathalla, M. Ben Said, R. Bennaceur: Phys. Status Solidi A **99**, 521-526 (1987)
5.60   F. Coriand, H.-G. Walther, E. Welsch: Thin Solid Films **130**, 29-35 (1985)
5.61   O. Göde, W. Heimbrodt, F. Sittel: Phys. Staus Solidi A **93**, 277-282 (1986)
5.62   E. Welsch, H.-G. Walter, H. J. Kuhn: J. de Phys. **48**, 419-424 (1987)
5.63   B. R. Weinberger, C. B. Roxlo, S. Etemad, G. L. Baker, J. Orenstein: Phys. Rev. Lett. **52**, 86-89 (1984)
5.64   M. D. Porter, D. H. Karwcik, T. Kuwana, W. B. Theis, G. B. Norris, T. O. Tiernan: Appl. Spectrosc. **38**, 11-16 (1984)

5.65   W. Knoll, H. Coufal: Appl. Phys. Lett. **51**, 892-894 (1987)
5.66   J. K. Dohrmann, U. Sander: Ber. Bunsenges. Phys. Chem. **90**, 605-609 (1987)
5.67   P. Helander, I. Lundström: J. Appl. Phys. **52**, 1146-1151 (1981)
5.68   J. Pelzl, R. Grygier, H. Coufal: In *Progress in Basic Principles of Imaging Systems*, ed. by F. Grazer, E. Moisar (Vieweg, Braunschweig 1986) p. 695
5.69   H. Coufal: Appl. Phys. Lett. **45**, 516-518 (1984)
5.70   C. Starr: Rev. Sci. Instrum. **8**, 61-64 (1937)
5.71   M. J. Adams, G. F. Kirkbright: Analyst **102**, 678-682 (1977)
5.72   H. Coufal, P. Hefferle: Appl. Phys. A **38**, 213-219 (1985)
5.73   P. K. John, L. C. M. Miranda, A. C. Rastogi: Phys. Rev. B **34**, 4342-4345 (1986)
5.74   C. C. Ghinzoni, L. C. M. Miranda: Phys. Rev. B **32**, 8392-8394 (1985)
5.75   O. Pessoa, Jr., C. L. Cesar, N. A. Patel, H. Vargas, C. C. Ghinzoni, L. C. M. Miranda: J. Appl. Phys. **59**, 1316-1318 (1986)
5.76   W. P. Leung, A. C. Tam: Opt. Lett. **9**, 93-95 (1984)
5.77   R. E. Imhof, F. R. Thornley, J. R. Gilchrist, D. J. S. Birch: J. Phys. D **19**, 1829-1841 (1986)
5.78   A. Skumanich, H. Dersch, M. Fathalla, N. M. Amer: Appl. Phys. A **43**, 297-300 (1987)
5.79   J. P. Roger, F. Lepoutre, D. Fournier, A. C. Boccara: Thin Solid Films **155**, 165-174 (1987)
5.80   G. Benedetto, R. Spagnolo: Appl. Phys. A **46**, 169-172 (1988)
5.81   T. Hashimoto, J. Cao, A. Takaku: Thermochim. Acta **120**, 191-201 (1987)
5.82   A. Mandelis, E. Siu, S. Ho: Appl. Phys. A **33**, 153-159 (1984)
5.83   M. E. Abu-Zeid, A. E. Rakhshani, A. A. Al-Jassar, Y. A. Youssef: Phys. Status Solidi A **93**, 613-620 (1986)
5.84   R. Florian, J. Pelzl, M. Rosenberg, H. Vargas, R. Wernhardt: Phys. Status Solidi A **48**, K35-K38 (1978)
5.85   C. Pichon, M. Le Liboux, D, Fournier, A. C. Boccara: Appl. Phys. Lett. **35**, 435-436 (1979)
5.86   M. A. A. Sigueira, G. G. Ghinzoni, J. I. Vargas, E. A. Menez, H. Vargas, L. C. M. Miranda: J. Appl. Phys. **51**, 1403-1406 (1980)
5.87   P. S. Bechthold, M. Campagna, T. Schober: Solid State Commun. **36**, 225-231 (1980)
5.88   P. Korpiun, J. Baumann, E. Lüscher, E. Papamokos, R. Tilgner: Phys. Status Solidi A **58**, K13-K16 (1980)
5.89   A. Mandelis, F. Care, K. K. Chan, L. C. M. Miranda: Appl. Phys. A **38**, 117-122 (1985)
5.90   A. L. Glazov, S. B. Gurevich, N. N. Il'yashenko, N. P. Kalmykova, K. L. Muratikov, N. A. Rogachev: Sov. Tech. Phys. Lett. (USA) **12**, 59-60 (1986)
5.91   H. Coufal, W. Lee: Appl. Phys. B **44**, 141-146 (1987)
5.92   C. B. Scruby, R. J. Dewhurst, A. A. Hutchins, S. B. Palmer: In *Research Techniques in Nondestructive Testing*, Vol. 5, ed. by R. S. Sharpe (Academic, New York 1982) pp. 281-327
5.93   G. Birnbaum, G. S. White: In *Research Techniques in Nondestructive Testing*, Vol. 7, ed. by R. S. Sharpe (Academic, New York 1984) pp. 259-365
5.94   D. A. Hutchins: *Physical Acoustics*, Vol. 18, ed. by W. P. Mason, R. N. Thurston (Academic, New York 1986)
5.95   A. C. Tam, H. Coufal: Appl. Phys . Lett. **42**, 33-35 (1983)
5.96   G. M. Sessler, J. E. West, G. Gerhard: Phys. Rev. Lett. **48**, 563-566 (1982)
5.97   H. Sontag, A. C. Tam: Appl. Phys. Lett. **46**, 725-727 (1985)

5.98   H. Sontag, A. C. Tam: Can. J. Phys. **64**, 1330-1333 (1986)
5.99   G. L. Eesley, B. M. Clemens, C. A. Paddock: Appl. Phys. Lett. **50**, 717-719 (1987)
5.100   C. Thomsen, H. J. Maris, J. Tauc: Thin Solid Films **154**, 217-223 (1987)
5.101   A. Rosencwaig: J. Photoacoust. **1**, 371-386 (1983)
5.102   R. C. Frye, J. J. Kumler, C. C. Wong: Appl. Phys. Lett. **50**, 101-103 (1987)
5.103   H. G. Walther, E. Welsch: Thin Solid Films **142**, 27-35 (1986)
5.104   S. Aithal, G. Rousset, L. Bertrand: Thin Solid Films **119**, 153-158 (1984)
5.105   H. Coufal, R. Grygier, D. Horne, J. Fromm: J. Vac. Sci. Technol A **5**, 2875-2889 (1987)
5.106   G. Li, Q.-R. Yin, W.-G. Luo, Z.-W. Yin: Jpn. J. Appl. Phys. Suppl. **24**, 425-426 (1985)
5.107   G. M. Sessler, R. Gerhard-Multhaupt: Radiat. Phys. Chem. **23**, 363-370 (1984)
5.108   H. Coufal, U. Möller, S. Schneider: Appl. Opt. **21**, 116-119 (1982)
5.109   H. Coufal, U. Möller, S. Schneider: Appl. Opt. **21**, 2339-2343 (1982)
5.110   J. C. Murphy, J. W. Maclachlan, L. C. Aamodt: In [5.10], pp. 529-541
5.111   L. D. Favro, P.-K. Kuo, R. L. Thomas: In [5.4], pp. 69-96
5.112   W. C. Mundy, R. S. Hughes: Appl. Phys. Lett. **43**, 985-987 (1983)
5.113   M. Guardalben, A. Schmid: Phys. Rev. **35**, 4026-4030 (1987)
5.114   J. I. Burov, D. V. Ivanov: J. de Phys. **47**, 549-552 (1986)
5.115   A. Rosencwaig, J. Opsal: In [5.10], pp. 516-528
5.116   A. Rosencwaig: In [5.4], pp. 97-135
5.117   H. Taalat, H. D. Dardy: In *1983x Ultrasonics Symposium Proceedings*, ed. by B. R. McAvoy (IEEE, New York 1983) pp. 700-703
5.118   B. Rothenhäusler, J. Rabe, P. Korpiun, W. Knoll: Surf. Sci. **137**, 373-383 (1984)
5.119   T. Inagaki, M. Motosuga, E. T. Arakawa, J. P. Goudonnet: Phys. Rev. B **32**, 6238-6245 (1985)
5.120   T. Inagaki, M. Motosuga, E. T. Arakawa, J. P. Goudonnet: Phys. Rev. B **31**, 2548-2550 (1985)
5.121   C. S. Jung; G. Park, Y. D. Kim: Appl. Phys. Lett. **47**, 1165-1167 (1985)
5.122   R. K. Grygier, W. Knoll, H. Coufal: Can. J. Phys. **64**, 1067-1069 (1986)
5.123   T. Inagaki, J. P. Goudonnet, P. Royer, E. T. Arakawa: Appl. Opt. **25**, 3635-3639 (1986)
5.124   V. N. Rai: Appl. Opt. **26**, 2395-2400
5.125   P. Royer, J. P. Goudonnet, T. Inagaki, G. Chabrier, E. T. Arakawa: Phys. Status Solidi A **105**, 617-625 (1988)
5.126   C. L. Cesar, H. Vargas, H. Pelzl: J. Appl. Phys. **55**, 3460-3464 (1984)
5.127   U. Netzelmann, U. Krebs, J. Pelzl: Appl. Phys. Lett. **44**, 1161-1162 (1984)
5.128   M. Davies: J. Phys. D **18**, 1655-1663 (1985)
5.129   O. v. Geisau, U. Netzelmann, J. Pelzl: IEEE Trans. MAG-20, 1954 (1988)

6. Photothermal Characterization of Surfaces and Interfaces

A.C. Tam

With 15 Figures

Photothermal (PT) generation, referring to the heating of a sample and its surroundings by electromagnetic irradiation, can produce many effects, including refractive-index gradients, acoustic emission, surface deformation, reflectivity changes, desorption, phase change, ablation, and "gray-body" infrared emission, providing various techniques for materials characterization. Many methods of detecting these PT effects are possible, using "probe" laser beams, transducers, microphones, pyrometers, or infrared detectors. Several PT effects often occur simultaneously, depending on whether there is mainly surface heating (opaque sample) or mainly bulk heating (transparent sample), and on the nature of the coupling medium, if any. The choice of a suitable PT effect and the appropriate detection scheme depend on the nature of the sample and its environment, the light source used, and the measurement desired. This chapter describes the experimental arrangements and detection schemes for PT surface heating applied to surface and interface characterizations, and provides recent examples of such studies utilizing pulsed photoacoustic detection for imaging of layered structures and for detecting laser ablation, photothermal refraction of a probe beam to monitor pulsed PT desorption from a surface, and photothermal radiometry following pulsed heating of a layered structure for measuring the subsurface thermal contact resistance.

6.1 Photoacoustic Generation and Transducer Detection

Pressure variation or modulation resulting from the absorption of modulated light by a sample is usually called "photoacoustic" (PA) or "optoacoustic" (OA) generation. Photoacoustic generation mechanisms include electrostriction [6.1,2], thermoelastic expansion [6.3,4], volume changes due to photochemistry [6.5], gas evolution [6.6], boiling or ablation [6.7-11], and dielectric breakdown [6.12,13], with the PA generation efficiency η (i.e., acoustic energy generated/light energy absorbed) generally increasing in this order. For the usual case of thermal expansion (also called thermoelastic mechanisms), η is very small, often less than 10^{-10}, while for breakdown mechanisms, η can be as large as 30% [6.12]. Photothermal generation via thermoelastic expansion is the most common mechanism used in PA spectroscopy. Photoacoustic generation can be classified as either direct or indirect. In direct PA generation, the acoustic wave is produced in the sample where the excitation beam is adsorbed. In indirect PA generation, the acoustic wave is generated in a coupling medium adjacent to the sample, usually due to heat leakage and sometimes also due to acoustic transmission from the sample; here, the coupling medium is typically a gas or a liquid, and the sample is a solid or a liquid. Extensive reviews of the PA generation and detection for spectroscopy and other measurements have been given in the literature, e.g., [6.3] and [6.4]. Most earlier work emphasized the use of the PA technique as a convenient spectroscopic tool for highly transparent, opaque, or light-scattering samples. Subsequently, extensive PA imaging and microscopy work was also developed. The particular advantage of the PA generation method in producing a single ultra-narrow acoustic pulse with a well-defined pulse shape and no ringing has received less attention up to now. This has been proposed as a mechanism to generate a standard acoustic profile, since the pulsed PA emission shape can be theoretically modeled in detail. Here, we shall discuss some specific applications of narrow PA pulses detected by fast piezoelectric transducers: mapping of structures of thin layers and the study of laser-induced ablation of a surface.

6.1.1 Pulsed PA Imaging of Thin Layered Structures

An experimental arrangement to detect a nanosecond PA pulse shape is shown in Fig. 6.1. The width of a PA pulse in an opaque sample with "end-on" detection is determined [6.14] by both the laser pulse duration and the optical penetration depth; both of these parameters must be short to produce a narrow PA pulse. When such a PA pulse is produced in a thin layered structure, multiple reflections of the pulse are produced [6.15] due to acoustic impedance changes at each interface. The use of a sufficiently narrow single PA pulse makes it possible to distinguish the reflections from the various layers (ambiguities occur for a broad acoustic pulse when the echoes overlap). Multiple PA pulses observed with this arrangement for a coated ceramic plate (25 μm alumina coating on a 4 mm thick plate) are shown in Fig. 6.2. We see that the narrow PA pulse permits the "coating echoes" to be time-resolved when each "substrate echo" is expanded in high time resolution. By measuring these signals at various locations of the sample, PA echo images can be obtained for the substrate and coating separately, as indicated in Fig. 6.3. Note that nonuniformities in the thin coating can be separately detected from those in the substrate by this technique of short-pulsed PA imaging.

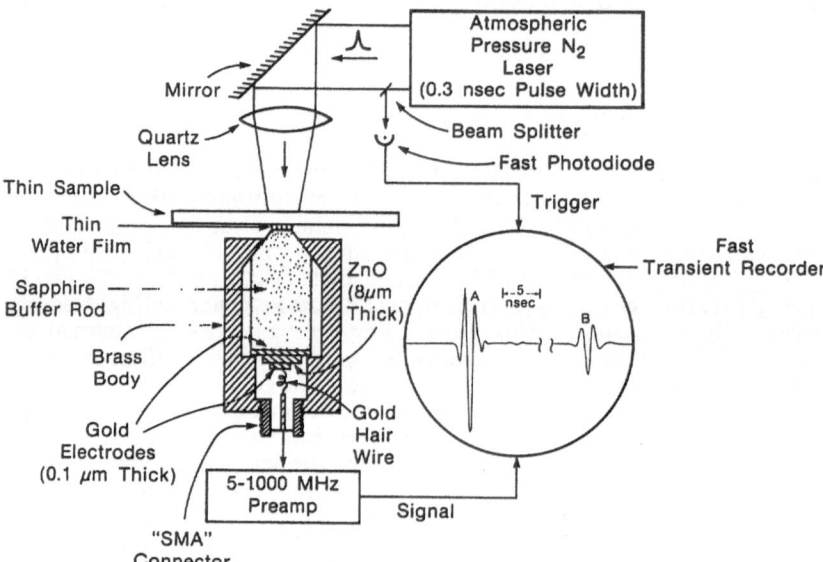

Fig. 6.1. Experimental setup for the generation and detection of single sub-nanosecond PA pulses in a sample. A ZnO transducer (bandwidth exceeding 2 GHz) for contact detection is shown. The transient recorder shows a signal of the PA pulse transmitted through a black ceramic plate of thickness 4 mm; the left peak is a delayed digitized signal of the first pulse A, and the right peak is a further-delayed digitized signal of the second pulse B (after a round trip reflection)

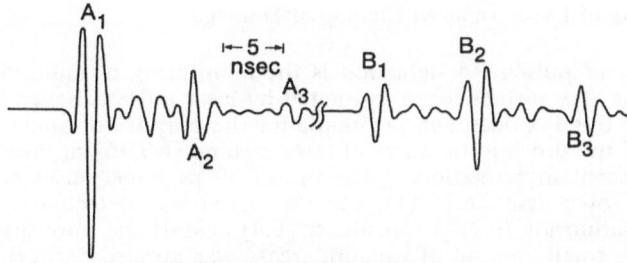

Fig. 6.2. Signal observed in the experimental arrangement of Fig. 6.1 where the sample is a coated 4 mm thick ceramic plate with a 25 μm alumina coating. The coating reflections cause each pulse in Fig. 6.1 to split into a sequence of "coating echoes" labeled A_1, A_2, A_3, ..., B_1, B_2, B_3, ..., etc.

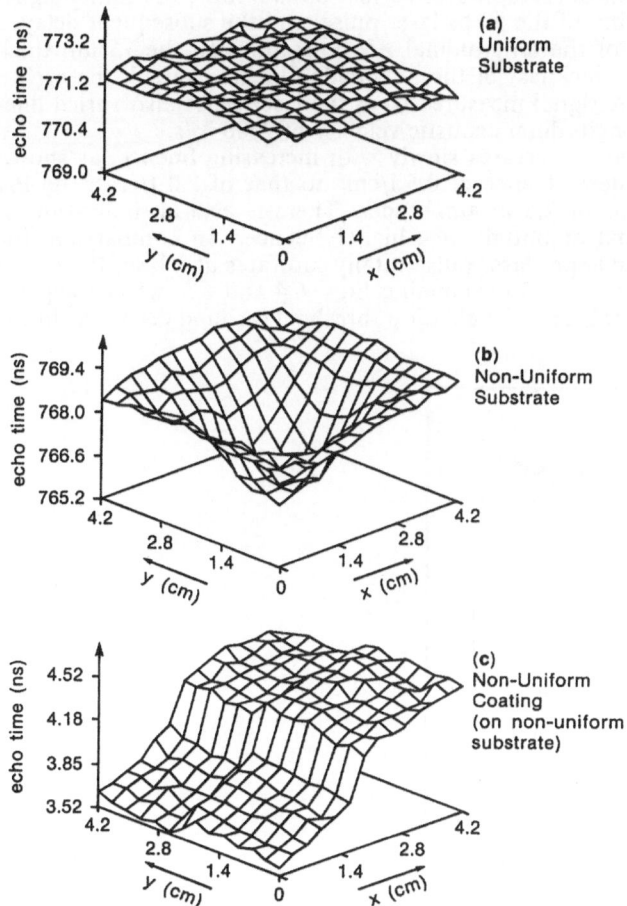

Fig. 6.3a-c. Maps of variation of longitudinal PΛ pulse echo times in nanoseconds (vertical axis) as a function of x, y location for coated black ceramic samples. (a) and (b) show the round trip echo times for the substrate only (unaffected by any nonuniformities in the coating), while (c) shows the single trip echo times for a nonuniform sputtered Al_2O_3 coating on the substrate in (b)

6.1.2 Pulsed PA Monitoring of Laser-Induced Etching or Damage

Another recent application of pulsed PA detection is the monitoring of induced etching or phase transitions at a surface when irradiated by laser pulses. Several authors [6.7-11] have used this PA detection technique for the purpose of understanding laser ablation and monitoring the onset of laser etching [6.7,16] or laser damage [6.10]. In our recent investigation of the use of 30 ps laser pulses at 355 nm to cause ablation of polyimide [6.11], the PA signal was detected by pressing a polyvinylidene difluoride (PVF_2) transducer [6.4] against the opposite side of the ablated spot. A small amount of vacuum grease was applied between the 75 μm thick polyimide film and the transducer to increase the acoustic coupling between them. The pressure between the two was also kept constant when different spots on the polyimide film were ablated. The acoustic signal detected by the PVF_2 transducer was amplified by a preamp (Trontech Model W1G2H), and integrated by a boxcar integrator (Stanford Research Systems Model SR 250). Figure 6.4 shows the fluence dependence of the photoacoustic signal as well as a real-time trace of the signal, observable from a fast oscilloscope; the initial signal spike occurs due to the firing of the 30 ps laser pulse, and the subsequent delayed peak is due to the arrival of the longitudinal PA pulse through the 75 μm thick polyimide film. When the thickness of the polyimide film was varied, the arrival time of this longitudinal PA signal measured from the laser pulse also varied linearly, corresponding to a longitudinal acoustic velocity of 2500 m/s.

The PA signal magnitude increases slowly with increasing fluence, as shown in Fig. 6.4. From an incident fluence of 0.5 J/cm^2 to that of 2.0 J/cm^2, the PA signal increases by a factor of 2.5 in amplitude. There is also an indication of leveling off in the PA signal amplitude at a higher fluence. In comparison, the ablation rate of the polyimide per laser pulse totally saturates at a laser fluence of ~ 1 J/cm^2, as shown in Fig. 6.5. By examining Figs. 6.4 and 6.5, we can suggest the following hypotheses: (a) near the ablation threshold (around 0.1 J/cm^2 in the

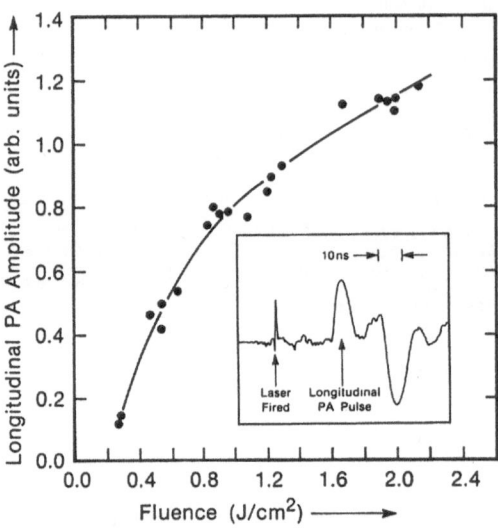

Fig. 6.4. Dependence of the longitudinal PA pulse magnitude on the fluence for a polyimide sample of thickness 75 μm excited by a 30 ps laser pulse at 355 nm. Each data point is derived from integrating the area of the longitudinal PA pulse signal detected by the PVF_2 transducer. A signal for a fluence of 2 J/cm^2 is shown in the inset

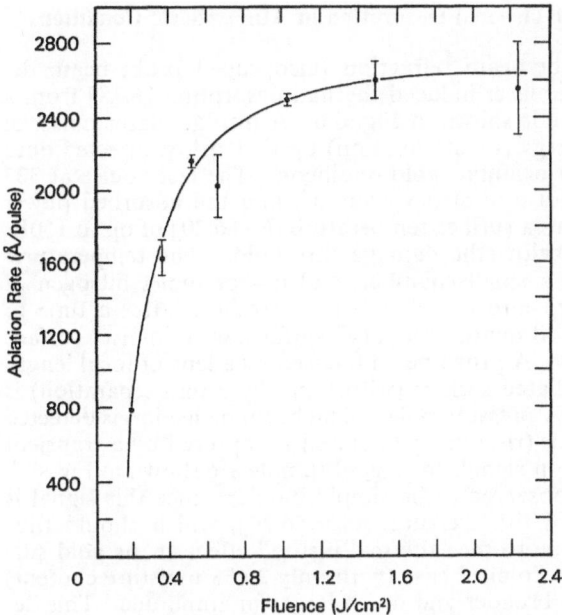

Fig. 6.5. Variation of the ablation rate of polyimide by laser pulses of 30 ps duration (full width at half maximum) at 355 nm. The curve drawn through the experimental point is for visualization only

present case), a large increase in the PA signal is produced due to the onset of ablation; (b) the etch rate saturates faster than the PA signal amplitude at a high fluence; assuming that the ablative PA signal magnitude is proportional to the total momentum of ablated materials, this suggests that the average emitted velocity increases while the emitted mass saturates at a higher laser fluence.

6.2 Photothermal Probe-Beam Refractions

Photothermal heating of a sample can produce a refractive index gradient (RIG) in the sample (direct effect) or in an adjacent "coupling fluid" (indirect effect), both of which can be monitored by a probe-beam refraction (PBR) scheme [6.17], which is also called mirage detection. Also, there are two types of RIG produced by the PT heating of the sample, namely, a "thermal RIG" and an "acoustic RIG." The thermal RIG (also called a "thermal lens") is produced by the decreased density of the medium (sample or coupling fluid) caused by the local temperature rise at the illuminated region of the sample. It decays in time following the diffusional decay of the temperature profile, and remains near the illuminated "source volume" for a stationary medium. The rapid expansion of the source volume also produces a propagating acoustic RIG (i.e., a PA wave) which travels at the acoustic velocity (for a stationary medium) away from the source volume and decays as it propagates due to the acoustic attenuation and angular spreading. By locating a probe-beam laser near a pulse-excited surface, certain specific surface properties or processes can be measured, for example, laser-induced surface effects like desorption [6.18], damage [6.10], or phase change. In addition, acoustic absorption and dispersion in a medium can be measured [6.19] by observing the probe beam deflection signal at various separations of the probe beam from the irradiated solid surface.

6.2.1 Detection of Laser-Induced Thermal Desorption in Atmospheric Conditions

An example of the use of probe-beam refraction (also called probe-beam deflection) schemes to detect pulsed laser-induced thermal desorption [6.18] from a solid surface in a gas atmosphere is shown in Fig. 6.6. A nitrogen laser pulse, of 8 ns duration and adjustable energy (by attenuation) up to 1 mJ, is directed onto a polished surface of a solid, for example, gold or silicon. The laser pulse at 337 nm is slightly focused into a spot size of 0.5×4 mm^2, and the adsorbed power density of up to 5 MW/cm^2 causes a surface temperature rise [6.20] of up to 150K. The pulse energy is kept well below the damage threshold. The temperature-controlled sample is situated in a small chamber, and dry or moist nitrogen at room temperature can be passed into the chamber to produce (after a time to reach steady state, typically several minutes) a "dry" surface, or a "moist" surface with adsorbed water, respectively. A probe beam focused by a lens of focal length 25 mm vertically above the irradiated surface (with typically 3 mm separation) is used to detect the shape of the PA pressure pulse. The beam deflection is detected by a fast avalanche photodetector (risetime <5 ns) and is captured by a transient digitizer. Some observed deflection signals for a gold sample are shown in Fig. 6.7; the signal for the "dry" case is observed to be simply bipolar, since this signal is proportional to the derivative of the pressure pulse [6.20], which should thus simply be a compression pulse due to the "thermal piston" effect at the gold surface. However, the signal for the "moist" case (with only 2.6% moisture content) in Fig. 6.7 is very different: it is broader and much larger in amplitude. This derivative signal suggests that the pressure pulse produced at the "moist" surface is composed of three components: (1) prompt compression due to a "thermal piston" effect of the gas adjacent to the gold surface; (2) prompt compression due to the desorption of the water molecules originally on the gold surface; (3) delayed rarefaction due to surface and gas cooling, and re-adsorption of water molecules, producing the delayed positive peak in Fig. 6.7.

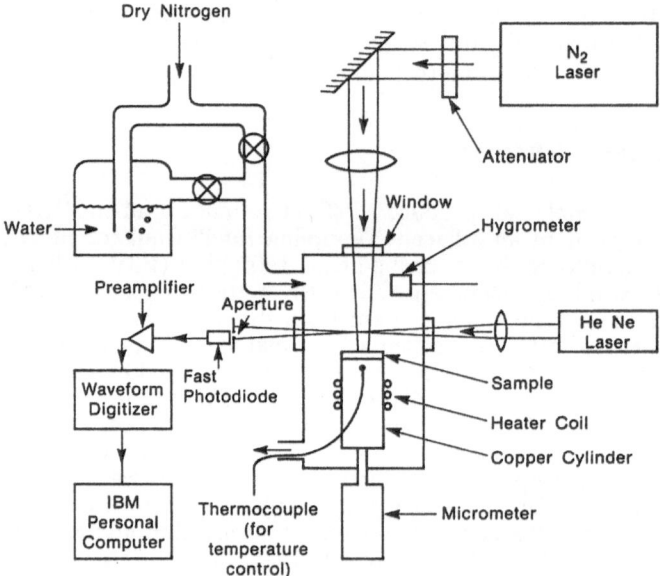

Fig. 6.6. Schematic experimental arrangement for measuring the photothermal desorption of water adsorbed on a solid surface in an ambient gas pressure of 1 atmosphere and room temperature

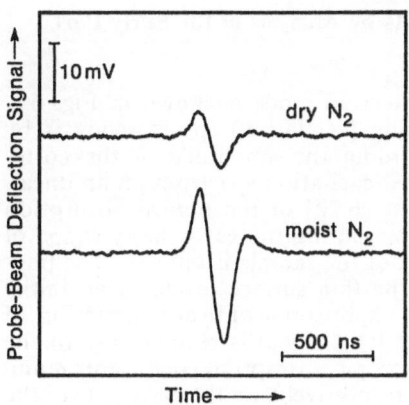

Fig. 6.7. The upper curve shows the PBR signal observed from a gold surface at 22°C exposed to 1 atmosphere of dry N_2 after steady state is reached; the lower curve shows the corresponding signal for moist N_2, i.e., 740 Torr N_2 mixed with 20 Torr of water vapor

6.3 Photothermal Radiometry

Photothermal radiometry (PTR) relies on the detection of variations in the infrared thermal radiation emitted from a sample excited by an electromagnetic "pump" beam (typically from a laser or from an arc lamp) with modulated intensity, wavelength, or position of illumination. A simple theory of PTR is given by *Nordal* and *Kanstad* [6.21]. The total radiant energy W emitted from a gray body of emissivity ε and absolute temperature T is given by the Stefan-Boltzmann law

$$W = \varepsilon\sigma T^4 , \tag{6.3.1}$$

where σ is the Stefan-Boltzmann constant. When a sample is irradiated by an optical pulse of energy E and wavelength λ that is absorbed by the sample with an absorption coefficient $\alpha(\lambda)$, resulting in a *small* temperature rise $\delta T(E, \alpha)$, the total radiant energy is increased by

$$\delta W(E, \alpha) = 4\varepsilon\sigma T^3 \delta T(E, \alpha) . \tag{6.3.2}$$

If $\delta T(E, \alpha)$ varies linearly with the absorbed energy αE, spectroscopic measurement is possible by defining the "normalized" PTR signal S as

$$S(\alpha) = \delta W(E, \alpha)/E . \tag{6.3.3}$$

An excitation spectrum called a PTR spectrum can be obtained by monitoring S for various excitation wavelengths λ.

In a typical PTR measurement, the excitation beam of photons (more generally, of some form of energy) is either continuously modulated, with about 50% duty cycle, or pulse modulated, with a low duty cycle and high peak power. The observation spot can, in principle, be anywhere on the sample; however, the IR emission is usually detected at the excitation spot in a backward direction (called "back-detection PTR") or from a spot that is "end-on" through the sample thickness from the excitation spot (called "transmission PTR"). Thus, there are four common modes of PTR in the literature [6.22], as classified according to the excitation mode (continuously modulated or pulsed) and the detection mode (transmission or back detection).

The mode of pulsed PTR with back-detection is of special interest here for surface and interface characterization because of its remote sensing capability and single-ended arrangement (where all instrumentations are on only one side of the sample), high detectivity, and quantitative depth-profiling capability.

6.3.1 Spectroscopy and Thickness Measurements by Analysis of the Early Part of the PTR Line Shape

A schematic of pulsed PTR with the back-detection mode is shown in Fig. 6.8. The most commonly known way of performing (excitation) spectroscopy is by scanning the excitation wavelength and recording the amplitude of the corresponding PTR signal. This can only provide an excitation spectrum on an uncalibrated scale. However, absolute measurement [6.22] of the optical absorption coefficient (or optical density) is possible by measuring the early decay shape of the pulsed PTR signal if the thermal diffusivity of the sample is known. The physical explanation is that the flash heating of the film surface produces an initial temperature gradient that corresponds to the exponential optical penetration of the opaque film; hence, the early decay of the PTR signal is related only to the spectroscopic properties. The effect of spectroscopic absorption coefficients α and α' at the excitation and detection wavelengths, respectively, on the early part of the pulsed PTR signal can be theoretically calculated as indicated in Fig. 6.9, and experimentally shown in Fig. 6.10. On the other hand, if the excitation penetration depth is constant for two samples of isolated opaque films with different thicknesses, the late decay shape of the pulsed PTR signals provides a single-ended measurement of the thickness, because the decay of the surface temperature of a thin isolated film is slowed down when the heat has diffused uniformly over the thickness of the film. This is shown in the theoretical calculation of Fig. 6.11 and experimentally demonstrated in Fig. 6.12. The pulsed PTR signal of an opaque isolated thin film ultimately decays back to zero due to convection, radiation cooling, and conduction along the film; this occurs on a time scale much longer than that in Fig. 6.12 for the conditions indicated.

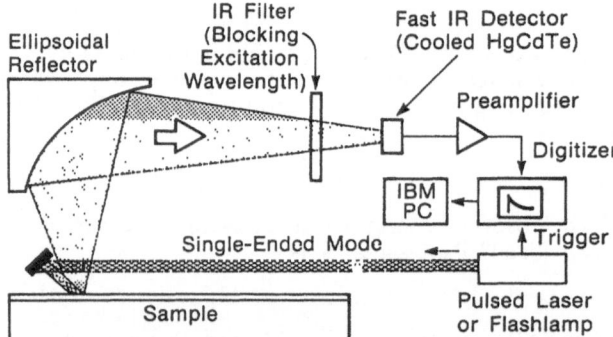

Fig. 6.8. Experimental scheme for pulsed PT radiometry of a sample with single-end back detection

6.3.2 Subsurface Air Gap and Contact Resistance Measurement by Analysis of the Late Part of the PTR Line Shape

If an opaque thin-film is not isolated, but is positioned near a substrate with a certain thermal contact resistance between the film and the substrate, the temperature decay of the flash-heated thin film will depend on this thermal contact resistance after a time delay corresponding to the thermal diffusion across the film, and can be much faster than is the case of the isolated film. This means that the pulsed PTR signal analysis after a suitable time delay will provide a quantitative measurement of subsurface thermal contact resistance [6.23] and air-gap thickness measurements [6.24]. (A *thin* air gap is a special case of an interface thermal

164

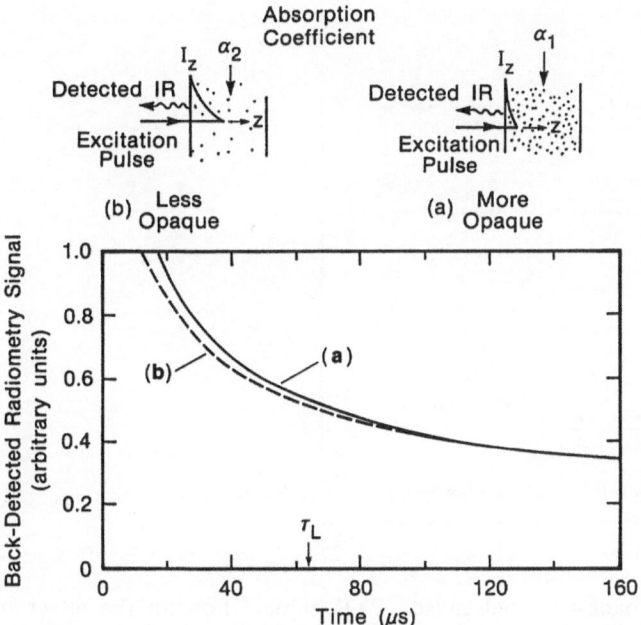

Fig. 6.9. Theoretical back-detected pulsed PTR signals for 2 samples of the same thickness $L = 50\ \mu$m and thermal diffusivity $D = 0.04\ \text{cm}^2/\text{s}$ but different optical absorption coefficients α and α' at the excitation and detection wavelengths, respectively: dashed line, $\alpha = 5 \times 10^3\ \text{cm}^{-1}$, $\alpha' = 2.5 \times 10^3\ \text{cm}^{-1}$; solid line, $\alpha = 2 \times 10^4\ \text{cm}^{-1}$, $\alpha' = 1 \times 10^4\ \text{cm}^{-1}$

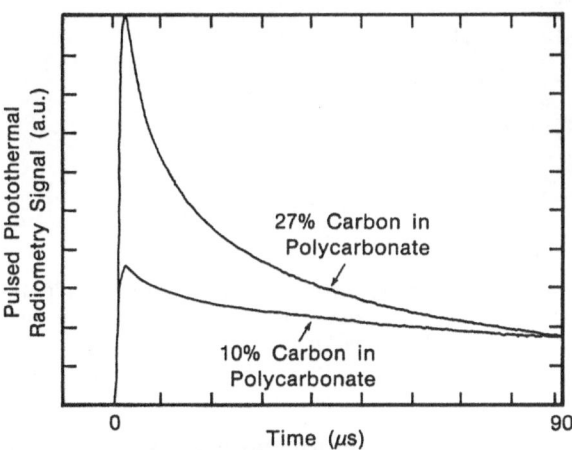

Fig. 6.10. Radiometry signals observed from two similar black polycarbonate films: one containing more amorphous carbon content (27%) and the other with less amorphous carbon content (10%). The difference in amplitude is mainly due to the smaller optical depth in the former case

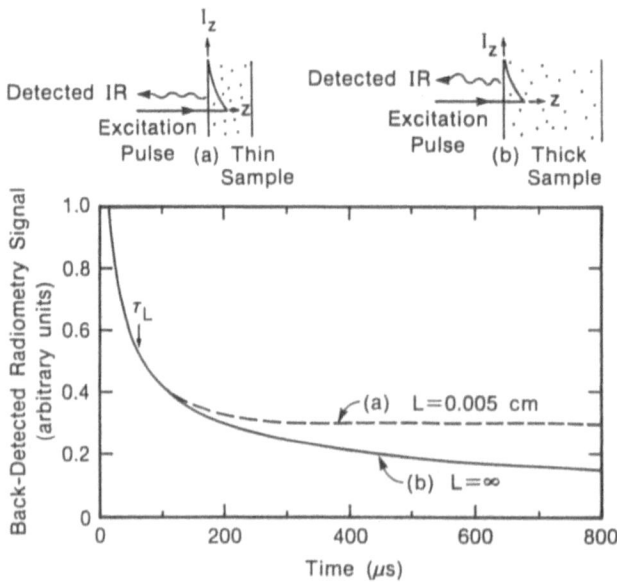

Fig. 6.11. Theoretical back-detected pulsed PTR signal showing the effect of sample thickness L: dash line $L = 50\ \mu m$; solid line $L = \infty$. In both curves, α, α' and D are taken as $2 \times 10^4\ cm^{-1}$, $1 \times 10^4\ cm^{-1}$ and $0.04\ cm^2/s$, respectively (typical values for stainless steel).

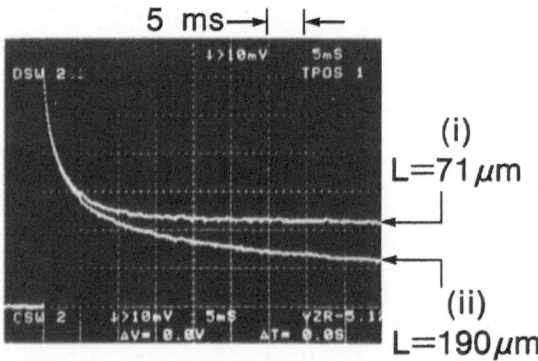

Fig. 6.12. Observed backscattering PTR signal for two black PVC films of identical composition but different thicknesses L. Curve (i), $L = 71\ \mu m$; curve (ii), $L = 190\ \mu m$. Horizontal time scale is 5 ms per division

contact resistance.) Specific applications of such measurements have been made by various authors [6.25,26]. It should be noted that subsurface thermal contact measurements have long been actively pursued by the PA and PT community, and various PA/PT techniques have been developed to monitor such thermal contacts and the related adhesion measurements. Earlier PA studies mainly relied on the use of a microphone detection of the PA signal. For example, *Helander* et al. [6.27] have examined the PA signals from a multilayered photographic emulsion excited by a sinusoidal-modulated laser beam, assuming perfect thermal contact;

this means that the PA signal (magnitude and phase) would change if the excitation beam is scanned over a region of imperfect thermal contact. The dependence of the PA signal on the "thermal resistance" between layers has been studied by *Monchalin* et al. [6.28]. For the simplest case of a thin film on a thick substrate, the thermal contact can be monitored either by a modulated light beam absorbed only by the thin layer as performed by *Nordal* and *Kanstad* [6.29]) or by a light beam absorbed only by the substrate as demonstrated by *Kirkbright* and co-worker [6.30]).

The advantages of the present decay-shape analysis of pulsed PTR signal for subsurface thermal resistance measurement compared to earlier PA techniques are the following: (a) The sample need not fit in a PA cell so that large or inaccessible samples can be measured. (b) The pulsed PTR measurement, done in the time domain, can be performed faster and easier than the earlier PA measurement, done in the frequency domain. (c) Pulsed PTR measurement provides a *quantitative* determination of an air gap thickness or contact resistance value.

An illustrative description of pulsed PTR for measuring a subsurface air gap or thermal contact resistance between an opaque film and a substrate is given in Fig. 6.13. At a delay time much longer than the thermal diffusion time across the film, the surface temperature of the film is sensitive to the subsurface thermal contact resistance. At such delay times, the temperature across the film is almost uniform, while there is a substantial temperature drop across the air gap that now controls the signal decay from the front surface of the film. This problem can be solved mathematically and the results for a one-dimensional case with flash heating by a laser pulse of a spatial width large compared to the film and gap thicknesses is shown in Fig. 6.14. The experimental pulsed PTR signal observed from the front surface is also shown for comparison. It is seen that for an air gap thickness exceeding 15 μm the theoretical shape and the experimental shape agree well. Below 15 μm, and especially at "contact," the theory neglecting film roughness and thermal contact resistance does not agree with experiment, since these parameters dominate the signal decay shape.

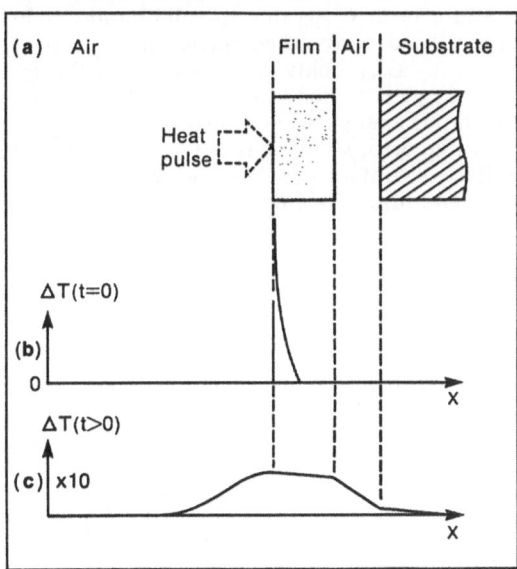

Fig. 6.13. Temperature distributions in the vicinity of an opaque film at time $t = 0$ (i.e., immediately after the short excitation laser pulse) and at a later time when the temperature is uniform throughout the thickness of the film

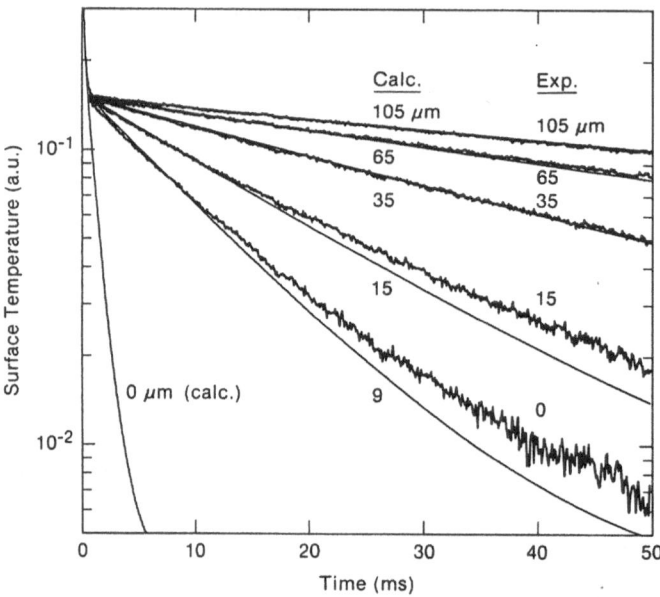

Fig. 14. Comparison between the calculated and experimental PPTR signal I_R for several airgap thicknesses. The maximum surface temperature is normalized to be unity

This leads us to try to understand what really happens when a film "makes contact" with a substrate due to a certain "contact pressure P." In this situation, the interfacial contact conductance is composed of two contributions, a term C_{gas} due to the interfacial gas conduction and a term C_{solid} due to solid contact conduction. Fortunately, C_{gas} and C_{solid} can be deconvoluted since only C_{gas} is affected by a gas change at constant P, with C_{gas} being proportional to the gas thermal conductivity for most cases, and C_{solid} is varied only when P is changed.

The pulsed PTR signals for various interfacial gases at constant contact pressure (hence constant C_{solid}) are shown in Fig. 6.15, showing that C_{gas} is significant; by repeating these measurements at different contact pressures, we can derive the contributions C_{gas} and C_{solid} for various conditions of interfacial gas and contact pressure.

6.4 Conclusions

While PA and PT techniques have many applications in numerous fields of science, engineering and medicine, one specific area covered here is the application for the characterization of surfaces and interfaces. Examples are given here using contact detection with transducers and noncontact detection with probe beams or radiometry. Measurements exemplified include the mapping of thin layered structures (i.e., locating acoustic impedance mismatch interfaces), studying laser-induced surface desorption or etching, spectroscopy and thickness measurement of an opaque surface film, and quantitative determination of subsurface air gap or contact resistance. This chapter is only meant to be indicative, not exhaustive. More information on the use of PT and PA techniques for surface and interface characterization can be found in the research papers and reviews cited here, as well as in other contributions to this volume, especially Chap. 5. Other significant

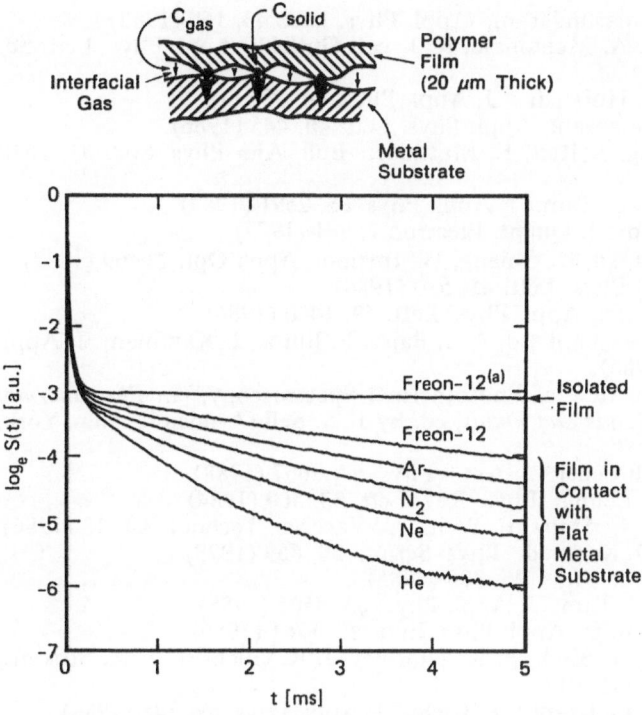

Fig. 6.15. Typical pulsed PTR signal $S(t)$ for a carbon-loaded polycarbonate film pressed against a polished brass surface at an applied contact pressure of 55 kPa in various gas atmospheres, except for the top signal, labelled (a), which is for the case of an isolated film (obtained when Freon-12 is used and the film is approximately 2 mm away from the substrate)

examples in this field include probe deflection monitoring of PT surface deformation for surface spectroscopy [6.31], transient thermal-reflectance and piezo-reflectance of a surface after PT heating with picosecond time resolution [6.32], photoacoustic generation of surface elastic waves [6.33] and its spectroscopic application [6.34], photothermal beam-deflection microscopy [6.35], and PA characterizations of plasma-modified surfaces [6.36]. Many quantitative calculations of PT heating of a surface are also reported in the literature, for example [6.37 and 6.38].

Acknowledgments: The author greatly appreciates the valuable contributions and suggestions of Drs. Wing-Pun Leung, Holger Schroeder, Heinz Sontag, Mei-Chen Chuang, Frank Träger, Eckart Matthias, and Hans Coufal.

References

6.1 H. M. Lai, K. Young: J. Acoust. Soc. Am. **72**, 2000 (1982)
6.2 J. M. Heritier: Opt. Commun. **44**, 267 (1983)
6.3 A. Rosencwaig: *Photoacoustics and Photoacoustic Spectroscopy* (Wiley, New York 1980)
6.4 A. C. Tam: Rev. Mod. Phys. **58**, 381 (1986)
6.5 G. J. Diebold, J. S. Hayden: Chem. Phys. **49**, 429 (1980)

6.6   P. Ganguly, T. Somasundaram: Appl. Phys. Lett. **43**, 160 (1983)
6.7   R. W. Dreyfus, F. A. McDonald, R. J. von Gutfeld: Appl. Phys. Lett. **50**, 1491 (1987)
6.8   A. P. Ghosh, J. E. Hurst, Jr.: J. Appl. Phys. **64**, 287 (1988)
6.9   P. E. Dyer, R. Srinivasan: Appl. Phys. Lett. **48**, 445 (1986)
6.10   S. Petzoldt, A. Elg, J. Reif, E. Matthias: Bull. Am. Phys. Soc. 33, 1638 (1988)
6.11   M. C. Chuang, A. C. Tam: J. Appl. Phys. **65**, 2591 (1989)
6.12   V. S. Teslenko: Sov. J. Quant. Electron 7, 981 (1977)
6.13   A. C. Tam, W. Zapka, K. Chiang, W. Imaino: Appl. Opt. **21**, 69 (1982)
6.14   A. C. Tam: Appl. Phys. Lett. **45**, 510 (1984)
6.15   A. C. Tam, G. Ayers: Appl. Phys. Lett. **49**, 1420 (1986)
6.16   J. H. Brannon, J. R. Lankard, A. I. Baise, F. Burns, J. Kaufman: J. Appl. Phys. **58**, 2036 (1985)
6.17   A. C. Tam: "Overview of Photothermal Spectroscopy," in *Photothermal Investigations of Solids and Fluids*, ed. by J. A. Sell (Academic, New York, 1988) Chap. 1
6.18   A. C. Tam, H. Schroeder: J. Appl. Phys. **64**, 3667 (1988)
6.19   A. C. Tam, W. P. Leung: Phys. Rev. Lett. **53**, 560 (1984)
6.20   D. Burgess, Jr., P. C. Stairs, E. Weitz: J. Vac. Sci. Technol. A4, 13 (1986)
6.21   P. E. Nordel, S. O. Kanstad: Phys. Scripta **20**, 659 (1979).
6.22   A. C. Tam: Infrared Phys. **25**, 305 (1985)
6.23   W. P. Leung, A. C. Tam: J. Appl. Phys. **63**, 4505 (1988).
6.24   A. C. Tam, H. Sontag: Appl. Phys. Lett. **49**, 1761 (1986)
6.25   R. E. Imhof, D. J. S. Birch, F. R. Thornley, J. R. Gilchrist, T. A. Strivens: J. Phys. E 17, 521 (1984)
6.26   D. L. Balageas, J. C. Krapez, P. Cielo: J. Appl. Phys. **59**, 348 (1986)
6.27   P. Helander, I. Lundstrom, D. McQueen: J. Appl. Phys. **52**, 1146 (1981)
6.28   J. P. Monchalin, J. L. Parpal, L. Bertrand, J. M. Gagne: Appl. Phys. Lett. **39**, 391-3 (1981)
6.29   S. O. Kanstad, P. E. Nordal: Applic. Surf. Sci. **5**, 286 (1980)
6.30   M. J. Adams, G. F. Kirkbright: Analyst (London) **102**, 678 (1977)
6.31   M. A. Olmstead, W. M. Amer: Phys. Rev. Lett. **52**, 1148 (1984)
6.32   G. L. Eesley, B. M. Clemens, C. A. Paddock: Appl. Phys. Lett. **50**, 717 (1987)
6.33   R. E. Lee, R. M. White: Appl. Phys. Lett. **12**, 12 (1968)
6.34   S. R. J. Brueck, T. F. Deutsch, D. E. Oates: Appl. Phys. Lett. **43**, 157 (1983)
6.35   N. J. Dovichi, T. G. Nolan, W. A. Weimer: Anal. Chem. **56**, 1700, 1704 (1984)
6.36   B. K. Bein, J. Pelzl: In *Plasma Diagnostics: Surface Analysis and Interaction* (Academic, New York 1988)
6.37   R. J. Dewhurst, D. A. Hutchins, S. B. Palmer, C. B. Scruby: J. Appl. Phys. **53**, 4064 (1982)
6.38   U. Schleichert, K. J. Langenberg, W. Arnold, S. Fassbender: *Review of Progress in Quantitative NDE*, ed. by D. O. Thompson, D. E. Chimenti, Vol. 8A (Plenum, New York 1989) p. 489

7. Spectroscopic Depth Profiling Using Thermal Waves

Richard M. Miller

With 16 Figures

Spectroscopy can be used to gain information on the identity and concentration of chemical species present in a sample. Unfortunately, it is rarely possible to obtain the equivalent information on the spatial distribution of a chromophore. There are many samples where such information would be extremely valuable, ranging from the study of deposition processes, through chemical modification of surfaces to membrane transport. Spatial information in the x-y plane normal to the incident radiation can be readily obtained by the use of imaging detectors [7.1,2] or scanned spot systems [7.3,4]. Whether information can be obtained in the z direction through the sample thickness depends on the scale over which information is required. If selective information is required on the surface layers over a range of angstroms to nanometers, then techniques from conventional surface science can be used; for example photoelectron spectroscopy and high resolution electron energy loss spectroscopy [7.5]. For penetration depths of a micrometer or so, evanescent wave methods such as attenuated total reflection (ATR) may be suitable [7.6]. For scales of greater than a few hundred micrometers, the only practical techniques are destructive, involving sectioning of the sample to convert it into a thin planar geometry which can be examined by the methods described above. In the interesting region of a few micrometers to a few hundred micrometers, photoacoustic and photothermal methods offer the possbility of performing real nondestructive depth profiling on a wide range of sample types.

The ability of thermal wave techniques to provide information on the variations of chromophore concentration or thermal properties with depth arises naturally from the mechanism of generation of the thermal waves. In a condensed phase sample, the energy source pumping the generation of the thermal waves is the decay of spectroscopically excited states through internal conversion. Since this process is extremely fast compared to the rate of propagation of the thermal wave, the spatial distribution of the heat source arising within the sample from the incident radiation is governed by the optical properties of the sample, and the temporal profile of the heat source is governed by the amplitude modulation of the incident radiation. The relaxation of the system towards equilibrium will be governed by the thermal properties of the sample.

Figure 7.1 illustrates the effect of the optical properties of a sample on the generation of thermal waves. The figure represents a one-dimensional view of a sample being illuminated from the direction indicated by the arrow. The solid curve shows the optical density of the sample at the wavelength of illumination, and the dotted curve the amount of energy dumped into the sample per unit time at that point, as a result of absorption of the incident radiation. In Fig. 7.1a the optical

171

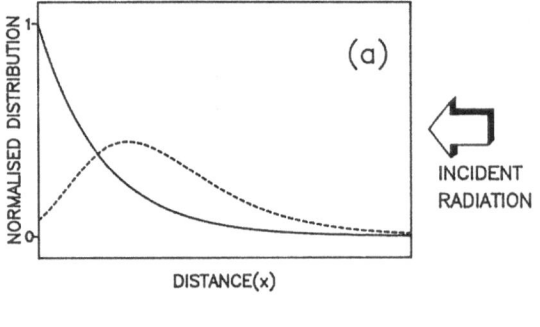

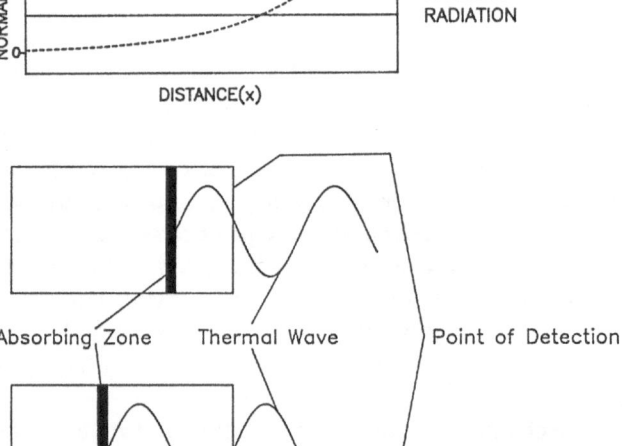

Fig. 7.1a,b. Effect of variations in chromophore distribution on heat source distribution: (—) optical density, (· · ·) heat input per unit time. (a) Optical density increasing exponentially with depth. (b) Isotropic optical density with same average value as (a)

Fig. 7.2. Detection of optical features at different depths in samples using the propagation delay of the thermal waves

density increases exponentially in the direction away from the radiation source. In Fig. 7.1b the optical density is isotropic, and is equal to the average optical density in Fig. 7.1a. Although the amount of energy released in the sample is the same, the spatial distribution of the heat sources are clearly different in the two cases, and the resulting thermal wave signature should reflect these differences. Thus, by carefully examining the thermal wave response of a sample, it should be possible to infer the spatial distribution of the chromophores in the sample.

A very simple example of depth profiling and depth discrimination in thermal wave spectroscopy is shown in Fig. 7.2. Two samples contain strongly absorbing features at different depths below the surface, illuminated by the amplitude modulated radiation. The thermal source produced by these features will track the modulation of the incident radiation accurately as the excited state lifetimes in condensed phases are very short. We therefore have two identical sources of thermal waves displaced in space by different amounts relative to the illuminated surface of the samples. If we are detecting the thermal waves at the illuminated surface, for example with a gas microphone photoacoustic cell or a photothermal beam deflection system, then there will be a detectable time delay between generation of the thermal waves and their detection, due to their propagation through the intervening material (for clarity, the damping of the thermal waves has been omitted). This propagation time will be detected as a phase shift if the experiment is being carried out in the frequency domain, and as a time delay if the experiment is being carried out in the time domain. Since the thickness of material through which the thermal waves must propagate is different for the two samples, the detected thermal wave signals will show a difference in phase shift or time delay. If the thermal properties of the matrix are known, then for such highly localised sources, a location in space can be quantitatively determined with respect to the point at which the thermal waves are detected.

The possibility of carrying out depth profiling using thermal wave spectroscopy has always appeared attractive, and many different applications have been proposed or demonstrated, (see Sect. 7.3). Potentially, the most useful feature of the technique is its capability for performing nondestructive measurements. The sample can retain its integrity after examination. This means that it is possible to apply other measurement tools to the sample to further characterise it without having to worry about its properties changing during the thermal wave experiment. Samples can also be analysed using thermal wave spectroscopy and then subjected to further treatments before re-analysis to investigate how those treatments have changed the sample. This is of particular interest in areas such as the weathering of coatings [7.7].

The technique is also particularly suitable for the characterisation of physically complicated systems. Thermal wave spectroscopies are capable of characterising materials of both very low and very high optical density, light scattering and specularly reflecting materials. The technique is therefore particularly applicable to studying the structure of materials such as layered composites, semiconductors, plant and animal tissues and membranes.

Finally, because of the nondestructive nature of the technique, it can be used to monitor systems that are slowly changing with time. This feature has been of particular interest in looking at the diffusion of chromophores through membranes, and the study of chemical reactions which modify the chromophore distribution.

It has frequently been claimed that depth profiling is a unique characteristic of thermal wave spectroscopies. Although it is largely true that other spectroscopic methods of characterisation do not offer the same degree of both spectral and spatial discrimination, there is now at least one other technique which has the potential to challenge thermal wave spectroscopies in the area which they have claimed as their own: confocal microscopy.

Confocal microscopy uses a novel optical arrangement to provide information from a shallow image plane, which can be located within a transparent or semitransparent sample [7.8,9]. In conventional microscopy, it is possible to produce an in-focus image plane which is very thin; however, light originating from points outside the focal plane can still enter the optical train and contribute to the overall image produced. In micriscopy, this is seen as a clouding of the image, which reduces its

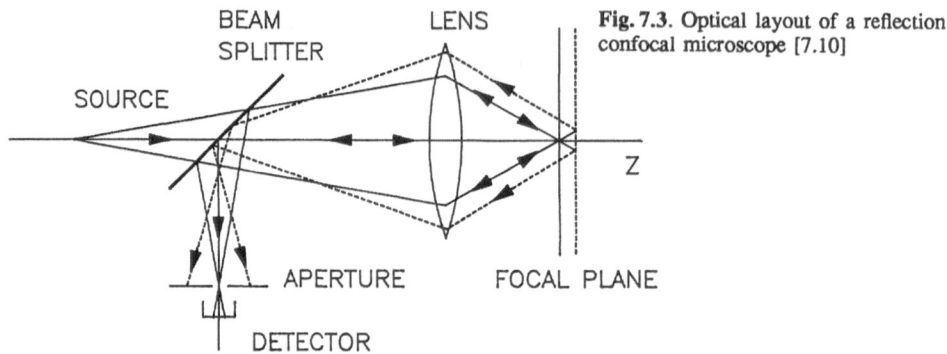

Fig. 7.3. Optical layout of a reflection confocal microscope [7.10]

SOURCE

BEAM SPLITTER

LENS

APERTURE

DETECTOR

FOCAL PLANE

Z

contrast and eventually causes it to disappear. In micro-spectroscopy, the inability to discriminate light from the image plane from light from other parts of the sample has made it impossible to obtain quantitative information from complex systems. Confocal microscopy gets round this problem in an elegant way, as illustrated in Fig. 7.3 [7.10].

This diagram shows the configuration for reflectance microscopy, but the principle is very similar for transmission microscopy. Light from a point source is focused onto the image plane by the objective lens. Reflected light is gathered by the same lens, and directed by a beam splitter onto a small aperture confocal with the focal spot in the image plane. Light scattered or reflected from planes above or below the image plane will not focus onto the aperture and will be selectively rejected. By scanning the incident beam or the sample, a two-dimensional image can be built up.

Confocal microscopes have a very high level of discrimination against light from outside the image plane, and have shown themselves to be capable of providing high quality images from significant depths below the surface of highly light scattering materials. Initially developed in the field of material science, confocal microscopy is now finding its widest application in the biological sciences. By coupling the confocal microscope with a spectrophotometric detection system, it may be possible to construct wavelength selective images of any thin slice through a semitransparent sample. By collecting and analysing a series of such slices, it should be possible to reconstruct the pattern of absorption at any particular wavelength in three dimensions.

Although confocal microscopy has a great deal of potential in spectroscopic depth profiling, it has yet to be extensively applied. In the future we may expect to be able to directly compare confocal microscopy with thermal wave methods.

This chapter reviews the theoretical treatments available for thermal wave depth profiling, the experimental methods, and the applications that have been reported in the literature. Throughout the chapter certain assumptions have been made to reduce the range of options that must be described. It has been assumed that the optical and thermal properties of the sample are static on the time scale of the thermal wave measurements. Dynamic systems will be treated, but only those where the rate of change is slow compared to the time taken to make a thermal wave observation. The same restriction also limits consideration to solid and semisolid samples. Any mixing within the system, for example as a result of convection, should be minimal on the time scale of the experiment. Similarly, the thermal properties are assumed to be constant within the time scale of the experiment. For example, systems operating very close to a phase transition have not been considered, as the extra thermal input from the absorption of the incident radiation can cause dramatic structural changes, with consequent changes in the physical properties.

A whole class of very interesting thermal wave studies have therefore been excluded, but the remaining systems cover the vast majority of sample types where depth profiling studies can provide useful information.

7.1 Theory

In order to obtain the maximum information from a thermal wave depth profiling experiment, it is necessary to construct a theoretical model that predicts the thermal wave response from a sample with optical and thermal properties which vary arbitrarily with spatial location in the sample. Ideally, the theoretical model should be capable of providing a full three-dimensional treatment of the generation and detection of the thermal wave signal. However, for many samples of practical importance, the sample diameter is very much greater than the sample thickness. Under these circumstances, the edge effects are relatively small, and the system may be adequately modelled using a one-dimensional formulation. The bulk of the analyses carried out have therefore concentrated on the one-dimensional case, with extensions to three dimensions. Comparison of theoretical with experimental data for the high sample diameter to thickness ratios typical of spectroscopic experiments suggests that the errors introduced by the one-dimensional formulation are small.

The problem to be solved is to calculate the variation in temperature with time for a sample containing a heat source defined by the optical absorption properties of the sample and the modulation waveform applied to the incident radiation. The heat sources in the sample can be simply described from a knowledge of the sample optical properties, heat capacity and the modulation waveform. To this must be added the relaxation of the perturbed system towards equilibrium by conduction into the rest of the system from the heat sources. Heat conduction in the sample is controlled by the Fourier heat conduction equation [7.11]:

$$J = -k\nabla T \,, \tag{7.1.1}$$

where k is the thermal conductivity, J the heat flux and T the temperature.

7.1.1 Analytical Approaches

One of the most widely used and sucessful theories for describing the generation of the thermal wave signal in the specific case of photoacoustic detection is the *Rosencwaig-Gersho* (RG) theory [7.12,13]. This model assumes a sample that is isotropic, both in thermal and in optical properties. The model assumes sinusoidal modulation of the incident radiation, and calculates the sinusoidal component of the fluctuation in temperature at the sample gas boundary, and from this the ac acoustic signal. The physical pressure variation $\Delta p(t)$ in the cell is given by

$$\Delta p(t) = Q_1 \cos\left(\omega t - \frac{\pi}{4}\right) - Q_2 \sin\left(\omega t - \frac{\pi}{4}\right) , \qquad (7.1.2)$$

where Q_1 and Q_2 are the real and imaginary parts of a function Q, which specifies the complex envelope of the sinusoidal pressure variation. The expression for Q is relatively complex, and analytical solutions for the variation of amplitude and phase with frequency of the photoacoustic signal have been obtained for several special cases [7.13]. These special cases have found very wide application in phenomenological explanations of experimental data in photoacoustic studies. The special cases can also be applied to depth profiling studies, but are limited to certain simple geometries. An early example of this was a demonstration by *Adams* and *Kirkbright* [7.14] that it is possible to measure the thickness of a thin nonabsorbing layer on an opaque substrate by measuring the increased phase lag introduced by the coating.

Using the full version of the RG theory, *Rosencwaig* [7.15] showed how the photoacoustic signal depends on the relationship between the thermal diffusion length and the sample thickness for transparent samples, or the optical path length for opaque samples. In particular, a log-log plot of signal magnitude versus normalised length shows a distinct kink as the thermal diffusion length passes through the optical path length. The phase angle of the detected signal shows a dramatic shift at the same point.

As it is conventionally formulated, the RG theory cannot be applied to spatially nonhomogeneous optical and thermal properties. However, by assuming that the sample is composed of a number of homogeneous layers of finite thickness, it is possible to predict the photoacoustic signal by obtaining the solution from RG theory for each of the layers and then matching the optical and thermal boundary conditions. By the use of a sufficiently large number of layers, any practical system could be modelled. However, the calculations would become extremely difficult to perform, and solutions using more direct time and space discretisation may be preferred, as will be shown below.

Using the same initial approach, *Afromowitz* et al. [7.16] addressed the problem of modelling a system where the optical absorption varies continuously with distance from the surface, but where the sample is thermally homogeneous. The authors defined a thermal forcing function $H(x)$, which could be related to the absorption coefficient function $\beta(x)$ by

$$\beta(x) = H(x) \left(\frac{(1-R)}{2k} I_0 \eta_0 - \int_0^x H(y) dy\right)^{-1} , \qquad (7.1.3)$$

where R is the optical reflection coefficient of the solid, I_0 the incident power density, k the thermal conductivity, and η the conversion efficiency of the de-excitation process.

Since the photoacoustic signal at a particular frequency can be directly related to the ac variation of the sample surface temperature at that frequency [7.12], if the spatially dependent thermal forcing function $H(x)$ can be deduced from the surface temperature fluctuations, then the absorption coefficient function $\beta(x)$ can be calculated from the frequency dependence of the photoacoustic signal. From the analysis, it emerges that the frequency-dependent surface temperature response of a solid is directly related to the thermal forcing function through the Laplace transform. If the photoacoustic signal for an unknown sample is recorded at a number of modulation frequencies, it should be possible to estimate the optical absorption distribution by inverting the Laplace transform. The inversion is accomplished by making a polynomial fit through the data points available. This process is prone to errors in the presence of noise; however, the authors reported that variations of approximately 1 % in the magnitude of the photoacoustic signal and 1° of the phase did not appear to affect the inversion results appreciably.

A completely different approach to the problem of providing a theoretical framework for the understanding of thermal wave depth profiling has been taken by *Mandelis* [7.17]. This model is formulated from the analogy between classical mechanical plane wave propagation and thermal wave motion. This analysis shows that a Hamilton–Jacobi formulation of thermal wave physics can be produced, which explains the thermal wave field behavior in terms of a thermal harmonic oscillator subject to a restoring conservative force generated by the effective harmonic potential field. Further, by replacing all the classical variables of the Hamilton–Jacobi theory with thermal wave quantum mechanical operators analogous to those in tradition quantum theory, a thermal wave quantum theory can be produced that yields equations for macroscopic thermal wave phenomena similar to those derived by classical theory.

A number of insights into the thermal wave process can be obtained from this formulation. For example, there is a constant that relates the energy of the thermal wave packets to their angular frequency in an analogous fashion to Planck's constant for electromagnetic radiation. Also, the thermal wave vector k_{th} has an imaginary component, which is responsible for the exponential attenuation of thermal waves propagating in the continuous medium. Such a medium can be described as thermally lossy in the same way that a medium can be optically lossy in the propagation of electromagnetic radiation. By similar argument, it is possible to show that there is an uncertainty associated with either the thermal wave temperature excursion, or its momentum. This uncertainty decreases with increasing modulation frequency. This uncertainty principle limits the precision with which the heat centroid can be located at a given modulation frequency, and provides the theoretical limit to the depth resolution in thermal wave depth profiling.

In principle, this analysis provides a method for quantitative analysis of depth profiling experiments in media where the thermal properties and the optical properties are arbitrarily varying in space. Although cast as a one-dimensional formulation, it has been claimed that this analysis can readily be extended to three dimensions if required. This analysis provides an interesting theoretical framework with which to study problems in thermal wave physics, and the author has shown how observable macroscopic phenomena can be derived from this formalism; however, this theory has not yet been applied to practical problems of materials characterisation, and it may be that the analysis of real systems by this method is technically difficult.

7.1.2 Digital Simulation

The difficulty of working with analytical treatments of thermal wave generation and propagation has prompted a number of researchers to adopt a different approach. Instead of attempting to solve the governing parabolic differential equations directly, the continuous functions are replaced by piecewise approximations, usually polynomials. Space and time are divided up into discrete elements, each space element representing a control volume. If the properties of each element are known, and the interactions between adjacent elements can be described by simple polynomial expressions, then the evolution of the state of an ensemble of such elements can be calculated through successive time steps.

Various approaches to the solution of continuum problems by discretisation have been developed over the years. Mathematicians have developed general techniques applicable to the differential equations covering these problems, an approach which is typified by the use of finite difference approximations [7.18,19]. The engineering approach has been to look at the problem intuitively by making a direct analogy between real discrete elements of the system being studied, and small regions of the continuum domain employed [7.20]. This "finite element" approach has been extensively used in structural engineering. As the mathematical treatment of the finite element approach has developed, it has converged with the approximation methods developed for differential equations, and a unified approach to the analysis of continuum problems by discretisation and approximation has developed [7.20,21]. Despite the convergence of these two approaches, there are still practical differences in their application. Finite difference methods can be applied in a relatively simple fashion to the governing differential equations of thermal wave methods, allowing relatively simple programmes to be developed for the analysis of practical systems. Finite element methods in general require the use of commercial packages and are relatively less accessible. They can cope with very complex shapes, but this adds significantly to the complexity of the analysis. For general applications in spectroscopic depth profiling, the finite difference simulation is quite adequate [7.22–25], whereas for thermal wave imaging applications, or the use of piezoelectric detection, where deformation of the sample is significant, the finite element approach may be more effective [7.26,27].

Since the application of the finite difference approach can be more readily demonstrated, it will be discussed in some detail. Figure 7.4 illustrates the way in which the representation of a photoacoustic cell used for the *Rosencwaig–Gersho* one-dimensional theory [7.12] can be broken down into control volumes. The generalised finite difference formulation for the approximate solution of heat flow in an isotropic region in this model is given by the second-order difference equation [7.19]

$$a'_j T'_j = b_j [f T'_{j+1} + (1 - f)T_{j+1}] + c_j [f T_{j-1} + (1 - f)T_{j-1}]$$
$$+ [a_j - (1 - f)b_j - (1 - f)c_j]T_j , \qquad (7.1.4)$$

$$a_j = \frac{\varrho_j C_j \Delta x_j}{\Delta t} , \qquad a'_j = f a_{j+1} + f a_{j-1} + a_j ,$$

$$b_j = \frac{k_{j+1}}{\Delta x_{j+1}} , \qquad c_j = \frac{k_{j-1}}{\Delta x_{j-1}} ,$$

T_j is the temperature in element j at time t, T'_j is the temperature in element j at time $t + \Delta t$, ϱ is the density, C is the specific heat, k_{j+1} is the thermal conductivity in

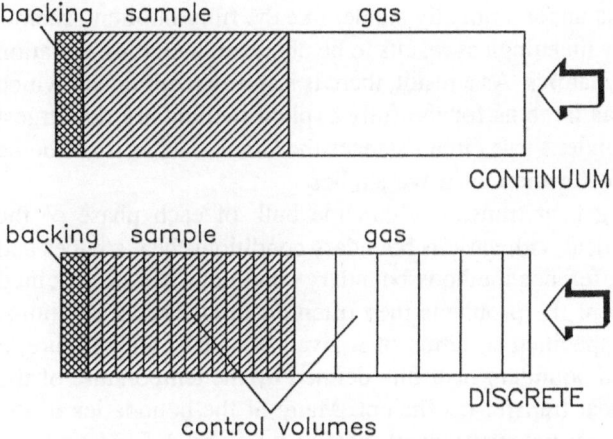

backing   sample   gas

CONTINUUM

backing   sample   gas

DISCRETE

control volumes

Fig. 7.4. Continuum and discrete one-dimensional models of a photoacoustic cell

the direction of the next element, k_{j-1} is the thermal conductivity in the direction of the previous element, x is the displacement, and f is a weighting factor ($0 < f < 1$).

There are three schemes for the solution of the second order of difference equations depending on the value assigned to the weighting factor f in (7.1.4). If $f = 0$, then only the temperatures at time t are used to calculate the final temperature at time $t + \Delta t$. This is equivalent to assuming that the starting temperature in the control volumes prevails throughout the time step. This model is known as the fully explicit or forward difference scheme, and produces a particularly simple solution, which can readily be programmed on any small computer. However, for the solution to be stable, the following criterion must be met:

$$\frac{k}{\varrho^C} \frac{\Delta t}{(\Delta x)^2} < 0.5 \ . \tag{7.1.5}$$

From this, it is clear that as the scale of the spatial discretization is varied, particularly as the grid becomes finer, the time step may also need to be changed to maintain stability.

In contrast, if $f = 1$, the new temperature of the grid point is calculated from the old temperature of the grid point and the new temperatures of the adjoining elements. This is equivalent to assuming that the temperature changes from its old value to the new value at the beginning of the time step. This is the implicit, or backward difference scheme. Since the difference equation (7.1.4) becomes an implicit equation, it is not possible to calculate the temperature distributions without solving a set of $n - 1$ simultaneous equations. The computational overhead is therefore significantly greater than for the fully explicit scheme. However, the method does have the advantage of being unconditionally stable, allowing much more flexible definition of spatial element size and time step.

If $f = 0.5$, the temperature is assumed to vary linearly during the time step, and both the old temperatures and new temperatures are given equal weight in calculating the new temperature of the grid point [7.28]. This is known as the Crank–Nicholson scheme. Since this is still an implicit method, a similar set of simultaneous equations must be solved. The method is unconditionally stable, like the fully implicit scheme, but it is possible for physically meaningless results to be obtained, including violation of the second law of thermodynamics. As a result, there is stability requirement which means that time steps twice as large as for the fully explicit method are the largest that can be tolerated [7.29]. Under these circumstances the Crank–Nicholson scheme does not seem to be attractive for thermal wave studies.

In addition to considering heat transfer within the bulk of each phase of the system, it is also necessary to deal with various boundary conditions, heat sources and sinks. Fortunately, in finite difference methods boundary conditions can be specified very simply, and do not present the problems they often do in analytical solutions. Boundary conditions can be specified in terms of a given boundary temperature, a given boundary heat flux, or a boundary heat flux defined by the temperature of the surrounding medium and a heat transfer coefficient. Many of the boundaries in the type of system used for the thermal wave depth profiling can be defined as fixed boundary temperatures, for example the thermal conductivity and heat capacity of the cell walls are usually very much greater than that of either the sample or the gas. Other boundaries can be simply specified by defining the thermal conductivity of the interface between the two phases. A frequently used simple assumption has been that the interface thermal conductivity is the arithmetic mean of the conductivities of the two adjacent control volumes. However, it has been pointed out that a much more realistic estimate is obtained by taking the harmonic mean of the thermal condictivities [7.19]. The effect of heat sources and sinks can be separately calculated for each control volume involved during a particle time step, and added to the effects of heat transfer to give an overall change in temperature for each control volume in the system.

Most of the published work on digital simulation of thermal wave spectroscopy has used the explicit finite difference scheme because of its ease of implementation [7.22–24]. It has also used a one-dimensional formulation of the problem to reduce computation time. This relies on the same argument as the analytical studies in that the typical width-to-thickness ratio of a sample used in thermal wave spectroscopy permits this simplification. Recently, a study has been published of the digital simulation problem for thermal wave experiments that uses the fully implicit formulation [7.30]. This study used a number of elegant methods derived from the extensive literature on the solution of partial differential equations in engineering to reduce the computation time required for solving the $n - 1$ simultaneous equations. The computation time was reduced to the point where the model could be run on a conventional personal computer. The fact that the fully implicit scheme is unconditionally stable allowed a wider latitude in the selection of dimensions of the control volumes, further reducing the computation time. This not only allowed the more reliable fully implicit scheme to be used in the simulation, but also permitted the model to be extended to two and three dimensions. The simulation model produced results which agreed with a wide range of experimental data obtained by various authors, and also was consistent with the predictions of the RG theory. In its three-dimensional form, it showed that the difference between the three-dimensional formulation and the one-dimensional formulation due to edge effects was small, in agreement with both theory and experimental observation.

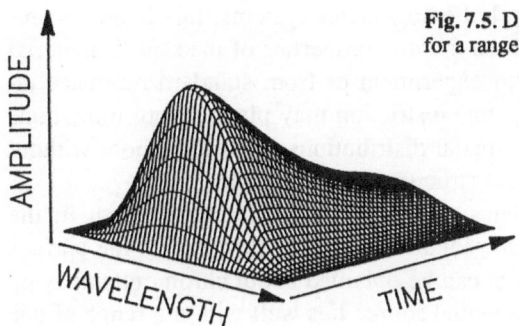

Fig. 7.5. Digital simulation of the thermal impulse response for a range of wavgelengths across an absorption maximum

AMPLITUDE

WAVELENGTH   TIME

As an example of the capabilities of simulation methods, Fig. 7.5 shows the digital simulation of the thermal impulse response obtained for a spectroscopic sample at a series of wavelenghts across an absorption maximum. This is typical of the type of results which can be obtained by digital simulation and, in terms of the response shape, is in good agreement with experimental observation [7.23].

7.1.3 Information Content of Depth Profiling Experiments

A number of theoretical approaches exist to predict the response of a given example in a thermal wave depth profiling experiment. Each method has its own advantages and limitations, but in general they all provide broad agreement with the main features of the body of experimental data. Given a theoretical basis for predicting the sample response, it is now possible to explore the information content of the experiment. How much can we expect to learn from a thermal wave depth profiling experiment?

Each sample can be described by a set of thermal properties and a set of optical properties, each of which can vary spatially. Each three-dimensional location within a sample can therefore be described in terms of a set of optical properties and a set of thermal properties. The optical properties involve such parameters as absorption and scatter, which can be grouped together as the amount of energy deposited per unit time at that location for given illumination conditions. For very simple systems, this can be treated as a pure absorption function. The thermal properties can be lumped together as the thermal diffusivity of the system at that location. The system can therefore be defined in terms of two parameters that are spatially variant. Further simplification can be achieved by taking a one-dimensional view of the system, which then reduces the description to two parameters that vary with one spatial parameter.

From a consideration of some of the underlying physical processes involved in thermal wave methods, certain simple exclusions can be made about the information gathering power of the experiment. For example, the time delay or phase shift of a signal is related to the thermal diffusivity of the medium, and the distance from the source to the point of detection. Without prior knowledge, or some alternative probe of the system properties, it is not possible to distinguish between the effects of thermal diffusivity and source location from an inspection of the sample frequency response or impulse response. For simple homogeneous systems, this is not too restrictive, since a reasonable estimate of the thermal properties of the sample material will frequently be available, either from experiment or from standard reference tables [7.68]. For more complex systems, this restriction may place severe limitations on the ability to obtain information on spatial distributions of chromophore without a thorough understanding of the physical structure of the sample.

The second area of restriction relates to the optical penetration depth in the sample. It is obvious that if the incident radiation is totally absorbed in the surface layers of the sample, then no information can be obtained about chromophores lying below the blocking layer, even if the potential source lies well within a range of one thermal diffusion length from the detector.

Apart from such obvious limitations on the ability to provide a full description of the sample characteristics from the thermal wave experiment, there are also practical limitations associated with signal-to-noise ratio in real experiments. In reality, the experiment does not have infinite discriminating power in the exact shape of the thermal wave response. Weak components of the response will be relatively uncertain due to the influence of experimental noise, and the impact of this uncertainty on the overall estimate of the sample properties will depend on the type of sample being probed. High-frequency components of the response are relatively weak in amplitude due to the smaller temperature excursion in the source. This is partially compensated for by the fact that due to the damping of the thermal wave only high-frequency components originating close to the point of detection can be observed at all. In contrast, source regions being pumped at low frequencies but which are some distance from the point of detection will produce very weak signals due to dissipation in the bulk of the sample. The interaction of these various factors has not been thoroughly explored despite its importance for understanding the capabilities and limitations of the method. For example, it is not known whether an arbitrary distribution of optical density with depth in a sample produces a unique characteristic response that can be discriminated from all other arbitrary distributions. Intuitively, we would believe that this might only be true for a perfect noise-free system. In a real system, noise would blur subtle differences. However, it is not known what difference between two alternative profiles will produce an observable difference in the experimental response for a given experimental regime.

A simple simulation experiment can be carried out to illustrate the effect of noise on the discriminating power of the thermal wave experiment, and the impact of the sample properties on that discrimating power. Consider an isotropic homogeneous sample of known thermal diffusivity and optical density. We can simulate the theoretical impulse response that would be obtained from such a sample using a finite difference method. If we now make a series of simulations allowing the absorption coefficient to vary by $\pm 50\%$ from the notional value used in the reference simulation, we can calculate how the two simulations deviate with variation in optical absorption coefficient, using the sum of the squared errors as a measure of the difference. Providing a simulation is sufficiently accurate, we would predict a theoretical discrimination power which is of a very high order. Certainly, a 1% variation in optical density produces a clear difference between the two simulations. However, as soon as we add in experimental noise, the situation changes. Figure 7.6

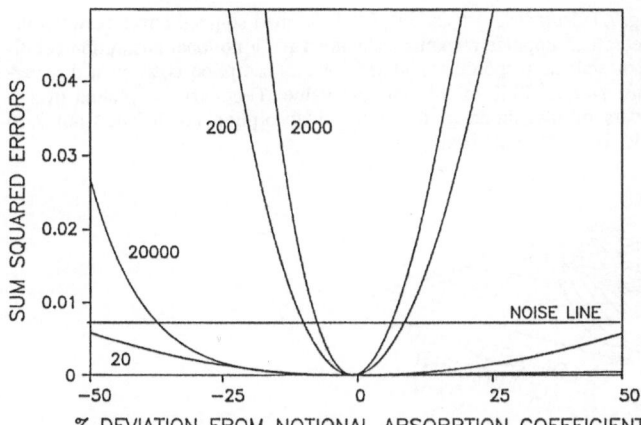

Fig. 7.6. Effect of deviation from a notional value of absorption coefficient on error in response simulation

shows the sum of the squared errors versus deviation from the notional absorption coefficient for a range of notional absorption coefficients. Overlaid on this is a line representing the typical experimentally observed level of noise for a single-pulse experiment using a gas microphone cell.

We can make the simplistic argument that the area below the noise line in the diagram represents an area of confusion where the noise prevents us from discriminating between curves for different optical densities. Although not statistically valid, this assumption gives a simple visual representation of the effects of noise. For this particular set of examples, at extremely low levels of absorption coefficient the entire curve lies below the noise line. For that level of noise we would therefore be unable to describe the absorption coefficient of a weak absorber to within a precision of $\pm$ 50%. As the absorption coefficient increases, the error between the notional absorption coefficient and deviation from that value increases dramatically, and the curve dips more briefly below the noise line. For such a system, we might expect to be able to determine the absorption coefficient to better than $\pm$ 10%. As the extinction coefficient continues to increase, we find that the curve flattens and becomes more asymmetric. This increases the length of the curve below the noise line, and decreases the precision. What we are seeing is the onset of photoacoustic saturation. For a strong absorber, virtually all of the light can be absorbed in a relatively thin section of the sample. As a result, small changes in the absorption coefficient do not produce as large a change in the initial distributed heat source as is seen at lower absorption coefficients. The error curve is asymmetric, as increasing saturation at higher absorption coefficients leads to relatively smaller changes in the initial heat source distribution. This means that there is a continuous variation with absorption coefficients in the precision with which one can assign that assorption coefficient from its thermal wave response. There is an optimum absorption coefficient for the sample at which the precision is at its maximum. At all other levels of absorption coefficient, the precision which can be achieved in the experiment is lower.

The effect of optical density on the discriminating power of the experiment is shown more graphically in Fig. 7.7. This plots deviation from the notional absorption coefficient on the x-axis, the logarithm of the absorption coefficient on the y-axis,

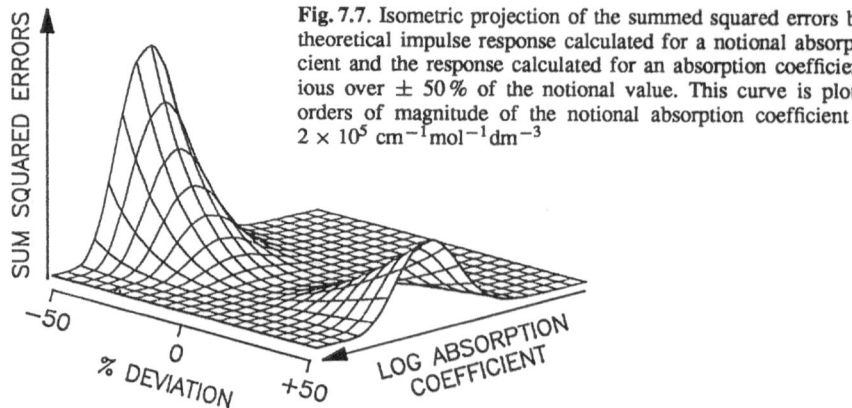

Fig. 7.7. Isometric projection of the summed squared errors between the theoretical impulse response calculated for a notional absorption coefficient and the response calculated for an absorption coefficient that various over ± 50% of the notional value. This curve is plotted over 5 orders of magnitude of the notional absorption coefficient from 2 to 2×10^5 cm^{-1}mol^{-1}dm^{-3}

over five orders of magnitude, and the sum of the squared errors on the z-axis. The continuous variation in discriminating power is clearly shown, and the existence of a peak discriminating power at a particular absorption coefficient is also evident.

Despite the importance of a full understanding of the information gathering capabilities of a thermal wave experiment in performing a quantitative depth profiling analysis, little work on this subject has been published. The above discussion shows how a very simple system can give rise to quite a complex pattern of errors. This analysis was made for an isotropic distribution of chromophore, and as yet there has been no attempt to evaluate how reliably the shape of a distribution can be inferred from the thermal wave response in the presence of realistic experimental noise. Evidence from other theoretical studies [7.16] suggests that even relatively small amounts of noise can have a significant impact on the ability to obtain information from deep within the sample. Further work in this area is urgently required to provide a means of evaluating whether the prior knowledge available concerning the sample, together with data obtainable from the thermal wave experiment, is capable of providing an answer to the questions being posed in the experimental study.

7.2 Experimental Methods

In carrying out thermal wave depth profiling experiments, the problem reduces to one of finding an efficient method of probing the system so as to produce a good estimate of the response of the sample. This can then be compared with the theoretical models to yield an estimate of the spectroscopic properties. Since we are deliberately dealing with sytems where the properties are spatially dependent, the overall system is a distributed parameter rather than a lumped parameter system. Although there is extensive literature in the field of distributed parameter systems, there appears to be no optimal approach to their identification and characterization [7.31]. However, a wide range of tools have been developed within the field of system identification which can be applied to thermal wave experiments. Only a limited range of these have so far been exploited in practice, and they have many features in common. It is important to appreciate that the choice of the strategy for probing the system is as much defined by the available equipment, and the type of experiment which is planned, as by the individual merits of each approach.

From the underlying physical properties of thermal wave generation and propagation in condensed phases, we would expect a sample to behave as a linear system, provided the temperature excursion is small. From the point of view of trying to characterise a linear system, its major properties are frequency preservation and superposition. The output or response of a linear system to any input contains only those frequencies represented in the input waveform. No new frequencies are generated within the system, and all the system does is modify the amplitude and relative phase of the various input frequency components. The principle of superposition states that the output or response of the system to a complex input containing several components applied simultaneously is equal to the sum of the responses of the system to each individual input component applied separately.

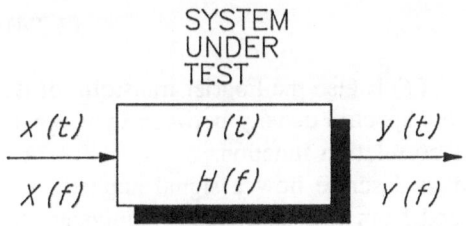

Fig. 7.8. Relationship between input and output signals for a linear system

The experimental problem is to devise a method to probe the signal processing characteristics of a "black box" by applying external perturbation to the system, and monitoring the response. This is illustrated diagrammatically in Fig. 7.8 [7.32]. A perturbation signal of frequency distribution $X(f)$ is applied to the system, where it is modified by the transfer function of the sample $H(f)$, yielding the output spectrum $Y(f)$. The system transfer function describes how each frequency present in the input waveform is changed in amplitude and phase by the system under test. From the definition of a linear system, we know that the only effects that a system can have on an individual frequency component are to change its amplitude and to phase shift it with respect to the input. The output spectrum can therefore be obtained simply by multiplying the input spectrum $X(f)$ by the system transfer function $H(f)$,

$$Y(f) = H(f)X(f) . \tag{7.2.1}$$

In the time domain, the situation is slightly more complicated. An input waveform or time series $x(t)$ is applied to the system to perturb it. This input waveform is modified by the system weighting function or impulse response $h(t)$ to produce the output $y(t)$. Instead of a simple multiplication, the relationship between the input waveform, the output waveform and the system impulse response is given by the convolution integral

$$y(t) = \int_{-\infty}^{\infty} x(\tau)h(t - \tau)d\tau , \tag{7.2.2}$$

where τ is an auxiliary time variable. Fortunately, convolution in the time domain is equivalent to multiplication in the frequency domain, and if $y(t)$ is the Fourier transform of $Y(f)$ and $x(t)$ is the Fourier transform of $X(f)$, then $h(t)$ is the Fourier transform of $H(f)$.

More generally, the effect of a linear system on an arbitrary or random input can be expressed in terms of power spectral density or correlation functions. A power spectral density function is the distribution of the variance of a time series over frequency, and a correlation function measures the similarity between a time series and a delayed version of that or another series. The important correlation functions are the autocorrelation function R_{xx}, which compares a series with a delayed version of itself, and the cross-correlation function R_{xy}, which compares two signals:

$$R_{xx}(\tau) = \frac{1}{T} \int_0^T x(t - \tau)x(t)dt , \tag{7.2.3}$$

$$R_{xy}(\tau) = \int_0^T x(t - \tau)y(t)dt . \tag{7.2.4}$$

The power spectral density of a signal $S_{xx}(f)$ is also the Fourier transform of its autocorrelation function $R_{xx}(\tau)$, and the cross spectral density between two signals $S_{xy}(f)$ is the Fourier transform of the cross-correlation function.

The cross spectral density can be used to describe how a signal applied to a linear system is modified by that system, and from this an equation equivalent to (7.2.1) can be derived for an arbitrary input:

$$S_{xy}(f) = H(f)S_{xx}(f) . \tag{7.2.5}$$

Similarly, in the time domain, the effect of passing an arbitrary time series through a linear system can be derived from the correlation function

$$R_{xy}(\tau) = \frac{1}{T} \int_0^T h(t)R_{xx}(\tau - t)dt . \tag{7.2.6}$$

In the special case where $R_{xx}(\tau)$ approximates to a Dirac pulse, the input–output cross-correlation function is an approximation to the system weighting function:

$$R_{xy}(\tau) \approx h(t) . \tag{7.2.7}$$

These relationships give a considerable degree of flexibility in probing a sample. Experiments can be carried out in either the time or the frequency domain, and the results expressed as either an impulse response or a transfer function, whichever is appropriate to the model with which it is being compared. Further, since there is no specification of the type of perturbation waveform in these relationships, the experiment can be tailored to give the information required.

In a real experimental system, there are noise contributions from various sources and effects of the various elements of the detector and data acquisition system on the overall response. To obtain a good estimate of the sample response, calibration studies must be carried out to characterise the detection system, and various experimental strategies employed to reduce the influence of noise on the final estimate.

The types of perturbation experiments carried out in thermal wave depth profiling can be classified into three groups:

1. single-frequency measurements
2. pulse measurements
3. multifrequency measurements.

Each experimental method has its own characteristics, and will be discussed separately.

7.2.1 Single-Frequency Measurements

Single-frequency measurements were the earliest used to perform depth profiling [7.33–36]. The method relies on the fact that at a given modulation frequency a signal is derived from the sample only within approximately one thermal diffusion length of the point of detection. By controlling the modulation frequency it is possible to control the thickness of sample interrogated, and thus to obtain information about the distribution of chromophores. It is also possible to obtain similar information by examining the phase of the detected signal. For a given thermal wavelength, the location of a source can be identified over the range of a thermal wavelength from the point of detection by its relative phase shift.

The attraction of this method is that it can be easily used with a conventional photoacoustic spectrometer of the kind shown in Fig. 7.9. This is simply a light source, mechanical chopper, monochromator, detection cell and lock-in amplifier. At a given modulation frequency, both the phase and amplitude of the photothermal signal can be directly monitored from the lock-in amplifier, and by scanning the wavelength of the incident radiation an absorption spectrum at a given modulation frequency can be recorded.

Despite the availability of a theoretical basis for determining the chromophore distribution from measurements of the thermal wave magnitude spectrum made at a number of discrete modulation frequencies [7.16], there has been little attempt to use this method in practice. The demonstration of depth profiling capabilities has been limited either to simple two layer systems, where a high modulation frequency

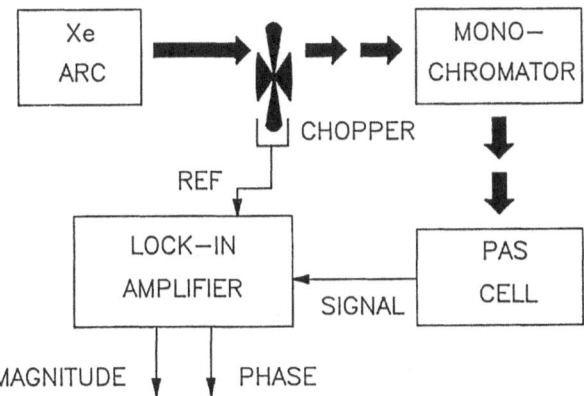

Fig. 7.9. Block diagram of a conventional single-frequency photoacoustic spectrometer

gives access to the upper layer, and a low modulation frequency gives a combined signal from both upper and lower layers [7.35], or to a simple qualitative assessment of the characteristics of a sample [7.37]. This latter study demonstrated the difficulty of interpreting data obtained by discrete frequency measurements without extensive prior knowledge of the sample characteristics.

Using the phase information to discriminate between upper and lower layers has proved more popular, as it does not require that spectra are repeatedly recorded at different modulation frequencies. In addition, since it is easy to make reliable measurements of the phase angle, it is usually easier to make a quantitative assessment of the location of a chromophore within the system. This has been demonstrated for a sample of color photographic film [7.38]. More recently, a phase resolved method has been explicitly developed for two-layer systems. This method was developed by *Moore* and co-workers for biological materials [7.39,40] and by *Vargas, Miranda* and co-workers for other solid samples [7.41–44]. The method is illustrated in Fig. 7.10. The lock-in amplifier records the in-phase and quadrature components of the thermal wave signal. For a two-layer system there are two signals contributing to the overall spectrum: a signal arising from the surface layer, and a signal from the lower subsurface layer, which will show a phase lag relative to the surface signal. The resultant detected signal is a vector sum of the two components, and will be recorded as a two-phase signal based on its projection onto the in-phase and quadrature axes. From (7.2.8), the spectrum at a particular phase angle θ can be calculated from the in-phase (S_0) and quadrature (S_{90}) signals:

$$S_\theta(\lambda) = S_0(\lambda) \cos \theta + S_{90}(\lambda) \sin \theta .\tag{7.2.8}$$

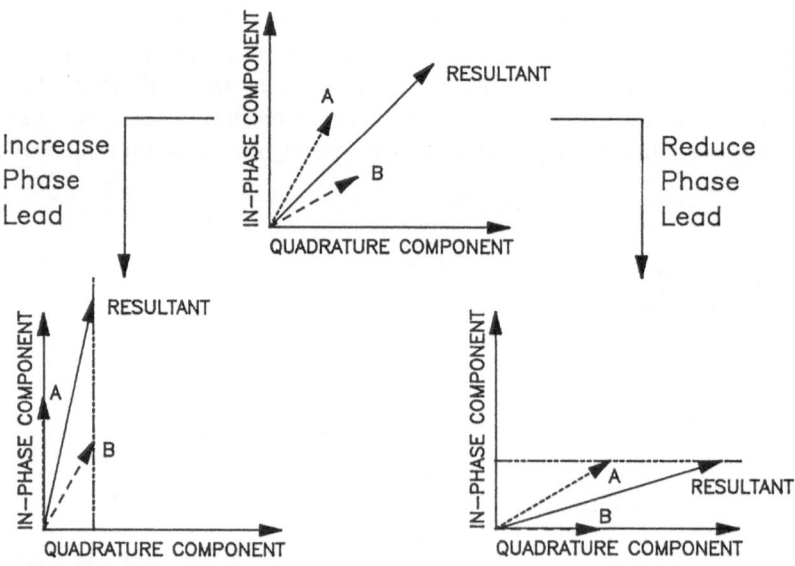

Fig. 7.10. Phase resolved method for depth profiling [7.40]: (A) surface component; (B) subsurface component; RESULTANT: vector sum of A and B

By calculating the spectra at different phase angles, it is possible to null out the contribution from either the surface or subsurface signal by rotating the resultant vector until one component of the vector pair has been rotated to a nodal point. At this point it has no projection on the perpendicular axis, and does not contribute to a spectrum plotted using that axis projection. In practice, the method requires that a spectroscopic feature be present that is attributable to either the surface or the subsurface layer. By monitoring the amplitude of this signal, the appropriate rotation can be carried out. It is an empirical method, most applicable where clear spectroscopic differentiation exists between the two layers, and the two layers themselves are highly differentiated, for example skin [7.40] and plant tissues [7.43].

The method cannot provide information on continuous chromophore distributions, since the phase angle at a particular wavelength will be a vector sum of all thermal wave sources within the thermal diffusion length range of the detector. Even if the sample does decompose into two discrete layers, it is not possible to monitor the distribution of a particular feature within that layer. In principle, the method can be extended to a greater number of layers by selectively nulling a particular component, then subtracting that from the other spectral components to allow a gradual process of decomposition. However, there are no reports of such studies, and this is probably due to the difficulty of selectively nulling out the various components in a complex system.

The use of spectra recorded at a number of different modulation frequencies to provide information on the spatial distribution of a chromophore is theoretically possible, but if only a limited number of discrete frequencies are available, the confidence with which the distribution can be assigned is low. To obtain reliable information it would be necessary to record the spectra at a large number of modulation frequencies, and with conventional instrumentation this would be a very time consuming process. There has therefore been considerable interest in finding ways of multiplexing different modulation frequencies together to give a more efficient data collection process, for example by using pulse or multifrequency perturbation signals.

7.2.2 Pulsed Methods

The impulse response of a sample can be obtained by applying an impulse perturbation and directly monitoring its response in the time domain. Methods using pulsed light sources have been widely used in photoacoustic and photothermal spectroscopy, but generally not for depth profiling studies. The major application has been in the area of the spectroscopy of very weak absorption using pulsed lasers [7.45–47]. The high peak power of the laser pulse produces a good signal-to-noise ratio for samples of low optical density. The short laser pulse also allows the use of a time-gated detection system to discriminate the direct signal arising from the spectroscopic process from other spurious signals from sources such as absorption in the windows, since the spurious signals will arrive at the detector at a different time. The high optical power density also enables nonlinear spectroscopy to be carried out, and this has shown particular gains in selectivity for gas phase samples through multiphoton excitation [7.45,48–50]. Another area of application is the determina-

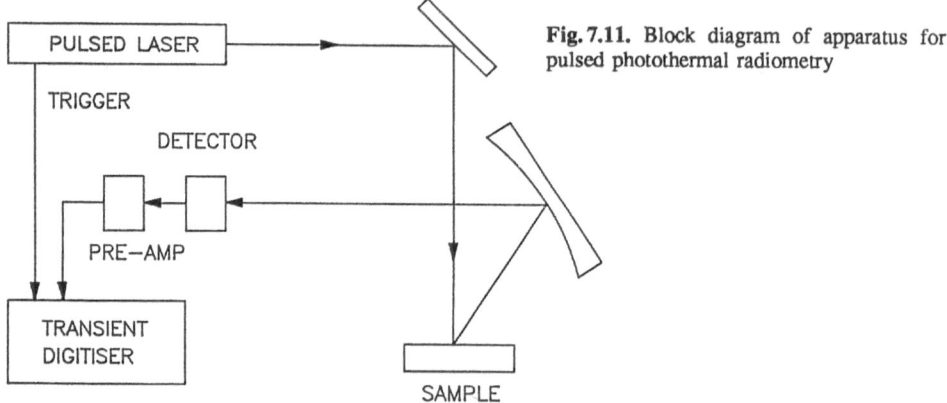

Fig. 7.11. Block diagram of apparatus for pulsed photothermal radiometry

PULSED LASER

TRIGGER

DETECTOR

PRE—AMP

TRANSIENT DIGITISER

SAMPLE

tion of the kinetics of photo-induced processes. This can provide information on quantum yields [7.51] and excited states' lifetimes [7.52], and has been used in the characterization of flowing systems [7.53].

Of more relevance to depth profiling studies has been the application of pulsed methods for nondestructive evaluation of solid samples. Here, variations in thermal properties rather than optical properties are being probed, but as can be seen from the relevant theory, these are two faces of the same problem. Pulsed radiation can be used to generate acoustic waves within a material, which can then be detected by a piezoelectric transducer or other appropriate scheme [7.54]. Alternatively, heat flux from the sample can be detected using a pyroelectric calorimeter [7.55,56], or by infrared radiometry [7.57–61].

A block diagram for a typical experimental apparatus for nondestructive evaluation using infrared radiometry is shown in Fig. 7.11. The sample is illuminated with a short pulse of radiation from a pulsed laser, and infrared emission from the sample surface is detected using a fast detector. This is usually a quantum device such as a mercury-cadmium-telluride detector. The resulting signal is then captured using a transient digitizer. Typical transients may last up to 100 milliseconds, although for many practical applications a great deal of important information is contained in the first few microseconds.

The experiment is usually carried out using broad band detection of the infrared radiation, but recently, information has been published on the use of wavelength-selective detection to provide information on the infrared emission spectroscopic profile of the sample, and hence the material characteristics [7.62]. The experimental geometry is basically the same as for Fig. 7.11, except that a wavelength-selective element such as a grating monochromator or an interference filter is interposed between the sample and the detector. This experimental arrangement has been successfully used to look at variations in the emission from liquid crystal systems at different temperatures, and to investigate the effects of sunscreens on skin [7.63].

Given the wide application in the field of nondestructive evaluation, why have pulsed excitation methods not been used for spectroscopic depth profiling? The reason probably lies in the properties of the sources themselves. The almost universal source used for fast pulsed measurements is the laser. Unfortunately, lasers are not tunable over a very wide wavelength range, without major additional complication. In nonspectroscopic applications such as nondestructive evaluation, the lack of wavelength tunability is not a problem. It is only necessary that radiation be strongly absorbed by the sample, which is frequently opaque. In spectroscopy, lack of tunability severely limits the applications. Nitrogen- or excimer-pumped dye lasers provide limited tunability, and these have been used for spectroscopic studies of weakly absorbing samples [7.46,47]. However, they do not appear to have been used for depth profiling studies.

Another problem may be the way in which the photoacoustic or photothermal signal strength is proportional to the amount of energy deposited in the sample. To obtain a reasonable time resolution in the impulse response, it is necessary to use a narrow pulse with a relatively long inter-pulse spacing. The temptation is therefore to use the maximum laser power possible to improve the signal-to-noise ratio of the individual transients, and to reduce the data acquisition time. Many of the samples of interest in nondestructive testing are sufficiently rugged to withstand relatively intense thermal transients. For other condensed phases systems, the optical power density during the pulse may be sufficient to cause damage to the sample, particularly in the case of relatively fragile materials, such as polymers and biological materials. Indeed, good use is made of the ablation of material by intense laser pulses for the generation of ultrasonic pulses in samples for relatively nondestructive evaluation [7.54].

Lower pulse powers can be used in spectroscopic studies of biological systems without causing damage, particularly with the use of thin film pyroelectric detectors [7.67]. However, the detector geometry generally used is inconvenient for depth profiling studies, and no use appears to have been made of this combination. A study has been published comparing single pulse methods with multifrequency methods at low laser power [7.65]. Although acceptable results could be obtained for the single pulse experiment by ensemble averaging of a series of transients, the number of transients that had to be averaged to achieve an acceptable signal-to-noise ratio meant that the method was unacceptably slow compared to multiplex methods.

At least one report has been published of the use of a xenon arc source directed through a variable speed shutter into a monochromator, allowing a range of pulse widths from 14.3 to 69.4 ms to be generated across the entire UV-visible spectral range [7.66]. The relatively wide pulse widths used precluded high quality depth profiling, although they did aid the signal-to-noise ratio. However, the authors only reported experiments at a single wavelength, and showed no spectroscopic results. This work does not seem to have been followed up, and, we must assume, has no practical applications.

From this we can conclude that the poor duty cycle obtained using pulse modulation of white light sources at low peak powers will always produce a signal-to-noise ratio too low for practical depth profiling studies. Pulsed lasers show more promise where a wide spectral range is not required. They would prove particularly useful where it is desirable to use a time-gated detector to reject unwanted signals. They would also show advantages where very fast transients must be studied, since fast time domain experiments are often cheaper to undertake than the equivalent frequency domain experiments. In other circumstances, multiplex methods are likely to be much more effective.

7.2.3 Multifrequency Methods

From a frequency domain viewpoint, pulse methods of probing the sample are successful because a narrow pulse contains a wide range of frequencies. However, there are limitations on the quality of information that can be obtained, due to the poor duty cycle of the excitation. Other modulation waveforms have the same spectral content as a pulse, but are generated by a time series with a much more efficient duty cycle. This significantly increases the amount of energy absorbed by the sample per unit time, and therefore the magnitude of the thermal wave response.

For example, Fig. 7.12 compares the spectral power density of a single pulse, with a multifrequency modulation signal with the same peak amplitude and autocorrelation function. The duty cycle of the single pulse was 0.8 %, and that of the multifrequency signal, 50 %. The improvement in energy deposited in the sample per unit bandwidth is clear. From (7.2.6) we know that if a multifrequency modulation has the same autocorrelation function as a pulse perturbation, then the input–output cross-correlation function is a good estimate of the response of the system to a pulse. Where the perturbation is a unit impulse, then the input–output cross-correlation function is a good estimate of the system impulse response.

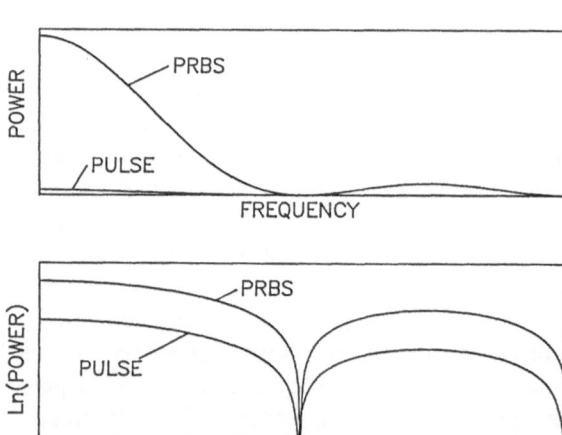

Fig. 7.12. Comparison of the power spectral density for a pulse and a multifrequency modulation sequence with equivalent autorcorrelation functions

A number of experimental strategies have been developed that use a perturbation time series with a wide frequency content. A special case of multifrequency modulation is found in the use of photoacoustic or photothermal detection with Fourier transform infrared (FTIR) spectrometers. The Michelson interferometer is the normal geometry used in most FTIR spectrometers. For this geometry, transmission maxima at a given wavelength are evenly spaced throughout the travel of the interferometer moving mirror, at intervals of optical retardation equal to the wavelength. For a constant mirror velocity, the modulation frequency imposed on the radiation transmitted through the interferometer is wavelength dependent. At low wavenumbers, the effective modulation frequency is low, giving a long thermal wavelength and large penetration depth. At high wavenumbers, the effective modulation frequency is high, leading to short thermal wavelengths favoring surface features. An FTIR photoacoustic spectrum is therefore distorted by the change in thermal diffusion length from one end of the scan to the other. For a homogeneous material whose thickness significantly exceeds the longest thermal diffusion length, and which does not contain any saturated peaks within the scan range, the distortions to the relative peak heights introduced by the variation in modulation frequency can easily be corrected for. Where saturation occurs, or when there are variations in spectroscopic features with depth, correction is much more difficult.

Fortunately, most FTIR spectrometers can scan at a range of mirror velocities, and it is therefore possible to alter the range of modulation frequencies covered within the scan. Recording the spectra at different mirror velocities allows information to be obtained from a range of different probe depths, and a crude depth profile of the sample can be constructed. Compared with the methods employed with dispersive spectrometers in the UV and visible, where a very wide range of modulation frequencies can be applied, the number and range of different modulation frequencies is fixed by the interferometer design for each wavelength in the spectral scan. However, careful experimentation can still allow a useful amount of depth profiling information to be obtained. Figure 7.13 shows the wavenumber dependence of the thermal diffusion length for a range of mirror scan velocities in a Michelson interferometer. The data covers a typical spectral scan from 400 to 4000 cm^{-1}. Thermal diffusion lengths were calculated for a typical polymeric material [7.67].

Although the experiment is carried out using multifrequency modulation, FTIR photoacoustic spectroscopy shares many similarities with conventional single-frequency methods. Depth profiling studies have been carried out, but the method is relatively tedious, and does not yield the kind or quality of information obtainable from other kinds of multifrequency modulation methods.

Turning to more flexible multifrequency experiments, a wide range of perturbation signals can be used. Because of the simplicity of implementation, there has been much interest in the use of multifrequency test signals that have only two levels. The most popular signal of this type is the pseudorandom binary sequence (PRBS) [7.68]. The PRBS modulation signal has the following properties:

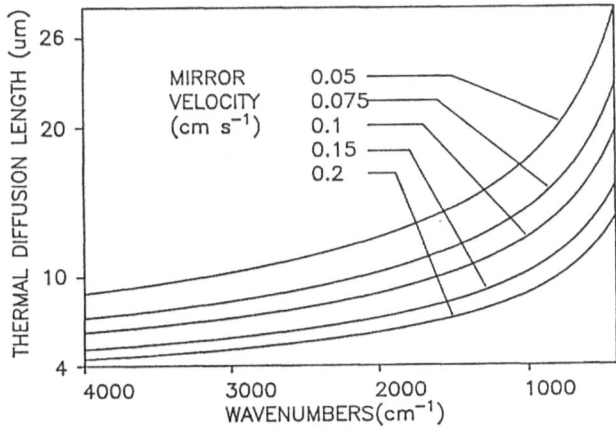

Fig. 7.13. Effects of mirror velocity on thermal diffusion length in an FTIR-PAS experiment

1. The signal has only two levels and can switch from one level to the other at multiples of the basic clock cycle Δt.
2. The probability of a level change at each clock cycle is 0.5.
3. Whether or not the signal changes level at any particular time interval is predetermined. The PRBS is deterministic and periodic, with period $T = N\Delta t$ where N is the number of bits in the sequence.

Provided that Δt is sufficiently short, the autocorrelation function for a PRBS signal is the same as the autocorrelation function of a pulse. The impulse response of the system can therefore be obtained from the convolution integral relationship between the input auto-correlation function R_{xx} and the input–output cross-correlation function R_{xy} (7.2.7).

Where the system is contaminated by additive noise $n(t)$, it is not possible to measure R_{xy} directly, but only R_{xz} where $z(t)$ includes the effect of noise. The relevant equation now becomes

$$R_{xz}(\tau) = \frac{1}{T_s} \int_0^{T_s} h(t) R_{xx}(t - \tau)dt + R_{xn}(\tau) , \qquad (7.2.9)$$

where T_s is the settling time of the system and R_{xn} is the cross-correlation function between the input signal and noise in the system. If the sequence length is greater than the system settling time, then the noise contribution from R_{xn} can be reduced by averaging R_{xz} over an integral number of sequences. Provided Δt is sufficiently small and N sufficiently large, the system weighting function $h(t)$ is directly proportional to the input–output cross-correlation function R_{xz}. A typical PRBS signal is shown in Fig. 7.14 with its associated power spectral density.

A number of groups of workers have exploited this type of experiment to measure photoacoustic and photothermal impulse respones [7.66,69–74]. In the earliest reported study [7.69], a mechanical chopper was cut to emulate a 31 bit PRBS. This mechanical chopper was used in a conventional photoacoustic spectrometer (Fig. 7.9) with the lock-in amplifier replaced by an analogue cross-correlator and time delay device. This instrument was capable of producing a plot of response versus delay time at a particular wavelength, which was called a correlation spectrum of the first

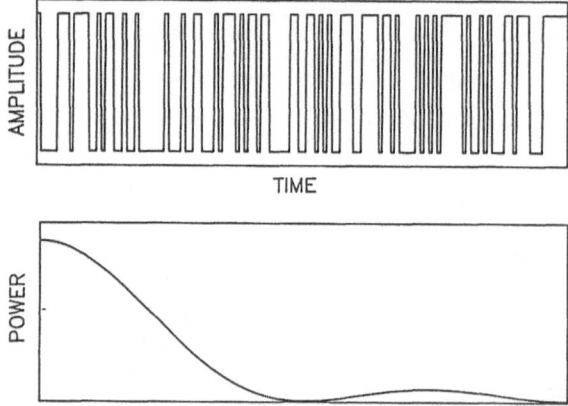

Fig. 7.14. Pseudorandom binary sequence and its power spectral density

194

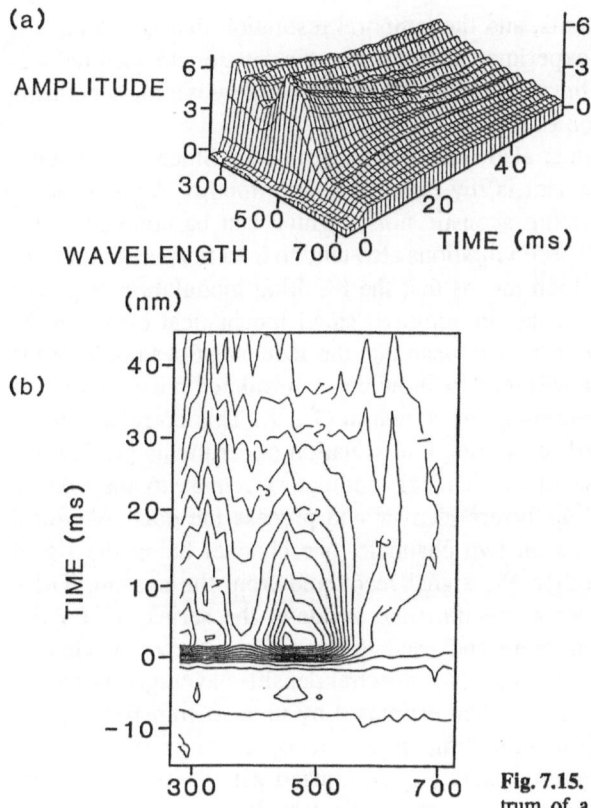

(a)

AMPLITUDE

WAVELENGTH (nm)

TIME (ms)

(b)

TIME (ms)

WAVELENGTH (nm)

Fig. 7.15. Impulse response photoacoustic spectrum of a bulk dyed polymer film: (a) isometric projection, (b) contour map [7.72]

kind, or a plot of intensity versus wavelength at a fixed delay time, a correlation spectrum of the second kind. The short period of the PRBS gave coarse time resolution, and the results were dramatically improved by a 63 bit sequence and finally a 127 bit sequence [7.70–73]. A considerable improvement in the technique was obtained by replacing the analogue correlator and time delay unit with a digital signal analyzer [7.72,73]. This permitted full impulse response photoacoustic spectra (IMPAS) to be obtained by recording the sample impulse response at a series of discrete wavelengths, and then combining these to form a two-dimensional data set, which could be represented as an isometric projection or as a contour map (Fig. 7.15).

The principle difficulty with this approach lies in the use of a mechanical chopper with a specially cut disk to produce the PRBS modulation. The slit width of the spectrometer provides an ultimate limit on the length of the sequence that can be accommodated on a practical size chopper wheel. This is compounded by the fact that reducing the slit width on the monochromator to allow the use of narrower PRBS segments will reduce the throughput of the monochromator, reducing the signal-to-noise ratio, and losing some of the benefits of the multiplex modulation scheme. The practical limit for mechanical choppers of a usable size is probably a sequence 255 bits long. This means that the time window that can be studied can only be divided

195

up into a maximum of 255 elements, and the temporal resolution that is available is not sufficient for many types of experiments. Increasing the rotational speed reduces the time step in the time series, but also reduces the time window width, decreasing the length of transient that can be observed.

The use of a mechanical chopper also creates noise problems. Since the sequence is pseudorandom, the chopper wheel is, by definition, asymmetric. As a result, it may vibrate when rotated, producing acoustic noise, which can be coupled to the photoacoustic detection system. These vibrations also tend to lead to speed variations in the rotation of the chopper, which means that the resulting modulation sequence does not have the expected properties. In addition, small mechanical errors in the manufacture of the chopper wheel can also mean that the sequence does not have the expected properties. These errors will tend to show up as correlated noise within the system that is not removed by averaging, as shown in (7.2.9). This correlated noise has been experimentally observed in multiplex modulation experiments [7.72,73].

The final problem in the use of mechanical modulators relates to the way in which commercial digital signal analyzers recover and process the data. A typical method is to capture a time series on two channels, one channel being the signal going into the system, and the other, the signal recovered from the system. These two time series are acquired over a specific time window, the series are Fourier transformed into the frequency domain and used to form a cross spectral density function. The process is repeated, and the cross spectral densities averaged to reduce noise. The cross-correlation function is then estimated by inverse–transforming the cross spectral density function. It is quite difficult in practice to ensure that the time window of the digital signal analyzer is exactly synchronized with the pseudorandom binary sequence. If it captures slightly more or slightly less than one full sequence, then once again the sequence does not have the expected statistical properties. This creates noise in the cross spectral density function. This can be eliminated by repeated averaging, but significantly reduces the quality of the system weighting function estimates that can be obtained from this experiment for a given measurement time. With a mechanical modulator it is particularly difficult to ensure synchronization between the modulation clock period and the data capture clock period.

All of these problems have made it difficult in practice to achieve high clock rates and long pseudorandom sequences using mechanical modulators. A number of possible alternatives are being explored, although none have proved totally satisfactory.

If a cw laser is being used at the source, modulation can easily be applied using an acousto-optic or electro-optic modulator. These have the capability of applying any desired modulation waveform to the laser with frequencies as high as 100 MHz. Where tunability is not required and a cw laser is available, this is clearly the best choice. Acousto-optic and electro-optic modulators can also be applied to white light sources, although with much lower efficiency. Their main problems are that they require polarized light, the modulation is wavelength dependent, and they usually have a small acceptance angle. All of this leads to a low throughput, and has made them impractical for the white light sources currently available.

Acousto-optic tunable filters combine wavelength selection and modulation, and by removing the need for a monochromator should improve the total throughput of the system. However, the acceptance angle is rather small, and although it is possible to produce photoacoustic signals, the signal-to-noise ratio is not high enough for practical studies [7.75].

Another possible type of modulation is the liquid crystal light valve or light shutter. These devices are much slower and still require polarized light but have a much bigger acceptance angle and aperture than the acousto-optic and electro-optic devices. Although no practical experiments have been carried out with these devices, they show considerable promise for future development.

A final option is the use of direct modulation of the current in a xenon arc. The depth of modulation that can be achieved is severely limited by stability requirements within the arc plasma. If the current is modulated too deeply, then the stability of the lamp goes down dramatically and the noise content also increases. The frequency range is severely limited, with a maximum frequency of 5 kHz for typical commercial devices. However, preliminary experiments have shown that despite the poor modulation depth, the optical throughput of the system is far superior to that achieved using opto-electronic modulation devices [7.75]. This approach is currently showing the most promise for future development.

The appropriate autocorrelation function for the input time series can be achieved by any waveform with a relatively flat power spectral density over a wide frequency range. For a sine wave modulation of the input, it is clearly possible to achieve this by slowly sweeping the frequency from a low to a high value. This is the basis upon which conventional sine wave testing of electronic circuits is carried out using lock-in amplifier signal recovery. It is possible to demonstrate that the same effect can be achieved by rapidly sweeping the modulation frequency over the same range in a short frequency chirp in a technique known as Frequency Modulated Time Delay Spectrometry (FM-TDS). This technique has been pioneered for photothermal applications by *Mandelis* and co-workers [7.77–81], based on earlier work in the field of acoustics [7.82].

A typical linear frequency sweep signal is shown in Fig. 7.16, together with the power spectral density. The power spectral density function is flat across a frequency window Δf equal to the frequency sweep range, but suffers some ringing at each end of the sweep due to the effects of the finite sweep width.

As for the cross-correlation function in the time domain experiment, the cross spectral density $S_{xy}(f)$ is a good approximation to the system transfer function provided the input power spectral density $S_{xx}(f)$ is equivalent to that of a Dirac pulse. This would require a flat power spectral density function over an infinite frequency range. This is not achievable in practice, but the approximation will be satisfactory provided certain criteria are met; specifically, the linear system being tested should not pass frequencies higher than the maximum of the frequency sweep f_{max}, and the sweep time should be very much greater than the delay introduced by the linear system. This technique is dealt with in considerable detail in Chap. 8 of this volume.

The technique is claimed to give a considerable improvement in the quality of the derived system transfer function or system weighting function compared to other modulation types such as random noise or PRBS. Whilst advantages can be clearly seen over the use of random noise, due to the higher energy content of the

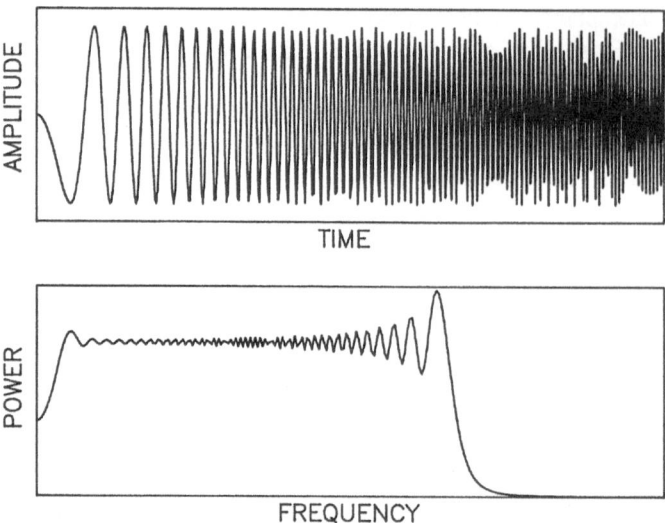

Fig. 7.16. Linear frequency sweep and its power spectral density

modulation, one would intuitively expect that similar results would be obtained as for the use of PRBS. However, since the performance achieved with these experiments critically depends on the way in which the experiment is set up, and on the apparatus used to perform the measurements, it is difficult to be sure that experiments are being carried out under exactly identical conditions, and it may well be that the apparent advantages of FM-TDS are an artefact of the way in which the experiments have been carried out.

A difficulty experienced with FM-TDS is that the requirement for sinusoidal modulation precludes the use of white light sources with mechanical choppers to provide the multiplex modulation. All studies published to date have used laser excitation of the thermal wave, coupled with acousto-optic modulation of the laser power. For many nondestructive testing applications, the restrictions introduced by the use of a laser source do not cause many difficulties; however, for true spectroscopic applications it will be necessary to find some other means of applying the modulation. It is possible that this could be achieved using direct modulation of the lamp current, but no work in this area has been published to date.

From the above discussion, it should be clear that the highest information content is obtained when a multiplex modulation scheme is used to give the widest excitation bandwidth possible in the experiment, and thus the best estimate of the system response obtainable for a given observation time. Pulse methods offer a wide bandwidth, but suffer from poor duty cycle and therefore poor signal-to-noise ratio. Single-frequency modulation techniques offer the poorest information gathering power for a given observation time, but can be applied to very simple instrumentation. As will be demonstrated in the next section, the limitations on the single-frequency methods are not necessarily fatal, provided the samples are well chosen and sufficient other information is available about the sample.

7.3 Applications

7.3.1 Biological Samples

Depth profiling in biological systems has been of interest since the early days of photoacoustic spectroscopy. The ready availability of naturally occurring systems with layered or distributed structures, and the need for a nondestructive method of examination, has led to many demonstrations of the depth profiling capabilities of thermal wave spectroscopies, but fewer applications. The very earliest studies employed the single-frequency technique, and used control of the thermal diffusion length and phase sensitive detection to discriminate the spectrum of the cuticular layer of a leaf or fruit rind from the underlying structures [7.33–35]. These early studies were essentially demonstrations of the practicality of depth profiling using photoacoustic spectroscopy. A more interesting study was reported by *Moore* and co-workers in 1979 [7.36]. This group made an in situ study of the distribution of chromophores with depth in an intact lobster shell. Coloration in the lobster shell is caused by a carotenoid pigment, astaxanthin, which is protein bound. The specific carotenoprotein complexes produced have different absorption spectra. The spectra were recorded at four modulation frequencies ranging from 26 to 326 Hz. The spectra showed a progressive bathochromic shift of absorption maximum as the thermal diffusion length increased. This was assigned to an anisotrophic distribution of different pigment complexes within the endocuticle of the lobster shell. This was confirmed by thermal denaturing of the protein complexes, which eliminated the bathochromic shift with modulation frequency. This study represented a genuine contribution to the understanding of the structural organisation of lobster carapace, and is therefore one of the earliest real applications of the depth profiling capability of thermal wave methods.

Another biological material for which depth profiling using thermal wave spectroscopy has been proposed is skin. The *stratum corneum* is the outer horny layer that governs much of the appearance, texture and barrier properties of skin. The thickness of this layer matches the typical thermal diffusion lengths available in thermal wave spectroscopy, and it thus represents an ideal sample material. In addition, there is a strong need for nondestructive and in situ tools for probing the properties of skin to study both its normal properties, pathological conditions, the impact of cosmetic treatments, transcutaneous drug delivery systems and barrier compounds. Several early studies were carried out using single-frequency methods to monitor water content and distribution, and the effects of topically applied tetracycline [7.83,84], but the information gained was of limited usefulness, as there was no ability to analyze any anisotropies in chromophore distribution.

A more quantitative analysis was reported by *Anjo* and *Moore* [7.40], who used the phase resolved method to carry out a depth profile of beta-carotene in the skin of mouse ears. An untreated sample showed a relatively featureless surface spectrum until the UV absorption of the aromatic amino acids was reached at 380 nm. The subsurface spectrum clearly showed the Soret bands of hemoglobin in the capillary bed of the mouse ear. This hemoglobin-containing layer was phase separable at frequencies up to 500 Hz, suggesting an epidermal thickness of approximately 8 μm.

Samples from mice who had been fed on beta-carotene were then examined. The spectra showed that the bulk of the beta-carotene lay above the hemoglobin in the skin sample. To further locate the beta-carotene, methylene blue was applied to the skin sample to delineate the surface, and eosin-y was injected into the mice 4 h before killing. The eosin-y is water soluble and infiltrates the dermis and viable epidermis. Analysis of the spectra from these samples showed that the beta-carotene could not be separated from the methylene blue spectrum on the surface of the sample, but was readily separable from the eosin-y spectrum. They therefore concluded that the beta-carotene was strongly associated with the surface layers of the skin, and that very little, if any, was associated with viable dermis and epidermis. However, the beta-carotene did not effectively block light reaching the hemoglobin layers, although this had been suggested as a possible mode for its action in alleviating photo-induced skin disorders. In contrast, the melanin layer from a pigmented mouse ear had a strong filtering effect, allowing little light to reach the hemoglobin layers. This elegant series of experiments clearly shows the quality of information that can be obtained using relatively simple experimental procedures.

Later studies on skin have concentrated on the penetration of topically applied drugs into human skin using an in vivo photoacoustic cell [7.85,86]. These studies represent the only experiments that combine in vivo experimentation with depth profiling. It is a tribute to the experimental capabilities of this group that results of such high quality have been obtained under very difficult circumstances. Of most interest was the study of the topical application of the muscle relaxant Tizanidine in different vehicles. The drug was applied incorporated into a gel or a cream in a 2 μm thick layer, and the decay of the photoacoustic signal recorded over a period of 4 days. Measurements were made at modulation frequencies of 1100 Hz and 180 Hz. At 1100 Hz the signal is localized in the delivery system and the surface regions of the *stratum corneum*. The 180 Hz signal depends on the amount of drug within the vehicle and total *stratum corneum*. For the gel vehicle, the 1100 Hz signal decays very much faster than the 180 Hz signal, suggesting that the rate-determining step for penetration through the skin is the rate of drug diffusion in the *stratum corneum*. For the cream, both signals decay at the same rate, indicating that the rate-determining step is the release of the drug from the cream. As with the earlier skin studies, no attempts have been made to provide a complete depth profile, but valuable information can still be obtained from a carefully chosen experiment.

Much of the work on skin has employed externally introduced chromophores. In a recent study, the water spectrum in the near infrared was obtained for samples of human cadaverous skin at different modulation frequencies [7.87]. At a low modulation frequency of 13 Hz, the absorption peak due to the first outcome of the hydroxyl stretch in water was observed at 1456 nm, indicating association between hydroxyls through hydrogen bounds. However, at 1500 Hz the peak was observed at 1367 nm, suggesting a monomeric hydroxyl. The conclusion drawn was that in the outer *stratum corneum*, water molecules are absorbed to the dense keratin matrix as a monomeric species, whereas deeper in the skin structure water is associated.

The other type of biological sample that has received considerable attention in depth profiling studies has been plant tissues. The highly segregated structure coupled with strong chromophores has made them an ideal sample for demonstrating the capabilities of thermal wave depth profiling, although few real applications have been reported.

The earliest work was the discrimination of the cuticle from the colored cellular layer using single-frequency methods [7.34,35]. A more thorough study using the same technique has recently been reported by *Nagel* and *Lichtenthaler* [7.88], in which they compared the spectra from leaves of the green radish (*raphanus sativus*) and maize leaves (*zea mays*) before and after treatment with the bleaching herbicide norfluorazon, which destroys the photosynthetic pigment. Leaves from these plants have different morphologies, but the experiment showed that between 400 and 700 nm the photoacoustic spectrum obtained is primarily defined by the photosynthetic pigments, and that leaf structure plays a less important role.

A reverse experiment was carried out by *Helander* et al. [7.38], who investigated India rubber leaves before and after treatment with alcohol to remove the waxy surface layer from the leaf. All samples showed a strong near infrared absorption at wavelenghts above 750 nm. This was assumed to be absorption at the leaf surface. An examination of the phase spectra showed that the strong chlorophyll absorption spectrum that was observed occurred at a much greater depth than the infrared absorption. It was estimated that the waxy layer was approximately 25 μm thick, and that the center of the chlorophyll absorption was 43 μm below the wax/leaf interface. As the leaf fades, the chlorophyll absorption decays, and a new strong blue absorption develops, possibly due to carotenes. An examination of the phase spectra for the different components suggests that absorptions in the blue region and the yellow/red region of the spectrum are located at different depths within the leaf, and are therefore due to different molecules.

Phase resolved spectroscopy has been carried out on a variety of plant materials, including those attacked by the contact herbicide paraquat [7.43]. The phase resolved method was used to separate the cuticle signal from the pigment layer in intact leaves, and the pericarp and endosperm from a corn kernel. Comparison of the spectra with those obtained from mechanically isolated pericarp and endosperm showed good agreement. In the herbicide test, the plants where sprayed with a solution of paraquat under normal sunny field conditions. Samples were taken at regular intervals and the phase resolved spectra obtained. For soybean, coffee and datura leaves, a decrease in phase shift between the cuticle and the pigment layers of the leaf was observed as a function of time. A high correlation was found between the changes in phase shift and changes in leaf thickness as a result of paraquat action. The intensity of the pigment bands was also observed to decrease due to the toxic effects of the paraquat, and this was assigned to the generation of hydrogen peroxide from the interaction of paraquat with photo-system I. The possibility of applying this method to the study of dynamic systems is extremely attractive.

The method of impulse response photoacoustic spectroscopy using multifrequency modulation and cross-correlation has been applied to plant materials by two groups [7.10,72,89,90]. *Sugitani* and co-workers [7.70,89,90] plotted signal amplitude versus either delay time at a specific wavelength or wavelength at a specific delay time, to form correlation spectra of the first or second kind for a variety of plant materials. *Kirkbright* et al. [7.73] recorded the full sample impulse response over a range of wavelenghts, allowing a complete response surface to be plotted. From this, spectra of the first or second kind could be extracted at will, and it was shown that information could be obtained similar to that available from phase resolved spectra [7.38,43]. Spectra were obtained from maturing laburnum leaves, which showed the development of the cuticle, with the peak of the chlorophyll moving to longer time delays as the cuticle thickened. An increase in concentration of the photosynthetic pigments could also be observed. Treatment of a lettuce leaf with dilute acid showed destruction of the surface cuticle, and rupture of the subsurface cells. The effects of the contact herbicide paraquat could also be detected, although a different analysis of the changes in signal was proposed than had been given by *Nery* et al. [7.43].

7.3.2 Polymeric Samples

A significant amount of depth profiling work has been published on polymeric samples, both as demonstrations of the capability of the technique and as practical applications. Because a large amount of spectroscopic information for polymeric materials is found in the infrared, the bulk of the work on polymeric matrices has been carried out using FTIR-PAS. However, a number of studies have also been published on chromophores distributed within a polymeric matrix.

A good example of the application of FTIR-PAS to polymeric matrices was reported by *Urban* and *Koenig* [7.91]. The sample they studied was a poly(ethylene terephthalate) (PET) film, coated with different thicknesses of poly(vinylidine fluoride) (PVF$_2$). The PET substrate has a strong carbonyl resonance at $1736 \, cm^{-1}$ in a region where there is no strong absorption due to PVF$_2$. It was therefore possible to study the effect both of varying the thickness of the PVF$_2$ layer and of controlling the thermal diffusion length by varying the mirror speed. For the mirror speeds available, the thermal diffusion length was calculated to vary between 3 and 9 μm. Increasing the PVF$_2$ thickness caused a decrease in the intensity of the PET carbonyl band at a given mirror speed. At the slowest mirror speed, a thermal diffusion length of 9 μm was calculated, and a 12 μm film led to the complete disappearance of the carbonyl band. Analysis of the data for all the available mirror speeds at each film thickness showed that the results were in agreement with RG theory, and it would be therefore possible to quantitatively determine the thickness of a PVF$_2$ overcoat from the FTIR-PAS spectra.

Fateley and co-workers [7.92–94] have applied the technique to a wide variety of polymeric materials, particularly natural and snythetic fibres. By control of the thermal diffusion length it was possible to separate spectra of the bulk fibre from a surface finish, size or chemical treatment. Sensitive work was also carried out on the distribution of chemical additives in cellulose fibres. Cotton yarn and fabric treated with a polyacrylate sizing agent shows an intense peak at $1730 \, cm^{-1}$ due to the carbonyl groups of the polyacrylate. Two methods of applying the size were compared, and it was found that a foam finishing treatment gave a much more uniform distribution than the alternative padding treatment. This evidence was used to explain the observation that wrinkle recovery was generally better for the foam finished cotton fabrics because of increased cross-linking between the fibres due to more even distribution of finishing agents. Differences in penetration into the yarn were observed for polyacrylate and polyurethane sizing agents [7.93]. This set of studies represents a good example of the potential application of depth profiling techniques in real industrial problems.

Another type of sample often studied are packaging foils, which consist of an aluminum foil/polymer composite. In a recent example, a foil was examined consisting of an aluminum substrate coated with 14 μm of polyamide, overcoated with 16 μm of polyethylene [7.96]. Although no quantitative results were presented, the study was again able to distinguish spectroscopic signals from the various layers.

The bulk of the work using FTIR spectrometers has been carried out using gas microphone photoacoustic cells. In recent years, there has been extensive exploration of the potential of photothermal beam deflection as a detection scheme for thermal wave spectroscopy using FTIR spectrometers. A much wider range of sample geometries can be readily handled, and there is the potential for higher signal-to-noise ratios. *Varlashkin* and *Low* have evaluated the application of FTIR-PBDS for infrared spectral depth profiling [7.95]. The samples studied were a 20 μm nitrocellulose layer coated onto commercial polyethylene sheet, a section of over-printed commercial polyethylene wrapping material, and a short section of human hair. Only three mirror scan speeds were used in this study, so that the results obtained were limited in so far as quantitative analysis is concerned. However, they did indicate that similar results can be obtained using photothermal beam deflection and the gas microphone cell. However, the signal-to-noise ratio for PBDS was disappointing, particularly at higher modulation frequencies. This was attributed in part to the fact that the photothermal beam deflection detector is more microphonic at the higher frequencies. One observed advantage was the absence of cell resonances. With the gas microphone cell, these can severely limit the range of modulation frequencies that can be used in a commercial system, and the photothermal beam deflection detector does not suffer from this problem. However, care has to be taken that the signal input channel on the spectrometer has the appropriate frequency response for the experiments being carried out and does not introduce extraneous effects.

The problem of wavelength-variable modulation frequency and fixed mirror scan speeds has prompted some groups to look at an alternative form of interferometer to the conventional rapid scan devices. An obvious alternative approach is to use a step-and-integrate interferometer, which would allow complete control of the modulation frequency through an external modulation device [7.97]. Recently, this has been applied in the mid infrared with both gas microphone and photothermal beam deflection [7.98]. The capability of accurately controlling the modulation frequency across the full wavelength range has been demonstrated, but as yet not true depth profiling studies have been reported. Since so much valuable spectroscopic information is available in the infrared, this innovation may lead to the application of depth profiling techniques across a broad range of science and technology. However, the fact that many simple depth profiling studies have been carried out using FTIR-PAS owes much to the wide availability of the instrumentation, and if the step-and-integrate techniques require a specialized apparatus in addition to the conventional type of FTIR instrument common in laboratories around the world, take up of this concept may be restricted.

The majority of work on bulk polymer matrices has been carried out in the mid infrared region of the spectrum. In a recent report, *de Oliveira* et al. studied low density polyethylene in the near infrared over the spectral range 1200–2800 nm. The spectra were obtained over a range of sampling depths from 11 to 56 μm, and shifts in relative amplitude of assignable peaks recorded. From these, it was possible to determine that the thermal conductivity of the polyethylene slab was nonuniform, and was a function of depth. In addition, the concentration of methyl, vinyl, and hydroxyl groups was much greater at the surface of the sample than in the bulk. This was assumed to be due to the presence of low molecular weight, waxy, noncrystalline material at the polymer surface.

An example of an application of thermal wave depth profiling of a chromophore in a polymeric matrix is the determination of water content and distribution in polymer dielectrics [7.100]. Phase resolved spectroscopy was used on samples of ethylene–propylene rubber and cross-linked polyethylene. Spectra were recorded in the 2000–4000 nm range. This covers strong absorption peaks at 2660 and 2740 nm for water vapor, a large absorption band between 2800 and 3200 nm due to associated water, and strong CH_2 peaks at 3420 nm and 3510 nm. Both the surface and subsurface spectra showed a correlation between the relative humidity at which the samples were maintained and the associated water peak. The bulk of the water is absorbed onto the sample surface, where it associates and forms a thin film. This layer is strongly dependent on ambient humidity. A small but significant amount of water is found in the bulk of the samples and seems to be associated with water clusters originating from the manufacturing process, in which high pressure water vapor is used in the cross-linking process. Ambient humidity has a smaller effect than for the surface layer, suggesting that water may be diffusing in and out of the sample.

Helander et al. [7.38] used phase resolved spectroscopy to examine a sample of exposed color reversal film. An examination of the phase spectra allowed spectroscopic features associated with each of the three absorbing layers of the film to be identified. By assuming that the gelatine coating of the film had the same thermal properties as water, an estimate was made of the predicted thickness of these layers from the observed phase shifts. The predicted overall thickness of the blue layer was 6.2 μm and for the blue and green layers together, 11.4 μm. These are in agreement with the nominal values for the film used. A similar analysis of a color reversal film has been carried out using impulse response photoacoustic spectroscopy [7.89]. *Kirkbright* et al. [7.72] studied a range of polymer samples with both uniform and nonuniform chromophore distributions by impulse response photoacoustic spectroscopy. The results obtained were consistent with the known structures of the samples, and showed the sensitivity of the technique to small variations in the position of the heat source centroid. However, the time resolution that was available in this experiment was not sufficient to allow thin film samples such as color reversal films to be studied, and the results could not directly be compared with those of phase resolved spectroscopy [7.38]. This method did, however, offer an advantage of phase resolved methods in decomposing the signal into a number of sources in different layers, rather than just into a bulk and surface signal. Examples of the application of this were reported in the paper [7.72].

7.3.3. Other Solid Phase Samples

Although the bulk of the depth profiling work reported in the literature has been carried out on biological or polymeric samples, a number of papers deal with applications on other types of solid materials.

An interesting example of FTIR-PAS depth profiling was the study of coal weathering by *Zerlia* [7.101]. Intact coal samples were either aged naturally in air for about 6 months, or submitted to accelerated ageing by heating in air to 200°C for different periods of time. Spectra were recorded at each available mirror velocity and differenced with the spectrum at the slowest mirror velocity. Since this represented a spectrum dominated by the bulk of the sample, it was argued that at increasing mirror velocities the difference spectrum would show spectral features associated with a gradually shallower probe depth. The different spectra showed negative features corresponding to OH and CH absorptions that were enhanced at higher mirror velocities. These were seen as evidence of oxidation of the surface. Such negative features were not observed on a freshly cleaved coal sample cut under nitrogen in a dry box. In the accelerated ageing studies, the changing intensities of the features associated with oxidation as a function of ageing time showed the gradual penetration of oxidation from the surface of the sample into the bulk. Further analysis of the results suggested that there was evidence of different oxidation mechanisms in the surface and bulk zones of the sample. The absence of ether or carboxyl absorption bands increasing in intensity to compensate for the decrease in the CH band supported the model that oxidation occurs by atmospheric oxygen uptake and the formation of peroxides as transient intermediates.

In one of the early papers developing the concept of phase resolved spectroscopy, *Cesar* et al. [7.41] reported experiments on ferromagnetic layered samples. A 50 μm nickel film was bonded to am 50 μm iron film, and placed in a cell exposed to amplitude-modulated X-band microwave radiation. The signal was detected using a conventional gas microphone. The entire ensemble was then placed in a variable magnetic field. By an analysis of the phase resolved spectra at different modulation frequencies and different fields, it was possible to separate the contributions from the iron and nickel films. This work was later extended [7.42] for a sample of computer tape supported on an aluminum foil. Again, it was possible to discriminate between the diamagnetic aluminum foil and the ferromagnetic oxide coating. Analysis of the phase shift between them also provided information about the polymer backing of the computer tape, and the bond coat attaching the film sample to the aluminum substrate.

Another research group has applied phase resolved spectroscopy using a tunable CO_2 laser to examine a 100 Å SiO coating evaporated onto a 3 mm thick CaF_2 substrate [7.102]. By careful measurement of an uncoated sample, the photothermal signal was converted to an absolute absorption coefficient. From this, the absolute absorption properties of the coating could be calculated. The method has considerable potential for accurately determining the properties of optical thin film coatings, providing sufficient information can be gained about the supporting substrate.

The bulk of reported work using impulse response photoacoustic spectroscopy has been on polymer and biological materials, but in 1986 *Mandelis* and *Dodgson*

[7.74] published a theoretical study of a method that used powdered samples of holmium oxide as a reference material, against which they tested the theory which had developed. Even for such a highly saturated sample, a full response surface was obtained, which agreed with the theoretical predictions. Earlier work by *Kirkbright* et al. [7.72] had also shown that impulse response methods can be used for optically saturated samples, although they did not present such a thorough theoretical analysis of the results.

7.3.4 Dynamic Processes

Using thermal wave spectroscopy it is possible to obtain information about changes in chromophore distribution in samples relatively rapidly. Although it may not be possible to carry out a full analysis, measurements capable of demonstrating a redistribution trend can be obtained in a few seconds. This suggests the possibility of studying systems where the chromophore distribution is changing slowly as a function of time. There are many such dynamic systems of potential interest. For example, the diffusion of molecular species through separation or barrier membranes, controlled drug release, transcutaneous delivery systems, photochemical reactions in solid samples, curing processes in polymers, solid-state chemical reactions, etc. A wide range of specific areas of application can be envisaged.

The simplest possible study of this type would be to measure the photoacoustic spectrum as a function of time at a fixed modulation frequency, monitoring changes in the concentration of chromophore within the thickness of the sample probed by the thermal waves. This would not provide a full profile of the changes in concentration with depth in the sample, but could give information about average changes within a thin layer. A good example of this would be the measurement of the flux of a chromophore diffusing into or out of a zone in a sample defined by the thermal diffusion length. This method has been applied to the in vivo monitoring of the diffusion of topically applied muscle relaxant in human skin [7.85,86]. This application has been described in Sect. 7.3.1. By modelling the diffusion of the drug in skin, and combining that with a model for the generation of the photoacoustic signal, the authors were able to make estimates of the diffusion coefficients for the drug in the *stratum corneum* and the carrier vehicle, and also to estimate partition coefficients within the vehicle.

A better analysis of chromophore distribution can be obtained using phase resolved spectroscopy. *Nery* et al. [7.43] examined the effects of the herbicide paraquat on leaf samples for a period of 6 h after application of the herbicide. The most significant change observed was a decrease in phase shift between the cuticle signal and the subsurface pigment signal. This was ascribed to progressive dehydration of the leaf sample as a result of the toxic effects of paraquat. This was confirmed by measuring the thickness of a series of samples as a function of time under similar experimental conditions. In parallel with the changes in phase shift, there were significant changes in the intensity of both the cuticle and pigment signals, and substantial shifts in the spectral features of the pigment layer. In particular, the chlorophyll absorption was dramatically reduced in intensity after paraquat treatment. Although no analysis was made of the kinetics or mechanism of these changes, the potential is there for the study of quite complex biological systems whose characteristics vary with time.

The method of impulse response photoacoustic spectroscopy has also been applied to the characterization of dynamic systems. *Miller* et al. [7.24,103–105] have studied both the diffusion of a dye through a polymer membrane, and the action of paraquat on plant tissues. In the dye diffusion studies, a strongly absorbing dye was allowed to diffuse through a thin polymer membrane, and the photoacoustic impulse response obtained from the polymer surface opposite the dye reservoir recorded at intervals up to 30 min after the start of the experiment. No attempt at complete spectral characterization was made, and all responses were measured at a fixed laser wavelength. At the beginning of the experiment, the impulse response showed a small surface absorption feature due to the polymer film and a broad low amplitude subsurface absorption due to the chromophore. As the experiment progressed, the surface feature film remained constant, and the subsurface peak became sharper, more intense, and moved to shorter time delays, paralleling the movement of the chromophore through the membrane. Analysis of a number of experiments showed a strong linear relationship between chromophore absorption peak amplitude and diffusion time, and a reproducible nonlinear decay of peak delay for the chromophore absorption with diffusion time. A theoretical analysis of these data using a one-dimensional digital simulation of the photoacoustic impulse response [7.24] showed that the observed results could be explained by assuming classical Fickian diffusion through the membrane, and that many alternative diffusion profiles did not agree with the experimental data. To date, it has not been possible to extract real diffusion coefficients from these experiments, but an improved understanding of the theoretical model should lead to a better mapping between the model and the experiment, opening the way for useful information about diffusion profiles to be obtained.

The same group has also carried out experiments on the action of the herbicide paraquat on plant tissues [7.103,105]. In these experiments the impulse response for excised leaf samples was recorded after exposure to paraquat. In order to simplify the description of the experiment, the centroid of the photoacoustic impulse response was calculated for each observation and plotted against time. An analysis of untreated control samples showed a reduction in the overall intensity of response and a shift in the positions of surface and subsurface signals compatible with the dehydration process that had also been observed by *Nery* et al. [7.43]. However, *Nery* et al. did not report results obtained from control samples, so it is not clear whether the dehydration effects in the presence of paraquat that they reported were indeed due to the herbicide. If the response changes observed for the control are subtracted from the results for samples treated with paraquat, a further series of changes in the impulse response profile can be observed. These appear to suggest changes in the distribution of the photosynthetic pigments as well as changes in their spectral profile. It is possible that the toxic effects of the paraquat lead to a rupturing of the chloroplast membranes, and a diffusion of the photosynthetic pigments within the bulk of the leaf tissue. However, this has not been confirmed.

Acknowledgement: The author would like to thank his friends and colleagues in the photoacoustic community for the many hours they have spent with him discussing the issues presented in this review.

References

7.1   P.M. Epperson, J.V. Sweedler, R.B. Bilhorn, G.R. Sims, M.B. Denton: Anal. Chem. **60**, 282A (1988)
7.2   P.M. Epperson, J.V. Sweedler, R.B. Bilhorn, G.R. Sims, M.B. Denton: Anal. Chem. **60**, 327A (1988)
7.3   D.S. Burgi, N.J. Dovichi: Appl. Opt. **26**, 4665 (1987)
7.4   G.F. Kirkbright, M. Liezers, R.M. Miller, Y. Sugitani: Analyst **109**, 465 (1984)
7.5   H. Oechsner: *Thin Film and Depth-Profile Analysis*, Topics Curr. Phys., Vol. 37 (Springer, Berlin, Heidelberg 1984)
7.6   P.R. Griffiths, J.A. de Haseth: *Fourier Transform Infra-Red Spectrometry*, (Wiley, New York 1986)
7.7   J. Hodson, J.A. Lander: Polymer **28**, 251 (1987)
7.8   T. Wilson, C.J.R. Sheppard: *Theory and Practice of Scanning Optical Microscopy* (Academic, London 1984)
7.9   G.J. Brakenhoff, P. Blom, P. Barends: J. Microsc. **117**, 219 (1987)
7.10 T. Wilson: Lab. Practice **36** (12), 53 (1987)
7.11 M.D. Mithailov, M.N. Ozisik: *Unified Analysis and Solutions of Heat and Mass Diffusion* (Wiley, New York 1984)
7.12 A. Rosencwaig, A. Gersho: J. Appl. Phys. **47**, 64 (1976)
7.13 A. Rosencwaig: *Photoacoustics and Photoacoustic Spectroscopy* (Wiley, New York 1980)
7.14 M.J. Adams, G.F. Kirkbright: Analyst **102**, 678 (1977)
7.15 A. Rosencwaig: J. Applied Phys. **49**, 2905 (1978)
7.16 M.A. Afromowitz, P.S. Yeh, S. Yee: J. Appl. Phys. **48**, 209 (1977)
7.17 A. Mandelis: J. Math. Phys. **26**, 2676 (1985)
7.18 J. Crank: *The Mathematics of Diffusion* (Clarendon, Oxford 1975)
7.19 S.V. Patankar: *Numerical Heat Transfer and Fluid Flow* (McGraw-Hill, London 1980)
7.20 O.C. Zienkiewicz: *The Finite Element Method*, 3rd ed. (McGraw-Hill, London 1977)
7.21 A.J. Davies: *The Finite Element Method: A First Approach* (Clarendon, Oxford 1980)
7.22 V.A. Fishman, A.J. Bard: Anal. Chem. **53**, 2034 (1981)
7.23 R.M. Miller: Can. J. Phys. **64**, 1053 (1986)
7.24 R.M. Miller: Spectrochim. Acta B **43**, 687 (1988)
7.25 J. Fromm, H. Coufa: In *Photoacoustic and Photothermal Phenomena*, ed. by P. Hess, J. Pelzl, Springer Ser. Opt. Sci., Vol. 58 (Springer, Berlin, Heidelberg 1988) p. 464
7.26 D. Levesque, G. Rousset, L. Bertrand: Can. J. Phys. **64**, 1030 (1986)
7.27 M. Kasai, M. Ishioka, M. Kaihara, S. Fukushima, T. Sawada, Y. Gohshi: In *Photoacoustic and Photothermal Phenomena*, ed. by P. Hess, J. Pelzl, Springer ser. opt. sci. Vol. 58 (Springer, Berlin, Heidelberg 1988) p. 346
7.28 J. Crank, P. Nicholson: Proc. Camb. Philos. Soc. **43**, 50 (1947)
7.29 A.M. Clausing: In *Advanced Heat Transfer*, ed. by B.T. Chao (University of Illinois Press, Champaign, IL 1969) pp. 156–216
7.30 J.T. Jones: Ph.D. Thesis, Victoria University of Manchester (1989)
7.31 S.G. Tzafestas (ed.): *Distributed Parameter Control Systems: Theory and Application* (Pergamon, Oxford 1982)
7.32 P.A. Lynn: *An Introduction to the Analysis and Processing of Signals*, 2nd ed. (Macmillan, London 1982) Chap. 7
7.33 M.J. Adams, B.C. Beadle, A.A. King, G.F. Kirkbright: Analyst **101**, 553 (1976)
7.34 M.J. Adams, G.F. Kirkbright: Analyst **102**, 281 (1977)
7.35 A. Rosencwaig: Adv. Electron. Electron Phys. **46**, 207 (1978)
7.36 M.L. Mackenthun, R.D. Thom, T.A. Moore: Nature **279**, 265 (1979)
7.37 G.F. Kirkbright, R.M. Miller, D.E.M. Spillane, Y. Sugitani: Anal. Chem. **56**, 2043 (1984)
7.38 P. Helander, I. Lundstrom, D. McQueen: J. Appl. Phys. **52**, 1146 (1981)
7.39 E.P. O'Hara, R.D. Thom, T.A. Moore: Photochem. Photobiol. **38**, 709 (1983)
7.40 D.M. Anjo, T.A. Moore: Photochem. Photobiol. **39**, 635 (1984)
7.41 C.L. Cesar, H. Vargas, J. Pelzl, L.C.M. Miranda: J. Appl. Phys. **55**, 3460 (1984)
7.42 C.L. Cesar, H. Vargas, L.C.M. Miranda: J. Phys. D. **18**, 599 (1985)
7.43 J.W. Nery, O. Pessoa Jr., H. Vargas, F. de A.M. Reis, A.C. Gabrielli, L.C.M. Miranda, C.A. Vinha: Analyst **112**, 1487 (1987)
7.44 G.A.R. Lima, M.L. Basseo, Z.P. Arguello, E.C. Da Silva, H. Vargas, L.C.M. Miranda: Phys. Rev. B **36**, 9812 (1987)
7.45 C.K.N. Patel, A.C. Tam: Rev. Mod Phys. **58**, 381 (1981);
      A.C. Tam: Rev. Mod. Phys. **58**, 381 (1986)
7.46 G. Buchau, R. Stumpe, J.I. Kim: J. Less-Common Met. **122**, 555 (1986)

7.47   R. Klenze, R. Stumpe, J.I. Kim: In *Photoacoustic and Photothermal Phenomena*, ed. by P. Hess, J. Pelzl, Springer Ser. Opt. Sci., Vol 58 (Springer, Berlin, Heidelberg 1988) p. 139
7.48   S.E. Bialkowski, G.R. Long: Anal. Chem. **59**, 873 (1987)
7.49   S.L. Chin, D.K. Evans, R.D. McAlpine, W.N. Selander: Appl. Opt. **21**, 65 (1982)
7.50   Y. Bae, J.J. Song, Y.B. Kim: Appl. Opt. **21**, 35 (1982)
7.51   K. Heihoff, S.E. Braslavsky, K. Schaffner: Biochem. **25**, 1422 (1987)
7.52   R.W. Redmond, S.E. Braslavsky: In *Photoacoustic and Photothermal Phenomena*, ed. by P. Hess, J. Pelzl, Springer Ser. Opt. Sci., Vol. 58 (Springer, Berlin, Heidelberg 1988) p. 95
7.53   H. Sontag, A.C. Tam: Can. J. Phys. **64**, 1121 (1986)
7.54   D.A. Hutchins: Can. J. Phys. **64**, 1247 (1986)
7.55   H. Coufal: IEEE Trans UFFC 33, 507 (1986)
7.56   H. Coufal, P. Hefferle: Appl. Phys. A **38**, 213 (1985)
7.57   W.P. Leung, A.C. Tam: Opt. Lett. **9**, 93 (1984)
7.58   P. Cielo: J. Appl. Phys. **56**, 230 (1984)
7.59   S.O. Kanstad, P.E. Nordal: Can. J. Phys. **64**, 1155 (1986)
7.60   R.E. Imhof, D.J.S. Birch, F.R. Thornley, J.R. Gilchrist, T.A. Strivens: J. Phys. E **17**, 521 (1984)
7.61   D.L. Balageas, A.A. Deon, D.M. Boscher: Mater. Eval. **45**, 461 (1987)
7.62   R.E. Imhof, C.J. Whitters, D.J.S. Birch, F.R. Thornely: J. Phys. E **21**, 115 (1988)
7.63   R.E. Imhof: Private communication
7.64   S.E. Braslavsky: In *Photoacoustic and Photothermal Phenomena*, ed. by P. Hess, J. Pelzl, Springer Ser. Opt. Sci., Vol. 58 (Springer, Berlin, Heidelbeg 1988) and references therein
7.65   G.F. Kirkbright, R.M. Miller: Anal. Chem. **55**, 502 (1983)
7.66   M.F. Cox, G.N. Coleman: Anal. Chem. **53**, 2034 (1981)
7.67   C.Q. Yang, R.R. Bresee, W.G. Fateley: Appl. Spectrosc. **41**, 889 (1987)
7.68   Y.S. Touloukian, R.W. Powell, C.Y. Ho, M.C. Nicolasn: *Thermal Diffusivity* (Plenum, New York 1973)
7.69   K. Kato, S. Ishino, Y. Sugitani: Chem Lett. 783 (1980)
7.70   Y. Sugitani, A. Uejima, K. Kato: J. Photoacoust. **1**, 217 (1982)
7.71   Y. Sugitani, A. Uejima: Bull. Chem. Soc. Jpn. **57**, 2023 (1986)
7.72   G.F. Kirkbright, R.M. Miller, D.E.M. Spillane, Y. Sugitani: Anal. Chem. **56**, 2043 (1984)
7.73   G.F. Kirkbright, R.M. Miller, D.E.M. Spillane, I.P. Vickery: Analyst **109**, 1443 (1984)
7.74   A. Mandelis, J.T. Dodgson: J. Phys. C **19**, 2329 (1986)
7.75   R.M. Miller, C.T. Tye, I.J. Flynn: Unpublished data
7.76   A. Mandelis: Rev. Sci. Instrum. **57**, 617 (1986)
7.77   A. Mandelis, L.M.L. Borm, J. Tiessinga: Rev. Sci. Instrum. **57**, 622 (1986)
7.78   A. Mandelis, L.M.L. Borm, J. Tiessinga: Rev. Sci. Instrum. **57**, 630 (1986)
7.79   J.F. Power, A. Mandelis: Rev. Sci. Instrum. **58**, 2018 (1986)
7.80   J.F. Power, A. Mandelis: Rev. Sci. Instrum. **58**, 2024 (1986)
7.81   J.F. Power, A. Mandelis: Rev. Sci. Instrum. **58**, 2033 (1986)
7.82   R.C. Heyser: J. Audio Eng. Soc. **15**, 370 (1967)
7.83   S.D. Campbell, S.S. Yee, M.A. Afromowitz: J. Bio. Eng. **1**, 185 (1977)
7.84   S.D. Campbell, S.S. Yee, M.A. Afromowitz: IEEE Trans. BE-26, 220 (1979)
7.85   K. Giese, A. Nicolaus, B. Sennhenn, K. Kolmel: Can. J. Phys. **64**, 1139 (1986)
7.86   B. Sennhenn, M. Rohr, K. Giese, K. Kölmel: In *Photoacoustic and Photothermal Phenomena*, ed. by P. Hess, J. Pelzl, Springer Ser. Opt. Sci., Vol. 58 (Springer, Berlin, Heidelberg 1988) p. 548
7.87   U. Haas, J. Franz, F. Nimmerfall: In *Photoacoustic and Photothermal Phenomena*, ed. by P. Hess, J. Pelzl, Springer Ser. Opt. Sci., Vol. 58 (Springer, Berlin, Heidelberg 1988) p. 552
7.88   E.M. Nagel, H.K. Lichtenthaler: In *Photoacoustic and Photothermal Phenomena*, ed. by P. Hess, J. Pelzl, Springer Ser. Opt. Sci., Vol. 58 (Springer, Berlin, Heidelberg 1988) p. 568
7.89   A. Uejima, Y. Sugitani, K. Nakashima: Anal. Sci. **1**, 5 (1985)
7.90   A. Uejima, F. Itoga, Y. Sugitani: Anal. Sci. **2**, 113 (1986)
7.91   M.W. Urban, J.L. Koenig: Appl. Spectrosc. **40**, 994 (1986)
7.92   C.Q. Yang, W.G. Fateley: Anal. Chim. Acta **194**, 303 (1987)
7.93   C.Q. Yang, R.R. Bresee, W.G. Fateley, T.A. Peremich: *The structures of cellulose*. ACS Symp. Ser. **304**, 214 (1987)
7.94   C.Q. Yang, W.G. Fateley: Polym. Mater. Sci. Eng. **54**, 404 (1986)
7.95   P.G. Varlashkin, M.J.D. Low: Infra-red Phys. **26**, 171 (1986)
7.96   J. Philippaerts, E. Vanderleyden, E.F. Vansamt: In *Photoacoustic and Photothermal Phenomena*, ed. by P. Hess, J. Pelz, Springer Ser. Opt. Sci., Vol. 58 (Springer, Berlin, Heidelberg 1988) p. 33
7.97   D. Debarre, A.C. Boccara, D. Fournier: Appl. Opt. **20**, 4281 (1981)

7.98   R.A. Palmer, M.J. Smith, C.J. Manning, J.L. Chao, A.C. Boccara, D. Fournier: In *Photoacoustic and Photothermal Phenomena*, ed. by P. Hess, J. Pelzl, Springer Ser. Opt. Sci., Vol. 58 (Springer, Berlin, Heidelberg 1988) p. 50
7.99   M.G. DeOliveira, O. Pessoa Jr., H. Vargas, F. Galenbeck: J. Appl. Polym. Sci. **35**, 1791 (1988)
7.100 A. Bordeleau, G. Rousset, L. Bertrand, J.P. Crine: Can. J. Phys. **64**, 1093 (1986)
7.101 T. Zerlia: Appl. Spectrosc. **40**, 214 (1986)
7.102 B. Mongeau, G. Rousset, L. Bertrand: Can. J. Phys. **64**, 1056 (1986)
7.103 R.M. Miller, G.R. Surtees, C.T. Tye, I.P. Vickery: Can. J. Phys. **64**, 1146 (1986)
7.104 R.M. Miller, G.R. Surtees, C.T. Tye: Analyst **114**, 547 (1989)
7.105 D.A. Adesida, R.M. Miller, M.L. Waller: Unpublished data

8. Frequency-Modulated Time-Delay-Domain Photothermal Spectrometry: Principles, Instrumentation and Applications to Solids

A. Mandelis

With 27 Figures

The origins, development and state of the art of the photothermal wave frequency-modulated (FM) spectrometry, a technique intermediate between the frequency and time domains, are examined in this chapter in terms of correlation and spectral functional analysis. The nature of FM excitation is discussed in the context of photothermal signal generation and detection. The instrumentation pertinent to the FM thermal wave experimental techniques is further described, including signal recovery methods and instrumental detection limitations posed by the spectral requirements of FM optical excitation and impulse response analysis. A review of photothermal applications to date is then presented which includes three major photothermal detection schemes and their application to solids and surfaces: photothermal beam deflection spectrometry (FM-PDS) and its application to the measurement of thermal diffusion delay times associated with the photothermal impulse response of thin solid layers; thin-film photopyroelectric spectrometry (FM-P^{2}ES) and its application to thermal diffusivity and thermal conductivity measurements of metals and insulating solids; and photothermal reflectance spectrometry (FM-PTR), which exhibits the widest frequency response bandwidth of the available photothermal techniques, as well as instrumental limitations arising from high frequency contributions to the photothermal impulse response.

The experimental and theoretical framework of the photothermal FM techniques developed to date is shown to be indicative of the capabilities of these techniques to perform frequency-domain photothermal measurements of superior dynamic range via fast Fourier transformations of the input data, in a very short time compared to the conventional lock-in detection, and simultaneous time-domain-equivalent (time-delay spectrometric) impulse response measurements, which are capable of replacing pulsed laser sources with cw lasers plus the mathematical equivalent of an optical pulse (the Green's Function). The latter feature holds very good promise for use in non destructive evaluation of delicate or high technology materials with low damage threshold to optical pulses.

8.1 Introduction and Conceptual Building Blocks

8.1.1 Nature of FM Excitation

The concept of the frequency-modulated (FM) photothermal excitation and signal detection technique is intermediate between the time and frequency domains. The most widely used application of this method is, perhaps, the frequency modulation (FM) signal transmission and detection of communications systems. It was introduced by *Heyser* [8.1] in 1967 in the field of acoustical measurements of loudspeakers and was named time delay spectrometry (TDS) by the same author. Through its implementation and long-term use in acoustic engineering, TDS has been shown to out-perform any other time selective technique with respect to noise rejection and nonlinearity suppression from measurements of systems with linear behavior [8.2]. The TDS technique, which is based on a linear frequency sweep of the excitation function, is akin to chirp modulation [8.3,4]. It has been specifically compared to the impulse response transformation and the wideband (pseudo) random-noise methods and has been proven to have superior measurement dynamic range properties [8.5]. The reason for the excellent dynamic range characterizing TDS can be understood by regarding the measurement system to which the frequency-swept signal is applied as a transmission line. An important parameter of such a system is the total time delay τ_D required for the transmission of information from the input to its output. Primary consideration, therefore, must be given to minimizing the effective duration of a transmitted pulse. A relevant measure of this duration is T_D, the second moment of the power of the impulse response $h(\tau)$ about a suitably chosen origin τ_0 [8.1]:

$$T_D = \min_{-\infty < \tau_0 < \infty} \left[\int_{-\infty}^{\infty} \tau^2 \, |h(\tau - \tau_0)|^2 \, d\tau \right]. \tag{8.1.1}$$

It can be shown [8.6] that T_D is minimized when

$$t_g(\omega_i) = \frac{d\phi_i(\omega_i)}{d\omega_i} = \text{constant} \tag{8.1.2}$$

where $t_g(\omega_i)$ is a group time delay in the classical wave mechanical sense and expresses the relative time shift of instantaneous signal frequency components $\omega_i = 2\pi f_i$ adjacent to a reference frequency [8.7]; and $\phi_i(\omega_i)$ is the instantaneous signal phase angle. Using the minimum transit time condition (8.1.2), the rate of change of the phase angle

$$\frac{d\phi_i(t)}{dt} = \frac{d\phi_i}{d\omega_i} \frac{d\omega_i}{dt} \tag{8.1.3}$$

is seen to involve the instantaneous signal frequency dependence on time, a concept incompatible with conventional Fourier transform theory, where the variables frequency f and time t are taken to represent physical phenomena in two mathematical domains, which are Fourier transforms of each other. Nevertheless, (8.1.2) and (8.1.3) show that as long as

$$\frac{d\phi_i(t)}{dt} = t_g \frac{d\omega_i(t)}{dt}, \tag{8.1.4}$$

the effective duration of a transmitted pulse will be minimized. This network has been called a minimum phase system [8.8]. The minimization of the pulse duration (width) thus results in an extended dynamic range of the system transfer function, defined as the Fourier transform of the impulse response, $h(\tau)$, of the network:

$$H(f) = \int_{-\infty}^{\infty} h(\tau) e^{-2\pi i f \tau} d\tau. \tag{8.1.5}$$

This important characteristic of minimum phase networks was first exploited in photothermal applications by *Mandelis* et al. [8.9,10]. A minimum phase system can be shown to be a linear phase system [8.2], i.e.,

$$d\omega_i(t)/dt = \text{constant}. \tag{8.1.6}$$

This implies that a sweep input signal can be any function of time $x(t)$ modulated between the two extreme deviation frequencies, such that the instantaneous value of the frequency-like quantity $f_i(t)$ is given by [8.1]

$$f_i(t) = \left[\frac{\Delta f}{T} \right] t + \tfrac{1}{2}(f_2 + f_1) \equiv \left[\frac{\Delta f}{T} \right] t + f_c. \tag{8.1.7}$$

In (8.1.7), $\Delta f = f_2 - f_1$ is the carrier signal modulation bandwidth, $f_c = (f_2 + f_1)/2$ is the average carrier frequency, and T is the total sweep period.

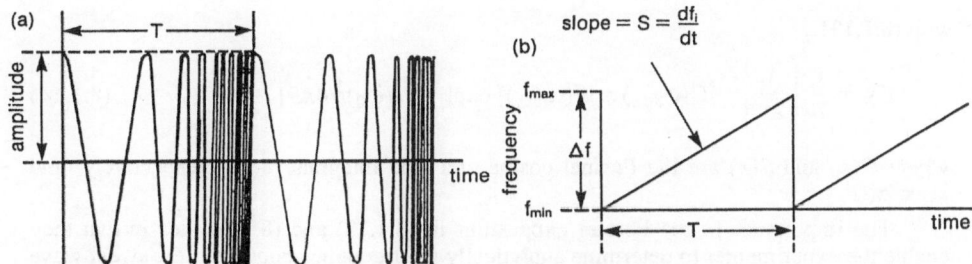

Fig. 8.1. Schematic representation of linear frequency sweep concepts: **(a)** amplitude-time profile; **(b)** frequency-time profile corresponding to sweep (a)

The sweep rate S is defined as the time derivative of the frequency-like quantity $f_i(t)$:

$$S = \frac{df_i(t)}{dt} = \frac{\Delta f}{T} \text{ [Hz/s]}. \tag{8.1.8}$$

The sweep rate S is independent of time, a general feature of linear sweep modulation systems that results directly from the linear dependence on time of (8.1.7). Assuming the excitation function to be a cosinusoidal carrier wave [8.1,2,5], a time-delay photothermal system input will be given by (Fig. 8.1a)

$$x(t) = A(t) \cos\left[\phi_i(t)\right] \tag{8.1.9}$$

where $A(t)$ is the amplitude modulation (AM) function, usually chosen to be constant. Taking t_g to be a time delay commencing at the initiation of the input signal [8.11], i.e., $t_g = t$ in (8.1.4), we find upon integration

$$\phi_i(t) = (\pi S)\, t^2 + \omega_c t + \phi_o \tag{8.1.10}$$

where $\phi_o \equiv \phi_i(0)$ is the input phase at $t = 0$.

The experimental conditions chosen for photothermal wave measurements are [8.9,10] $\phi_o = 0$ and $f_1 = 0$ and $\phi_i(T + \delta t) = \phi_i(\delta t)$, for $\delta t \to 0$, where A is constant. These correspond to a linear saw-tooth frequency sweep between dc and $f_2 = f_{\max}$ with multiple repetitions of the sweep process every period T as shown in Fig. 8.1b. Under these conditions, (8.1.9) can be conveniently written in the form [8.1]

$$x(t) = x_+(t) + x_-(t) ; \quad 0 \le t \le T \tag{8.1.11}$$

where

$$x_+(t) = \frac{A}{2} \exp[i(\pi S t^2 + \omega_c t)] \tag{8.1.12a}$$

and

$$x_-(t) = \frac{A}{2} \exp[-i(\pi S t^2 + \omega_c t)]. \tag{8.1.12b}$$

It can further be shown [8.1] that the frequency swept signal component functions $\exp[\pm i(\pi S t^2)]$ in (8.1.12) can be expanded in conventional Fourier series

$$\exp[\pm i(\pi S t^2)] = \sum_{N=-\infty}^{\infty} C_N^{\pm}\, e^{iN\omega_0 t} \tag{8.1.13}$$

213

with [8.1,12]

$$C_N^\pm = \frac{1}{T} \left[\frac{1}{2S} \right]^{\frac{1}{2}} [C(\omega_{max}) \pm iS(\omega_{max})] \exp[\mp i(N\omega_0)^2/4\pi S] \qquad (8.1.14)$$

where $C(x)$ and $S(x)$ are the Fresnel cosine and sine integrals [8.13], respectively, and $\omega_0 \equiv 2\pi/T$.

The importance of the Fourier expansions in (8.1.13) and (8.1.14) lies in that they enable the experimenter to determine analytically the frequency content of the swept wave along with the weighting factors $C_N^\pm$, which ultimately give the frequency bandwidth of the Fourier transform of the input signal.

8.1.2 Classification of Correlation Functions and Photothermal Spectral Analysis

The measurement of photothermal signals using FM-TDS methods relies on the validity of the assumption that the photothermal system may be modeled as a general linear system, or "black box" (Fig. 8.2) with impulse response $h(\tau)$, input wave train $x(t)$ and output response wave train $y(t)$. The system transfer function is $H(f)$, defined by (8.1.5), and the frequency contents of input and output wave trains are given by $X(f)$ and $Y(f)$, respectively. The photothermal wave information, which is contained in the output signal response of the system, can be recovered using correlation and spectral analysis techniques [8.14]. The frequency swept input signal can be expressed analytically as a complex function of time, (8.1.11) and (8.1.12). The time-dependent signals obtained via correlation and convolution analysis carry information related to the delay of the output signal with respect to the input. It is, therefore, customary [8.1,15] to use the variable τ instead of t for time and to refer to the signal temporal evolution process in the time delay domain. In the time domain, the complex output of the photothermal system $y(t)$ is given as the convolution integral between the complex FM input signal and the unit impulse response function $h(t)$ of the system, provided the latter is assumed to be time invariant

$$y(t) = \int_0^\infty h(\tau)x(t-\tau)d\tau \quad [V] \qquad (8.1.15a)$$

$$\equiv h(t) * x(t). \qquad (8.1.15b)$$

Equation (8.1.15a) assumes a causal relationship between input and system response, so that the lower limit of the integration is set to zero, rather than $-\infty$. The three time delay domain functions of importance to time delay photothermal wave spectrometry are the autocorrelation function $R_{xx}(\tau)$ of the input; the cross-correlation function $R_{xy}(\tau)$ between input and output; and the unit impulse response function of the system $h(\tau)$:

$$R_{xx}(\tau) = \lim_{T\to\infty} \frac{1}{T} \int_0^T x^*(t)x(t+\tau)dt \quad [V^2], \qquad (8.1.16)$$

$$R_{xy}(\tau) = \lim_{T\to\infty} \frac{1}{T} \int_0^T x^*(t)y(t+\tau)dt \quad [V^2], \qquad (8.1.17)$$

and

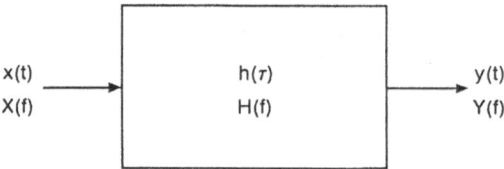

Fig. 8.2. Generalized "black-box" representation of frequency response $H(f)$ for a photothermal system

$$h(\tau) = \int_{-\infty}^{\infty} H(f)e^{2\pi i f \tau} \, df. \tag{8.1.18}$$

In (8.1.16) and (8.1.17) starred quantities indicate complex conjugation. The following important relationship can be proved [8.12] from (8.1.15-17):

$$R_{xy}(\tau) = h(\tau) * R_{xx}(\tau) \ . \tag{8.1.19}$$

One advantage of FM-TDS over conventional frequency and time domain techniques is its ability to generate information and signals in both domains simultaneously with one set of measurements. Therefore, the time delay domain functions described above must be supplemented by frequency domain counterparts. These are complex functions carrying amplitude and phase information and are defined as Fourier transforms of the time delay domain functions. Due to the fact that it is experimentally impossible to measure negative frequencies, as expected by the definition of the Fourier transform, the one-sided input autospectral density and the input-output cross-spectral density are given by

$$G_{xx}(f) = 4\int_{0}^{\infty} R_{xx}(\tau)\cos(2\pi f\tau)d\tau \quad [\text{V}^2/\text{Hz}], \tag{8.1.20}$$

$$G_{xy}(f) = 4\int_{0}^{\infty} R_{xy}(\tau)\cos(2\pi f\tau)d\tau \quad [\text{V}^2/\text{Hz}] \tag{8.1.21}$$

$$\equiv C_{xy}(f) - \mathrm{i}\,Q_{xy}(f). \tag{8.1.22}$$

Separating out real and imaginary parts defines the cross-spectral magnitude and phase:

$$G_{xy}(f) = [C_{xy}^2(f) + Q_{xy}^2(f)]^{\frac{1}{2}} \quad [\text{V}^2/\text{Hz}] \tag{8.1.23}$$

and

$$\theta_{xy}(f) = \tan^{-1}[Q_{xy}(f)/C_{xy}(f)] \quad [\text{rad}]. \tag{8.1.24}$$

The coherence function is defined as

$$\gamma_{xy}^2(f) = \frac{|G_{xy}(f)|^2}{G_{xx}(f)G_{yy}(f)} \ ; \quad 0 \le \gamma_{xy}^2(f) \le 1 \tag{8.1.25}$$

and constitutes a powerful indicator of the strength of the relation between the input signal and output response of the system. In (8.1.25), $G_{yy}(f)$ can be determined from the output autocorrelation function $R_{yy}(\tau)$ in a manner analogous to the one used for the determination of $G_{xx}(f)$. A very useful related function is the coherent output power spectrum

$$\gamma_{xy}^2(f)G_{yy}(f) = |G_{xy}(f)|^2/G_{xx}(f). \tag{8.1.26}$$

This function is a measure of the output power spectrum caused by the input excitation in the presence of noncoherent noise. The ability of the system to respond to the input signal frequency spectrum is determined by the complex transfer function or frequency response function $H(f)$ defined in terms of spectral densities by

$$H(f) = G_{xy}(f)/G_{xx}(f). \tag{8.1.27}$$

If $H(f)$ is written in polar coordinate form

$$H(f) = |H(f)| e^{-i\phi(f)}, \tag{8.1.28}$$

it can be shown from (8.1.23,24) and (8.1.28) that

$$|H(f)| = |G_{xy}(f)|/G_{xx}(f)$$ (8.1.29)

and

$$\phi(f) = \theta_{xy}(f).$$ (8.1.30)

The transfer function $H(f)$ and the unit impulse response function have been previously defined to be Fourier transforms of each other.

8.2 Experimental FM-TDS Recovery Techniques, Dynamic Range and Limitations

The above considerations show that, in theory, FM-TDS signals require recovery through the correlation and spectral analysis presented here. In practice, however, in cw broadband-modulation experiments, the use of a FFT signal analyzer such as the HP 3562A enables the recovery of all correlation functions by inverse transformation of averaged spectral density functions. This is made feasible by means of the Weiner-Khinchine relations [8.14] in which successive Fourier transforms of the input $X(f)$ and output $Y(f)$ are averaged over a measurement period T:

$$G_{xx}(f) = \lim_{T \to \infty} (1/T) <X^*(f)X(f)>,$$ (8.2.1)

$$G_{yy}(f) = \lim_{T \to \infty} (1/T) <Y^*(f)Y(f)>,$$ (8.2.2)

$$G_{xy}(f) = \lim_{T \to \infty} (1/T) <X^*(f)Y(f)>,$$ (8.2.3)

for the autospectra of the input and output, $G_{xx}(f)$ and $G_{yy}(f)$, respectively, and for the cross spectrum, $G_{xy}(f)$.

The cross and autocorrelation functions, $R_{xy}(\tau)$ and $R_{xx}(\tau)$, are recovered by inverse transformation of (8.2.1-3) to give the appropriate correlation functions:

$$R_{xx}(\tau) = \int_0^\infty e^{2\pi i f \tau} G_{xx}(f) df,$$ (8.2.4)

$$R_{yy}(\tau) = \int_0^\infty e^{2\pi i f \tau} G_{yy}(f) df,$$ (8.2.5)

$$R_{xy}(\tau) = \int_0^\infty e^{2\pi i f \tau} G_{xy}(f) df.$$ (8.2.6)

Similarly, frequency-response (transfer function) data are directly recovered from the spectral density functions via (8.1.27) and (8.1.28) from cross-spectral density data and

$$|H(f)| = [G_{yy}(f)/G_{xx}(f)]^{1/2}$$ (8.2.7)

from autospectral data.

The strategy of broadband-modulated cw techniques, in general, is to excite the photothermal system with $G_{xx}(f) = 1$. Equivalently, the input autocorrelation function must satisfy $R_{xx}(\tau) \sim \delta(\tau)$ on the time scale of the experiment, where $\delta(\tau)$ is the symbol for the

216

Dirac delta function. Consideration of (8.1.18) and (8.1.27) shows that a flat (unity) G_{xx} has the consequence that

$$H(f) \sim G_{xy}(f) \tag{8.2.8}$$

while (8.1.19) shows that an infinitesimally narrow input pulse (measured on the time scale of the photothermal experiment) results in the same functional form between the system impulse response and the cross-correlation function:

$$R_{xy}(\tau) \sim h(\tau). \tag{8.2.9}$$

It is relatively easy to visualize the application of these correlation techniques in experiments which use stochastic excitation. By using random noise of sufficiently wide bandwidth as the excitation waveform, the flatness of the input autospectrum is maintained. By contrast, it is important to verify that the condition $G_{xx}(f) = 1$ is valid for excitation by deterministic signals. In particular, the linear sine sweeps used for signal excitation in FM-TDS techniques possess flat autospectra only in the limit of sufficiently long sweep times. The consequences of any spectral approximations impact directly on the quality of the spectral functions and frequency-response information obtained.

For the linear FM sweeps described by (8.1.11) and (8.1.12)

$$R_{xx}(\tau) = (1/T)[R_{x_+x_+}(\tau) + R_{x_-x_+}(\tau) + R_{x_+x_-}(\tau) + R_{x_-x_-}(\tau)], \tag{8.2.10}$$

where [8.16]

$$R_{x_+x_+}(\tau) + R_{x_-x_-}(\tau) = \frac{1}{4} \frac{T - |\tau|}{T} \left[\frac{(e^{i2\pi S\tau(T - |\tau|)} - 1)}{i2\pi S\tau(T - |\tau|)} \right.$$

$$\left. - \frac{(e^{-i2\pi S\tau(T - |\tau|)} - 1)}{i2\pi S\tau(T - |\tau|)} \right] \tag{8.2.11}$$

and

$$R_{x_+x_-}(\tau) + R_{x_-x_+}(\tau) = \frac{1}{2T} \int_{|\tau/2|}^{T - |\tau/2|} \cos\left[2\pi S \left[t^2 + \frac{\tau^2}{4} \right] \right] dt. \tag{8.2.12}$$

For large T, (8.2.12) approximates to $1/[8\sqrt{(\Delta f)T}]$ so that its contribution is suppressed with long sweep times T and wide modulation bandwidths Δf.

It is possible to attain the condition $R_{xx}(\tau) \approx \delta(\tau)$ if the cross terms in (8.2.12) are zero so that

$$R_{xx}(\tau) \approx R_{x_+x_+}(\tau) + R_{x_-x_-}(\tau).$$

Even with the cross terms suppressed, it is still not obvious why and when $R_{x_+x_+}(\tau) + R_{x_-x_-}(\tau) \approx \delta(\tau)$. This may be clarified by considering the desired condition $G_{xx}(f) = 1$. Ideally, this condition is only met as $\Delta f \to \infty$. Infinite modulation bandwidth is neither achievable nor actually required in practice as long as the output response of the photothermal system, $Y(f) \sim 0$ for $f > f_{max}$.

For long enough sweep times $T >> |\tau|$, the "autopolarized" terms in (8.2.11) become

$$R_{x_+x_+}(\tau) + R_{x_-x_-}(\tau) = \frac{1}{4}\frac{e^{i2\pi(\Delta f)\tau} - 1}{2\pi i S\tau T} - \frac{1}{4}\frac{e^{-i2\pi\Delta f\tau} - 1}{2\pi i S\tau T} \tag{8.2.13}$$

$$= \sin(2\pi\Delta f\tau)/4\pi\Delta f\tau.$$

This result has the identical form predicted for the real part of the inverse Fourier transform of the idealized square frequency window depicted in Fig. 8.3. As the modulation bandwidth Δf approaches very large values, the real part of $R_{xx}(\tau)$ approaches the Dirac delta function $\delta(\tau)$.

The imaginary part of the time-domain expression is shown in Figs. 8.3b and c, however, it constitutes redundant information, since it is readily obtainable from the real part by Hilbert transformation [8.2].

The form of the input autocorrelation function is thus demonstrated to be mathematically equivalent to a Dirac delta function provided the sweep time is long. If the sweep time is insufficiently long, i.e., $T \leq 1/2\pi S\tau$, the approximation to the Dirac delta function, which is recovered by standard correlation techniques, no longer applies.

An important advantage of the FM time-delay domain technique over both the impulse-response transformation (pulsed) method and the wideband random noise methods [8.17] is its superior dynamic range. For a system or device in which the frequency response exhibits a resonance with a characteristic bandwidth Δf_R centered at f_0, the maximum value of the Fourier transform of the transient time response of the system to an impulsive type of excitation (theoretically, a Dirac delta-function) is given by [8.5]

$$[F_{IR}(f)]_{max} = F_{IR}(f_0) = \frac{1}{\pi}X_{peak}\left[\frac{\Delta f_M}{\Delta f_R}\right] \tag{8.2.14}$$

where X_{peak} is the maximum amplitude of the transient response, and $\Delta f_M = 1/T_M$ with T_M the pulse recirculation time. For a typical fast-Fourier-transform (FFT) signal-analyzer frequency resolution of, e.g., $3\Delta f_M = \Delta f_R$, (8.2.14) gives

$$F_{IR}(f_0) = \frac{1}{3\pi}X_{peak}. \tag{8.2.15}$$

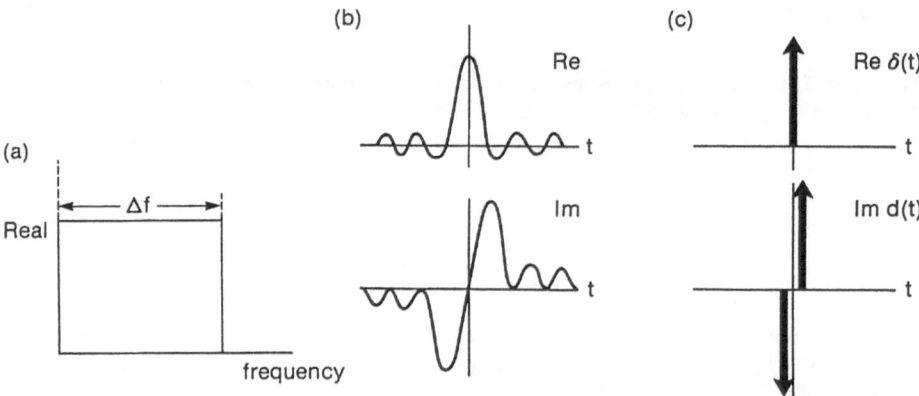

Fig. 8.3. Frequency-time schematics showing (**a**) idealized square frequency window $u(f - f_a)$ and (**b**) its Fourier transform (IFT); (**c**) Fourier transform of (a) in the limit $f_{max} \to \infty$; $d(t)$ stands for the doublet symbol (i.e., the imaginary component of the inverse Fourier transform of the infinitely broadband frequency window)

If the input signal is of the frequency swept type and given by (8.1.9) with $A(t) = X_{peak}$, then at the resonant frequency f_0, the maximum value of the Fourier transform of the FM response is

$$F_{FM}(f_0) = X_{peak}.$$ (8.2.16)

From (8.2.15) and (8.2.16) it can be seen that, for an instrumental frequency resolution requirement of $\Delta f_M = \Delta f_R/3$, the difference in available dynamic range between a frequency sweep measurement and an impulse response transformation measurement on the same singly resonant test system is of the order of

$$20 \log_{10}\left[\frac{F_{FM}(f_0)}{F_{IR}(f_0)}\right] \approx 20 \text{ dB}.$$ (8.2.17)

In the FM mode, Δf_M is defined as the inverse of the sweep period: $\Delta f_M = 1/T$. Figure 8.4 shows a comparison between frequency response functions of a multiresonant acoustic system measured by the two techniques, where the superior dynamic range of the FM technique is clearly demonstrated. If wideband random noise is used as the input signal to the test system, the maximum value of the Fourier transform can be shown to be [8.5]

$$F_{RN}(f_0) = \frac{1}{3}\left[\frac{\Delta f_M}{\Delta f_R}\right]^{\frac{1}{2}} X_{peak}$$ (8.2.18)

where Δf_M now is the maximum root-mean-square (rms) value of the measured noise bandwidth centered at f_0. For $\Delta f_M/\Delta f_R = \frac{1}{3}$, the loss in dynamic range compared to the FM technique is

$$20 \log_{10}\left[\frac{F_{FM}(f_0)}{F_{RN}(f_0)}\right] \approx 15 \text{ dB}.$$ (8.2.19)

This value corresponds quite closely to actually measured results. For a linear frequency sweep, a requirement for good spectral resolution is that the sweep rate, (8.1.8), through a resonance f_0 with bandwidth Δf_R, must satisfy

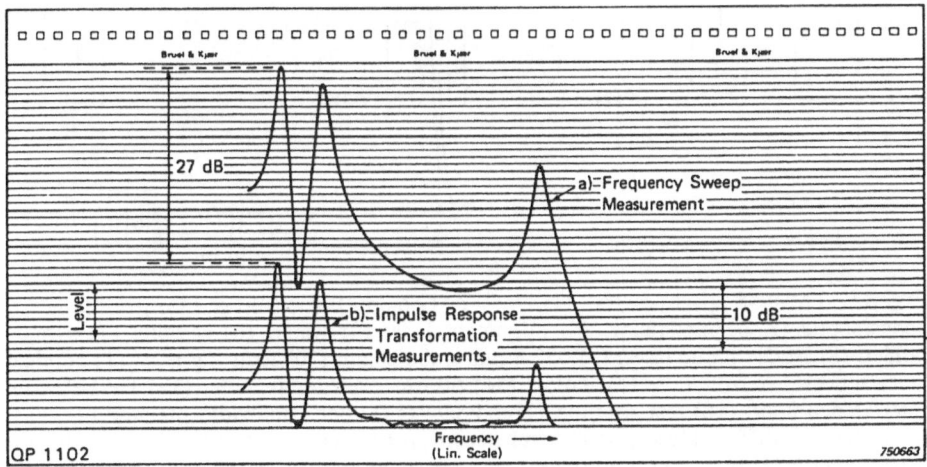

Fig. 8.4. The frequency response function of a multiresonant acoustic device [8.5]

$$S \le (\Delta f_R)^2. \tag{8.2.20}$$

This relation is tantamount to an instrumental "uncertainty principle" of the form

$$(\Delta f_R)(\Delta t) \ge 1. \tag{8.2.21}$$

Hence

$$S = \lim_{\Delta t \to 0} \frac{\Delta f_R}{\Delta t} \le (\Delta f_R)^2. \tag{8.2.22}$$

8.3 Photothermal Wave Applications

8.3.1 Photothermal Beam Deflection FM Spectrometry

(a) Instrumentation and Performance

The first reported photothermal wave system with FM time-delay-domain optical excitation is the photothermal deflection spectroscopic apparatus of *Mandelis* et al. [8.9] shown in Fig. 8.5. These authors investigated the performance of the PDS apparatus using a

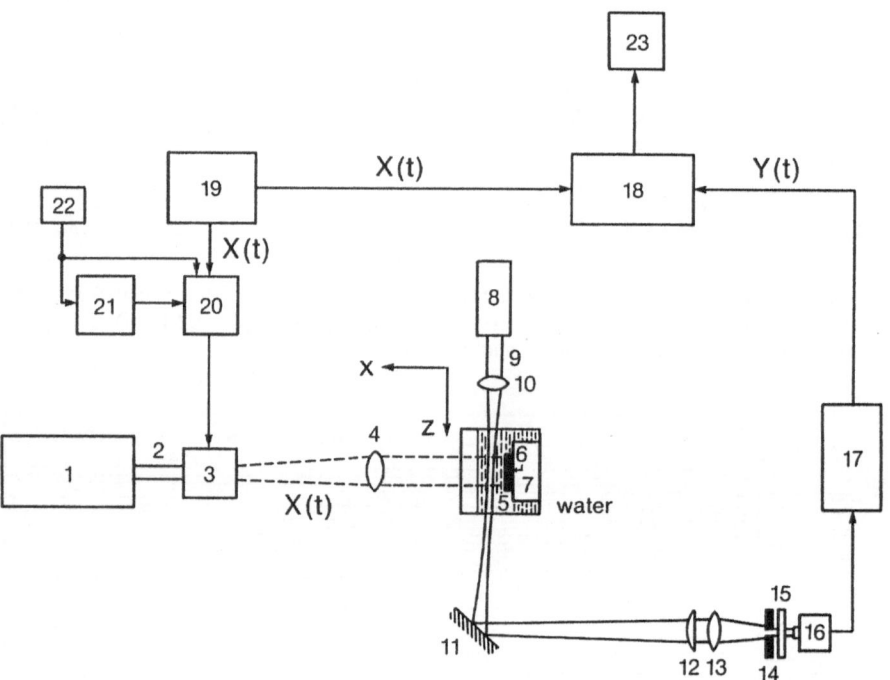

Fig. 8.5. FM-PDS spectrometer. (1) Nd^{3+}:YAG pump laser; (2) cw 1.06 µm beam; (3) acoustooptic (A/O) modulator; (4) alignment lens; (5) water; (6) sample; (7) sample holder; (8) He-Ne probe laser; (9) 632.8-nm probe beam; (10) focusing lens; (11) optical lever reflector mirror; (12, 13) lenses; (14) 50-µm diameter pinhole; (15) He-Ne beam interference filter; (16) fast risetime photodiode; (17) wide bandwidth preamplifier; (18) dual channel FFT analyzer; (19) synthesizer/function generator; (20) A/O modulator driver; (21) A/O driver power amplifier; (22) A/O modulator power supply; (23) computer memory storage

220

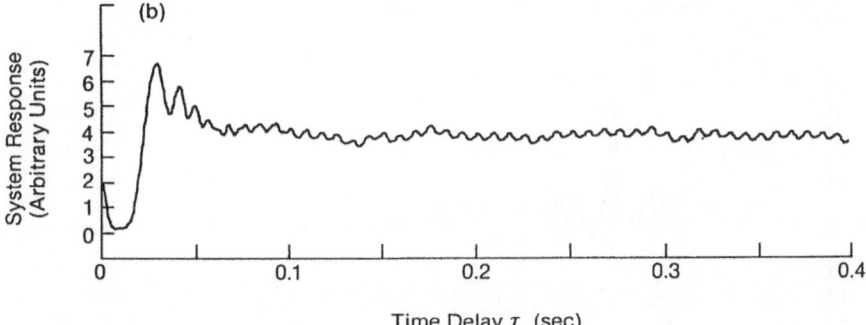

Fig. 8.6. FM excitation swept square-wave function (**a**) and PDS response (**b**) of an anodized aluminum/water interface

blackbody reference sample (anodized aluminum) in water. A fast beam position detector was fabricated using a pinhole-photodiode arrangement with a 34 ns response time. The excitation beam from a 2 W Nd^{3+}:YAG pump laser was expanded over the sample in order to facilitate the (one-dimensional) theoretical interpretation of the data. Frequency modulation of the pump beam intensity was effected using an HP 3325A Synthesizer/Function Generator. The system output was registered as a photovoltage whose amplitude was proportional to the spatial deflection of the He-Ne probe beam due to the mirage effect [8.18]. All the necessary frequency and time-delay-domain functions were calculated via a Nicolet Scientific Corp. Model 660A dual channel FFT analyzer. Figure 8.6 shows typical FM-PDS excitation and response sweeps using square-wave modulation of the acousto-optic (A/O) modulator.

Figure 8.7 shows the impulse response $h(\tau)$ and the cross-correlation $R_{xy}(\tau)$ functions generated with a beam offset of ~10μm, averaged over 1000 frequency sweeps with 1024 data points per sweep. The total amount of time required for each curve in Fig. 8.7 was approximately 6-7 min. This time could easily be reduced by half or more, as the quality of the spectral functions remained essentially unaltered after 200-300 averages were obtained. Both $h(\tau)$ and $R_{xy}(\tau)$ have essentially the same peak delay time τ_0 and full-width at half-maximum time τ_{FWHM}, with a somewhat longer delay time of the minimum (negative) cross-correlation τ_{min} than the impulse response τ_{min}. This small discrepancy may be due to the oscillatory nature of the cross-correlation function about the zero level on either side of the peak [8.14]. Otherwise, the curves of Figs. 8.7a and b are essentially identical. Figure 8.8 shows the autocorrelations of the input (a) and output (b) signals corresponding to the data of Fig. 8.7. It can be observed from Fig. 8.8a that the input autocorrelation $R_{xx}(\tau)$ is extremely narrow on the time scale of the experiment and can, therefore, be approximated by the Dirac delta function

$$R_{xx}(\tau) \cong \delta(\tau). \tag{8.3.1}$$

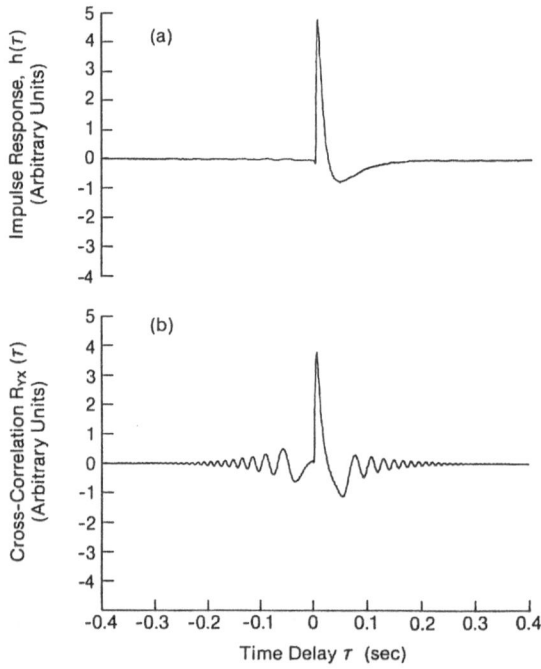

Fig. 8.7. (a) Impulse response of anodized aluminum/water interface at beam offset $x_0 \cong 50 \, \mu m$. Peak delay time $\tau_0 = 2.34$ ms; $\tau_{FWHM} = 5.13$ ms; $\tau_{min} = 23.44$ ms. **(b)** Cross-correlation of the same system. Peak delay time $\tau_0 = 2.34$ ms; $\tau_{FWHM} = 5.2$ ms; $\tau_{min} = 26.56$ ms

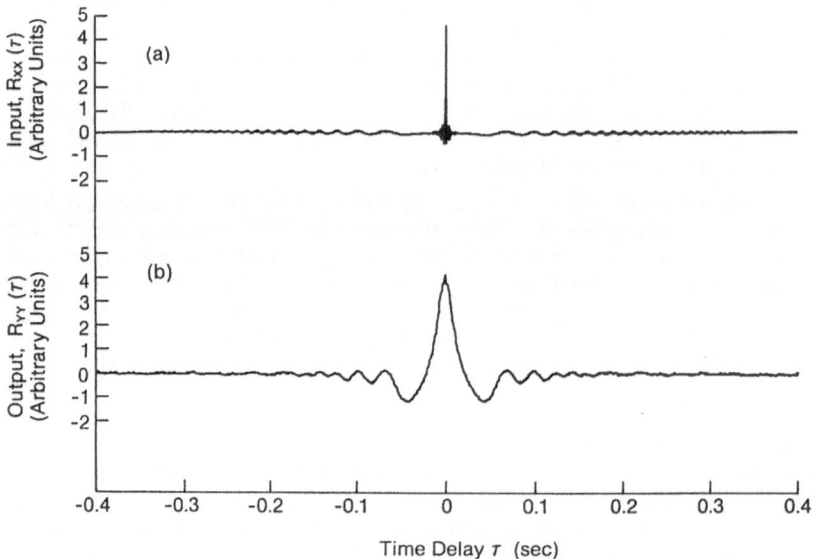

Fig. 8.8. Autocorrelation functions of (a) the input and (b) the output PDS signal waveforms. Swept wave excitation between dc and 1280 Hz

222

From (8.1.19) we can write

$$R_{xy}(\tau) \cong h(\tau) * \delta(\tau) = h(\tau) \tag{8.3.2}$$

in agreement with Fig. 8.7. The impulse response of Fig. 8.7a can be essentially understood in terms of thermal wave conduction from the blackbody surface into the water mass, after excitation by a thermal pulse of infinitesimal duration at time $\tau = 0$. In that case the spatial and temporal profile of the temperature rise in the fluid is given by the Green's function corresponding to the one-dimensional heat conduction equation. The solution is [8.19]

$$T_f(x,\tau) = \frac{Q k_s (\alpha_f/\alpha_s)^{1/2}}{(k_s \alpha_f^{1/2} + k_f \alpha_s^{1/2})} \frac{\exp(-x^2/4\alpha_f\tau)}{(\pi\tau)^{1/2}}, \tag{8.3.3}$$

where a plane instantaneous heat source of strength Q [J/cm^2] was assumed in the solid at its surface, the solid-liquid interface, at $x' = 0$. Using the paraxial ray approximation in the geometry of Fig. 8.5, we can write the equation of motion of the probe beam intensity centroid deflection [8.20]:

$$\frac{d^2 x_a}{dz^2}(\tau,z) + \frac{F}{\tau^{3/2}} x_a(\tau,z) \exp[-x_a^2(\tau,z)/4\alpha_f\tau] = 0, \tag{8.3.4}$$

where

$$F \equiv \frac{1}{2} Q \left[\frac{\partial n}{\partial T} \right]_{T=T_0} \frac{k_s}{n_0(\pi\alpha_s\alpha_f)^{1/2}(k_s\alpha_f^{1/2} + k_f\alpha_s^{1/2})}. \tag{8.3.5}$$

In (8.3.3-5), the symbols α and k stand for thermal diffusivity and conductivity, respectively, for the solid (s) and the fluid (f), while n_0 is the fluid refractive index at ambient temperature. Equation (8.3.4) is nonlinear and can be solved analytically only via a perturbation analysis [8.20] in the time delay domain. The analysis predicts a peak delay time

$$\tau_0 = x_0^2/6\alpha_f. \tag{8.3.6}$$

Using $x_0 = 50$ μm and $\alpha_f = 1.4 \times 10^{-3}$ cm^2/s, one finds $\tau_0 = 2.97$ ms in good agreement with the experimental τ_0, Fig. 8.7. Equation (8.3.6) can be also used to compute the FWHM in Fig. 8.7

$$x_a(\tau_{FWHM},z) = \frac{1}{2} x_a(\tau_0,z). \tag{8.3.7}$$

It can be shown that

$$\tau_{FWHM} \cong x_0^2/(4 \ln 2)\alpha_f, \tag{8.3.8}$$

provided that $\tau_0 << (e^{-3/2} F z^2/2)^{2/3}$. Numerically, (8.3.8) gives $\tau_{FWHM} = 4.99$ ms, in good agreement with the experimental τ_{FWHM} in Fig. 8.7. The trough observed in the impulse response and the cross-correlation function past the zero-crossing delay time cannot be explained by heat diffusion considerations alone. *Jackson* et al. [8.18] have observed a similar negative response in a PDS pulsed laser experiment, in which the absorbed optical pulse from a pulsed dye laser in 0.1% benzene in distilled CCl$_4$ acted as an instantaneous heat source in the fluid. The trough has also been observed by other workers photoacoustically [8.21, 22]. It has been explained theoretically in terms of a pressure wave rarefaction following a strong compression in the fluid due to a pulsed heat source of short duration [8.23].

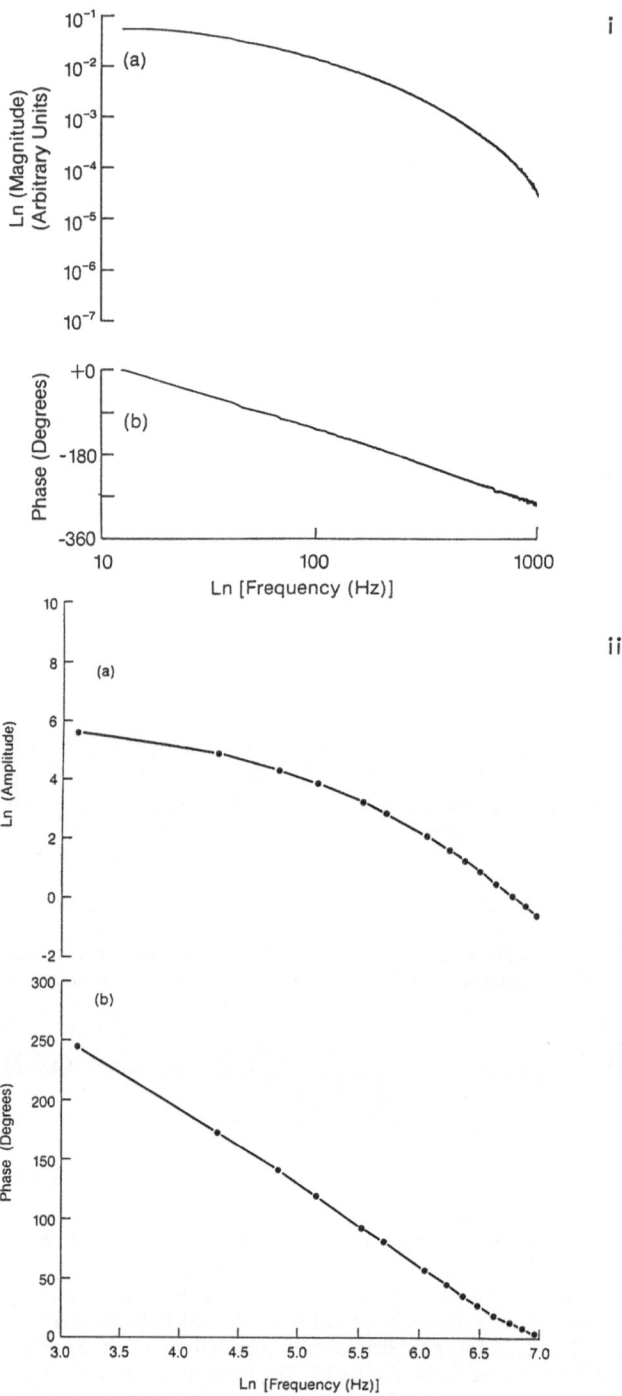

Fig. 8.9.(i) (a) Magnitude and (b) phase of the complex transfer function $H(f)$ of the blackbody/water interface at 10-μm probe beam offset, using the FM Time Delay Domain PDS apparatus. (ii) (a) Amplitude and (b) phase of PDS signal with same data obtained using lock-in detection

The system response obtained in Figs. 8.7 and 8.8 indicates that the FM time delay mirage effect apparatus generates the equivalent of a thermal impulse relaxation during the delay times of the measurements. The signal profiles $h(\tau)$ and $R_{xy}(\tau)$ are, mathematically, the one-dimensional heat diffusion Green's functions of the blackbody/water system with a fluid pressure wave superposed at $\tau >> \tau_0$. Another important feature of the spectrometer is its ability to calculate and plot on the screen of the FFT analyzer the magnitude and phase of the complex transfer function of the system in an amount of time equal to a frequency sweep time T. The total experimental time required for such a display [Fig. 8.9(i)] can be as low as 1 min, corresponding to a minimum number of ~200 sweeps/average. This time is far shorter than the time required to obtain the same information dispersively using lock-in detection, Fig. 8.9(ii). This characteristic is by no means unique to FM time delay systems, but rather common to random noise signals.

(b) PDS Impulse Response of Thin Solid Layers

The spectrometer in Fig. 8.5 was further used to measure the response from thin microscope quartz slide layers in direct contact with the backing material (anodized aluminum support). The back of the slides was painted with black paint, which was allowed to dry out after contacting the backing, in order to avoid interfacial thermal resistances due to air gaps, water seepage, etc. Such resistances could alter the temporal profile features of the response [8.24]. A single slide cut into many pieces was used for these experiments, to assure material uniformity. Each piece was etched in 50% HF:50% H_2O down to the desired thickness. Figure 8.10 shows a superposition of the impulse responses for two different thicknesses, 30 μm (curve a) and 100 μm (curve b). The cross-correlation functions show similar features, i.e., an increased peak delay time, a broadened FWHM, and an increased trough time delay τ_{min} with increasing thickness. In each case data were taken at beam offset positions which maximized the PDS output at the detector.

The secondary oscillations on both wings of the main pulse in Fig. 8.10 are consistent with thermal energy arrivals at the sample surface after multiple reflections at the sample-backing interface. The delay time $\Delta\tau$ between two successive peaks corresponds roughly to twice the thermal transit time $\tau_{transit} = l^2/\alpha_2$ through the bulk of the sample (α_2

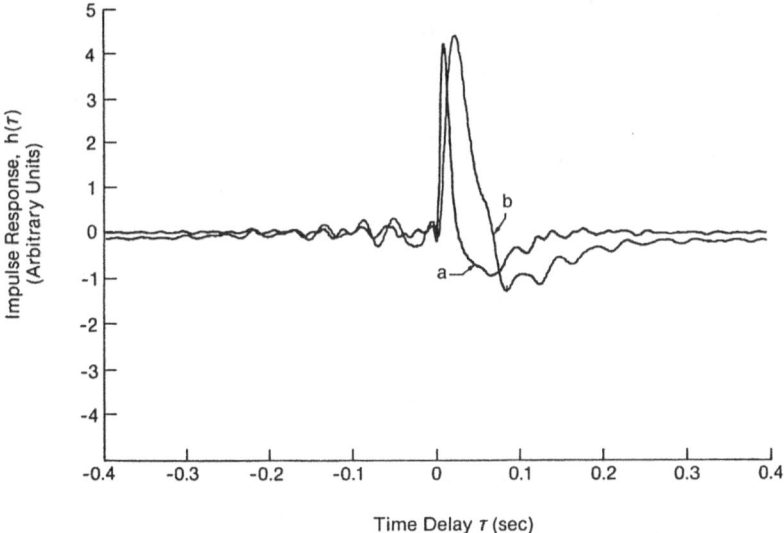

Fig. 8.10. Impulse response functions from quartz layers of thickness 30 μm (a) and 100 μm (b) . $\tau_0^a = 4.69$ ms, $\tau_{FWHM}^a = 6.04$ ms, $\tau_{min}^a = 32.81$ ms; $\tau_0^b = 10.94$ ms, $\tau_{FWHM}^b = 28.6$ ms, $\tau_{min}^b = 42.97$ ms

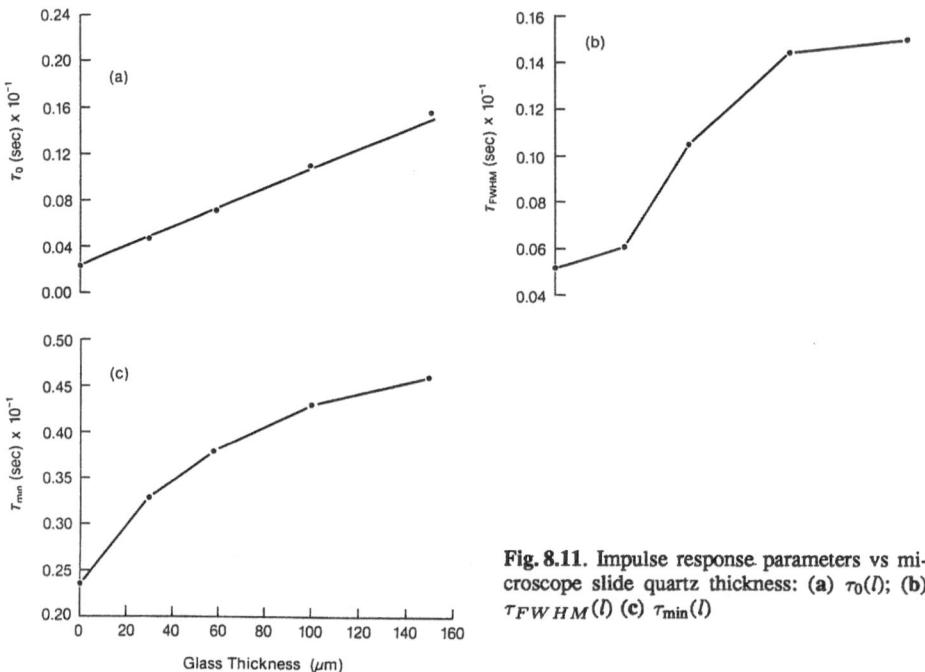

Fig. 8.11. Impulse response parameters vs microscope slide quartz thickness: (a) $\tau_0(l)$; (b) $\tau_{FWHM}(l)$ (c) $\tau_{min}(l)$

Glass Thickness (μm)

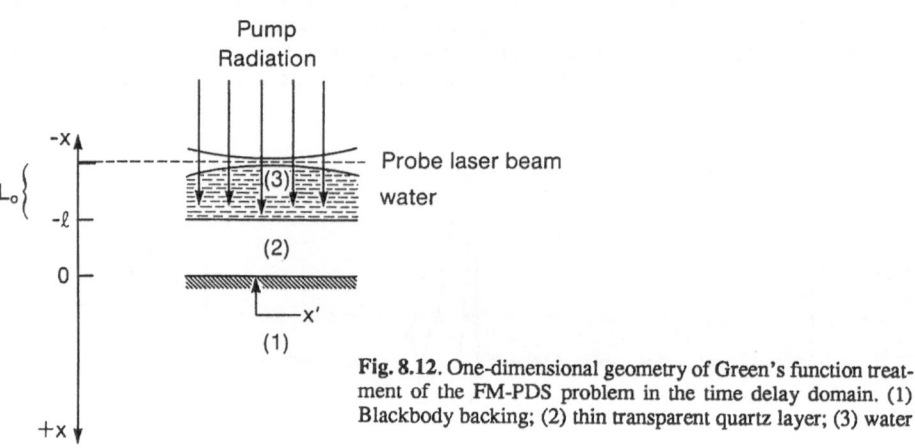

Fig. 8.12. One-dimensional geometry of Green's function treatment of the FM-PDS problem in the time delay domain. (1) Blackbody backing; (2) thin transparent quartz layer; (3) water

is the quartz thermal diffusivity). Similar effects have been predicted theoretically by *Burt* [8.25] in fluids excited by pulsed lasers and have been observed experimentally in liquids and solids by *Tam* et al. [8.26,27].

Figure 8.11 shows plots of τ_0, τ_{FWHM}, and τ_{min} for the system impulse response as functions of glass layer thickness. A theoretical model based on an extension of the solid-liquid interface heat conduction model [8.19] has been developed [8.9] for the geometry of Fig. 8.12. According to that model, the temperature increase in region (2), following an instantaneous photothermal impulsive excitation of unit strength at $x = x'$ at $\tau = 0$, is given by the Green's function for the system:

$$T_3(x,\tau) = \frac{2k_1k_2(\alpha_2\alpha_3/\alpha_1)^{\frac{1}{2}}}{(k_1\alpha_2^{\frac{1}{2}} + k_2\alpha_1^{\frac{1}{2}})(k_3\alpha_2^{\frac{1}{2}} + k_2\alpha_3^{\frac{1}{2}})}$$

$$\times \frac{\exp\{-[(\alpha_3/\alpha_2)^{\frac{1}{2}}l + |x+l|]^2/4\alpha_3\tau\}}{(\pi\tau)^{\frac{1}{2}}}. \tag{8.3.9}$$

Equation (8.3.9) is similar in form to (8.3.3) and, therefore, when subjected to the paraxial ray approximation perturbation analysis which follows (8.3.3), it predicts the following peak delay time for the system of Fig. 8.12:

$$\tau_0 = [(\alpha_3/\alpha_2)^{\frac{1}{2}}l + |L_0|]^2/6\alpha_3 \tag{8.3.10}$$

where

$$L_0 \equiv -x - l, \tag{8.3.11}$$

i.e., the sample (2) surface lies at $L_0 = 0$ and, thus L_0 is the beam offset above the sample surface. For geometries such that

$$(\alpha_3/\alpha_2)^{\frac{1}{2}}l << |L_0|, \tag{8.3.12}$$

(8.3.10) gives the approximate expression

$$\tau_0(l) \cong [1 + 2(\alpha_3/\alpha_2)^{\frac{1}{2}}(l/L_0)](L_0^2/6\alpha_3). \tag{8.3.13}$$

In these experiments, assuming the validity of relation (8.3.13), a linear dependence of τ_0 on l was predicted. From a least-squares fit of Fig. 8.11a to a straight line, the intercept $\tau_0(0)$ gives $L_0 = 44.3$ μm and the slope

$$\frac{\partial\tau_0(l)}{\partial l} \equiv \Delta = \frac{L_0}{3(\alpha_2\alpha_3)^{\frac{1}{2}}} \tag{8.3.14}$$

can be used to calculate α_2

$$\alpha_2 = (1/\alpha_3)(L_0/3\Delta)^2 \cong 4 \times 10^{-3} \text{ cm}^2/s \tag{8.3.15}$$

in good agreement with the published value [8.28] of 4.4×10^{-3} cm^2/s for the thermal diffusivity of quartz. Using the experimental values for α_2 and L_0, a check on relation (8.3.12) showed that this condition is approximately satisfied for sample thicknesses smaller than 100 μm, which is thus consistent with the observed linear dependence of τ_0 on l. No attempt was made at interpreting Figs. 8.11b and c due to the fluid pressure fluctuation phenomena which are likely [8.23] to dominate beyond the peak delay time regimes, on which τ_{FWHM} and τ_{min} depend strongly.

Further tests of the sensitivity of the FM-PDS spectrometric apparatus were performed on a silicon wafer of ~300 μm thickness. Half of the wafer had a 1 μm thick SiO$_2$ layer grown. Figure 8.13 shows the impulse responses obtained from both sides of the Si/SiO$_2$ interface. Trace (a) is broader than trace (b) and corresponds to the crystalline Si response. The sensitivity of the spectrometer is thus shown to be high enough to resolve the 1 μm SiO$_2$ layer. Qualitatively, the SiO$_2$ layer absorbs the 1.06 μm incident radiation more efficiently than the essentially transparent crystalline Si layer. Therefore, the SiO$_2$/Si system is expected to generate the equivalent of a pulsed thermal source closer to the surface than the Si wafer. The thermal pulse released to the overlying water medium from the SiO$_2$/Si structure will thus be narrower and will peak earlier than that from Si, in agreement with Fig. 8.13.

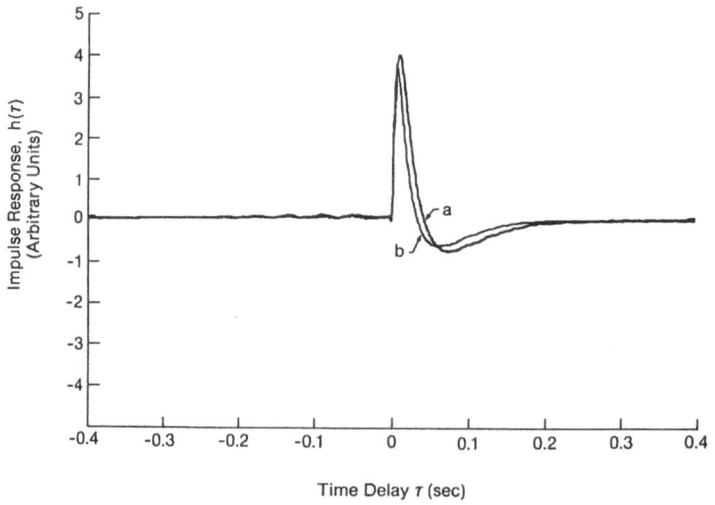

Fig. 8.13. Impulse responses from (a) Si and (b) SiO_2/Si. $\tau_0^a = 4.3$ ms; $\tau_{FWHM}^a = 10.42$ ms; $\tau_{min}^a = 37.5$ ms; $\tau_0^b = 3.12$ ms; $\tau_{FWHM}^b = 7.64$ ms; $\tau_{min}^b = 31.64$ ms. Traces (a) and (b) are the averages of 1000 responses

8.3.2 Photopyroelectric Thin Film FM Spectrometry

(a) Instrumentation and Nature of Recovered FM-P^2ES Signals

The ability of thin pyroelectric polyvinylidene fluoride (PVDF or PVF_2) films to generate and yield photothermal signals following optical excitation has been well established in both the frequency and the time domains [8.29]. Based on the photopyroelectric effect [8.30], the design of the time-delay-domain P^2E detection system is summarized in Fig. 8.14 [8.31]. The excitation source was a Coherent Innova 90 cw argon ion laser operating single line at 488 mm. The excitation power used in these experiments ranged from 100 to 200 mW depending on the sample thickness. The central excitation/detection component in the system was a HP 3562 fast Fourier transform analyzer equipped with an internal frequency synthesizer. The synthesizer was capable of generating fixed sine and

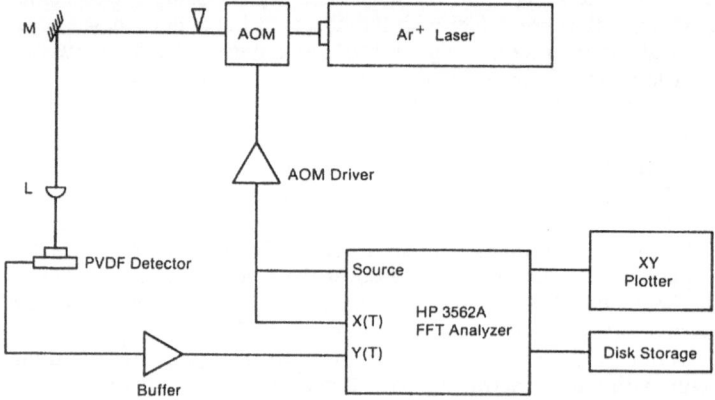

Fig. 8.14. FM-P^2E spectrometric apparatus [8.31]

random noise waveforms, as well as linear frequency sweeps, with modulation bandwidths up to 100 kHz. The sweep source was used to drive an acousto-optic modulator (Isomet 1201E) to modulate the intensity of the Ar$^+$ beam using a knife edge. The output of the acousto-optic modulator was optimized for modulation depth and minimum distortion by adjusting the peak-to-peak voltage applied to the driver circuit to a value of 400 mV. A reference beam was obtained by recording a fraction of the excitation signal with a beam splitter (BS) and a photodiode (PD). The photodiode input was used as a measure of the excitation waveform $x(t)$. In some experiments, the output of the HP 3562A internal synthesizer was used to supply the $x(t)$ waveform. The main excitation beam was directed through a microscope objective lens (magnification: 32×) (L) to simulate a point source, and focused onto the surface of a thin-film pyroelectric detector (PED) or a sample. The detector consisted of a 28-μm film of PVDF (Penwalt Corporation "KYNAR" piezo film) [8.32] supported by a stainless-steel backing. The detector was enclosed in an assembly supplied by Inficon to provide electrical contact and to ensure shielding from electromagnetic rf interference. Thermal contact between the sample and the detector was optimized by using a thin layer of thermally conducting grease (No. 120-8, Wakefield). The output of the detector was fed to a high-impedance buffer preamplifier (Comlinear, CLC-B-600, 3-pF parallel 3-MΩ input impedance; 600-MHz bandwidth), and applied to the $y(t)$ system input of the HP 3562A FFT analyzer. The recovery of all frequency and time-delay-domain functions was achieved via the analyzer computations.

In the case of FM-P^{2}ES, where the excitation wave train is a linear frequency sweep (Fig. 8.15), the criterion for flatness of the input autospectrum, $G_{xx}(f)$, is met when the sweep time T is much longer than the longest time-delay component τ in the system response, [8.2,12,33] and when the product of the modulation bandwidth, Δf and T, is very large, well above the limit of the instrumental uncertainty principle:

$$(\Delta f)T \gg 1. \tag{8.3.16}$$

Typical excitation and response signals for a 28-μm PVDF pyroelectric thin-film detector in an FM-P^{2}ES experiment are shown in Fig. 8.15. Swept waves were generated in the HP 3562A internal source with frequency spans ranging from 0-25 Hz (for thermally thick samples with slow time-delay-domain response) of up to 0-25 kHz (for thermally thin samples, with risetimes approaching that of the pyroelectric film itself). The swept waves generated within the HP 3562A were used to drive the acousto-optic modulator. Measurements carried out with the HP 3562A FFT analyzer were made in the linear resolution mode. In that mode the sweep time was related to the frequency span by the relationship [8.34]

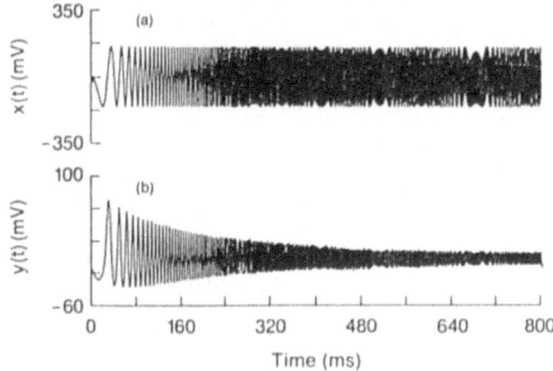

Fig. 8.15. FM time-delay swept wave excitation and response signals (a) input wave train $x(t)$; (b) PVDF thin-film detector response $y(t)$. $\Delta f = 0–1$ kHz. Sweep rate: 1.25 kHz/s

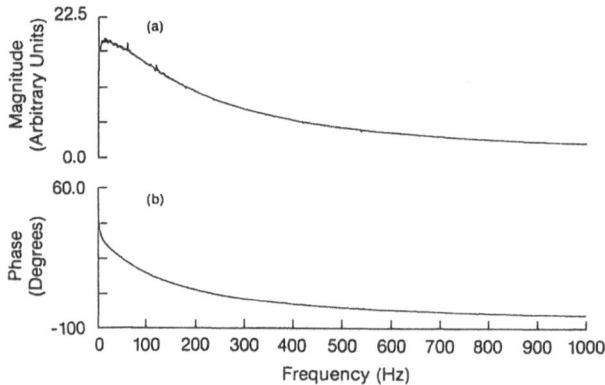

Fig. 8.16. FM-P^{2}ES measurements of frequency response for thin-film (28-μm) PVDF detector: magnitude and phase. $\Delta f = 0$–1 kHz; Sweep time: 800 ms. Result is the average of 300 measurements

$$T[\text{s}] = 800/\Delta f[\text{Hz}], \tag{8.3.17}$$

where the frequency span Δf is the modulation bandwidth of the carrier ($\Delta f = f_2 - f_1$) with $f_1 = 0$, and f_2 the maximum frequency in the sweep. The product ($\Delta f)T$ approaches 10^3, which ensures flatness of the excitation autospectrum, $G_{xx}(f)$ [8.1,2,33]. In all cases, the sweep was unidirectional from f_1 to f_2, and the resolution of the input data was 2048 points/sweep, with the sampling rate set above the aliasing frequency at 2.56 f_2.

The magnitude and phase response of the PVDF detector signals, recovered by FM-P^{2}ES (Fig. 8.16) were shown to compare favorably with standard $H(f)$ data obtained by lock-in measurements and very slow sine sweeps. The FM data were recorded in a fraction of the time required for the lock-in measurement, and at far greater resolution (2048 data pts/sweep). The latter type of measurement, in turn, limits the resolution of any impulse-response information that would be available from inverse transformation of frequency-domain data acquired via lock-in detection.

From the data of Fig. 8.16, it can be seen that the frequency response obtained from the pyroelectric thin film is weighted towards the lowest-frequency components. The features of the film response underline a general feature encountered in P^{2}E signal analysis typical of the diffusive nature of thermal waves: the strong response at low frequencies contains little information related to the transit time of thermal signals transmitted through the film. As shown in the impulse responses of Fig. 8.17 for the PVDF thin film detector, the thermal transit time information is observed at short time delays ($\tau < 2$ ms) which correspond to high frequencies under Fourier transformation. The high-frequency information is of low magnitude relative to the lower frequencies.

As already discussed in Sect. 8.2, the success of the FM-TDS measurement depends on the approximation of the input autocorrelation function to a Dirac delta function in the time-delay domain, or alternatively, on the flatness of the input autospectrum, $G_{xx}(f)$. These relationships are illustrated for the FM-P^{2}ES technique in Fig. 8.18. A comparison of $G_{xx}(f)$ and $G_{yy}(f)$ recorded for the PVDF detector [Figs. 8.18a and b] indicates that $G_{xx}(f)$ is nearly flat over the frequency range of 0-1 kHz. The oscillatory components at the edge of the span arise from the finite sweep time used in these experiments, which reduces the approximation of $R_{xx}(\tau)$ to a time delta function [8.2]. The output power spectrum $G_{yy}(f)$ appears to be of very low magnitude at the top of the frequency span.

This situation was interpreted cautiously, however [8.31], because in the frequency domain the output amplitude $Y(f)$ is the square root of the power and may be significantly

230

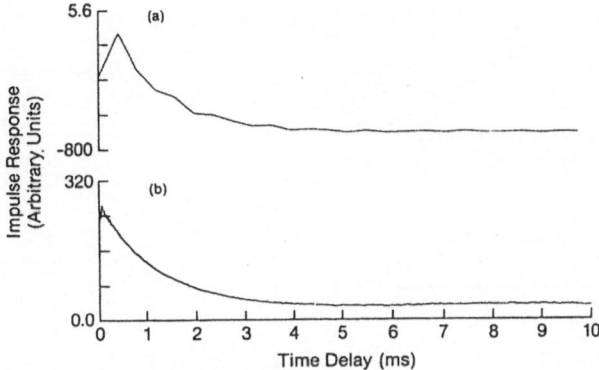

Fig. 8.17. Impulse response information recovered for thin-film PVDF detector with different modulation bandwidths: (a) $\Delta f = 0$–1 kHz, (b) $\Delta f = 0$–25 kHz. Each figure was the result of 100 averages. Peak delay information as follows: (a) $\tau_d = 390.6$ μs; $\Delta\tau_d$, 976.5 μs, (b) $\tau_d = 62.5$ μs, $\Delta\tau_d = 843.75$ μs

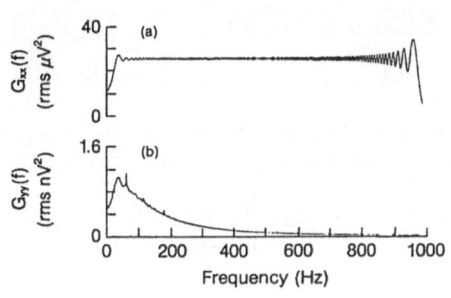

Fig. 8.18. Power spectra and autocorrelation functions for PVDF thin-film detector. (a) $G_{xx}(f)$ and (b) $G_{yy}(f)$ recorded with $\Delta f = 0 - 1$ kHz, (c) $R_{xx}(\tau)$ and (d) $R_{yy}(\tau)$ recorded with $\Delta f = 0 - 5$ kHz, (e) $R_{xx}(\tau)$ and (f) $R_{yy}(\tau)$ recorded with $\Delta f = 0 - 25$ kHz

greater than is indicated by a casual inspection of $G_{yy}(f)$. While the condition $G_{xx}(f) \cong 1$ appeared to be satisfied at $\Delta f = 0$–1 kHz (Fig. 8.18a), the corresponding impulse response recorded at this frequency span (Fig. 8.17a) shows a poor resolution at early time delays. The 0-1 kHz frequency span was therefore insufficient to resolve the rise time of the detector element, since the much weaker, high-frequency components associated with these early time delays lay outside the modulation bandwidth of the sweep. This point is also illustrated in Figs. 8.18c and d, which compare the input and output autocorrelation function $R_{xx}(\tau)$ and $R_{yy}(\tau)$, for the PVDF detector at $\Delta f = 0$–5 kHz. While $R_{xx}(\tau)$ approximates a delta function to within the time-delay linewidth defined by the modulation bandwidth of the sweep, the width of $R_{xx}(\tau)$ is not negligible compared with $R_{yy}(\tau)$. Increasing the modulation bandwidth of the sweep to $\Delta f = 0$–25 kHz enabled the resolution of the detector risetime information in the impulse response (Fig. 8.17b). The improved approximation of $R_{xx}(\tau)$ to a delta function relative to $R_{yy}(\tau)$ is indicated in Figs. 8.18e and f.

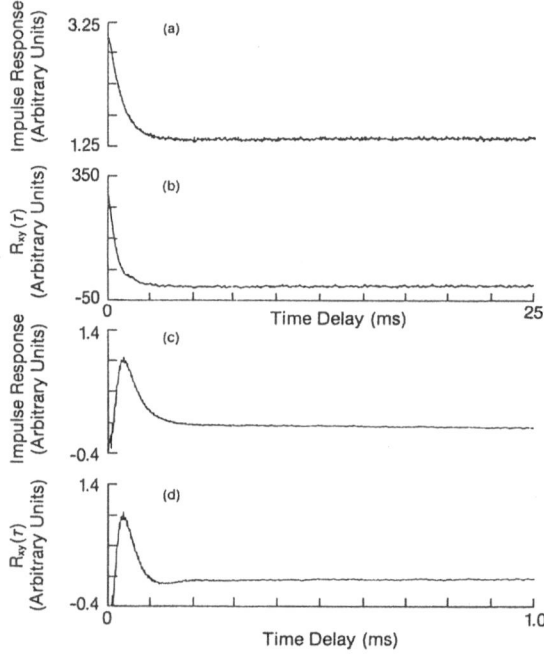

Fig. 8.19. Impulse response $h(\tau)$ and cross-correlation functions $R_{xy}(\tau)$ compared for **(a)** and **(b)** thin-film PVDF with $\Delta f = 0$–25 kHz; **(c)** and **(d)** 400–μm quartz sample and $\Delta f = 0$–200 Hz. $N = 300$ averages. Time-delay information was: PVDF **(a)** $h(\tau), \tau_d = 62.5$ μs, $\Delta\tau_d = 844$ μs; **(b)** $R_{xy}(\tau)$, $\tau_d = 62.5$ μs, $\Delta\tau_d = 470$ μs. 400–μm quartz **(c)** $h(\tau)$, $\tau_d = 35.16$ ms, $\Delta\tau_d = 29.5$ ms; and **(d)** $R_{xy}(\tau), \tau_d = 35.16$ ms, $\Delta\tau_d = 25.9$ ms

A further indication of the quality of an FM-P^{2}ES measurement is the degree of agreement between $h(\tau)$ and the cross-correlation function $R_{xy}(\tau)$ in accordance with (8.2.9). Figure 8.19 makes this comparison for the pyroelectric thin-film detector and for a 400-μm glass sample, coated with a thin layer of ink acting as a surface thermal source. Agreement between the two time-delay-domain functions was excellent with respect to peak-delay information, indicating the success of the FM-P^{2}ES measurement strategy. The peak widths of the cross-correlation functions, however, were consistently narrower than the corresponding impulse-response peaks. This feature has been attributed to nonuniformities in the input autospectrum $G_{xx}(f)$ at the edges of the frequency span.

(b) Thermal Diffusivity, Conductivity and Thickness Measurements of Solids

The capabilities of the FM-P^{2}ES spectrometric apparatus have been demonstrated in measurements of thermal and geometric properties of solids, by comparing the responses of a series of well-characterized samples to impulse response (Green's function) theoretical predictions. The samples consisted of a series of quartz microscope slides and cover slips of varying thickness. Each series of samples was prepared from a single slide, cut into many smaller pieces, to ensure material uniformity. The pieces were etched in 50% HF/50% H$_2$O to the desired thickness. The front surfaces of these samples were coated with a thin layer of water soluble ink, which served as a blackbody layer. The thinness of the ink layer and its large absorption coefficient ensured that the laser radiation absorbed at the sample surface acted as an ideal plane heat source.

The impulse responses of these samples may be readily predicted from a Green's function model of transient heat conduction in a four-layer system. In Fig. 8.20, the heat

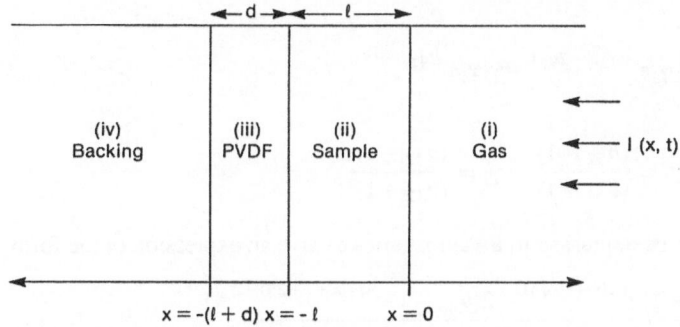

Fig. 8.20. Schematic of a four-layer theoretical model for the FM-P^{2}ES system

conduction equation was solved for each of layers (i)-(iv), assuming excitation of the sample with a unit impulse heat source at $x = x'$ and $t = t'$:

gas:
$$\frac{\partial^2 T_1}{\partial x^2} - \frac{1}{\alpha_1} \frac{\partial T_1}{\partial t} = 0, \tag{8.3.18}$$

sample:
$$\frac{\partial^2 T_2}{\partial x^2} - \frac{1}{\alpha_2} \frac{\partial T_2}{\partial t} = -\delta(x - x')\delta(t - t'), \tag{8.3.19}$$

pyroelectric thin film:
$$\frac{\partial^2 T_3}{\partial x^2} - \frac{1}{\alpha_3} \frac{\partial T_3}{\partial T} = 0, \tag{8.3.20}$$

backing:
$$\frac{\partial^2 T_4}{\partial x^2} - \frac{1}{\alpha_4} \frac{\partial T_4}{\partial t} = 0, \tag{8.3.21}$$

where α_i is the thermal diffusivity of the ith layer.

The boundary conditions associated with (8.3.18-21) were those of heat flux and temperature continuity. The impulsive heat source was located at the solid/gas interface, at $x' = 0$. The inhomogeneous solution in region (ii) for $t' = 0$ was given by the source Green's function [8.19,35]. Under these conditions, the Laplace transform of the temperature distribution in the pyroelectric film is given by

$$\overline{T}_3(x,s) = [2(b_{43} + 1)e^{q_3(x + l + d)} - 2(b_{43} - 1)e^{-q_3(x + l + d)}]$$

$$\times (\alpha_2 q_2 \{(b_{43} - 1)e^{-q_3 d}[(b_{32} - 1)(b_{12} + 1)e^{q_2 l} + (b_{32} + 1)(b_{12} - 1)e^{-q_2 l}]$$

$$+ (b_{43} + 1)e^{q_3 d}[(b_{32} + 1)(b_{12} + 1)e^{q_2 l} - (b_{12} - 1)(b_{32} - 1)e^{-q_2 l}]\})^{-1}, \tag{8.3.22}$$

where the thermal transport coefficients b_{ij} are defined as $b_{ij} = k_i \alpha_j^{1/2}/k_j \alpha_i^{1/2}$, i and j being adjacent layers in the model of Fig. 8.20; q_i contains the Laplace domain variable s: $q_i \equiv (s/\alpha_i)^{1/2}$.

The denominator may be expanded and expressed in the form

$$\alpha_2 q_2 (b_{43} + 1)(b_{12} + 1)(b_{32} + 1)e^{q_3 d}e^{q_2 l}(1 + \xi),$$

where

$$\xi = \gamma_1 \gamma_2 e^{-2q_3 d} + \gamma_1 \gamma_3 e^{-2q_3 d - 2q_2 l} - \gamma_2 \gamma_3 e^{-2q_2 l},$$

and

$$\gamma_1 = \frac{(b_{43} - 1)}{(b_{43} + 1)}, \quad \gamma_2 = \frac{(b_{32} - 1)}{(b_{32} + 1)}, \quad \gamma_3 = \frac{(b_{12} - 1)}{(b_{12} + 1)}.$$

The factor $1/(1 + \xi)$ may be expressed in a Taylor series to give an expression of the form

$$\bar{T}_3(x,s) = \frac{[2(b_{43} + 1)e^{q_3(x+l) - q_2 l} - 2(b_{43} - 1)e^{-q_3(x+l+2d) - q_2 l}]}{\alpha_2 q_2 (b_{43} + 1)(b_{12} + 1)(b_{32} + 1)}$$

$$\times \sum_{n=0}^{\infty} (-1)^n \xi^n. \tag{8.3.23}$$

Equation (8.3.23) formed the nucleus of the theoretical four-layer treatment. Instead of inverting (8.3.23) directly, the special case of discontinuous thermal properties of backing and pyroelectric was considered [8.36]. In that limit $b_{12} << 1$ and $b_{43} >> 1$, realistic assumptions for the air-quartz and PVDF-stainless steel interfaces. Under these interfacial mismatch conditions, (8.3.23) simplifies considerably and can be inverted to yield a spatially averaged temperature change in the pyroelectric

$$<T_3(x,t)> = \frac{1}{d} \int_{-(l+d)}^{-l} T_3(x,t)dx, \tag{8.3.24}$$

in the thermally thick limit:

$$q_2 l >> q_3 d. \tag{8.3.25}$$

This limit was found valid for thin PVDF film detectors ($d = 28 \ \mu m$) and yielded the expression

$$<T_3(x,t)> = \frac{2(\alpha_2 \alpha_3)^{1/2}}{\alpha_2 (b_{32} + 1)} \sum_{n=0}^{\infty} \gamma_2^n (-1)^n$$

$$\times \left[\text{erfc} \sqrt{\frac{\tau_{1n}}{4t}} - 2 \, \text{erfc} \sqrt{\frac{\tau_{2n}}{4t}} + \text{erfc} \sqrt{\frac{\tau_{3n}}{4t}} \right], \tag{8.3.26}$$

where

$$\tau_{1n}^{1/2} = \frac{2nd}{\alpha_3^{1/2}} + \frac{l}{\alpha_2^{1/2}}, \quad \tau_{2n}^{1/2} = \frac{(2n+1)d}{\alpha_3^{1/2}} + \frac{l}{\alpha_2^{1/2}},$$

and

$$\tau_{3n}^{1/2} = \frac{2(n + 1)d}{\alpha_3^{1/2}} + \frac{l}{\alpha_2^{1/2}}.$$

The exact form of the signal extracted from a pyroelectric measurement depends on whether the current or voltage response of the thin-film element is recovered. For the pyroelectric detector under the experimental conditions of Fig. 8.14, i.e. under load, the current response is recovered [8.29,37]:

$$I(t) = \frac{pd}{\varepsilon} \frac{\partial}{\partial t} <T_3(x,t)> \tag{8.3.27}$$

where p is the pyroelectric coefficient of PVDF and ε is its dielectric constant. The pyroelectric impulse response can be extracted from (8.3.26) and (8.3.27):

$$V(t) = \frac{KA}{t^{3/2}} \sum_{n=0}^{\infty} (-1)^n \gamma_2^n (\tau_{1n}^{\frac{1}{2}} e^{-\tau_{1n}/4t} - 2\tau_{2n}^{\frac{1}{2}} e^{-\tau_{2n}/4t} + \tau_{3n}^{\frac{1}{2}} e^{-\tau_{3n}/4t}), \quad (8.3.28)$$

$V(t)$ is the voltage recorded at the output of the buffer preamplifier, and K is a constant which incorporates the electrical properties of the pyroelectric film.

The factor A is a constant which incorporates the static thermal properties of the sample/pyroelectric system:

$$A = -2(\alpha_3 \alpha_2)^{\frac{1}{2}}/\alpha_2(b_{32} + 1)4\sqrt{\pi}. \quad (8.3.29)$$

Because the absolute intensity of the recovered signals is a function of instrumental factors such as irradiation power, amplifier gain, and excitation geometry, the impulse responses were normalized to give $V(t) = 1$ at the peak of the time-delay response.

Figure 8.21 shows the comparison between experiment and theory for a 500-μm sample of quartz. The impulse response profile was fitted assuming $\alpha_2 = 4.0 \times 10^{-7}$ m^2/s for quartz, which shows excellent agreement with literature values [8.38]: $\alpha_2 = 4.0 \times 10^{-7}$ m^2/s and [8.28] $\alpha_2 = 4.4 \times 10^{-7}$ m^2/s. The value of the parameter γ_2 which gave the best fit to experiment was $\gamma_2 = 0.7$. From this parameter, it was possible to calculate the thermal conductivity for the sample as $k_2 = 2.2$ W/mK.

Agreement between theory and experiment is excellent in the time-delay domain from 0 to 1 s. The theoretical expression of (8.3.28) required approximately five terms for convergence. The poor agreement at early times was due to light leakage through the sample blackbody layer onto the surface of the film, which causes simultaneous excitation of the film response. The effect was found to be generally more pronounced for thick samples with long time delays, since their peak intensities are weak relative to the early signals. Samples with earlier peak delays were generally less susceptible to this phenomenon because the signal intensity is much greater with the thinner samples.

The response of an entire set of samples is summarized in Fig. 8.22. For the series of quartz slides, plots were made of the time delay τ_d and peak width $\Delta\tau_d$ versus the square of the sample thickness. The plots shown in Fig. 8.22 indicate excellent agreement

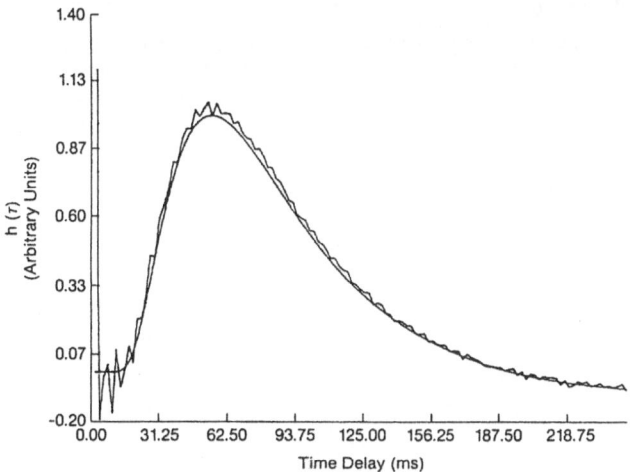

Fig. 8.21. Theoretical and experimental impulse-response profiles obtained for a 500-μm quartz sample

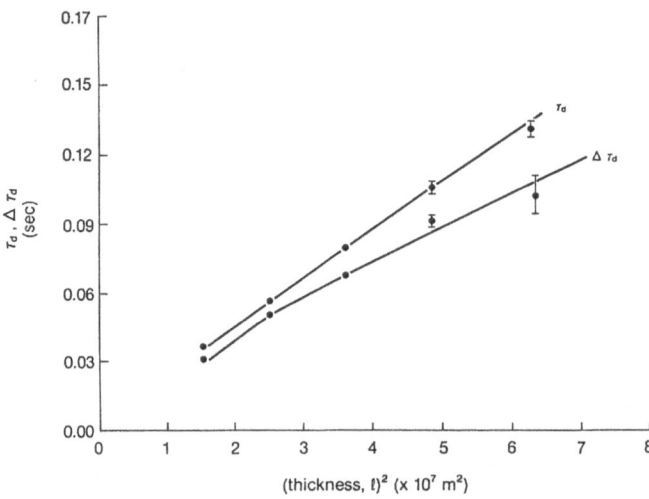

Fig. 8.22. Impulse-response parameters vs sample thickness for a series of well-characterized (quartz) samples. Solid lines represent theoretical responses

between theory and experiment with respect to both parameters over the full range of samples studied. It is of further interest to note that the relationship between τ_d and l^2 predicted from theory is nearly linear ($R > 0.9999$ where R is the correlation coefficient) over a reasonable range of sample thicknesses. This relationship suggests that the photopyroelectric methodology described could be used for precision thickness measurements on thin layers of glasses, by preparing calibration curves of τ_d (or $\Delta\tau_d$) vs l^2 and thereby standardizing the photopyroelectric system. The FM-P^{2}ES methodology was further tested using metallic samples (aluminum, stainless steel) which were contacted to the pyroelectric (PVDF) detector using thermal compound. In the case of the metal samples, $b_{32} << 1$, so that $\gamma_2 = 1$ in (8.3.28). The expression is valid provided $l^2/4\alpha_2\tau >> d^2/4\alpha_3\tau$ [relation (8.3.25)]. Thermal diffusivity measurements on these samples are shown in Table 8.1, which also reports the sample thickness and time-delay ranges for which the measurements were made. These conditions fall well within the approximations required by the model.

The results of Table 8.1 show a reasonable agreement between measured and literature values of α_2 for a wide range of different materials. The main factor controlling the agreement between theory and experiment for these results is the thermal contact resistance present at the sample/detector interface [8.40]. Flatness of the sample surface pro-

Table 8.1. Measured thermal diffusivity values for selected materials.

Material	α_2[m^2/s] Experimental, this work	α_2[m^2/s] Literature values	Reference	Measurement conditions
Quartz, fused	4.0×10^{-7}	4.0×10^{-7}	[8.38]	l_{range}:300 μm − 1 mm
		4.4×10^{-7}	[8.28]	τ_{range}:<1 s
Stainless steel	2.2×10^{-6}	3.5×10^{-6}	[8.39]	l_{range}:390 μm − 1 mm
				τ_{range}:<25 ms
Aluminum	38×10^{-6}	97.1×10^{-6}	[8.39]	l_{range}:500 μm − 2 mm
		68×10^{-6}	[8.39]	τ_{range}:<10 ms

motes a good thermal contact with the detector film, while warpage or other surface deformities cause poor contact and an overestimated value of the measured thermal diffusivity. A residual contact resistance persisted in the sample/pyroelectric system, even when thermal coupling compound was used, and contributed appreciable errors in α_2 at time delays earlier than 5 ms.

The FM-P^{2}ES results were contrasted with the work of *Sugitani* et al. [8.24], who reported a series of correlation photoacoustic experiments apparently showing the effects of significant contact resistance between sample and backing materials. The contribution of contact resistance was ascertained from the deviation of the time delay versus sample thickness data from the prediction of a model based purely on diffusion. The range of sample thickness used in that work was grouped towards a lower range of time delays so that susceptibility to these effects was greater. However, a gas-microphone cell was used as the detection system. Such systems have been known to contribute extraneous time delays to time domain photoacoustic signals due to variations in the flatness of the transducer frequency response [8.41] and to nonlinearities in response [8.42]. In FM-P^{2}ES, the response of the pyroelectric element is sufficiently fast to permit the observation of thermal signal transit through very thin films. The frequency response of the detector element was shown to be quite flat over the range of signal frequencies corresponding to thermal transit through the samples. It was noted that the risetime of the PVDF detector in these experiments was less than 50 µs, which is much less than the range of responses of the samples reported in Table 8.1.

8.3.3 Photothermal Reflectance FM Spectrometry

(a) Instrumentation, Limitations and Measurement Performance

In recent years, a very powerful methodology has emerged, in which thermal wave phenomena could be detected through noninvasive, noncontact means via optically induced thermoelastic deformations [8.43,44] and optical reflectivity changes [8.45] at the sample surface. Detection was achieved via a pump-probe configuration in which a highly focused heating beam was absorbed at the sample surface, inducing a thermal "bump" due to the thermoelastic effect, with simultaneous changes in sample reflectance due to variations of the surface temperature. Detection of thermal waves was achieved both thermoelastically, in which the thermal surface deformations produced deflections of the probe beam, or by changes in the sample's surface reflectance with temperature, which produced variations in the integrated intensity of the probe beam. Because of the availability of fast photodiodes and quad-cell detectors, the very wide bandwidth of the technique is therefore capable of resolving thermal images limited by optical rather than thermal diffraction. Time-domain (pulsed) laser schemes with wideband capabilities have been used with other photothermal wave systems, such as the flash radiometric technique developed by *Leung* and *Tam* [8.46], which is based on the noncontact detection of transient infrared (blackbody) radiation from a sample heated by a short optical pulse. These schemes, however successful they may be with materials tolerant to steep temperature excursions, cannot be used with delicate materials, such as semiconductor substrates and devices, without severe restrictions in the exciting laser beam parameters, as the high pump irradiances tend to alter (anneal or otherwise damage) the surfaces and device structures with which the beam interacts [8.47]. Furthermore, it has been shown that the greatest care must be exercised when using photothermal radiometry with some categories of samples, because the heat source is often difficult to define, due to the weak IR emissivity of the sample [8.48].

As an alternative to pulsed laser excitation, fast impulse-response FM photothermal wave excitation and detection techniques were recently coupled with PTR instrumentation of wideband frequency response, so as to avoid the destructive shortcomings of pulsed laser excitation while retaining desirable time-resolved advantages [8.49,50].

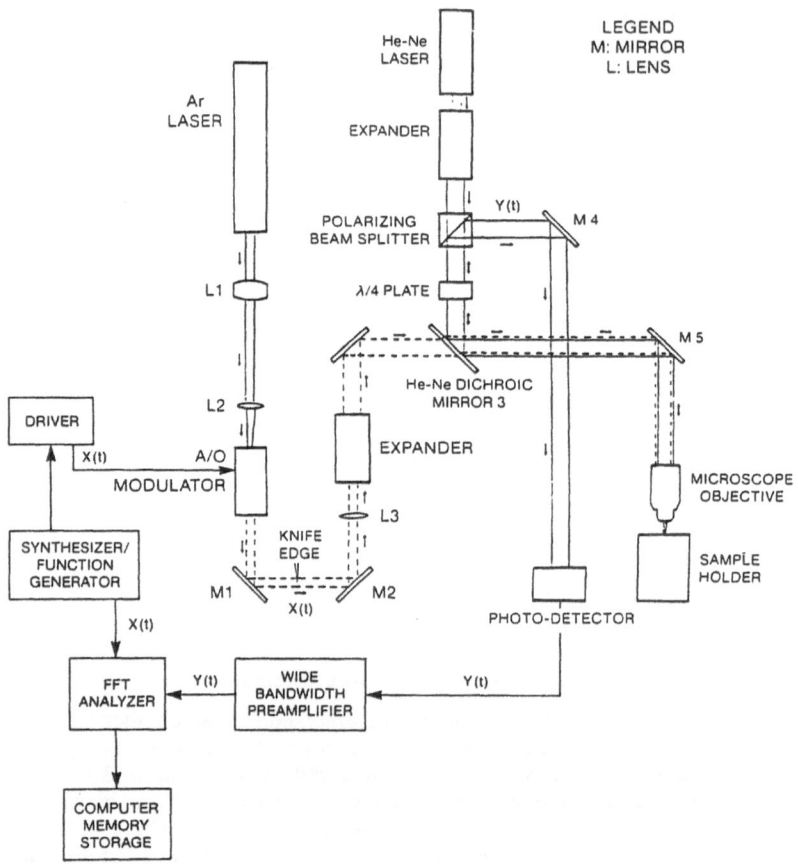

Fig. 8.23. Block diagram of wide-bandwidth FM-PTR instrumentation; $x(t)$ and $y(t)$ are real-time system input and output waveforms, respectively

Figure 8.23 shows a block diagram of the apparatus used for the detection of thermal waves by optical reflectance modulation. The basic optical design was previously reported by *Opsal* et al. in earlier modulated cw thermal wave experiments [8.43], whereas the associated FFT instrumentation formed the basis for the real-time Impulse Response PTR imager. Excitation was obtained from a Coherent Innova 90 Argon Ion Laser, which was operated single line at 488 nm. The Ar^+ beam was fed through an A/O modulator and thus its intensity was time varied by drive signals supplied by the internal function generator of an HP 3562A FFT analyzer. The drive waveforms were linear FM sweeps. Figure 8.24 shows typical excitation and response waveforms. The quad-cell signals were recorded in the sum mode. The quad-cell output was applied to the $y(t)$ input of the FFT analyzer, while the $x(t)$ input was obtained directly from the output of the internal sweep generator. A gating pulse, synchronous with the initiation of each frequency sweep was used to trigger the acquisition of time records at the $x(t)$ and $y(t)$ inputs to the HP 3562A. Time averaging of 300 to 1000 records was carried out to improve signal-to-noise ratios (SNRs). Impulse response information was recovered from the frequency sweep data by means of correlation and spectral functions computed by the HP 3562A waveform analyzer. The signal bandwidth of the instrument was 100 kHz.

The photoinduced thermoreflectance measurement has as its basis the monitoring of changes in a sample's reflectance, R, due to changes in temperature following light absorption. The reflectivity change has the form

238

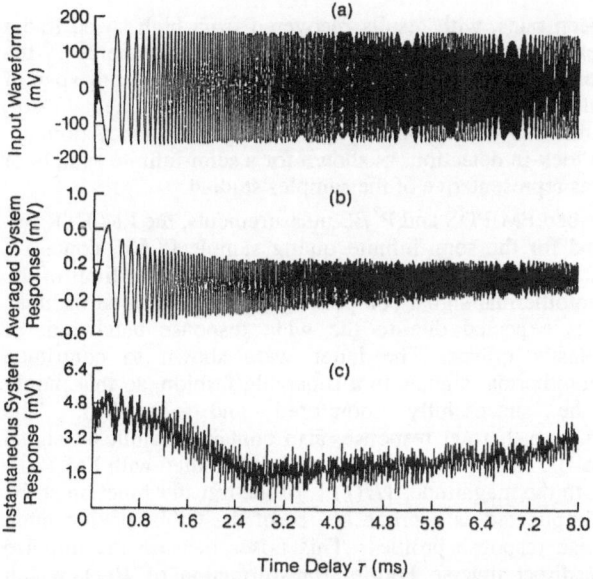

Fig. 8.24. Linear FM sweep wavetrains for photothermal reflectance response of a semi-infinite quartz sample. Sweep/measurement bandwidth: $\Delta f = 0$–100 kHz. Sweep rate, $S = 1.25 \times 10^7$ Hz/s; window: uniform; (**a**) excitation sweep, $x(t)$; (**b**) response waveform, $y(t)$; averaging $N = 1000$ records for a,b; (**c**) instantaneous $y(t)$ record with no averaging

$$R - R_0 = \Delta R = \left[\frac{\partial R}{\partial T} \right]_{T=T_0} \Delta T \tag{8.3.30}$$

where ΔR is the reflectivity change at the sample surface, R_0, is the unperturbed surface reflectivity and $\partial R/\partial T$ is the sample's temperature coefficient of reflectance. The temperature change, ΔT, is given by

$$\Delta T = T(r, z = 0, t) - T_0$$

where T_0 is the ambient temperature and $T(r,z,t)$ is the temperature distribution in the sample.

A direct proportionality between ΔR and $T(r,z,t)$ is predicted by (8.3.30), so that the reflectivity changes recorded in a typical experiment are a direct monitor of the surface temperature.

The detection of thermal waves by optical reflectance modulation is, in principle, a simple process, which suffers in practice from a number of interfering photothermal phenomena. These competitive effects include thermoelastic deformations of the sample and thermal blooming phenomena [8.51] which take place at the sample surface following light absorption. A careful optimization of the irradiation and detection geometry is required to minimize these effects. In addition, the effect of varying the relative sizes of the pump and probe beams, as well as the introduction of an offset between the beam centers, has significant repercussions on the recovered photothermal signals. The linear sweep excitation used to recover impulse and frequency response information has been known, in some cases [8.33], to depend on the sweep rates, S. This could potentially arise because of the slow speed of response of diffusive linear systems, typical of fast, wideband photothermal detection. By comparing point-by-point frequency response measurements, made using very low sweep rates, with results recovered from high speed linear FM chirps, it was possible to eliminate the sweep rate as a significant variable in the ranges of signals studied. Measurements recovered over the maximum frequency span of the analyzer (0-100 kHz), at a rate of 1.25×10^7 Hz/s agreed, to within experimental error, with frequency response measurements made point by point in the frequency domain, using both slow sine sweeps and lock-in detection, as shown for a semi-infinite sample of quartz, whose response profile was representative of the samples studied.

Unlike the previously described FM-PDS and P^2ES measurements, the FM-PTR frequency response profiles reported for the semi-infinite quartz sample [8.50], contain a significant magnitude component at 100 kHz, which corresponds to the upper limit of the analyzer bandwidth. Clearly, photothermal signal components extended beyond the span limit of the analyzer. This was expected due to the wide response bandwidth of thermoreflectance and thermoelastic effects. The latter were shown to contribute significantly to the measured photothermal signals in a separable fashion, so that purely photothermal effects could be successfully monitored and quantified. The thermoreflectance effect, itself a pure thermal response, also contains significant signal components at frequencies above 100 kHz for most of the samples studied with FM-PTR. The finite contribution of signals to the magnitude, $|H(f)|$, of the transfer function at the span edge (100 kHz) has been implicated as a prime source of the Gibbs phenomenon observed in the recovered impulse response profiles. This arises because the impulse response, $h(\tau)$, is recovered by direct inverse Fourier transformation of $H(f)$, which contains an effective truncation at 100 kHz. In some cases, multiplication of the frequency response by a logarithmic window could eliminate much of this truncation error while apparently preserving thermal signal information. The problems associated with the analyzer bandwidth limitation are further demonstrated in Fig. 8.25, which compares the input autocorrelation and output autocorrelation functions for the same quartz sample. Clearly, the approximation of $R_{xx}(\tau)$ to a Dirac δ function is poor since the input pulsewidth is not of very short duration relative to the pulsewidth of $R_{\hat{y}y}(\tau)$. Again, this phenomenon arises because of the presence of high frequency signal components at and above 100 kHz. Depending on the detailed beam geometries for irradiation and on the thermal properties of the sample studied, these high frequency components will contribute to a greater or lesser extent.

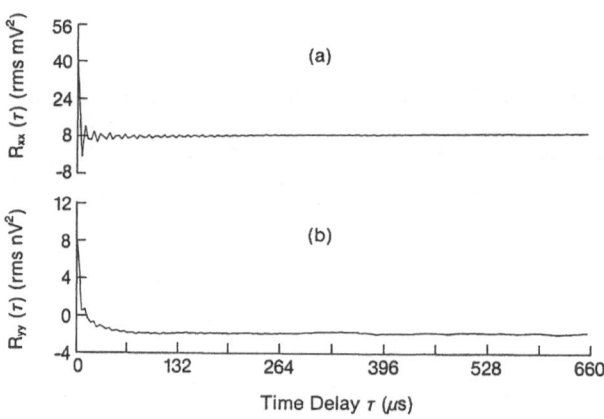

Fig. 8.25. Evaluation of measurement performance: (a) $R_{xx}(\tau)$, (b) $R_{yy}(\tau)$ for semi-infinite quartz sample. $\Delta f = 0$–100 kHz. $S = 1.25 \times 10^7$ Hz/s

(b) Quantitative Considerations of FM-PTR Irradiation Geometry

The geometry used for the theoretical interpretation of impulse response FM-PTR signals [8.49] via (8.3.30) is shown in Fig. 8.26. The sample solid [8.2] is assumed to be of infinite lateral extent, a valid approximation in view of the very tightly focused laser pump and probe beams (≈ 3 μm) in our experiments. The temperature distribution $T_2(r,z,t)$ can be obtained from the solution in cylindrical coordinates of homogeneous heat conduction equations of the form

$$\nabla^2 T_i(r,z,t) - \frac{1}{\alpha_i} \frac{\partial T_i}{\partial t}(r,z,t) = 0 \tag{8.3.31}$$

$i = 1$ (gas), 2 (solid) or 3 (backing), subject to homogeneous boundary conditions with an impulsive contribution (a Dirac δ-function in time) due to the cylindrical thermal source at $z = 0$. In (8.3.31), α_i is the thermal diffusivity of region (i). Taking Laplace transforms ($t \to s$) of the system of (8.3.31) subject to the initial conditions

$$T_i(r,z,0) = 0 \tag{8.3.32}$$

yields

$$\nabla^2 \bar{T}_i(r,z,s) - q_i^2 \bar{T}_i(r,z,s) = 0, \tag{8.3.33}$$

where $q_i(s)$ has been defined in conjunction with (8.3.22) and contains the Laplace domain variable:

$$q_i(s) \equiv (s/\alpha_i)^{\frac{1}{2}}. \tag{8.3.34}$$

The system of (8.3.33) can be solved using the conventional boundary conditions of temperature and heat flux continuity at the (1,2) and (2,3) interfaces [8.52]. Solutions of (8.3.33) can be written using Sommerfeld's method as applied by *Stratton* [8.53]:

$$\bar{T}_1(r,z,s) = \int_0^\infty J_0(kr) A(k) e^{-\sigma_1 z} dk \quad ; \quad z \geq 0 , \tag{8.3.35}$$

$$\bar{T}_2(r,z,s) = \int_0^\infty J_0(kr) \left[C(k) e^{\sigma_2 z} + D(k) e^{-\sigma_2 z} + \frac{ke^{-\sigma_2 |z|}}{4\pi\alpha_2\sigma_2} \right] dk \quad ;$$

$$0 \geq z \geq -l \tag{8.3.36}$$

Coupling Gas (1)

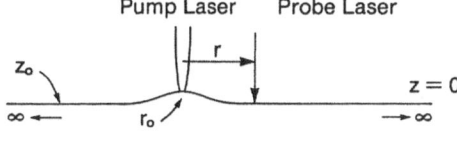

Solid (2)

Backing (3)

Fig. 8.26. Three-dimensional geometry for impulse response (Green's function) heat conduction generation due to an (effective) laser pulse $R^2 = r^2 + (z - z_0)^2$, where the exciting source is assumed to be at the sample surface ($z_0 = 0$)

and

$$\overline{T}_3(r,z,s) = \int_0^\infty B(k)J_0(kr)e^{\sigma_3(z+l)} \, dk \quad ; \quad z \le -l \, , \tag{8.3.37}$$

where A, B, C, and D are constants to be determined by the boundary conditions.

The last term in brackets of the integrand in (8.3.36) is due to the instantaneous source at $z_0 = 0$ in the $z \le 0$ half-space (Fig. 8.26). In (8.3.35-37) the definitions were made

$$\sigma_i(s) \equiv [k^2 + q_i^2(s)]^{1/2} \tag{8.3.38}$$

and J_0 is the Bessel function of zeroth order. Here it appears due to the cylindrical geometry imposed by the symmetry of the laser beams. Only solution for the Laplace transform of the temperature field at the gas-solid interface, $\overline{T}_1(r,0,s)$, was required [8.49], where probing of the temperature dependence of the reflectivity takes place. Determination of the constants A, B, C, and D yields for the Laplace transform of the gas temperature $\overline{T}_1(r,z,s)$

$$\overline{T}_1(r,z,s) = \frac{1}{2\pi\alpha_2} \int_0^\infty \left[\frac{1+e^{-2\sigma_2 l}}{\sigma_2} \right] \sum_{n=0}^\infty \zeta^n e^{-2n\sigma_2 l - \sigma_1 z} J_0(kr)k \, dk \, ;$$

$$z \ge 0 \tag{8.3.39}$$

where

$$\zeta \equiv \frac{(1-b_{32})(1-b_{12})}{(1+b_{32})(1+b_{12})}. \tag{8.3.40}$$

Because of the low thermal conductivity of gases relative to solid materials in general, the condition

$$k_2 >> k_1 \tag{8.3.41}$$

may be assumed to hold for all cases of experimental interest. Under this condition

$$\zeta \approx \frac{1-b_{32}}{1+b_{32}} \tag{8.3.42}$$

with

$$b_{ij}(s) \equiv k_i\sigma_i(s)/k_j\sigma_j(s). \tag{8.3.43}$$

Upon separating out the $n = 0$ term, (8.3.39) may be written as

$$\overline{T}_1(r,z,s) = \frac{1}{2\pi\alpha_2} \left\{ \int_0^\infty \frac{e^{-\sigma_1 z}}{\sigma_2} J_0(kr)k \, dk + \int_0^\infty \frac{e^{-2\sigma_2 l - \sigma_1 z}}{\sigma_2} J_0(kr)k \, dk \right.$$

$$+ \sum_{n=1}^\infty \int_0^\infty \left[\frac{\zeta^n}{\sigma_2} \right] e^{-2n\sigma_2 l - \sigma_1 z} J_0(kr)k \, dk$$

$$\left. + \sum_{n=1}^\infty \int_0^\infty \left[\frac{\zeta^n}{\sigma_2} \right] e^{-2(n+1)\sigma_2 l - \sigma_1 z} J_0(kr)k \, dk \right\} . \tag{8.3.44}$$

It is worthwhile noticing that the last three terms in (8.3.44) appear due to the presence of a finite solid boundary at $z = -l$. In the limit $l \to \infty$, the first term represents the thermal response in the gas due to a semi-infinite solid excited by a unit cylindrical thermal impulse of infinitesimal spatial extent in both r and z dimensions at $t = 0$.

Equation (8.3.44), which forms the basis for the three-dimensional FM-PTR point irradiation model, is a zero-order Hankel transform. It expresses the solution to the three-layer problem gas-solid backing in the Laplace domain as an integral function over the Bessel wavenumber k. It is of interest to point out that the expression (8.3.39) can be written equivalently in the Hankel transform form

$$\bar{T}_1(r,z,s) = \frac{1}{2\pi\alpha_2} \int_0^\infty F(z,s;k)J_0(kr)k \, dk \qquad (8.3.45)$$

where the kernel $F(z,s;k)$ becomes equivalent to the Laplace transform of the one-dimensional problem (uniform irradiation along the r-direction) [8.54] in the limit of $k = 0$. In that limit, the net effect on $F(z,s;0)$ is that the variables $\sigma_i(s)$ are replaced by the respective $q_i(s)$. This feature has also been pointed out by *Rosencwaig* and *Opsal* in a frequency-domain derivation of the expression for the surface temperature of a multilayered sample via thermoacoustic detection [8.55] and it reflects the formal equivalence between time- and frequency-domain formalisms.

Several special cases of the FM-PTR theory have been considered [8.50], regarding experimentally realistic case limits. For comparison with experiments, a simplified expression for $\bar{T}_1(r,z,s)$ was obtained in the limit of a semi-infinite solid ($l \to \infty$):

$$\bar{T}_1(r,z,s) = \frac{1}{2\pi\alpha_2} \int_0^\infty \frac{e^{-z\sqrt{k^2+q_1^2}}}{\sqrt{k^2+q_2^2}} J_0(kr)k \, dk. \qquad (8.3.46)$$

For the experimentally important case where the temperature field is required at the probe laser point on the sample surface, $z = 0$ and (8.3.46) becomes

$$\bar{T}_1(r,0,s) = \frac{1}{2\pi\alpha_2} \int_0^\infty \frac{J_0(kr)k \, dk}{\sqrt{k^2+q_2^2}}. \qquad (8.3.47)$$

This integral can be solved by use of a method presented by *Bellman* et al. [8.56] as applied to this particular integral to give

$$T_1(r,0,t) = \frac{1}{4(\pi\alpha_2 t)^{3/2}} e^{-r^2/4\alpha_2 t}. \qquad (8.3.48)$$

Equation (8.3.48) is the Green's function showing the explicit spatial and temporal dependence of the surface temperature of a semi-infinite three-dimensional solid, following an impulsive excitation by a point source at $r_0 = z_0 = 0$. The form (8.3.48) is valid for all locations r on the surface, in relation to the time-dependent thermal diffusion length [8.54]

$$\mu_s(t) \equiv 2\sqrt{\alpha_2 t} \qquad (8.3.49)$$

and is in agreement with other derivations of the Green's function predicted for the same semi-infinite geometry, the presence of the interface at $z = 0$ contributing twice the value of the Green's function for infinite geometry [8.19].

Equation (8.3.48) has taken no account of the size or location of the probe beam. In practice, the thermoreflectance measurement is made with a probe beam of spot size w_1 and Gaussian transverse intensity profile $\exp(-r^2/w_1^2)$ intercepting the heated region of the sample surface. Furthermore, the probe beam may be offset from the pump beam in alignment by a few micrometers, defined by a radial offset r_b. The quad-cell detector, operating in the sum mode, integrates the probe intensity over an aperture with dimensions large relative to w_1. This situation is described theoretically by integrating over the transverse coordinates:

$$h(t) = \int_0^\infty \int_0^{2\pi} T(\bar{r}, z = 0, t) e^{-r^2/w_1^2} r\, dr\, d\theta \tag{8.3.50}$$

where $h(t)$ is the sum signal observed from the quad-cell, $T(\bar{r}, 0, t)$ is the thermal profile at $z = 0$, and $\bar{r} = r - r_b$ where r_b is the center of the probe beam.

Assuming a radially distributed and temporally impulsive Gaussian beam profile impinging on an optically opaque solid of absorption coefficient β, the heat impulse produced per unit volume is given by

$$Q(r, z, t) \approx \beta P_0 e^{-r^2/w_0^2} e^{-\beta|z|} \delta(t) \tag{8.3.51}$$

where P_0 is the irradiance of the incident laser beam. For such a distributed thermal source, the temperature of the solid surface (semi-infinite case) can be written in terms of the Green's function, (8.3.48)

$$T_1^d(r, 0, t) = \int_0^\infty \int_{-l}^0 \int_0^{2\pi} \int_0^\infty T_1(r, 0, t; r_0, 0, t_0) Q(r_0, z_0, t_0) r_0\, dr_0\, dz_0\, d\theta_0\, dt_0$$

$$= \frac{\beta P_0}{4(\pi\alpha_2)^{3/2}} \int_0^\infty \int_0^\infty \int_0^{2\pi} \frac{e^{-(r-r_0)^2/4\alpha_2(t-t_0)} \delta(t_0)}{(t-t_0)^{3/2}}$$

$$\times e^{-r_0^2/w_0^2} r_0\, dr_0\, d\theta_0\, dt_0 \int_{-l}^0 e^{\beta z_0}\, dz_0$$

$$= \frac{P_0}{4(\pi\alpha_2 t)^{3/2}} \int_0^{2\pi} \int_0^\infty e^{-(r-r_0)^2/4\alpha_2 t} e^{-r_0^2/w_0^2} r_0\, dr_0\, d\theta_0 \tag{8.3.52}$$

where $\beta l \gg 1$ was assumed. Now in cylindrical coordinates

$$(r - r_0)^2 = r^2 + r_0^2 - 2rr_0 \cos(\theta_0 - \theta) \tag{8.3.53}$$

and using the result [8.57]

$$\int_0^{2\pi} e^{-(2rr_0/4\alpha_2 t)\cos(\theta_0 - \theta)}\, d\theta_0 = I_0 \left[\frac{rr_0}{2\alpha_2 t} \right] \tag{8.3.54}$$

where I_0 is the modified Bessel function of zero order, we can write

$$T_1^d(r, 0, t) = \frac{P_0 e^{-r^2/4\alpha_2 t}}{4(\pi\alpha_2 t)^{3/2}} \int_0^\infty e^{-r_0^2/\Omega_0^2} I_0 \left[\frac{rr_0}{4\alpha_2 t} \right] r_0\, dr_0 \tag{8.3.55}$$

with

$$\frac{1}{\Omega_0^2} \equiv \frac{1}{w_0^2} + \frac{1}{4\alpha_2 t}. \tag{8.3.56}$$

244

Equation (8.3.55) can be integrated using Weber's first integral [8.58] and gives

$$T_1^d(r,0,t) = \frac{P_0 w_0^2}{2\pi(\pi\alpha_2 t)^{1/2}(4\alpha_2 t + w_0^2)} e^{-r^2/(4\alpha_2 t + w_0^2)}.$$ (8.3.57)

Insertion of (8.3.57) in (8.3.50) in the form

$$T(\bar{r}, z = 0, t) = \frac{A_0}{(4\alpha t + w_0^2)t^{1/2}} \exp\left[-\frac{(r-r_b)^2}{4\alpha t + w_0^2}\right],$$ (8.3.58)

where

$$A_0 = \frac{P_0 w_0^2 \alpha^{3/2}}{2\pi^{3/2}},$$ (8.3.59)

P_0 is the irradiance of the incident laser beam, w_0 is the radius of the pump beam, and α is the thermal diffusivity of the sample, and integration over the radial coordinate, gives a quad-cell signal impulse-response of the form

$$h(t) = \frac{A_0 \tau_0}{2(t + \tau_1)t^{1/2}} \exp\left[-\frac{r_0^2}{4\alpha(t + \tau_1)}\right]$$ (8.3.60)

where

$$\tau_1 \equiv \frac{1}{4\alpha}(w_1^2 + w_0^2) \; ; \quad \tau_0 \equiv \frac{w_1^2}{4\alpha}.$$ (8.3.61)

The constants τ_0 and τ_1 have the dimensions of time and correspond to diffusional thermal transit times to the offset distances w_1 and $(w_1^2 + w_0^2)^{1/2}$, respectively, from the center of the optical excitation (pump beam). If the beam alignment is concentric, $r_b = 0$ and (8.3.58) becomes

$$h(\tau) = \frac{A_0 w_1^2}{8\alpha(\tau + \tau_1)\tau^{1/2}}$$ (8.3.62)

where the time-delay domain variable τ has been substituted for t.

The experimental effect of varying the relative sizes of w_0 and w_1 is illustrated in Fig. 8.27 and may be understood qualitatively from the forms of (8.3.58) and (8.3.62): Increasing the pump beam size produces a gradual transition from a pure three-dimensional to a one-dimensional response ($T \propto 1/t^{1/2}$), due to the reduction in the importance of the radial degrees of freedom of the conduction process. If the probe beam radius is kept small as w_0 is increased, this transition from three- to one-dimensional thermal decay is observed.

If, on the other hand, the pump beam radius is kept small, while the probe beam radius is increased, the effective aperture for integration of the signal increases, since the factor $\exp(-r^2/w_1^2)$, in (8.3.50), broadens with increasing w_1. Physically, we have a situation in which the probe beam profile samples an increasingly larger radial area. The increasing radial component of heat flow in the sample is weighted by the probe beam and integrated by the quad-cell as described in (8.3.50). This increased radial thermal contribution has the effect of broadening the impulse response profile, because the time required for thermal energy to propagate to a specified radial offset is given by $\tau_d = w_1^2/6\alpha$, as can be shown [8.49] by (8.3.50) and (8.3.62). Therefore, increasing the radial area integrated over produces a broadening or a delay contribution in the impulse response profile due to an increased weighting of these delayed radial components in the total response. As the probe beam size, w_1, approaches infinity, the time delay profile $h(\tau)$, approaches the form $1/t^{1/2}$, i.e., the form of a one-dimensional thermal decay in a semi-infinite medium.

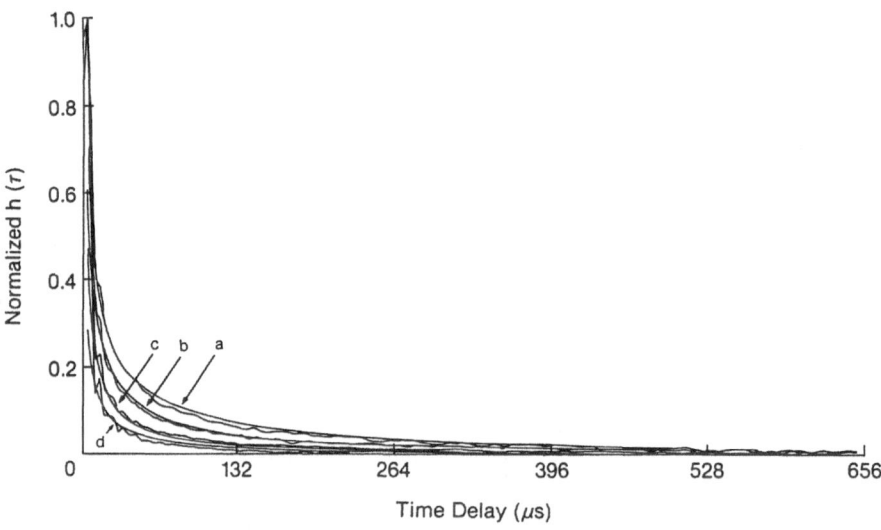

Fig. 8.27. Normalized impulse response profiles $[h(\tau = 0) = 1]$ for a semi-infinite quartz sample with various values of w_0 and w_1. Theory vs experiment. Upper curve: **(a)** $w_0 = 14 \, \mu m$; $w_1 = 12 \, \mu m$. **(b)** $w_0 = 10 \, \mu m$; $w_1 = 8 \, \mu m$. **(c)** $w_0 = 8 \, \mu m$; $w_1 = 2 \, \mu m$. **(d)** $w_0 = 2 \, \mu m$; $w_1 = 6 \, \mu m$ (bottom curve). These results were obtained with linear FM chirp excitation and $\Delta f = 0\text{--}100$ kHz; $S = 1.25 \times 10^7$ Hz/s. Theoretical curves (8.3.60) were fitted assuming that $\alpha = 5 \times 10^{-7}$ m^2/s for quartz. Experimental curves are the result of 1000 averages. Beam diameters were estimated independently

These trends may be noted from Fig. 8.27, which reports the experimental impulse response signals observed as w_0 and w_1 are varied from a semi-infinite sample of quartz. As either w_0 or w_1 are increased, the impulse response profile broadens, showing a greater contribution of the longer time delay processes associated with radial conduction. Experimentally, very fast time decay signals are observed for the condition $w_0 \approx w_1 \approx 5\mu m$. This case minimizes the radial flow effects just discussed, as well as the tendency for the thermal response to approach the one-dimensional condition as either w_1 or w_0 is increased.

An additional complication of the thermoreflectance signal is observed when the pump and probe beams are radially offset from each other. For metallic and most crystalline materials, this error in beam positioning is undetectable within the bandwidth of the instrumentation of Fig. 8.23, since the radial time delay components $\overline{r_b}/4\alpha$ are very small for beam offset errors of several micrometers. This is not always the case for highly insulating samples such as polymer films, with $\alpha \le 1 \times 10^{-7}$ m^2/s. An offset error as low as 2 micrometers can result in a significant radial delay component [8.49].

(c) Thermal Diffusivity Measurements

The contributions of the beam radii in (8.3.60) allow for the measurement of the thermal diffusivity on semi-infinite samples. For a given sample, variation of the beam profile characteristics gives rise to a family of impulse response curves as shown in Fig. 8.27. For the sake of convenience of reference, a 'half width' for the $h(\tau)$ profile may be defined as a means of estimating the width of the response curve. This half width is defined here as the time delay at which the thermal impulse response $h(\tau)$ has decreased to half the initial value observed in the time record, which was set at 3.9 µs. The variation in the half width of $h(\tau)$ with w_0 and w_1 is controlled by the thermal diffusivity, α, since this quantity determines the distribution of radial thermal transit times, which comprise the integrated quad-cell response. For the semi-infinite quartz sample, which has a moderately low thermal diffusivity (Fig. 8.27), the variation in the half width of the impulse response profile with beam radius is relatively large. For stainless steel, with one order of magnitude larger α, this variation of $h(\tau)$ with w_0 or w_1 is suppressed, since radial thermal transit times across the beam coordinate are greatly reduced. Nevertheless, in both cases there is good agreement between literature values of the thermal diffusivity and those obtained from the thermoreflectance measurement. Theoretical fits similar to those of Fig. 8.27 to experimental impulse responses of a semi-infinite stainless steel sample yielded the value 4×10^{-6} m^2/s for the thermal diffusivity of this material, in excellent agreement with the literature value [8.59] of 3.7×10^{-6} m^2/s. An extended detection bandwidth would be required, however, to resolve differences in impulse response profiles for highly conducting materials such as aluminum ($\alpha \geq 1 \times 10^{-4}$ m^2/s). In this latter case, the thermal transit times are so small that very high frequency measurements (corresponding to very early time delays) would be required to resolve differences in $w_1^2/4\alpha$ or $w_0^2/4\alpha$ in this regime.

In fitting the $h(\tau)$ data to (8.3.60), it was necessary to avoid the contribution due to background thermoelastic effects. For the quartz sample, the thermoelastic contribution appears as a delta function in the time delay domain. By fitting the data past the initial time delay period of $\tau = 10$ µs, it was possible to quantitatively evaluate the sample's thermal diffusivity independently of the thermoelastic contribution. This procedure is feasible for samples of moderately low thermal diffusivity. Highly conducting samples cannot be examined so successfully with the impulse response instrumentation of Fig. 8.23, because much more of the thermal information appears at these early time delays.

Highly insulating materials such as plastics and macromolecular polymer samples have a large variation in $h(\tau)$ with w_0 and w_1, and appear to be promising candidates for this type of measurement, within the instrumental limits of the apparatus in Fig. 8.23. Unfortunately, they usually have a larger coefficient of thermal expansion and a dominating thermoelastic response. Long time scale heating of thin films may cause deformations of the irradiated surface with an accompanying vertical translation of the surface due to the thermoelastic effect. As a case in point, FM thermoreflectance measurements made on thin PVDF films ($\alpha = 6 \times 10^{-8}$ m^2/s) showed accompanying cyclic changes in the thermal impulse response as the film heated and deformed over a time scale of minutes. These slow deformations were sufficient to move the sample surface in and out of the focal plane of the microscope objective lens. Due to the very large variation in response of $h(\tau)$ with changes in w_0 and w_1, surface motions of only a few micrometers over time could have been sufficient to produce these effects.

8.4 Conclusions – Future Directions

This review has shown that FM-TDS photothermal methods have become increasingly popular in recent years with investigators interested in depth-profiling, depth-resolution, and nondestructive analysis. The main advantages of these techniques are their superior dynamic range, speed of data acquisition, and information multiplexing, which is much richer in content and more reliable than frequency domain photothermal methods. The mathematical impulse-response equivalent obtained by the FM-TDS methods has the advantage of a higher damage threshold and higher SNR than conventional time-domain pulsed-laser-induced photothermal signals. Results to-date have shown that FM excitation holds excellent promise of favorably competing with the pseudorandom photothermal methods discussed by *Mandelis* [8.17], and by *Miller* in Chap. 7 of this volume. Detailed comparisons of FM-PDS with the pseudo-random-binary-sequence (PRBS) method of excitation revealed superior quality time delay and spectral information, when tested on a fast, flat frequency response mirage effect system [8.10]. Further comparisons of FM-P^{2}ES with wideband random-noise photothermal excitation were found to yield high-quality, band-limited impulse response information from both methods [8.60]. Random noise measurements, however, were found to be much less susceptible to distortions and nonlinearities in the excitation wave train. FM-P^2ES, on the other hand, demonstrated superior coherence and signal-to-noise ratio to random methods. This has a direct influence on the quality of spectral analysis and impulse response.

The experimental trend in FM-TDS photothermal techniques in the last few years since their introduction [8.9,10,12] has been toward the instrumentation of resolving earlier time delay domain information. The pseudorandom techniques have witnessed wide applicability in the millisecond range [8.17] and above. This range was also accessible by the FM-PDS technique, shown in Fig. 8.10, the time-delay limiting factor there being the transport of thermal information from the light absorbing surface to the probe laser beam position [8.9]. The sub-millisecond range was successfully accessed by thin-film FM-P^2ES instrumentation (Fig. 8.19) the limiting factor being the transport of thermal information from the front, optically excited, surface of the sample to the back surface where pyroelectric detection occurs [8.31].

The advent of FM-PTR spectrometry has pushed the limits of time-delay-domain resolution to the microsecond range [8.50], Fig. 8.27. With this technique, the limiting factor is essentially instrumental speed, i.e. the inability of commercial FFT analyzers to respond to signals with a frequency content higher than 100 kHz. Yet, many interesting and important applications of FM-PTR lie in the photothermal imaging of very shallow structures found in microelectronic devices, thin thermally conducting films and delicate optoelectronic materials. These applications require an increased detection bandwidth, so as to make it possible to extend the photothermal diagnostic analysis to very early time delays. Such early times, in the sub-microsecond range, will only become accessible with improved FFT instrumentation, effectively extending the bandwidth of present day FFT analyzers. Experimental efforts to build extended bandwidth instrumentation for very fast photothermal FM-PTR spectral analysis and impulse response recovery, and applications to NDE and physical systems exhibiting fast photothermal relaxations (e.g. semiconductors, laser materials) are currently well in progress in the author's laboratory.

References

8.1   R.C. Heyser: J. Audio Eng. Soc. **15**, 370 (1967)

8.2   H. Biering, O.Z. Pedersen: Brüel and Kjaer Tech. Rev. **1**, 5 (1983)

8.3   J.R. Klauder, A.C. Price, S. Darlington, W.J. Albersheim: Bell Syst. Tech. J. **39**, 745 (1960)

8.4   D.R. Bromaghim, J.P. Perry: IEEE Trans. MTT-**26**, 322 (1978)

8.5   J. Trampe Broch: Brüel and Kjaer Tech. Rev. **4**, 3 (1975)

8.6   A. Papoulis: *The Fourier Integral and its Applications* (McGraw-Hill, New York 1962)

8.7   R.C. Heyser: J. Audio Eng. Soc. **17**, 30 (1969)

8.8   R.C. Heyser: Monitor-Proc. IRE 67 (March 1976)

8.9   A. Mandelis, L.M.-L. Borm, J. Tiessinga: Rev. Sci. Instrum. **57**, 622 (1986)

8.10 A. Mandelis, L.M.-L. Borm, J. Tiessinga: Rev. Sci. Instrum. **57**, 630 (1986)

8.11 R.C. Heyser: J. Audio Eng. Soc. **17**, 130 (1969)

8.12 A. Mandelis: Rev. Sci. Instrum. **57**, 617 (1986)

8.13 E. Jahnke, F. Emde: *Tables of Functions* (Dover, New York 1945) pp. 35-88

8.14 J.S. Bendat, A.G. Piersol: *Engineering Applications of Correlation and Spectral Analysis* (Wiley, New York 1980)

8.15 S. Haykin: *Communication Systems* (Wiley, New York 1978) Chap. 4

8.16 H. Beiring, O.Z. Pedersen: Brüel and Kjaer Tech. Rev. **1**, 43 (1983)

8.17 A. Mandelis: IEEE Trans. UFFC-33, 590 (1986)

8.18 W.B. Jackson, N.M. Amer, A.C. Boccara, D. Fournier: Appl. Opt. **20**, 1333 (1981)

8.19 H.S. Carslaw, J.C. Jaeger: *Conduction of Heat in Solids*, 2nd ed. (Clarendon, Oxford 1959) Chap. XIV

8.20 A. Mandelis, B.S.H. Royce: Appl. Opt. **23**, 2892 (1984)

8.21 B. Sullivan, A.C. Tam: J. Acoust. Soc. Am. **75**, 437 (1984)

8.22 C.-Y. Kuo, M.F. Vieira, C.K.N. Patel: J. Appl. Phys. **55**, 3333 (1984)

8.23 H.M. Lai, K. Young: J. Acoust. Soc. Am. **72**, 2000 (1982)

8.24 A. Uejima, D. Curtis, Y. Sugitani: J. Photoacoust. **1**, 397 (1983)

8.25 J.A. Burt: J. Phys. D **13**, 1985 (1980)

8.26 A.C. Tam, C.K.N. Patel: Appl. Opt. **18**, 3348 (1979)

8.27 H. Sontag, A.C. Tam: Appl. Phys. Lett. **46**, 725 (1985)

8.28 A. Rosencwaig: In *Photoacoustics and Photoacoustic Spectroscopy*, Chemical Analysis, Vol. 57 (Wiley, New York 1980)

8.29 H. Coufal, A. Mandelis: In *Photoacoustic and Thermal Wave Phenomena in Semiconductors*, ed. by A. Mandelis (North-Holland, New York 1987) Chap. 7

8.30 A. Mandelis: Chem. Phys. Lett. **108**, 388 (1984)

8.31 J.F. Power, A. Mandelis: Rev. Sci. Instrum. **58**, 2024 (1987)

8.32 KYNAR^{TM} Piezo Film Applications Manual, Pennwalt Corp., King of Prussia, PA (1983)

8.33 H. Biering, O.Z. Pedersen: Brüel and Kjaer Tech. Rev. **2**, 28 (1983)

8.34 Hewlett-Packard Model 3562A Dynamic Signal Analyzer Operating Manual, Hewlett Packard, Everett, WA (1985) pp. 1-20

8.35 P.M. Morse, H. Feshbach: In *Methods of Theoretical Physics* Vol. 1 (McGraw-Hill, New York 1953) Chap. 9

8.36 J.F. Power, A. Mandelis: Rev. Sci. Instrum. **58**, 2018 (1987)

8.37 M.E. Lines, A.M. Glass: *Principles and Applications of Ferroelectrics* (Clarendon, Oxford 1977)

8.38 C.R.C. Handbook of Chemistry and Physics (Chemical Rubber Company, Cleveland, OH 1964)

8.39 F.P. Incropera, D.P. DeWitt: *Introduction to Heat Transfer* (Wiley, New York 1985) Appendix A, pp. 667-696

8.40 C.E. Yeack, R.L. Melcher, S.S. Jha: J. Appl. Phys. **53**, 3947 (1982)

8.41 A. Mandelis, B.S.H. Royce: J. Appl. Phys. **51**, 610 (1980)

8.42 A. Mandelis, J.T. Dodgson: J. Phys., C **19**, 2329 (1986)

8.43 J. Opsal, A. Rosencwaig, D.L. Willenborg: Appl. Opt. **22**, 3169 (1983)

8.44 A. Mandelis, A. Williams, E.K.M. Siu: J. Appl. Phys. **63**, 92 (1988)

8.45 A. Rosencwaig, J. Opsal, W.L. Smith, D.L. Willenborg: Appl. Phys. Lett. **46**, 1013 (1985)

8.46 W.P. Leung, A.C. Tam: Opt. Lett. **9**, 93 (1984)

8.47 T. Sawada, M. Kasai: In *Photoacoustic and Thermal Wave Phenomena in Semiconductors*, ed. by A. Mandelis (North-Holland, New York 1987) Chap. 1

8.48 S.O. Kanstad, P.-E. Nordal: Can. J. Phys. **64**, 1155 (1986)

8.49 A. Mandelis, J.F. Power: Appl. Opt. **27**, 3397 (1988)

8.50 J.F. Power, A. Mandelis: Appl. Opt. **27**, 3408 (1988)

8.51 J.P. Gordon, R.C.C. Leite, R.S. Moore, S.P.S. Porto, J.R. Whinnery: J. Appl. Phys. **36**, 3 (1965)

8.52 H.C. Chow: J. Appl. Phys. **51**, 4053 (1980)

8.53 J.A. Stratton: *Electromagnetic Theory* (McGraw-Hill, New York 1941) p. 573; A. Sommerfeld: Ann. Physik **28**, 665 (1909)

8.54 A. Mandelis, B.S.H. Royce: J. Appl. Phys. **50**, 4330 (1979)

8.55 A. Rosencwaig, J. Opsal: IEEE Trans. UFFC-**33**, 516 (1986)

8.56 R. Bellman, R.E. Marshak, G.M. Wing: Philos. Mag. **40**, 297 (1949)

8.57 H.S. Carslaw, J.C. Jaeger: *Conduction of Heat in Solids*, 2nd ed. (Clarendon, Oxford 1959) Sect.10.3

8.58 G.N. Watson: *A Treatise on the Theory of Bessel Functions*, 2nd ed. (Cambridge University Press, Cambridge 1944) p. 393

8.59 Y.S. Touloukian, R.C. Powell, C.Y. Ho, M.C. Nicolaou: *Thermal Diffusivity* (IFI/Plenum, New York 1973)

8.60 J.F. Power, A. Mandelis: Rev. Sci. Instrum. **58**, 2033 (1987)

9. Nondestructive Evaluation with Thermal Waves

G. Busse

With 29 Figures

The inspection of materials or components for very demanding applications is of considerable interest, since the costs due to a parts failure may exceed the value of the component by orders of magnitude. The detection and analysis of faults is a need that develops as more applications are found. Of course, one can analyze the quality of products by destructive evaluation and hence derive the probability of faults. However, statistical data obtained in this way can never guarantee a specific component.

This explains the increasing importance of nondestructive testing. One obtains information about the actual part that is going to be used, or even as it is being used. Unnecessary replacements can be avoided and faults (due to production or usage) are detected early enough. The necessary criteria have to be established from parallel investigations performed with both destructive and nondestructive evaluation (NDE).

The two best-known methods are ultrasonic and X-ray inspection, which both depend on the interaction of waves with matter, though the waves themselves are different. This chapter deals with another kind of wave that differs considerably from the two mentioned, although the general idea is the same: the interaction with boundaries or faults results in signal changes that can be used for quality criteria. The examples presented below should give some indication of which fields might profit from NDE using thermal waves.

9.1 Physical Background of Thermal Waves

Before presenting the applications, it is essential to understand the nature of thermal waves. Firstly, heat propagation is described by a parabolic differential equation that differs from popular "wave equations".

However, the "wave equation" in quantum mechanics does not fit into this scheme either. Wave-like heat propagation is not limited to interpretations where phonons are involved: under conditions of temperature modulation, e. g. at the surface of a solid (one-dimensional model), one finds that in the stationary situation this modulation propagates as a heavily damped wave. The wave vector is complex, where both the real and imaginary parts are equal to the inverse "thermal diffusion length μ",

$$\mu = \sqrt{\frac{2k}{\omega \varrho c}} \quad , \tag{9.1.1}$$

where k is the thermal conductivity, ϱ the density, c the specific heat, and ω the angular modulation frequency [9.1].

A typical value for μ is about 1 mm for metal and a tenth of that for polymers, if the modulation frequency is 10 Hz. Thus it is evident that thermal waves are slow, even though their phase velocity increases with the square root of frequency. Due to this dispersion, the group velocity is twice the phase velocity, and the shape of the pulse is distorted since the Fourier components in it propagate at different speeds. Propagation experiments – and amongst them NDE applications – are thus performed not with pulses but with stationary modulated waves. The stationary situation is achieved in a good approximation after about 10 cycles [9.2].

The ratio of the temperature amplitude required to generate a certain amplitude of heat flow density is named the "thermal wave impedance". This definition is useful to describe the boundary reflection, R, which is given by

$$R = \frac{Z_2 - Z_1}{Z_2 + Z_1} \tag{9.1.2}$$

for a thermal wave propagating from a medium with impedance Z_1 into one with impedance Z_2. Due to the superposition of the direct and reflected waves, it is obvious that reflections are detected from the resulting phase changes that they generate, where both signs of change are possible. From (9.1.2) and the definition of thermal conductivity, one finds that only the discontinuity of thermal properties determines the reflection coefficient of a boundary. For a crack in metal one finds $R \approx 1$ and about the same value for the paint/metal boundary. However, for very similar materials, e. g. paint on a polymer substrate, R can be quite small.

9.2 Experimental Arrangement

Although thermal waves can be generated by any periodic energy deposition, remote methods are of special interest since contact resistance is avoided. Optical absorption is about the most convenient method, with powers ranging from the milliwatt region (to generate thermal waves in polymers) to several watts (for experiments on metals).

For detection purposes, any device is suitable that is capable of monitoring the temperature modulation sensitively. The original opto- (or photo-)acoustic detection used by *Bell* [9.3] is still suitable for spectroscopy. However, it is not often used for NDE applications since the signal magnitude is inverse to the cell volume, which limits its applicability to the inspection of small components. A similar argument concerns the use of piezoceramic probes as solid-state microphones: although they remove the need for an airtight cell, they need physical contact with the part under investigation, and detection sensitivity is essentially limited to this small area. As an alternative to this stress/strain sensitive detector [9.4,5], one can use pyroelectric detection with a PVDF or PVF foil, or pyroelectric plate material [9.6,7].

Recently, true remote thermal wave detection has become more attractive with the use of the modulation of temperature-dependent optical properties for thermal wave detection. Examples are the mirage effect [9.8] or other methods of modulated optical deflection [9.9 - 11]. These methods may be restricted to certain geometries or optical properties, therefore they are not generally applicable to NDE. Modulated interferometry is also very sensitive, but it requires a reflecting surface [9.12].

The most versatile (although not the most sensitive) method is photothermal radiometry, where the modulated thermal infrared emission correlated with the thermal wave is analyzed [9.13]. Depending on the infrared optics in front of the detector, one can monitor a small spot of $40\,\mu$m in diameter in the thermal wave field, either on the illuminated front surface or on the rear surface after the wave has propagated through the sample (Fig. 9.1).

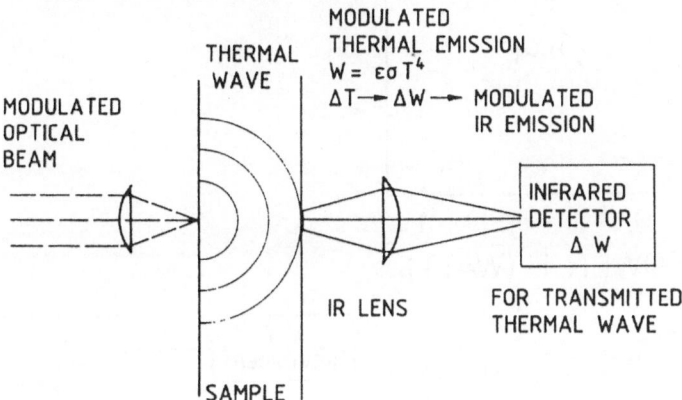

Fig. 9.1 Principle of photothermal infrared radiometry shown for a thermal wave transmission arrangement

If the temperature amplitude is small compared to 300 K, modulation of infrared emission is approximately proportional to the thermal wave amplitude [9.12]. At higher power densities this may not be true (Fig. 9.2); then one must be careful about signal distortions.

Between the temperature modulation in the focal spot and in the detector spot there is a phase shift (Fig. 9.3), from which thermal diffusivity can be determined. This is not a true local measurement, but an averaging process across the thermal wave path. If the detector is defocused, a large surface area is monitored. This thermal wave detector can therefore simulate the geometry of other detection principles. Here, it is essential to keep in mind that signal phase and amplitude provide quite different information [9.14]. In a raster scan, the photothermal magnitude image is the superposition of three images: the image of optical absorption, the image of the infrared emission coefficient (essentially similar to what thermography would show), and the thermal wave image [of local μ (9.1.1) and reflections]. The photothermal phase angle image, however, would contain almost exclusively the third image, which provides the most relevant information. This is true for low frequencies and opaque samples. It should also be noted that each of the three images has its own resolution [9.15].

A scan technique is generally required for NDE applications in order to detect inhomogeneities or to compare different sample regions. To do this, one can either

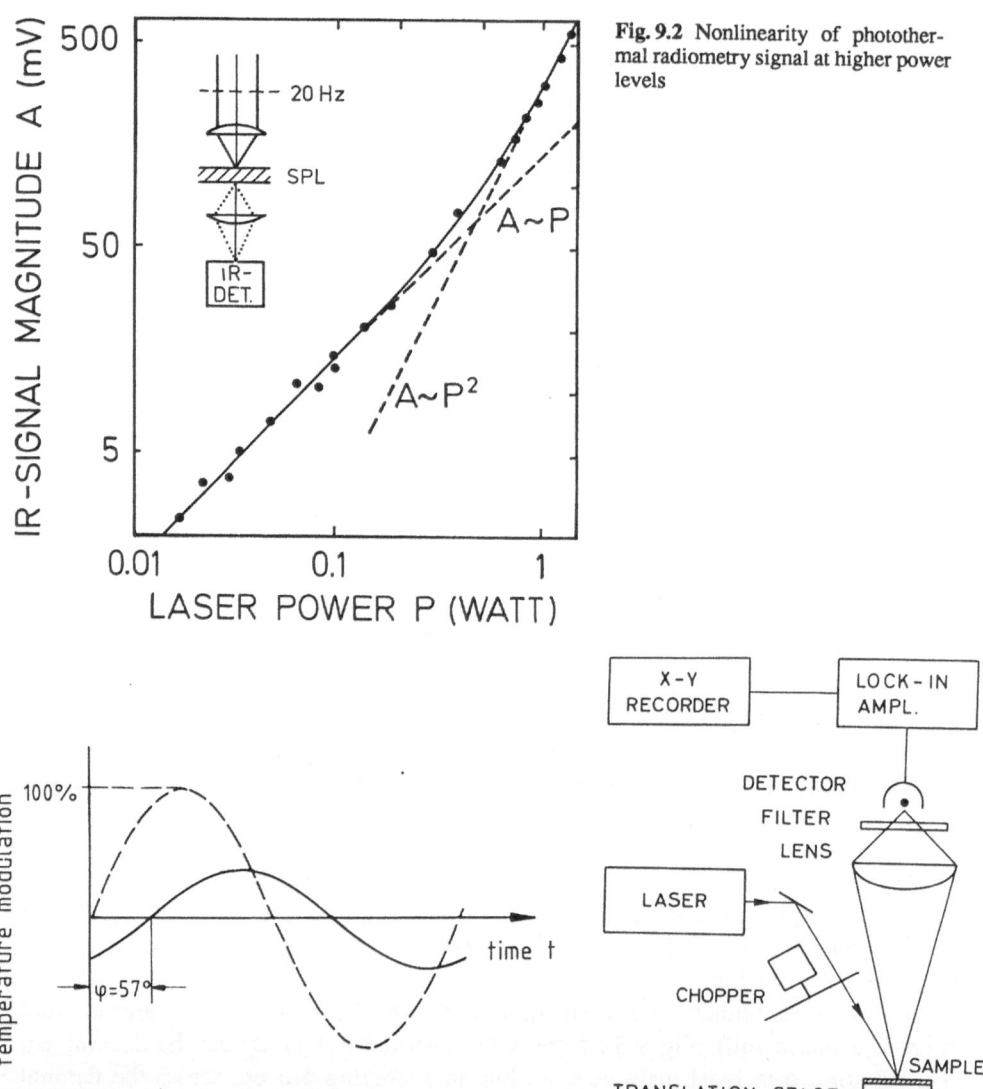

Fig. 9.2 Nonlinearity of photothermal radiometry signal at higher power levels

Fig. 9.3 Phase shift between temperature in the laser focus (dashed) and in the detector spot at a distance μ (solid line). A plane wave is assumed

Fig. 9.4 Experimental arrangement for remote photothermal nondestructive evaluation of the front surface [9.13]

scan the optical beam across the sample (if the detector is integrating over the whole surface), or the sample, if both the thermal wave generation and detection are spatially confined and need a constant relative position (Fig. 9.4). Usually, a computer is used to correlate the signal data with the sample coordinates where they were obtained. Data visualization is performed by signal surface representations, halftone images, or mixtures of both. An example of the various combinations is shown in Fig. 9.5.

254

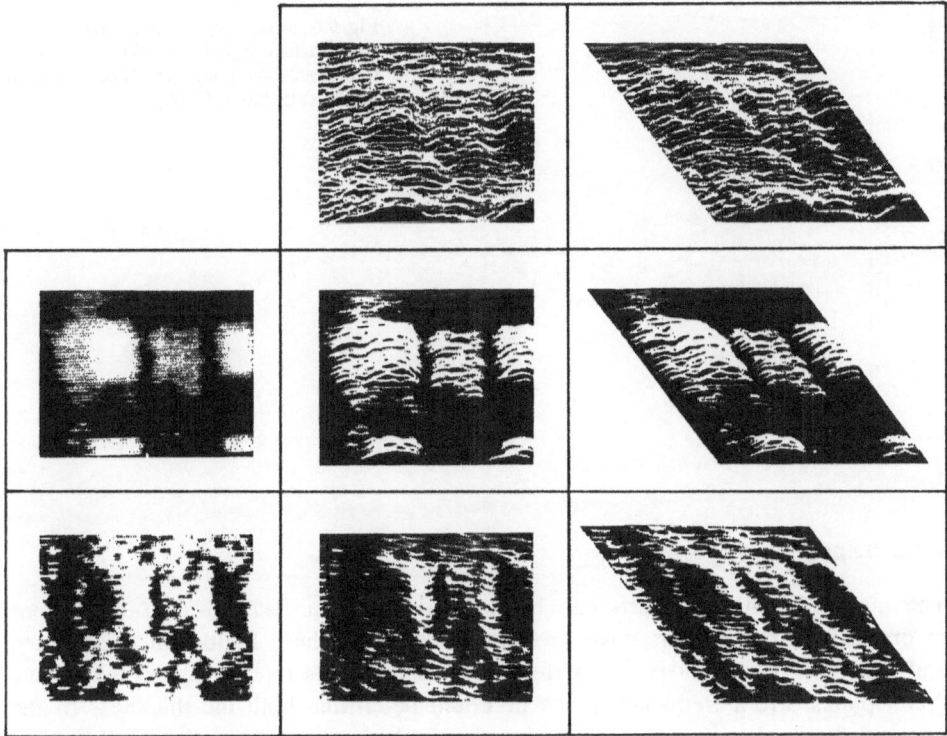

Fig. 9.5 Various kinds of image presentation (an integrated circuit)

9.3 Nondestructive Evaluation of Metals with Thermal Waves

For various reasons, much of the early thermal wave work was performed on metals. These samples can be machined in a reproducible way, are not as easily damaged as other materials during exposure to laser light, and the thermal diffusion length is about an order of magnitude larger than for most nonmetals. This last argument is crucial to propagation experiments, where one prefers larger distances and where model structures (e. g. holes) can be conveniently machined.

9.3.1 Depth Range Experiments

Work has been carried out to provide a better understanding of the depth range in thermal wave inspection. It was found that the depth range is about twice the thermal diffusion length, if one uses the phase angle [9.16 - 18] (Fig. 9.6). In a transmission arrangement, however, the range is noise-limited [9.19]. This is because, in front surface inspection, the signal change depends on thermal wave interference [9.20].

Stereoscopic thermal wave probing has also been applied to determine the depth of structures in metal [9.21,22]. However, the detection of faults was another major goal. Such faults can be simulated by subsurface holes or slots, or generated in a more realistic way.

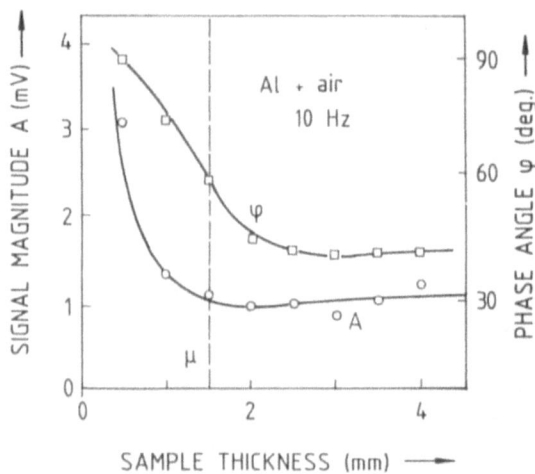

Fig. 9.6 Magnitude A and phase φ of front surface signal depending on sample thickness. Data points [9.19] compared to model calculation [9.16]

9.3.2 Inspection of Faults

The quality of metallic parts can be reduced by subsurface air gaps, inclusions or cracks. Various authors have investigated whether these faults can be detected and analyzed. *Lachaine* [9.23] performed measurements on air gaps between two metal plates. From frequency scans he could determine both the thickness of the gap and the thickness of the metal above it. Inclusions in an aluminum alloy have been investigated by *McDonald* [9.24] using a photothermal beam deflection method (Fig. 9.7).

Cracks may be produced either in the manufacturing process or while the component is in use. *Luukkala* and co-workers [9.25] performed thermal wave scans across a crack (generated in a crack opening device) that was almost invisible. However, the scan showed strong contrast (Fig. 9.8). Laser power was 50 mW in this experiment and the modulation frequency was 30 Hz. *Luukkala* also reports [9.26] that the

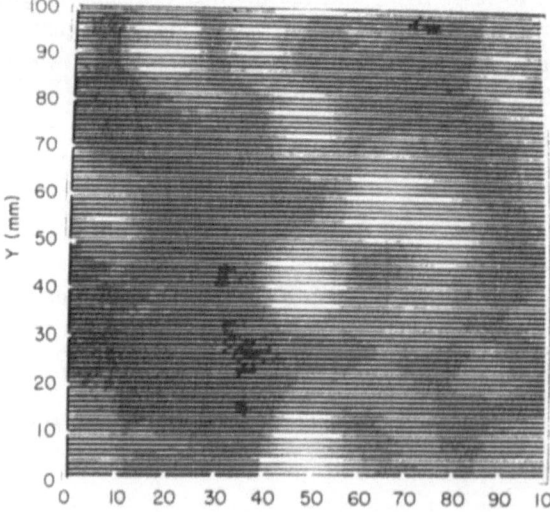

Fig. 9.7 Scan across an aluminum alloy sample using photothermal beam deflection. Structures are partly attributed to inclusions [9.24]

256

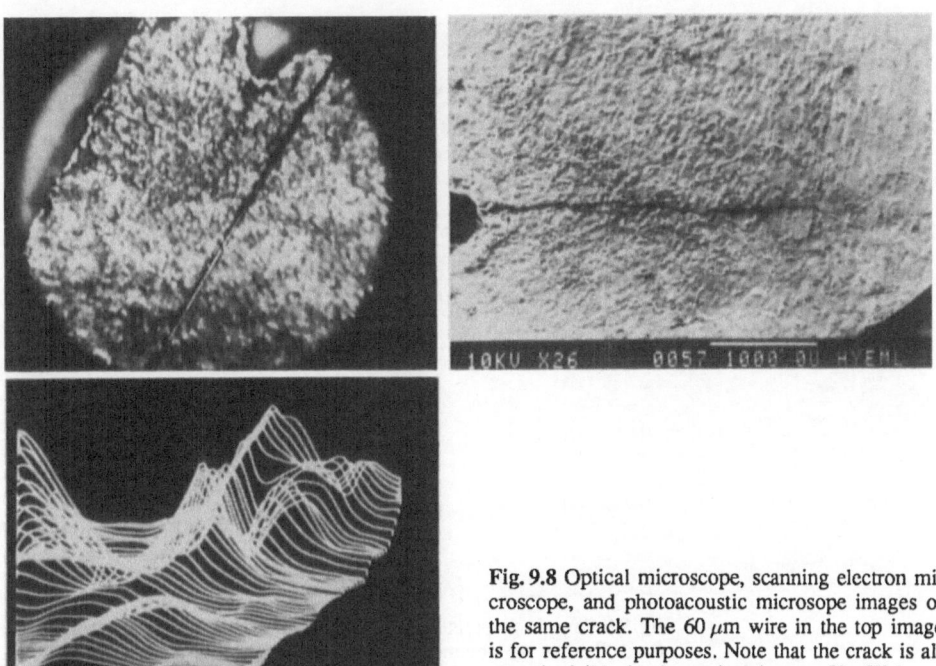

Fig. 9.8 Optical microscope, scanning electron microscope, and photoacoustic microsope images of the same crack. The $60\,\mu$m wire in the top image is for reference purposes. Note that the crack is almost invisible in the optical image. $50\,$mW laser, $30\,$Hz chopping frequency [9.25]

signal of old cracks is reduced by thermal short circuits, e. g. dirt. *Grice* et al. [9.27] investigated how the crack-correlated signal depends on geometry and signal detection method. They found that a vertical crack is not found with the gas cell method while it is easily seen with the optical beam deflection (OBD) method. For scans across vertical cracks, *McDonald* et al. [9.28] calculated the OBD signal assuming various geometries of the detection system. They found good agreement with experimental results. They also mention that a scan across a phase boundary gives a similar result.

9.3.3 NDE of Deformation, Seams and Hardening

The faults dealt with above have boundaries that should not exist in the ideal case. Another topic of interest is the use of thermal waves to monitor the influence of metal processing. An impressive example of this kind of NDE has been presented by *Luukkala* and *Askerov* [9.29]. They formed two grooves on their sample by striking it with a sharpened wedge (Fig. 9. 9a,b). Then they removed the grooves by grinding until the surface was smooth again. A scan across where the grooves had been clearly revealed the faults. Similar features were even observed during scans across the intact back surface of the 2 mm thick sample (Fig. 9. 9c - e). Upon annealing the sample, the signal structure disappeared. Later on, corresponding results were observed on deformed steel [9.30].

The inspection of welding seams was one of the early applications [9.31] where hidden discontinuities could be found (Fig. 9.10). By analyzing thermal wave prop-

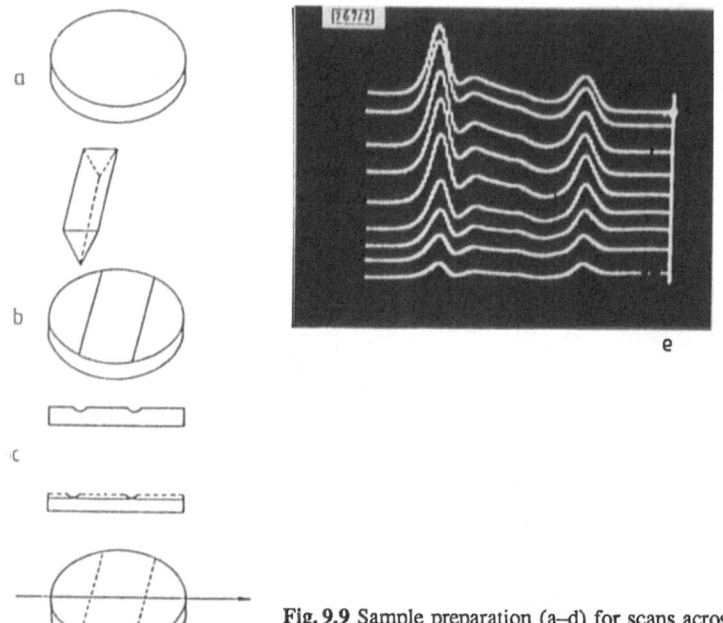

a

b

c

d

e

Fig. 9.9 Sample preparation (a–d) for scans across a deformed region in a metal. Results (e) obtained for scans across the rear surface. Modulation frequency ranges from 104 Hz (*top*) to 524 Hz (*bottom*) [9.29]

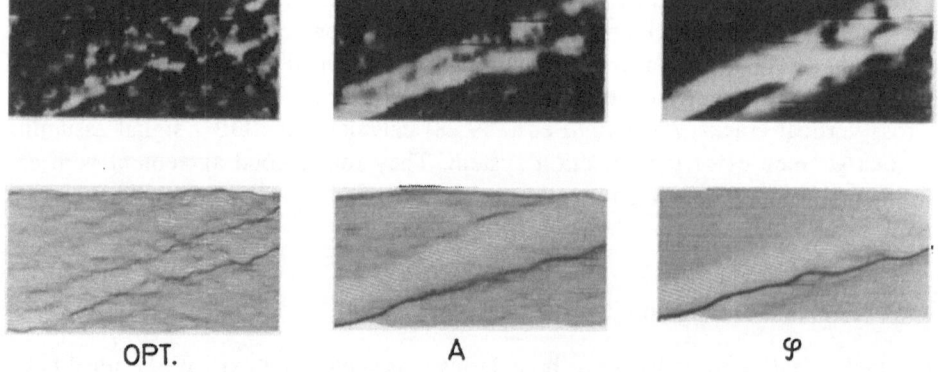

OPT.   A   φ

Fig. 9.10 Gray-scale (*top*) and perspective line drawings (*bottom*) of photothermal images of a weld seam in a stainless steel plate [9.31]

agation in or near the seam, one expects to probe the seam profile in depth and to detect faults. For thermal wave transmission a simple theoretical model has been used to determine the seam profile from a scan [9.30]. The agreement with values found from cut samples is acceptable. These investigations also indicated that seam quality and signal structure have some correlation. This would be in agreement with results published by *Rosencwaig* [9.32], who found that columnar grains and the transition zone in a weld region of an aluminum alloy affect the thermal wave signal (Fig. 9.11).

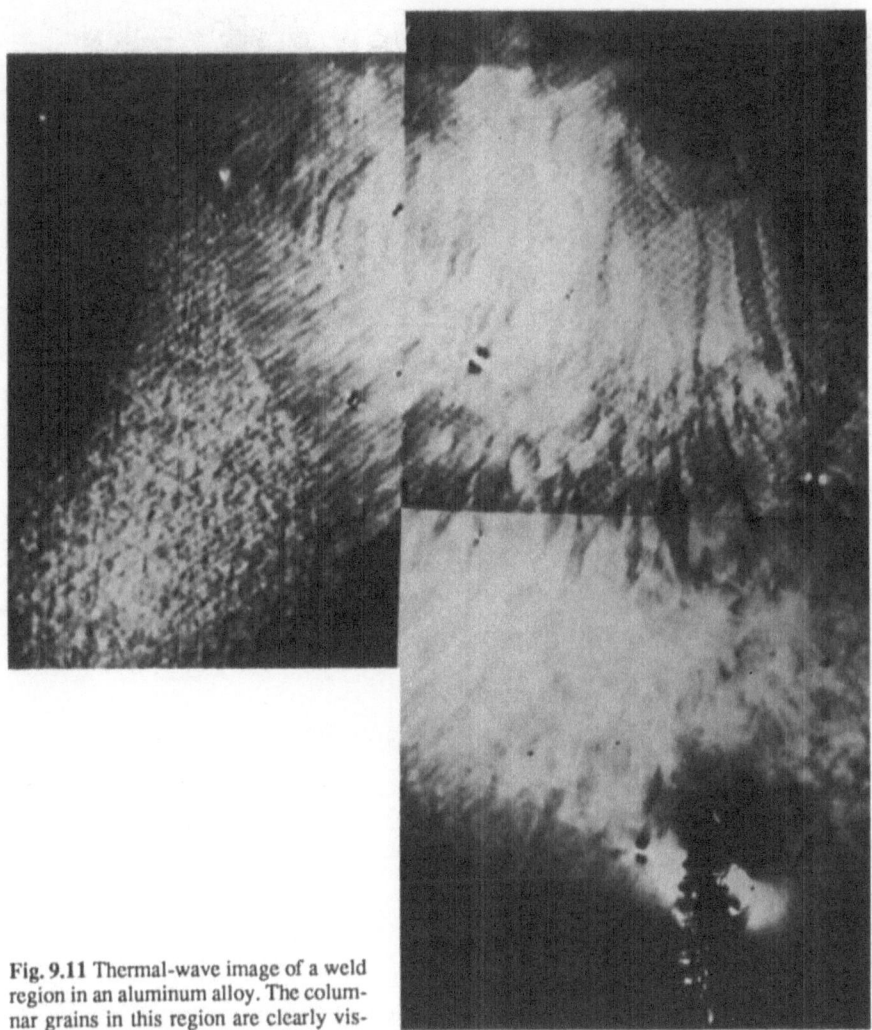

Fig. 9.11 Thermal-wave image of a weld region in an aluminum alloy. The columnar grains in this region are clearly visible. (Magnification 30×) [9.32]

Besides weld seams in metal plates, another potenial field is the inspection of ultrasonic bonded foils, as shown in Fig. 9.12. The thermal wave image shows an inhomogeneity that is not visible in the optical image [9.33].

Surface hardening of metals is highly relevant to increasing the performance and lifetime of components. Due to the carbon concentration, the hardened layer differs in its thermal properties from the material underneath. Analysis of thermal wave propagation can provide certain information. From experiments where signal phase was measured as a function of distance from the thermal wave source operated at 1.88 Hz, one obtains an effective diffusivity, which can be correlated to the thickness of the hardened layer [9.34]. The inverse determination of the thermal conductivity profile from the thermal wave surface data is possible according to a theoretical analysis [9.35]. If the sample is accessible from both sides, there seems to be a simpler way to determine the thickness of the hardened layer and the change of the

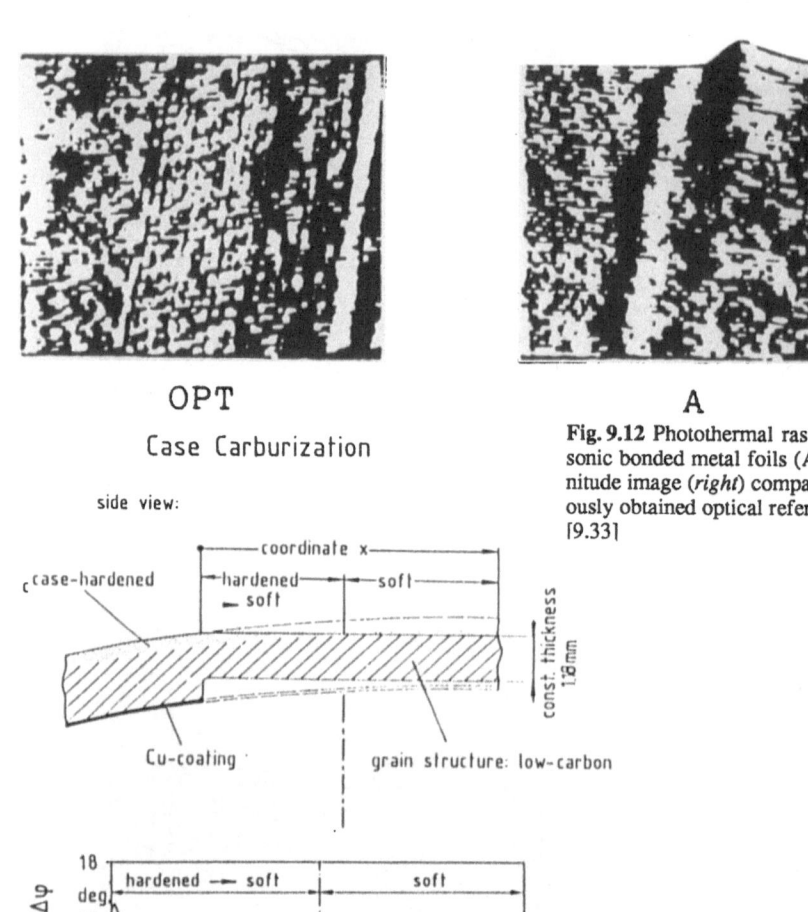

OPT

Case Carburization

Fig. 9.12 Photothermal raster scan of ultrasonic bonded metal foils (Al and Cu). Magnitude image (*right*) compared to simultaneously obtained optical reference image (*left*) [9.33]

side view:

coordinate x

c case-hardened

hardened — soft
— soft

soft

const. thickness 1.5 mm

Cu-coating

grain structure: low-carbon

A

phase angle Δφ

18 deg.

hardened — soft   soft

15

12

9

6

0   4   8   12   16   20 mm

coordinate x ⟶

Fig. 9.13 *Top*: Preparation of steel sample with a wedge-shaped hardened region. *Bottom*: Scan with thermal wave transmission arrangement (Fig. 9.1)

thermal parameters in depth (Fig. 9.13) [9.30]. However, in this case one is tied to a specific sample, so the applicability to general NDE is not obvious.

9.4 NDE of Nonmetals with Thermal Waves

Although thermal waves have successfully been used for probing metals, they have to compete with established NDE methods based on electrical conductivity or on magnetic properties. For probing nonconducting materials, there is less competition and also a greater need for new developments. This may be why there is such a variety of unanswered questions and so many (tentative or viable) solutions, some of which are reported in this section.

9.4.1 Semiconductors

Modern electronics depend on the reliability of doping and on defect-free materials intended to be used for a high density of tiny structures. Small undetected faults cause failure after many expensive production processes. Any early information on faults can significantly contribute to cost reduction. An example of this kind has been reported by *McClelland* et al. [9.36]: the alloy semiconductor $Hg_{1-x}Cd_xTe$ requires a very high compositional uniformity to ensure a homogeneous response in the focal plane of infrared imaging systems. It would be valuable to detect compositional variations prior to the fabrication of large arrays. A raster scan across dendritic and uniform-composition samples (Fig. 9.14) clearly reveals the difference. The compositional dependence of the thermal wave signal may allow faster quality inspection than electrofluorescence and luminescence methods.

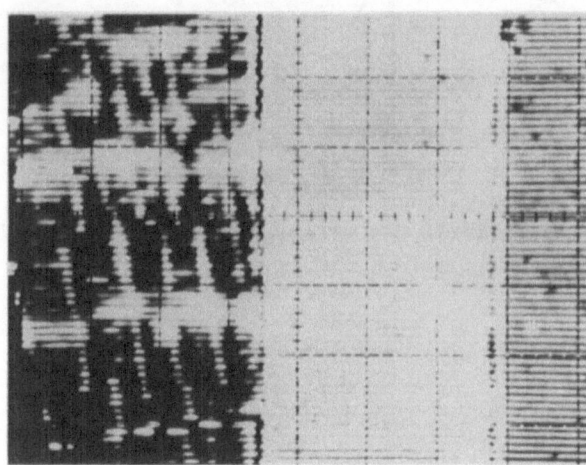

Fig. 9.14 Photoacoustic raster scan image of dendritic (*left*) and uniform composition $Hg_{1-x}Cd_xTe$ samples [9.36]

MacFarlane et al. [9.37] used photoacoustic detection to inspect silicon after both ion implantation and the subsequent process of laser annealing had been performed. The mechanism involved is a modified absorption in the implanted area.

Experiments performed by *Williams* [9.38] on boron-doped silicon indicated that the optical variation correlated with the doping is too small to account for the photothermal contrast seen in the image. To confirm this idea he coated the structured area with metal and could still distinguish the implanted areas (Fig. 9.15).

Although this chapter concentrates on optically generated thermal waves, it should be mentioned here that, in the area of semiconductors, the modulated electron beam competes with the laser. One reason for this is that this kind of sample is well suited to the vacuum conditions required for the electron beam. In fact, a lot of the early and recent work was done using this technique [9.39–41].

Microscopy using optically generated thermal waves has been tested on integrated circuits [9.42–44]. The image obtained with signal phase generally looks

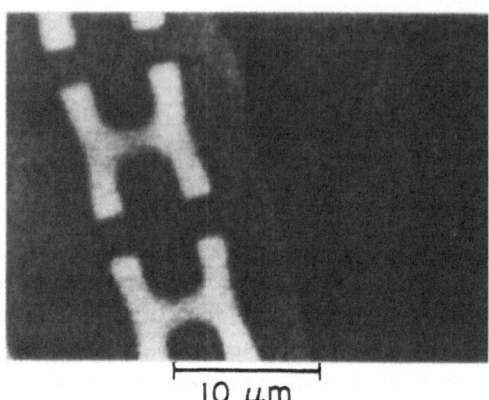

Fig. 9.15 Photothermal image of implanted silicon with 300 Å of titanium deposited on the silicon in the left half of the image [9.38]

10 μm

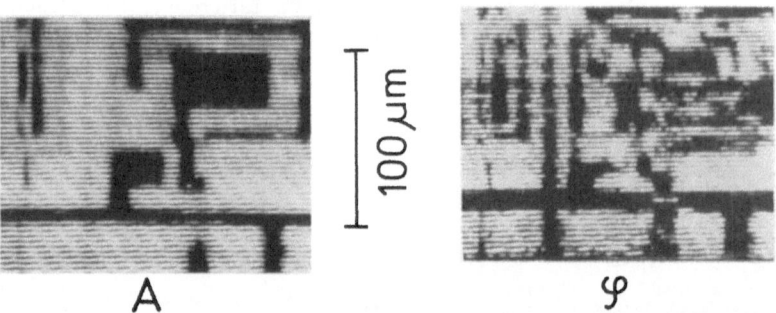

100 μm

A   φ

Fig 9.16 Thermal wave microscopy of an integrated circuit using signal magnitude (*left*) and phase (*right*) [9.31]

very different from the magnitude image containing a strong superposition of optical features (Fig. 9.16). The drawback of these point-by-point raster scans is their low speed, which makes in-line inspection of large numbers difficult.

9.4.2 Ceramics

The very first work on scanned thermal wave inspection was performed on ceramics [9.45]. The aim of this investigation was the detection of cracks and nonoptical features. There is a great deal of interest in this information since these surface defects provide an early failure mechanism. *Wong* [9.46] mentioned that the thermal wave signal depends on the sintering process and the grain size distribution. He transformed the line scan and interpreted the Fourier spectrum as the statistical distribution of grain size.

   Murphy and *Aamodt* [9.47] used the transverse optical deflection technique to improve the detection sensitivity for near-surface features (Fig. 9.17). The general problem with these investigations is the number of boundaries that are left after the sintering process, which provide a strong background to the possibly very similar dangerous microcracks.

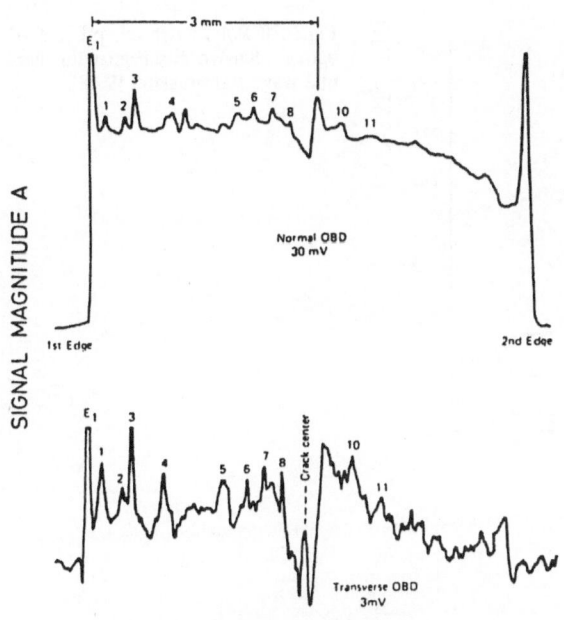

SIGNAL MAGNITUDE A

SAMPLE COORDINATE

Fig. 9.17 Transverse (*bottom*) and normal (*top*) beam deflection scan across a ceramic disk. Numbers indicate identified cracks [9.47]

9.4.3 Polymers

Though they were initially intended to be a cheap replacement for metals, polymers are used nowadays because they are convenient to manufacture and have excellent performance, even compared to metals. As an example, modern air and space systems depend on the high specific strength of carbon fiber reinforced materials to an ever increasing degree.

There is, meanwhile, a large variety of plastics tailored to fit the needs of specific applications. Common to all of them is the polymerization process, where smaller molecules form large units that are similar to chains. The length and the orientation of the chains are among the parameters that determine the performance. Therefore, there is a need to monitor these parameters.

The curing process is of interest for applications of adhesives. One early application was the measurement of the thermal wave transmission of an epoxy adhesive between two metal plates [9.48]. The initial exothermic reaction causes an increase in average temperature (measured simultaneously with the thermal wave), resulting in a steep change of signal phase. However, the phase keeps changing gradually over a long time, indicating that the polymerization process is still going on (Fig. 9.18). Alignment of molecules is produced by stretching polymers along one direction. Optical anisotropy induced in this way is suitable for monitoring the process in a remote way. However, not all materials are transparent and where this is the case thermal wave analysis might be an alternative [9.48].

Many technical applications require fiber-reinforced plastics. The high modulus of the fiber together with the matrix/fiber adhesion improve the strength significantly, depending on the amount and orientation of the fibers. It is thus of relevance to measure the amount of fibers and their orientation.

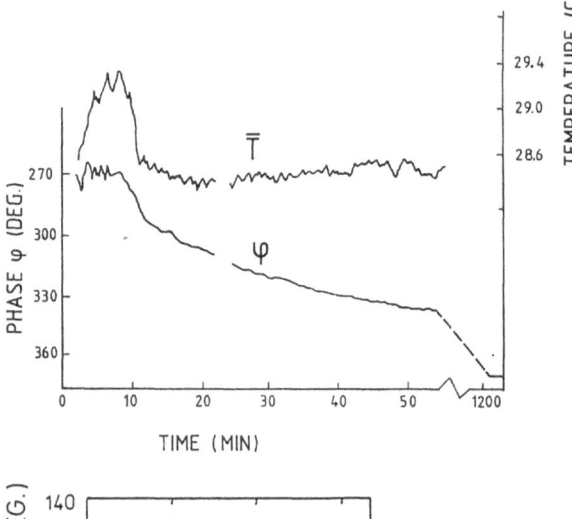

Fig. 9.18 Polymerization process of epoxy adhesive observed with thermal wave transmission [9.48]

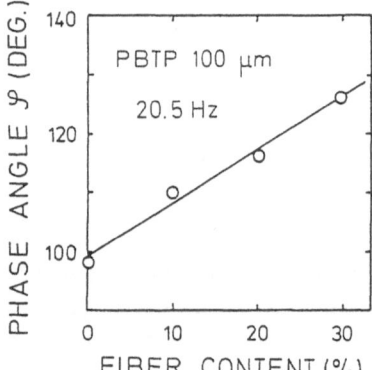

Fig. 9.19 Glass fiber reinforced PBTP: Influence of fibre content on signal phase in thermal wave transmission [9.49]

Measurements on glass fiber reinforced PBTP were performed in 1983 using remote thermal wave transmission [9.49]. Various samples of equal thickness and with a fiber content ranging from 0% to 30% were inspected. The result is shown in Fig. 9.19. With the present phase accuracy, it should be possible to determine the fiber content with about 0.1% accuracy. In the molding process, the fibers tend to be orientated along the flow direction, since for other orientations there is a torque resulting from the velocity gradient of the mold. To observe orientation effects, samples were cut along the flow direction or perpendicular to it. In the transmission arrangement, signal phase is linear in sample thickness with a slope depending on thermal diffusion length. From Fig. 9.20 it is obvious that at a small thickness the difference is too small to be detectable, while for thick samples (exceeding thermal diffusion length considerably) it is clearly established.

These transmission measurements have the advantage that the range is not limited by the thermal diffusion length. On the other hand, investigation requires accessibility from both sides and a thickness of significantly less than 1 mm. For both these reasons, single-ended inspection looks more attractive. Measurements with spatially resolved surface detection were performed on carbon fiber reinforced plastics. In this

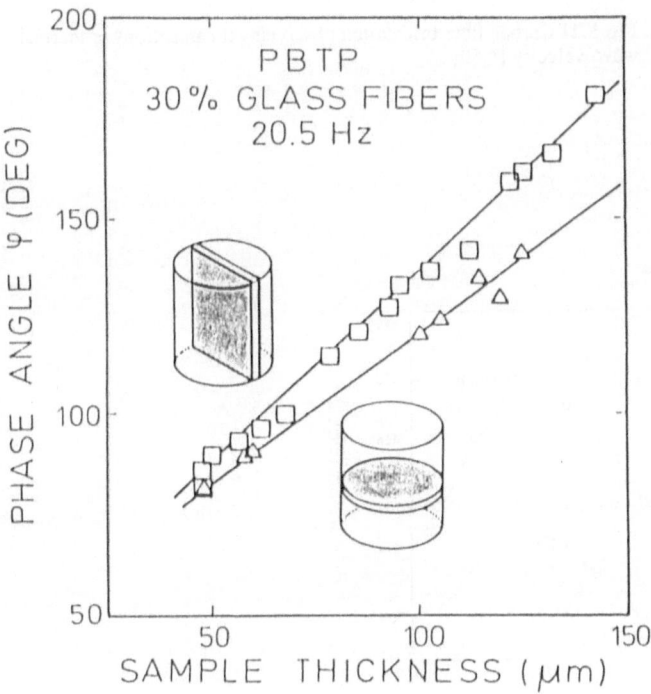

Fig. 9.20 Thermal wave anisotropy observed on an injection-molded glass fiber reinforced polymer. Directions A–A and B–B are along and perpendicular to the fibers, respectively [9.49]

case, the fiber differs much more from the matrix, so that thermal wave anisotropy due to fiber orientation is expected to be much stronger.

With a small spot imaged onto the infrared detector, the thermal wave propagation from the laser focus to this spot was observed as a function of the relative orientation with respect to the fiber direction. The dependence for two filled circles with different radii is shown in Fig. 9.21 [9.50]. The phase maximum indicates that the largest delay is observed when the thermal wave moves perpendicular to the carbon fibers, which are good thermal conductors. For an unknown orientation, such a curve would provide the local fiber direction, where "local" means an area of less than 1 mm in diameter. This resolution is good enough for most applications.

As a realistic example, two scans were performed across a seam in an injection molded component. When the two polymer streams meet each other to form the seam, flow direction and fiber orientation change in such a way that the fibers are orientated along the seam, while they are perpendicular to it initially. This is confirmed by the scan (Fig. 9.22). With the displacement between the laser focus and detector spot along the seam, signal phase is reduced in the seam region, while it increases if the displacement is perpendicular to the seam [9.51]. This result shows that spatially resolved thermal wave detection is suitable for scanning analysis of fiber orientation in seams. Its relevance is that such a scan reveals areas of hidden mechanical weakness. Any defect in fiber-reinforced materials is a potential source of mechanical failure. It has been shown previously that one can detect cracks (in the fibers or matrix) and inclusions in the near-surface region [9.52].

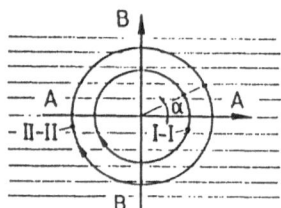

Fig. 9.21 Carbon fiber orientation observed with anisotropy of thermal wave velocity [4.50]

FIBER ORIENTATION ⟶

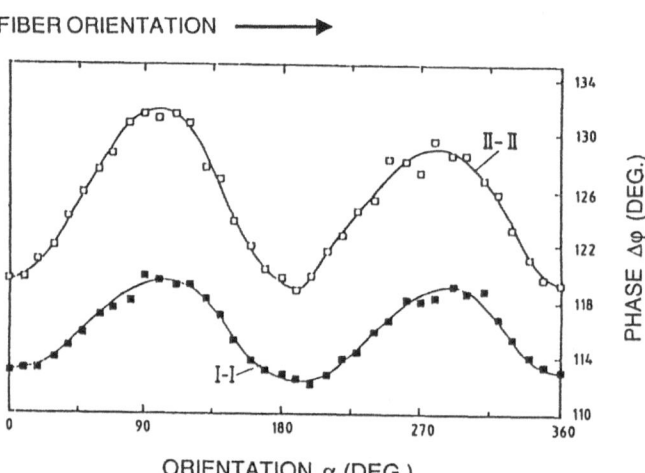

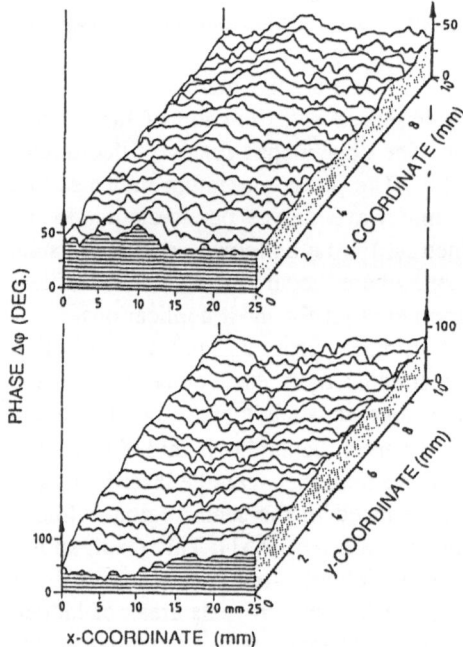

Fig. 9.22 Local fiber orientation in the seam of an injection-molded component. Seam direction along (*bottom*) or perpendicular (*top*) to offset direction between laser spot and detector spot [4.51]

9.5 Coatings

Surface modifications by coatings can improve material properties such as hardness, thermal resistance, conductivity or just color. The appearance of coatings may thus range from highly reflective to translucent or absorbing. The proper choice of detection may improve the result significantly. For semiconductor applications, one needs to monitor metal thickness on various materials and in this case optical detection of periodical surface deformation is a sensitive means of monitoring the thermal wave. Sensitivity is reported to be 10^{-4} Å/$\sqrt{\text{Hz}}$ [9.53]. Experimental and theoretical results show good agreement (Fig. 9.23), thereby indicating that the 3D model is correct. Three-dimensional calculations and measurements were performed for TiN layers on Si [9.54] using photothermal infrared radiometry.

While these were investigations of coated semiconductors, ceramics and other plasma sprayed coatings are applied to metal substrates to improve their hardness and thermal isolation.

Jaarinen and *Luukkala* measured thickness variations of tungsten carbide and detected simulated defects in the Cr_2O_3/metal boundary [9.55]. They point out that the reflection coefficient of the boundary changes from -1 to $+1$, thereby making thermal wave inspection of faults under the ceramic possible.

Almond et al. [9.56] mention that inspection of Al_2O_3 coatings is difficult due to their strong light scattering effects. Results for a scan are compared with ultrasonic

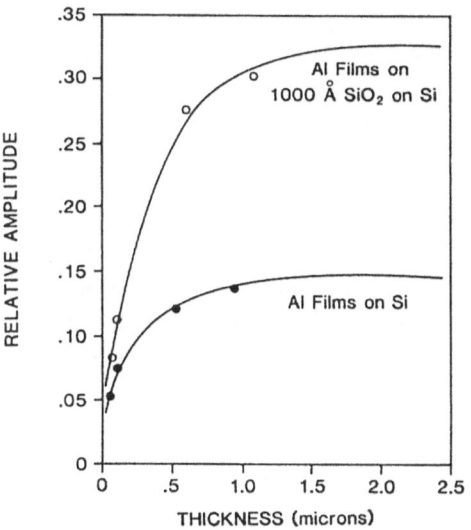

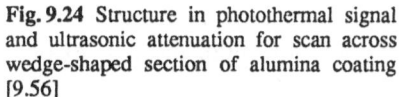

Fig. 9.23 Relative amplitude at 1 MHz of laser beam deflection signal for Al films on Si or SiO₂ on Si. Data points fit well to curves given by theory [9.53]

Fig. 9.24 Structure in photothermal signal ▶ and ultrasonic attenuation for scan across wedge-shaped section of alumina coating [9.56]

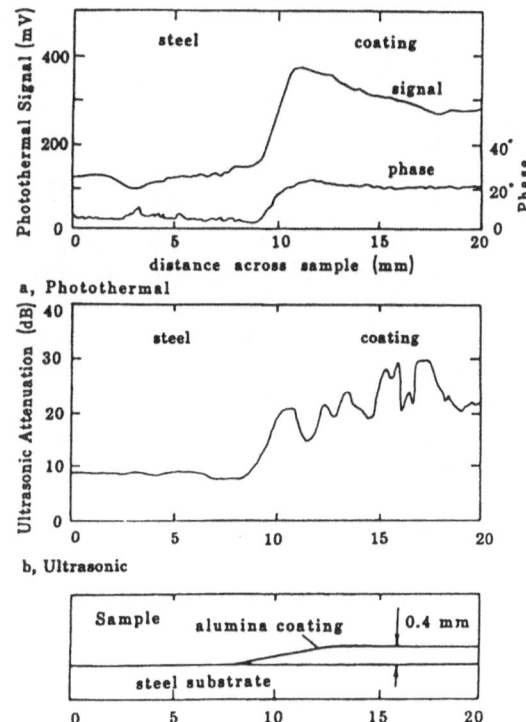

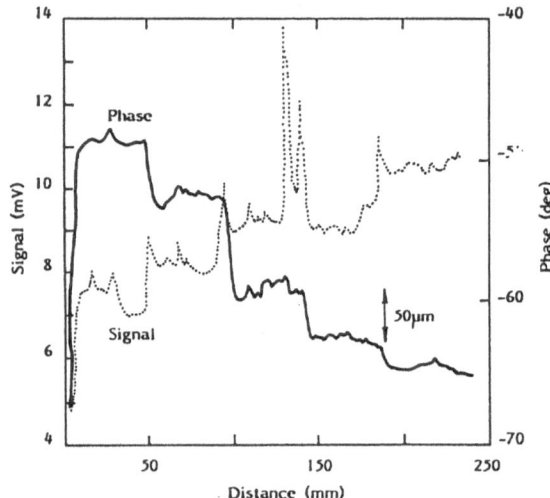

Fig. 9.25 Results for thickness measurements of a hard coating on mild steel [9.58]

Fig. 9.26 Enamel coating. Amplitude (×), phase (+), and theory (- - -) for a delaminated place. Amplitude (o), phase (•), and theory (—) for an undelaminated place [9.59]

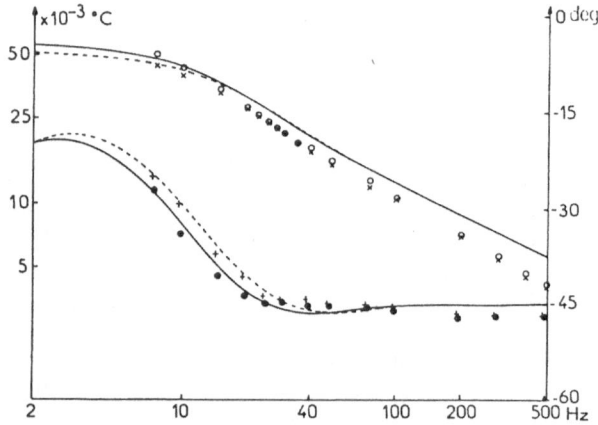

measurements (Fig. 9.24). Both the metal and the coating are sooted to achieve a homogenous infrared emission coefficient. *Liezers* and *Miller* [9.57] performed experiments where they detected features which may be correlated to defects in the coating above the bond coat layer. This results from the different frequency dependence.

Hard coatings on boiler tubes and reactor components were inspected to monitor thickness variations from 50 μm to 250 μm in 50 μm steps [9.58] (Fig. 9.25).

Egee et al. [9.59] performed theoretical and experimental investigations on semi-transparent ceramic coatings and found good agreement between the two (Fig. 9.26).

Organic coatings and paints are another field of interest. Protection of metal against corrosion is difficult if coating delamination occurs. Early experiments indicated that simulated delaminations can be detected with thermal waves [9.60] (Fig. 9.27). Inspection of these coatings is nondestructive only at reduced laser powers (usually less than 50 mW). Thickness variations are more difficult to observe on polymer substrates than on metals since the reflection coefficient at the

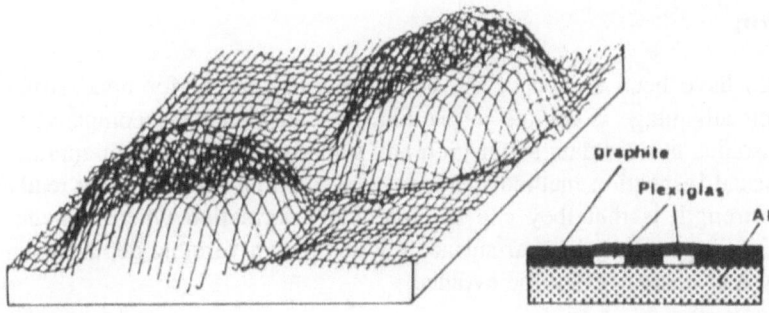

Fig. 9.27 Raster scan across two simulated delaminations using photoacoustic detection of thermal waves [9.60]

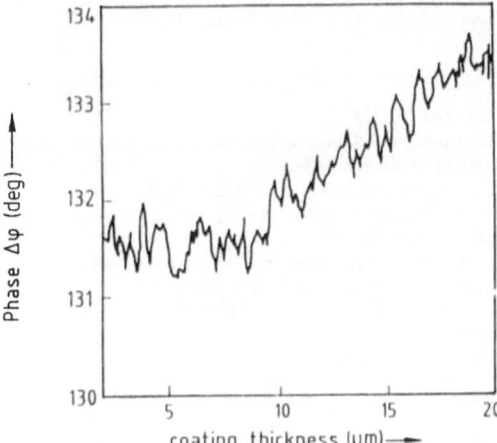

Fig. 9.28 Photothermal signal structure for scan across a wedge-shaped polyurethane (PUR coating on a PUR substrate). The gradual increase of phase indicates an increase of average thickness. The fine structure is due to local inhomogeneity of the coating [9.61]

paint/polymer boundary may be small [9.61]. Recently, it has even been possible to monitor polyurethane paint on a polyurethane substrate (Fig. 9.28) [9.62]. At low reflection, the influence of surface preparation or contamination may be probed through the coating [9.61]. The boundary situation may also be modified by deformation, as is shown in Fig. 9.29, where two coated polymer samples were scanned. The cross sections were reduced in the middle to increase strain locally. In this region the shaded area indicates a phase angle change that is opposite to the effect of reduced thickness.

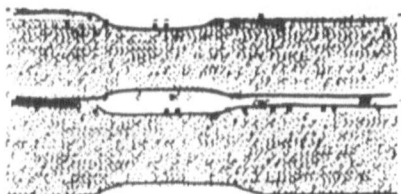
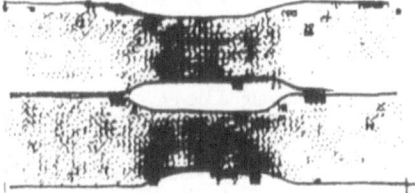

Fig. 9.29 Gray scale of local phase angle for scan across two unstrained painted polymer samples (*left*) and for scans after straining (*right*). Change of phase angle is 0.3° [9.62]

9.6 Conclusion

Various examples have been shown of the use of thermal waves for nondestructive testing. Their advantage is remote inspection, their disadvantage compared to X-rays and ultrasonics is slow data acquisition and sensitive equipment. In applications where classical inspection methods perform well, thermal waves cannot really compete. Their strength is that they can still provide information where conventional techniques fail: generally in near-surface regions with a variable depth range, especially if physical contact has to be avoided.

References

9.1   H. S. Carslaw, J. C. Jaeger: *Conduction of Heat in Solids* (Clarendon, Oxford 1959)
9.2   B. Rief: Can. J. Phys. **64**, 1303–1306 (1986)
9.3   A. G. Bell: Philos. Mag. **11**, 510–528 (1881)
9.4   R. M. White: J. Appl. Phys. **34**, 3559 – 3567 (1963)
9.5   A. Hordvik, H. Schlossberg: Appl. Opt. **16** 101–107 (1977)
9.6   M. Luukkala: "Photoacoustic Microscopy at Low Modulation Frequencies", in *Scanned Image Microscopy*, ed. by E. A. Ash (Academic, London 1980) pp. 273–289
9.7   C. R. Petts, H. K. Wickramasinghe: Proc. IEEE Ultrasonics Symp. 832–636 (1981)
9.8   A. C. Boccara, D. Fournier, J. Badoz: Appl. Phys. Lett. **36**, 130–132 (1980)
9.9   L. C. Aamodt, J. C. Murphy: J. Appl. Phys. **54**, 581–591 (1983)
9.10 W. B. Jackson, N. M. Amer, A. C. Boccara, D. Fournier: Appl. Opt. **20**, 1331–1344 (1981)
9.11 A. Rosencwaig, J. Opsal, D. L. Willenborg: J. de Phys. **44** (C6), 483–489 (1983)
9.12 S. Ameri, E. A. Ash, V. Neumann, C. R. Petts: Electron. Lett. **17**, 337–338 (1981)
9.13 P. -E. Nordal, S. O. Kanstad: Phys. Scr. **20**, 659–662 (1979)
9.14 A. Rosencwaig, G. Busse: Appl. Phys. Lett. **36**, 725–727 (1980)
9.15 A. Rosencwaig: American Lab. **11**, 39–49 (1979)
9.16 G. Busse: Appl. Phys. Lett. **35**, 759–760 (1979)
9.17 R. L. Thomas, J. J. Pouch, Y. H. Wong, L. D. Favro, P. K. Kuo, A. Rosencwaig: J. Appl. Phys. **51**, 1152–1156 (1980)
9.18 A. Lehto, J. Jaarinen, T. Tiusanen, M. Jokinen, M. Luukkala: Electron. Lett. **17**, 364–365 (1981)
9.19 G. Busse: Infrared Phys. **20**, 419–422 (1980)
9.20 C. A. Bennett, R. R. Patty: Appl. Opt. **21**, 49–54 (1982)
9.21 G. Busse, K. F. Renk: Appl. Phys. Lett. **42**, 366–368 (1982)
9.22 D. Fournier, F. Lepoutre, A. C. Boccara: J. de Phys. **44** (C6), 479–482 (1983)
9.23 A. Lachaine: "Photoacoustic Measurement of Subsurface Air Gaps" in *Photoacoustic and Photothermal Phenomena*, ed. by P. Hess, J. Pelzl, Springer Ser. Opt. Sci., Vol. 58 (Springer, Berlin, Heidelberg 1987) pp. 435–436
9.24 F. A. McDonald: Can. J. Phys. **64**, 1023 – 1039 (1986)
9.25 T. Pousi, M. Jokinen, A. Lehto, M. Luukkala: Proc. IEEE Ultrasonics Symp. 618–621 (1980)
9.26 M. Luukkala: "Photoacoustic Microscopy at Low Modulation Frequencies" in *Scanned Image Microscopy* ed. by E. A. Ash (Academic, London 1980) pp. 273–289
9.27 K. R. Grice, L. J. Inglehart, L D. Favro, P. K. Kuo, R. L. Thomas: J. Appl. Phys. **54**, 6245–6255 (1983)
9.28 F. A. McDonald, G. C. Wetsel, G. E. Jamieson: Can. J. Phys. **64**, 1265–1268 (1986)
9.29 M. Luukkala, S. G. Askerov: Electron. Lett. **16**, 84–85 (1980)
9.30 G. Busse, B. Rief, P. Eyerer: Can. J. Phys. **64**, 1195–1199 (1986)
9.31 R. L. Thomas, L. D. Favro, K. R. Grice, L. J. Inglehart, P. K. Kuo, J. Lohta, G. Busse: Proc. IEEE Ultrasonics Symp. 586–590 (1982)
9.32 A. Rosencwaig: J. de Phys. **44**, (C6), 437 – 457 (1983)
9.33 G. Busse: "Rasterbildverfahren mit optisch erzeugten Wärmewellen in der zerstörungsfreien Werkstoffprüfung", Habilitationsschrift, Universität Stuttgart (1984) p. 79
9.34 J. Jaarinen, M. Luukkala: J. de Phys. **44** (C6), 503–508 (1983)
9.35 H. J. Vidberg, J. Jaarinen, D. O. Riska: Can. J. Phys. **64**, 1178–1183 (1986)
9.36 J. F. McClelland, R. N. Kniseley, J. L. Schmit: *Photoacoustic Imaging of Compositional Variations in Hg $_{1-x}Cd_x Te$ Semiconductors*, in *Scanned Image Microscopy*, ed. by E. A. Ash (Academic, London 1980) pp. 353–364

9.37 R. A. McFarlane, L. D. Hess, G. L. Olson: Proc. IEEE Ultrasonics Symp. 628–632 (1980)

9.38 C. C. Williams: IEEE Trans. SU-**32**, 355–364 (1985)

9.39 A. Rosencwaig: "Thermal Wave Imaging and Microscopy", in *Scanned Image Microscopy*, ed. by E. A. Ash (Academic, London 1980) pp. 291–317

9.40 G. S. Cargill: "Electron-Acoustic Microscopy", in *Scanned Image Microscopy*, ed. by E. A. Ash (Academic, London 1980) pp. 319–330

9.41 L. J. Balk: Can. J. Phys. **64**, 1238 – 1246 (1986)

9.42 A. Rosencwaig, G. Busse: Appl. Phys. Lett. **36**, 725–727 (1980)

9.43 L. D. Favro, P. K. Kuo, J. J. Pouch, R. L. Thomas: Appl. Phys. Lett. **36**, 953–955 (1980)

9.44 S.-Y. Zhang, L. Chen: Can. J. Phys. **64**, 1316–1319 (1986)

9.48 Y. H. Wong, R. L. Thomas, G. F. Hawkins: Appl. Phys. Lett. **32**, 538–539 (1978)

9.46 Y. H. Wong: "Scanning Photo-Acoustic Microscopy", in *Scanned Image Microscopy*, ed. by E. A. Ash (Academic, London 1980) pp. 247–272

9.47 J. C. Murphy, L. C. Aamodt: Appl. Phys. Lett. **39**, 519–521 (1981)

9.48 G. Busse, P. Eyerer: Appl. Phys. Lett. **43**, 355–358 (1983)

9.49 G. Busse, P. Eyerer: J. de Phys. **44** (C6), 475–478 (1983)

9.50 G. Busse, B. Rief, P. Eyerer: Polymer Composites **8**, 283–286 (1987)

9.51 B. Rief: "Zerstörungsfreie Charakterisierung von kohlenstoffaserverstärkten Kunststoffen mittels Wärmewellenanalyse". Dissertation Universität Stuttgart (VDI, Düsseldorf 1988)

9.52 L. J. Inglehart, F. Lepoutre, F. Charbonnier: J. Appl. Phys. **59**, 234–240 (1986)

9.53 A. Rosencwaig, J. Opsal, D. L. Willenborg: J. de Phys. **44** (C6), 483–489 (1983)

9.54 M. Beyfuss, R. Tilgner, J. Baumann: "Photothermal Evaluation of Layered Samples with High Accuracy Based on 3-D Analysis of Thermal Waves", in *Photoacoustic and Photothermal Phenomena* ed. by P. Hess, J. Pelzl, Springer Ser. Opt. Sci., Vol. 58 (Springer, Berlin, Heidelberg 1988) pp. 392–395

9.55 J. Jaarinen, M. Luukkala: Proc. 3rd European Conf. on NDT, Florence (1984) pp. 128–138

9.56 D. P. Almond, P. M. Patel, H. Reiter: J. de Phys. **44**, (C6), 491–495 (1983)

9.57 M. Liezers, R. M. Miller: "Thermal Wave Imaging of Ceramic Thermal Barrier Coatings", in *Photoacoustic and Photothermal Phenomena*, ed. by P. Hess, J. Pelzl, Springer Ser. Opt. Sci. Vol. 58 (Springer, Berlin, Heidelberg 1988) pp. 437 - 439

9.58 J. Corbett, M. B. C. Quigley, B. Hart, B. L. Smith: "Laser-Generated Thermal Wave Interference for NDT of Hard Coatings on Boiler Tubes and Reactor Components", in *Photoacoustic and Photothermal Phenomena*, ed. by P. Hess, J. Pelzl, Springer Ser. Opt. Vol. 58 (Springer, Berlin, Heidelberg 1988) pp. 440-442

9.59 M. Egee, R. Dartois, J. Marx, C. Bissieux: Can. J. Phys. **64**, 1297–1302 (1986)

9.60 G. Busse, A. Ograbek: J. Appl. Phys. **51**, 3576–3578 (1980)

9.61 G. Busse, D. Vergne, B. Wetzel: "Photothermal Nondestructive Inspection of Paint and Coatings", in *Photoacoustic and Photothermal Phenomena*, ed. by P. Hess, J. Pelzl, Springer Ser. Opt. Sci., Vol. 58 (Springer, Berlin, Heidelberg 1988) pp. 427–429

9.62 D. Vergne, G. Busse: Infrared Phys. **29**, 839 – 849 (1989)

10. Surface Acoustic Waves in Solid-State Investigations

L. Konstantinov , A. Neubrand, P. Hess

With 13 Figures

The continuously growing interest in surface acoustic waves (SAWs) in the last decade arises from the fact that they form the basis of numerous signal processing devices (delay lines, frequency filters, resonators, convolvers, etc.) and because they can be used for studying thin films and surface properties of materials as well as for constructing various SAW sensors. These applications are motivated by the fact that, on the one hand, SAW parameters (amplitude, velocity, frequency, etc.) are affected in a very sensitive manner by a great number of physical and chemical processes and, on the other hand, that they can be realized in many cases on the basis of the same principles, methods and devices, as those employed for signal processing. Compared to its bulk acoustic wave (BAW) analog, a SAW device has the advantage that the signal, being concentrated at the substrate surface, is readily influenced, tapped, focused, reflected, etc., which leads to an extremely wide design flexibility. At the same time, the present SAW processors operate in a frequency range from a few megahertz up to above 1 GHz, allowing the investigation of processes scaled quite differently in time and dimension.

More surprisingly, very few review articles are available on the applications of SAWs for material evaluation, even in specialized periodicals. The main reason for this lack of articles is the difficulty in reviewing the subject adequately, because most of these applications stem from a quite heterogeneous community and exploit effects and phenomena from different fields which are more familiar to the specialists working therein.

The aim of this article is to review the present state of the art and to present some demonstrative results concerning the use of SAWs in the investigation of optical, electrical and elastic properties of materials, kinetic processes and parameters of electronic states in semiconductors. The review is not addressed to experts on SAW physics and devices but rather to specialists working in other fields, especially in surface and thin film physics and technology. To familiarize the reader with the fundamentals of SAWs we provide an introduction to the physical basis, main methods of generation and detection of SAWs, and the effects resulting from their propagation.

10.1 Fundamentals

10.1.1 Surface Acoustic Waves – Types, Properties and Main Characteristics

Surface acoustic waves are a class of waves that propagate along the interface of two elastic media, at least one of which is a solid, their amplitude decaying rapidly (on a scale of the wavelength) with distance away from the boundary surface [10.1]. There are various types of SAWs, differing in the motion of particles, velocity, dispersion, decay in depth, etc. Amongst the most important are Rayleigh SAWs on the plane surface of an isotropic half-space, their modifications in the presence of anisotropy (generalized Rayleigh waves), Love SAWs in a thin layer on the substrate surface (a wave type resembling horizontally polarized shear bulk waves but confined to the vicinity of the layer), Stoneley waves at the interface of two different solid media, Lamb wave modes in thin plates, and the Bleustein-Gulyaev type of SAW, a transversely polarized wave which penetrates much more deeply into the substrate and which would degenerate into a shear bulk wave in the absence of piezoelectricity.

SAWs are solutions of the set of wave equations for the displacement components which usually satisfy the boundary conditions for a mechanically free surface and, if the substrate is piezoelectric, appropriate electrical boundary conditions. SAW solutions can generally be presented as a linear combination of four partial waves with wave vectors k parallel to the surface, which propagate in the direction under consideration with the same phase velocity and decay in depth with a decay constant proportional to k. At the same time, the wave of the piezoelectric potential accompanying SAWs decays exponentially above the surface of propagation with a constant equal to k. It should be noted that in all but the simplest cases, an explicit determination of SAW solutions is not possible and numerical computations are necessary.

For the surface of a free isotropic substrate, the phase velocity of the Rayleigh SAW is of the order of 10 % less than the lowest bulk-wave velocity and cannot phase-match any bulk wave.

The depth dependence of the displacement components for this case is shown as an example in Fig. 10.1, revealing the most characteristic feature: the amplitude of

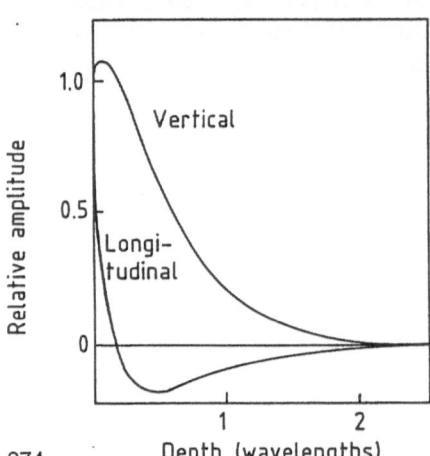

Fig. 10.1 Vertical and longitudinal displacement components for a SAW propagating in an isotropic substrate as a function of depth

the SAW disturbance becomes negligible for depths more than a few wavelengths from the surface. In addition, the displacement of particles is elliptically polarized, while the phase velocity is dispersionless.

If the depth of penetration of SAWs is comparable with the substrate thickness, the velocity of propagation depends on the plate thickness, i.e., there is dispersion, and different modes of waves with symmetric and antisymmetric distributions of displacements across the plate become possible. With an increase in the thickness to about 3-4 wavelengths, the phase velocities of the lowest modes asymptotically approach the velocity of the Rayleigh wave on a half-space of the same material.

While being of prime importance in the design of SAW devices, anisotropy and piezoelectricity are second-order effects as far as wave propagation is concerned. In anisotropic solids, the phase velocity and the form of the displacements depend on the crystal plane used as the free surface and on the direction of propagation, and generally the phase velocity is not parallel to the flow of wave energy or the group velocity. A feature often met in anisotropic cases consists in the SAW decay constant having a real part, which results in the appearance of oscillations in the distribution of the displacement components with depth, see Fig. 10.2. Nevertheless, about 90% of the energy of generalized Rayleigh waves remains concentrated at a distance of the order of one wavelength below the surface. Another difference from the isotropic case is that in certain directions of propagation the generalized Rayleigh waves can degenerate into shear bulk waves, which themselves satisfy the free-surface conditions.

One of the effects of piezoelectricity is to increase (stiffen) the effective elastic constants of the medium, and consequently the SAW velocity, by a factor less than 5% even for the strongest piezoelectrics. The principle distinguishing feature of the SAW propagation on piezoelectric substrates is, however, the accompanying electric field by means of which a SAW is usually coupled to external electric circuits, to other surface waves or to charge carriers in semiconductors. On the other hand, the

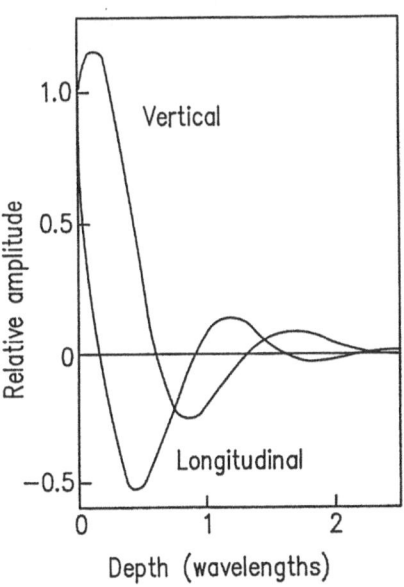

Fig. 10.2 Vertical and longitudinal displacement components for a SAW propagating along a cubic axis on the basal plane of Ni.(After [10.1])

presence of an external electric field above a surface with a propagating SAW makes it possible to easily influence the wave by conducting surfaces in close proximity or by changing the electrical boundary conditions.

SAWs owe much of their practical importance to the ease with which they can be manipulated and controlled by electrical or mechanical perturbations of the substrate surface. The most practical mechanical perturbation can be introduced by a solid layer of thickness h in intimate contact with the surface, which shifts the SAW velocity and makes it dependent on the frequency. At the same time, the medium becomes dispersive, with the phase velocity being a function of the ratio of the wavelength to the characteristic layer dimension h. If the perturbed velocity is greater than the unperturbed velocity, the layer is said to "stiffen" the substrate, while in the opposite case it is said to "load" the substrate. Which of these cases will be present depends predominantly on the relative magnitudes of the shear velocity of the layer material and the unperturbed Rayleigh velocity of the substrate. In addition, the depth distribution of displacements for the first Rayleigh mode changes and excitation of other higher modes in the layered structure is possible. As an example, Fig. 10.3 shows the vertical component of the displacement against depth for different values of kh in a ZnO layer on an isotropic silicon substrate. For $kh < 1$, the displacement is characteristic of the wave on the free surface of the substrate, while for large values of kh it corresponds to that of a SAW on the free surface of the layer and very little of the wave energy is carried into the substrate. Similar changes also occur in the profile for the longitudinal component of displacement.

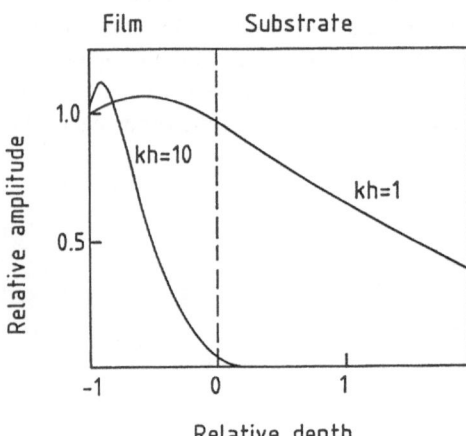

Fig. 10.3 Vertical displacement component for a SAW for different values of kh, with wave vector k and layer thickness h (After [10.1])

10.1.2 Methods for Generation and Detection of Surface Acoustic Waves

There are a variety of methods available, some of them reversible, for the generation and detection of SAWs, which can be classified according to the physical principles involved as

- methods based on the transformation of BAWs into SAWs or vice versa,
- methods with direct piezoelectric coupling,

- optical methods,
- other methods.

In the following, a brief survey is given of those methods which are currently most widely used in solid-state investigations. To stress its specific features, the optical generation of SAWs, referred to as surface wave photoacoustics because of its similarity to BAW photoacoustics, is considered in a separate section.

(a) Generation of Surface Acoustic Waves

The generation of continuous SAWs requires spatially periodic, time-varying mechanical deformations on the substrate surface. Due to the phase-synchronized superposition of elementary elastic displacements, the most effectively generated SAWs are those with the frequency optimized for correct timing and with a wavelength equal to the characteristic length of the spatial periodicity.

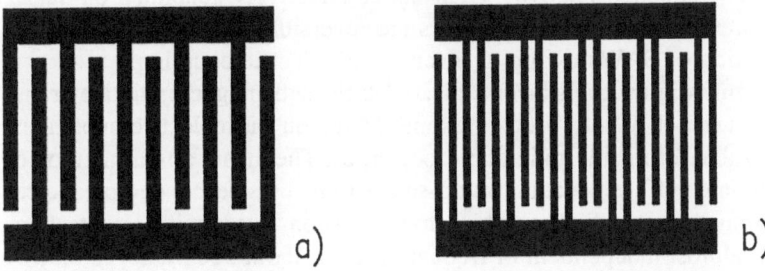

Fig. 10.4 Periodic interdigital transducers (a) with single electrodes, (b) with double electrodes ("split-finger geometry")

On piezoelectric substrates, SAWs are generated solely by means of transducers with direct piezoelectric coupling, the so-called interdigital SAW transducers (Fig. 10.4). This is one of the most effective and commonly used devices of this type. The transducers consist of a system of periodic metal electrodes ("fingers") deposited on a highly polished substrate surface. The periodic elastic deformation is produced in this case by the electrical field distribution arising in the substrate when a rf voltage is supplied to the transducer. With fields localized at the free surface, the coupling to SAWs can be made quite strong. The transducer possesses a maximum efficiency at the excitation frequency for which the surface wave propagates one transducer period in one rf voltage period (this is called the synchronous frequency, $f_0 = v/2L$, where L is the length of the electrode spatial periodicity). On the other hand, the efficiency possesses a maximum for equidistant transducer structures whose fingerwidth is equal to the gap between adjacent electrodes. The "split-finger" geometry of interdigital transducers (Fig. 10.4b) is recommended for eliminating the effect of SAW reflections from adjacent electrodes because these cancel in pairs due to a phase shift of π, while the electrical excitation periodicity is preserved and the elementary excitations continue to superimpose in phase at the synchronous frequency. By means of interdigital transducers with submicron peri-

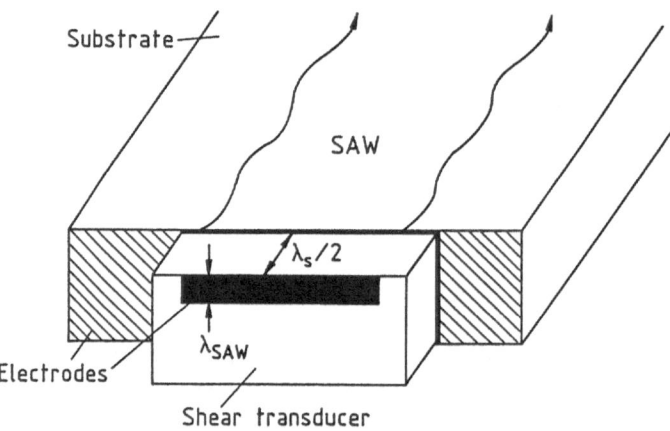

Fig. 10.5 Side-bonded thin piezoelectric shear-wave transducer for SAW generation

odicity, SAWs with frequencies above 1 GHz can be effectively generated on quartz or LiNbO$_3$ substrates. These SAW transducers are reversible, i.e., they can also be used for the detection of SAWs owing to inverse piezoelectric coupling.

Compared to interdigital transducers, piezoelectric and magnetostrictive transducers for BAWs are not widely used to excite SAWs on piezoelectric substrates, but they are preferable for metals and nonpiezoelectrics. The most popular method of transforming bulk into surface acoustic waves seems to be by side-bonded piezoelectric shear-wave transducers, which combine low insertion losses with a large fractional bandwidth, almost independent of frequency up to about 300 MHz [10.2,3]. A typical arrangement is shown in Fig 10.5. A LiNbO$_3$ shear-wave transducer of thickness d is attached with indium to the side face of the substrate, perpendicular to the surface of propagation, in such a way that their top surfaces are coplanar. To excite Rayleigh waves, the transducer should produce shear displacements perpendicular to the surface, while to excite Love waves, the displacements should occur parallel to the surface. As only the shear-strain energy located closer than one wavelength to the surface is efficiently converted into SAW energy, the back electrode of the transducer is only about one wavelength in height. The other electrode is usually a thin metal film deposited on the substrate side surface. The central resonant frequency of such a device is determined mainly by the thickness of the transducer, which is approximately half the shear wavelength at the center frequency. By means of a LiNbO$_3$ plate transducer side-bonded to a GaAs substrate, fractional bandwidths of above 90 % and an absolute conversion loss of 11 dB at the center frequency have been achieved [10.2]. Although ceramic or thin film PVDF foils can also be used as side-bonded transducers, LiNbO$_3$ is preferable due to its much higher piezoelectric constant compared to PVDF, and it can be more easily shaped into thin, flat and parallel plates than ceramics.

Concerning versatility, the use of piezoelectric and magnetostrictive transducers for the generation of SAWs has the advantage of being applicable to various substrates, that such transducers are movable and that, if properly chosen and designed, they allow one to cover a larger frequency range with one and the same

278

transducer. This method requires good acoustical contact to the sample, usually realized through liquids, gels, etc., which increase the energy losses and considerably lower the high-frequency limit of SAW generation compared to the methods with direct piezoelectric coupling. On the other hand, the main part of the acoustic energy injected into the sample by a BAW transducer is accumulated in bulk modes, diminishing the efficiency of SAW excitation and giving rise to parasite signals. Therefore, it is difficult to generate short SAW pulses of high amplitude and small risetime.

(b) Detection of Surface Acoustic Waves

Optical methods of SAW detection seem to be the most numerous as far as the principles of operation and realized schemes are concerned. The main incentive behind the intense effort to develop optical detection of SAWs stems from the wish to exploit the potential of laser generation of SAWs by amalgamating both techniques in order to make an "all-optical" contactless method for the inspection of materials. Optical SAW detection methods are very often based on various mechanisms of the interaction of SAWs with light (deflection, phase modulation or diffraction of light by SAWs). Nowadays, they form an important part of applied optics and acousto-optics and have been the subject of several reviews, see for example [10.4-7], so that they will not be discussed in detail in this chapter.

The detection of SAWs by means of transducers is closely related to the transducer generation of SAWs, as these two processes are most often the reverse of each other: SAW detection comprises transformation of acoustic into electric energy, while SAW generation exploits the opposite process. Thus, the statements on SAW generation by transducers in Sect. 10.1.2a apply in general to SAW detection as well.

However, transducers based on SAW transformation into BAWs are more frequently used for detection than for generation of SAWs. This is probably due to the fact that they require neither a reflecting surface of high quality and mechanical stability like the majority of optical detection schemes nor a polished piezoelectric substrate like the interdigital or other transducers with direct piezoelectric coupling. Therefore, detection with these transducers may be the most viable scheme in some cases. In the following, three important types of transducers used for the detection of SAWs will be described: electromagnetic acoustic transducers (EMATs), and thick and thin film piezoelectric transducers.

An EMAT consists of a suitable coil placed against a permanent magnet and mounted close to the sample surface without touching it. Motion of the surface perpendicular to the magnetic field will induce an eddy current provided the electrical conductivity of the surface is high enough. This current in its turn will induce a secondary current in the coil, which can be detected. In contrast to other schemes, it is not the surface displacement that is measured, but the surface velocity normal to both the magnetic field and the coil. In some cases, this may complicate the acoustic waveform, but it has the advantage that either the horizontal or the vertical motion can be recorded simply by rotating the magnetic field by 90°. The bandwidth of an EMAT can be varied by choosing an appropriate geometry. The largest bandwidth is achieved with a single conducting strip acting as a transducer. In this case, the bandwidth is limited by the width of the strip. Narrowband sensitivity can be realized using a meander line transducer, the center frequency of which is determined by its periodicity. More details can be found in [10.8], where the results for different EMAT detectors are compared.

Thick (or bulk) piezoelectric transducers can be prepared from a variety of materials of different shapes. There are two main groups of transducer materials, namely, ceramics (lead zirconate titanate, barium titanate, etc.) and single crystals (lithium niobate, quartz, bismuth germanium oxide, zinc oxide, etc.). Information on the properties and parameters of these materials can be found in [10.9]. Piezoelectric transducers can be operated in a resonant or nonresonant mode, the former being of higher sensitivity but sometimes giving rise to ringing effects [10.10]. Examples of the detection of SAWs with transducers of this type are reported in [10.8,11,12].

One disadvantage of thick piezoelectric transducers is that their bandwidth is naturally limited by their dimensions. This problem can be overcome by making use of very thin side-bonded piezoelectric disks [10.2,9,13]. However, such devices cannot readily be prepared for high frequencies as their lapping and bonding becomes increasingly difficult with the reduction in size, and therefore bandwidths substantially higher than 200 MHz are not easily achievable.

This limitation has led to the development of thin film transducers. The most widespread material for such transducers are CdS, ZnO and PVDF. Very thin films of these materials can be fabricated by evaporation (CdS) and sputtering (ZnO) techniques [10.9]. Metallized foils of polarized PVDF are commercially available with thicknesses down to a few micrometers. Although the coupling factors for films are somewhat lower than for bulk crystalline material, the use of such transducers is becoming increasingly popular, especially in connection with the detection of SAWs excited through short laser pulses, where fast response rather than very high sensitivity of the transducers is required. A typical example of the application of a thin film transducer for this purpose can be found in [10.14] where a bandwidth of about 1 GHz has been achieved by means of a 5 μm thick sputtered ZnO film. In this work, however, only BAWs have been detected. The use of PVDF foils for ultrasonic detection is described in [10.15,16]. Even with the quite simple bonding methods employed in these experiments (e. g. a spring pressing the foil to the sample via a thin layer of liquid for matching the acoustic impedances), response times of less than 4 ns have been achieved. The development of thinner foils and more elaborate bonding techniques should result in larger bandwidths, while new materials for thin film transducers may provide higher sensitivities.

10.1.3 Surface Wave Photoacoustics

Surface wave photoacoustics deals with the optical and, in particular, laser generation of SAWs and all related effects and applications. By means of laser radiation, one can excite SAWs in a wide range of frequencies, with varying spatial characteristics, both periodic and pulsed. There are several mechanisms of sound generation in solids caused by electromagnetic radiation. The most important of these are the thermoelastic and striction mechanisms, the transfer of recoil momentum when material or adsorbed species evaporate upon irradiation, and the material breakdown mechanisms which take place if a shock wave is produced as a result of a sudden phase transition in the irradiated substance. The latter two mechanisms occur only in very strong radiation fields typical of high intensity laser pulses. The striction mechanism, characteristic of materials with high lattice polarizability, is based on the polarization and dipole rearrangement in an electromagnetic field, which gives rise to elastic deformations and strains in the medium.

The most common and well known of these mechanisms of optical excitation of SAWs is the thermoelastic one, where the absorbed light energy transforms into heat and generates lateral elastic stresses due to the thermal expansion of the material in the near surface region. The dynamics of thermo-optical SAW generation is determined by the beam geometry and the time dependence of the light intensity as well as by the physical characteristics of the medium, such as the optical absorption coefficient, the heat capacity, the compressibility, and the thermal expansion coefficient.

A great number of experiments on laser excitation of SAWs have been performed so far, showing that this method is contactless, applicable to various materials, and quite flexible, as by making moderate changes in the geometry of a single experiment SAW sources of different shapes (point, line, annulus, etc.) and type (cw and pulsed) can be produced. The principle of such an experiment is shown in Fig. 10.6. It consists of an amplitude-modulated laser beam focused onto the surface of an absorbing specimen. The modulation can be performed either by chopping the output of a cw laser or by using a pulsed laser source. In the former case, a periodic SAW will be excited, with a frequency equal to the modulation frequency, while in the latter case, pulses of SAWs will be launched with a duration determined by the laser pulse length and spot size.

Laser excitation of SAWs was first reported in the pulsed mode [10.17] with the use of a Q-switched ruby laser to irradiate thin absorbing aluminum films on piezo- and nonpiezoelectric substrates. The frequencies of the resulting wideband SAW video pulses are Fourier components of the laser waveform and the authors

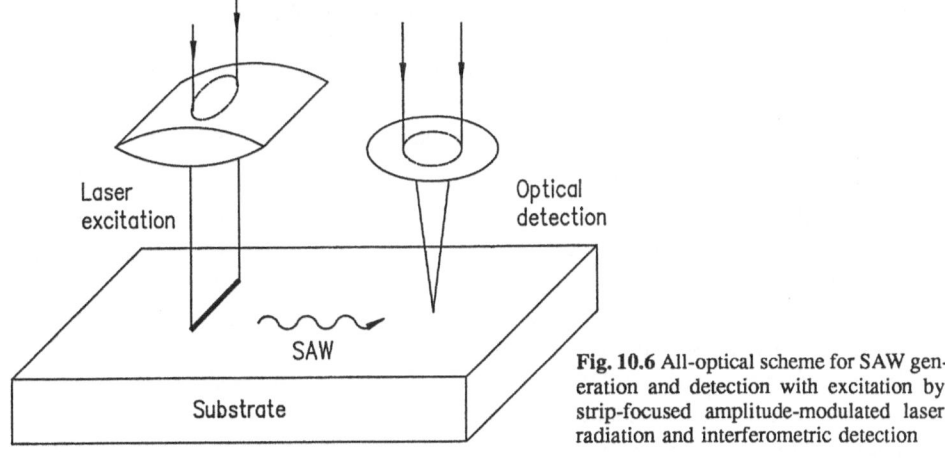

Fig. 10.6 All-optical scheme for SAW generation and detection with excitation by strip-focused amplitude-modulated laser radiation and interferometric detection

have shown that it is possible to enhance the effectiveness of SAW generation at a given frequency by using a spatially periodic mask with a periodicity equal to a corresponding SAW wavelength. A linear dependence of the wave amplitude on incident power density was observed, indicating the thermo-optical mechanism of excitation.

Later on, similar experiments were carried out by other authors [10.10,11,18-23] in order to study various aspects of pulsed-laser generation of SAWs both above and below the threshold intensity for plasma production. The thermoelastic strain and the stresses caused by the absorbed laser radiation excite not only surface, but longitudinal and transverse BAWs as well. Through an appropriate sample geometry, one can ensure that any reflected bulk waves arrive at the detector well after the SAW pulse so as not to confuse the registered signal [10.10,13,17]. In other cases, all three waves could be detected simultaneously in one single experiment by using different resonant piezoelectric transducers properly bonded to the specimen [10.11,18,19]. For example, in a study of the role of plasma production in pulsed-laser excitation of acoustic waves in metals, *Aindow* et al.[10.18] used narrowband longitudinal and shear piezoelectric ceramic transducers placed on the face opposite to the illuminated surface to detect the two bulk waves, while a resonant wedge-type transducer was employed on the illuminated face of a piezoelectric slab of the desired frequency, bonded at 45° to the sample edge, to record the amplitude of the Rayleigh SAW excited by a Q-switched Nd:YAG laser of 34 ns pulse duration. In the thermo-optical regime of generation, the amplitude of both the longitudinal and shear bulk waves was found to be proportional to the laser pulse energy in all samples studied. For a Rayleigh wave, the amplitude changes nonlinearly with pulse energy. At sufficient laser power densities, the surface can be heated to the boiling point, and recoil from the vaporized atoms may contribute to the generation of elastic waves. In this regime, a strong nonlinear increase of the wave amplitude on the power density was observed for the longitudinal bulk and the SAW pulses, while a more complicated behavior, with a drop in the amplitude above a power density threshold was characteristic of the transverse bulk acoustic pulse. The authors explained these dependences by assuming different contributions to the generation of longitudinal and shear deformations from the lateral and normal surface temperature gradients. The former are a major source of shear and surface wave pulses. This assumption also accounts for the experimentally observed fact that the effectiveness of SAW generation is strongly affected by the lateral dimension of the illuminated area, which determines the lateral gradient of the light intensity. On the other hand, the time duration τ of the Rayleigh SAW pulse was found to increase with an increase in the laser spot size, being given to a first approximation by $\tau = \sqrt{\tau_l^2 + \tau_h^2}$ [10.14], where τ_l is the duration of the laser pulse and $\tau_h = a/v$ is the time required for a SAW of velocity v to transverse the laser spot of diameter a. Similar results have also been reported by *Khodinskii* et al. [10.20] who, in contrast to *Aindow* et al.[10.18], observed that the SAW-pulse amplitude is proportional to the energy of the laser pulses up to intensities of the order of 10^8 W/cm^2.

In experiments exciting SAWs by short laser pulses, one basic difficulty is the recording of the resulting wideband Rayleigh SAW pulse. Its real waveform cannot be reproduced correctly by resonant narrowband transducers as the wave amplitude is measured, in fact, at the resonance frequency of the transducer. In principle, this can be done by ZnO thin film [10.14], PVDF foil [10.15] or capacitance transducers [10.24] or by using optical methods of SAW detection [10.21-24]. Waveforms of optically excited SAW pulses in various materials have been studied, both experimentally and theoretically, by many authors [10.14,15,19,21-24] in order to clarify the dominant mechanisms of photoacoustic generation and the type of the resulting acoustic phenomena. *Hutchins* et al.[10.23] have, for example, examined the waveform of laser induced transients in metal substrates with a plane surface. They used a Michelson interferometer of about 40 MHz bandwidth to measure the SAW amplitude with an accuracy of 0.2 Å. Owing to boundary conditions for thermoelastic expansion, forces acting normally to the surface are weak and the source is dominated by horizontal force dipoles. On the basis of such considerations, for the surface illuminated by radiation brought to a line focus, a waveform consisting of a bipolar Rayleigh pulse preceded by a much smaller longitudinal pulse was both predicted and experimentally observed. The same pulse shape can also be observed with a capacitance transducer [10.24] or with a PVDF foil transducer [10.25] at other excitation wavelengths and pulse durations. Figure 10.7 shows a pulse shape for thermoelastic excitation with a CO_2 laser pulse of 10.6 μm wavelength and 250 ns pulse duration focused to a strip of 5×1 mm; detection was performed with a PVDF foil detector and the sample consisted of an unpolished Al cube. Both the subsurface longitudinal pulse and the Rayleigh pulse can be clearly distinguished. The feature about 6 μs after the Rayleigh pulse is a reflection from the sample's edge. Ablation of material at higher laser power densities contributes an additional force normal to the surface, which becomes the dominant source under such conditions. A similar pulse-like source of even better reproducibility results if a thin surface coating evaporates on account of heating by the laser pulse, but now the polarity of the Rayleigh pulse is inverted. Waveforms of the same type for laser induced SAWs have also been reported by others [10.19,21,22,24]. *Jen* et al.[10.21] excited cylindrical SAWs by focusing laser pulses of 10 ns duration to a ring on the sample surface. Unipolar waveforms of converging SAW pulses were observed at a distance

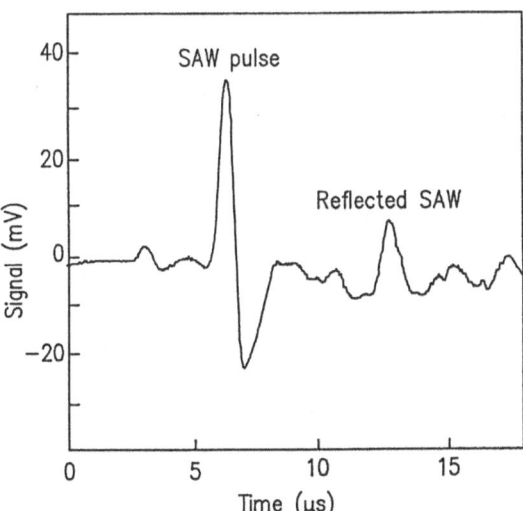

Fig. 10.7 Acoustic pulse shapes generated by pulsed CO_2 laser excitation in Al detected with a PVDF transducer

about 10 wavelengths away from the focus, while a bipolar form was recorded close to the focus of the ring source. This change of the waveform has been explained by a $\pi/2$ phase shift occurring when the cylindrical SAW pulse crosses the focus. The maximum surface displacement of about 15 nm was measured at the center. Similar results have also been obtained for SAW pulses optically excited on a sphere [10.22], where a π phase shift was observed when the spherical SAW pulse passed through the point diametrically opposite to the source. In samples of thickness on the order of one acoustic wavelength such as silicon wafers or steel foils, Lamb modes rather than Rayleigh waves have to be expected. *Sontag* and *Tam* [10.26,27] observed the propagation of strongly dispersive antisymmetric Lamb waves after irradiation of a Si wafer with a 1 mJ N_2-laser pulse using a probe beam deflection technique for detection. They also detected the reflection and refocusing of the Lamb waves from the circumference of the round wafer.

Generation of highly directional beams of SAWs is of prime importance for accurate location of flaws in the nondestructive testing of materials. Surface wave photoacoustics is one of the most convenient methods of solving this problem by focusing the laser beam to a line whose length is much greater and its width smaller than the acoustic wavelength [10.10,28]. Acoustic waves are emitted preferentially broadside from a line source excited in phase along its length, and the SAW output of such a thermoelastic source has been found to be strongly directional with well-defined maxima in the direction normal to the center of the line axis [10.10]. This means that SAWs propagating perpendicularly to the line focus are generated much more effectively than those in other directions. The directivity diagram of the photoacoustic source of Rayleigh waves was studied for various metals and angular full-widths-at-half-maximum of the order of 15° were readily achievable. It has been shown that the narrow-strip geometry is the optimum, both in efficiency of excitation and in beam spreading.

Laser generation of continuous SAWs can be performed, as mentioned previously, by a harmonically modulated laser beam directed onto the investigated surface through a periodic mask of alternating opaque and transparent strips or through a Fresnel biprism. In this case, the mask acts as a frequency filter and a sharp maximum occurs in the effectiveness of SAW generation at the modulation frequency, corresponding to a Rayleigh wavelength equal to the mask period. The use of a periodic structure, however, is not obligatory for excitation of periodic SAWs and just harmonic modulation of the output of a cw-laser source is sufficient to realize a wideband photoacoustic emitter of SAWs. For example, experiments have been reported [10.29,30] in which effective SAW generation is realized on $LiNbO_3$ covered with an optically absorbing coating by a mode-locked cw dye laser at a frequency of about 76.5 MHz. Using an interdigital transducer for detection purposes, with a resonance frequency equal to that of the laser mode synchronization, a signal-to-noise ratio of 45 dB has been achieved with laser intensities of the order of 100 W/cm^2 for a line focus, and 10 kW/cm^2 for a point focus. For a given geometry, the observed signal was proportional to the laser power, sharply decreasing when the width of the line focus exceeded half of the acoustic wavelength.

On the other hand, it is possible to excite a running SAW by scanning a cw-laser line (or band) focus along the surface with a velocity close to the SAW phase velocity [10.31]. The thermoelastic deformation is then accompanied by the moving source, increasing nonlinearly with the scanning sweep length. The SAW amplitude in this mode of generation is a strongly resonant function of the scan velocity, the resonance width being proportional to the laser spot size and inversely proportional to the scanning length. The SAW amplitude can reach substantial values (up to 0.1 μm), with quite weak surface heating of about 3 K [10.31], while remaining proportional to the laser intensity. The main advantage of the scanning mode of SAW generation lies in its ability to act on the surface in a mainly mechanical fashion, without heating it.

In all the examples considered so far, SAW pulses were excited by thermoelastic and momentum transfer mechanisms. A "strain" mechanism of SAW generation in semiconductors, based on photoexcitation of electron-hole pairs, has been proposed by *Avanesyan* et al.[10.12]. The experiment, consisting in illumination of a Si wafer by a Nd:YAG laser at 1.06 μm and recording of the longitudinal component of the resulting SAW pulse by a side-bonded piezoelectric transducer, is quite similar to that reported in [10.19], but the observed SAW pulses are of the opposite polarity. The authors suggest as an explanation that, in contrast to thermoelastic expansion, the production of an electron-hole plasma causes the lattice to contract. Further support of this proposal is provided by the observed change of the SAW pulse polarity when the silicon sample has been illuminated by the second harmonic of the laser radiation at 0.53 μm. This wavelength is absorbed in silicon in a very thin layer and the majority of photoexcited carriers recombine during the laser pulse, resulting in the dominant role of the thermoelastic mechanism and in the generation of SAW pulses with the same polarity as that reported in [10.19].

10.1.4 Effects Resulting from the Propagation of Surface Acoustic Waves

The variety of processes and effects resulting from the propagation of a SAW is so wide that it is difficult to discuss them in detail here. On nonpiezoelectric substrates, SAWs influence the medium only mechanically, periodically modulating the mass density, stiffness, refractive index and stress distribution on a subsurface region with a thickness of the order of the SAW wavelength. Thus, at the substrate surface the SAW forms a running "diffraction grating" capable of refracting and scattering incident light and giving rise to a set of acousto-optic interactions and effects which remain outside the scope of this discussion.

On the other hand, as the bandgap in semiconductors and semimetals depends on deformation, SAWs periodically modulate the potential energy of electrons and holes in the vicinity of band extrema, thus exciting a wave of the deformation potential and acting on charge carriers.

If the substrate on which a SAW propagates is piezoelectric, in addition to the effects mentioned above, a quasistatic electric wave is generated on account of the piezoelectric effect, which interacts in conductive media with carriers, both free and trapped. As a result of this interaction, which is generally nonlinear, several acousto-electric effects occur, all with potential for sensor and material testing applications. These acoustoelectric effects will be discussed in this section as they provide the basis for an experimental method to be elaborated on in Sect. 10.2. In the case of piezoelectric substrates, a substantial enhancement of the surface modulation of the refractive index by a SAW also becomes possible due to the electro-optic effect.

In ferro- and antiferromagnetic crystals where the elastic properties depend on the magnetic structure and parameters of the material, SAWs are accompanied by a wave of the spin orientation (or of magnetic moments), thus forming a coupled surface magneto-acoustic wave. Such waves result from nonlinear magneto-elastic interactions in the medium and can be used to transform SAWs into spin waves and vice versa. Effective magneto-elastic interaction occurs in the frequency range where the SAW wavelength is close to that of spin waves, i.e., at the intersection points of the corresponding dispersion curves. This is seen in the appearance of strong absorption of the SAW and dispersion of its velocity due to transfer of acoustic into spin wave energy. The spin wave frequency is usually between 1 and 10 GHz, while its wave vector is below $10 \, \mathrm{cm}^{-1}$, i.e., magnetoelastic interactions are characteristic of the high-frequency of SAWs that is difficult to access by BAWs. Effects similar to acoustoelectric effects, discussed below, can originate on the basis of the nonlinear magnetoelastic interaction, but they also lie outside the scope of the present chapter.

Acoustoelectric effects arise when an acoustic wave (bulk or surface) interacts with charge carriers in the medium in which it propagates. There are various mechanisms of such an interaction, but the piezoelectric mechanism is particularly important in practical applications. When the acoustic wave propagates in the bulk or along the surface of piezoelectric materials, it is accompanied by a quasistatic electric wave whose amplitude in some cases can reach up to thousands of volts per centimeter. The longitudinal and transverse components of this electric wave, which decay exponentially with distance away from the surface, are responsible for the SAW interaction with charged carriers. In this case, three different configurations are possible for the investigation of the acoustoelectric interaction:

(i) When a SAW propagates along the surface of a piezoelectric semiconductor.

(ii) When a SAW propagates along the surface of a piezoelectric substrate covered with a thin semiconductor (metal) film.

(iii) When a SAW propagates on the surface of a piezoelectric substrate with a semiconductor (metal) sample placed above it at a distance smaller than the SAW wavelength. This arrangement (Fig. 10.8), known as the separated-medium

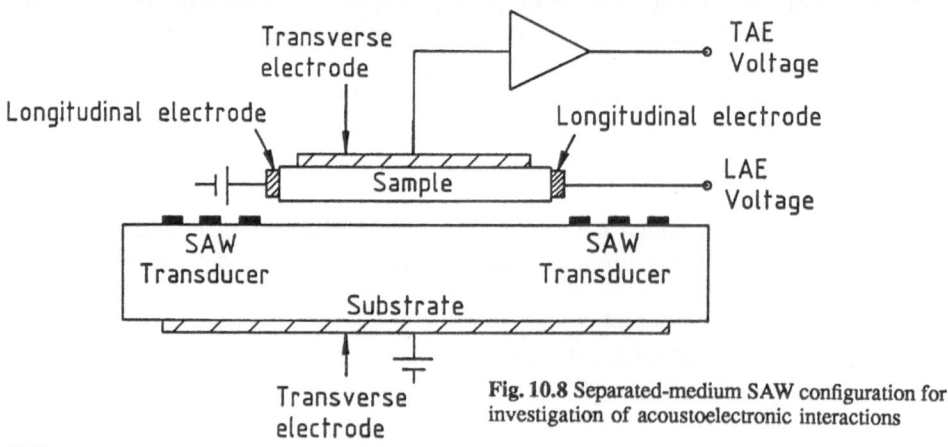

Fig. 10.8 Separated-medium SAW configuration for investigation of acoustoelectronic interactions

SAW configuration, has the advantage that the substrate and the sample can be independently chosen and easily replaced to meet various needs. The main requirement is that both surfaces have to be well polished in order to form a uniform air (vacuum) gap between them.

Due to the inhomogeneous spatial distribution of the acoustic and electric fields of SAWs in a piezoelectric medium, the acoustoelectric interaction with SAWs turns out to be much more complicated than that with BAWs. As a result of the interaction, which influences both the wave propagation and the distribution of charge carriers in the near-surface region, several acoustoelectric effects occur in the separated-medium SAW structure which will be considered shortly in the framework of a simple theoretical model in order to give some idea of their main characteristics.

Let us consider a structure consisting of a n-type semiconductor with energy bands not bending at the surface and a piezoelectric substrate along which a SAW is propagating parallel to the z-axis, the y-axis being directed normal to the surface into the piezoelectric substrate. The thickness of the semiconductors sample is d, while the air gap is assumed to be infinitesimal, i.e., the SAW electric field penetrates entirely into the semiconductor. The equations describing, in this case, the electrical properties of the semiconductor surface in the presence of a SAW are:

(i) The current density equation

$$j = \mu e n E + De \nabla n \quad , \tag{10.1.1}$$

where E is the electric field of the SAW, μ is the mobility of the electrons, D is their diffusion coefficient, j is the current density, e is the electron charge, and n is the nonequilibrium concentration of electrons in the SAW electric field.

(ii) Poisson's equation

$$\nabla \cdot E = -\frac{e}{\varepsilon}(n - n_0) \equiv -\nabla^2 \phi \quad , \tag{10.1.2}$$

where ϕ is the electrostatic potential and ε is the semiconductor dielectric permeability, and n_0 is the equilibrium concentration of electrons, which coincides in the case under consideration with the concentration of donor atoms N_d.

(iii) The continuity equation

$$\nabla \cdot j - e\frac{\partial n}{\partial t} = 0 \quad . \tag{10.1.3}$$

Equation (10.1.2) can be generalized for bands with surface bending, while (10.1.3) can be modified to account for additional optical generation and recombination of electrons.

Equations (10.1.1-3) have solutions of the surface-wave type:

$$E = E^{(0)} + E^{(1)}e^{i(\omega t - kz)} + E^{(2)}e^{2i(\omega t - kz)} + \cdots \quad , \tag{10.1.4}$$

and similarly for j, n, and ϕ, where ω and k are the SAW frequency and wave vector, respectively. Generally, all terms of these series solutions are functions of y and z. The first-order terms are proportional to the SAW amplitude, while those of zero and second order describe nonlinear interactions resulting in a change of the frequency.

Substituting (10.1.4) into (10.1.1-3) and combining the terms of the same order, one obtains sets of differential equations which, after some mathematical manipulation, can be transformed into separate equations for the successive-order components of the corresponding variables of interest. To preserve a reasonable level of clarity in such a treatment, usually only terms to the second order with respect to frequency harmonics are taken into account. For example, the resulting differential equations for the electric potential ϕ are of fourth order, having a general solution of the type

$$\phi^{(n)} = \sum_{l=1}^{4} A_l^{(n)} \exp\left\{\kappa_l^{(n)} y\right\} \quad , \qquad (n = 0, 1, 2) \quad . \tag{10.1.5}$$

The complex coefficients $\kappa_l^{(n)}$ are selectively related to the SAW wavenumber $2\pi/\lambda$ or to the quantity $2\pi/\lambda_D$, where

$$\lambda_D = \sqrt{\frac{\varepsilon}{\mu e D n}}$$

is the Debye length of electric field screening in the semiconductor. Owing to the exponential character of (10.1.5) and searching for surface wave solutions, terms decaying in depth of the semiconductor with the smallest decay constants dominate the solution. This means that the depth of penetration of the SAW electric field in the semiconductor sample, i.e. the acoustoelectronic interaction depth, is determined by the SAW wavelength or by the Debye length, whichever is smaller. The Debye length is known to depend on the free carrier concentration and can vary from hundreds of angstroms in low resistivity samples up to a few millimeters in semi-insulating materials.

The dependence of the variables (10.1.4) on z is periodic, with a period equal to the SAW wavelength, which describes the behavior of electrons in condensates under the action of the longitudinal periodic field of the wave.

The acoustoelectric effects resulting from the SAW interaction with electrons can be classified within the framework of the model under consideration as follows:

(a) Linear effects, consisting in the generation of an ac electric voltage with the frequency of the SAW, acoustoelectric extra losses of SAW energy (or SAW attenuation), and changes in SAW phase velocity. These effects are described by the first-order terms in (10.1.1-3), varying as $\exp(i\omega t)$, after linearization of the equations under the assumption of small amplitudes of the variables in (10.1.4). The amplitude of the resulting ac voltage is

$$V_\omega = \phi^{(1)}(-d) - \phi^{(1)}(0) \quad .$$

To calculate the change in the SAW velocity and attenuation one should take into account the influence of carriers on the wave. It can be shown, for example, that in materials with relatively weak electromechanical coupling the SAW attenuation is given by

$$\alpha = -\frac{1}{2}\kappa K^2 \frac{\varepsilon_p}{\varepsilon} \frac{\omega_c}{\omega} \left[\left(\frac{\varepsilon_p}{\varepsilon} + 1\right)^2 + \left(\frac{\omega_c}{\omega}\right)^2\right]^{-1} \quad , \tag{10.1.6}$$

where K^2 is the effective electromechanical coupling coefficient, ε_p is an effective dielectric constant for the structure, and $\omega_c = \mu n e / \varepsilon$ is the semiconductor conductivity frequency. This expression is, to a great extent, similar to that valid for the attenuation of BAWs in semiconductors with negligible diffusion of carriers. As an example of the order of magnitude of this effect, the acoustoelectric extra losses in a structure consisting of n-type Si with resistivity of $12\,\Omega\,\mathrm{cm}$ on LiNbO$_3$ exceed $1\,\mathrm{dB}/\mu s$.

Thus, changes in the conductivity of the layer penetrated by SAW may be determined by measuring the corresponding variations in either the SAW propagation losses or phase velocity.

When a dc current flows in the direction of SAW propagation with a velocity higher than that of the SAW, the charge carriers transfer energy to the wave, amplifying it. Such acoustoelectronic amplification (or "pumping") of SAWs by an electric current is of great interest, both theoretically and from a practical point of view.

(b) The nonlinear acoustoelectric effects described by the zero-order terms in (10.1.1-3) are those responsible for the appearance of dc currents and voltages. These are transverse and longitudinal, the latter being analogous to those arising in semiconductors when a BAW propagates due to the carriers being "dragged" by the longitudinal field of the wave. To calculate the longitudinal acoustoelectric current in the z direction, one should average over the semiconductor sample the product of its conductivity and the SAW electric field:

$$i_{\mathrm{AE}}^{\parallel} = \frac{1}{2}\mathrm{Re}\left\{\left\langle \sigma_{\mathrm{s}}^{(1)}(0)i\kappa\phi^{(1)*}(0) + \sigma_{\mathrm{s}}^{(1)}(-d)i\kappa\phi^{(1)*}(-d)\right\rangle\right\} \quad , \tag{10.1.7}$$

where σ_{s} is the surface conductivity. The total current which would flow in an external circuit is

$$I_{\mathrm{AE}}^{\parallel} = \frac{\mu S_0 W}{\nu l}\left(1 - e^{2\alpha l}\right) \quad ,$$

where W and S_0 are the width and the power density of the SAW flow respectively, and l is the length of the semiconductor sample in the direction of SAW propagation.

Analogously, one can also determine the transverse dc acoustoelectric current normal to the surface, which is produced by the y component of the SAW electric field:

$$i_{\mathrm{AE}}^{\perp} = \left\langle \sigma^{(1)}\frac{\partial\phi^{(1)*}}{\partial y}\right\rangle \quad . \tag{10.1.8}$$

This current charges the structure ·capacitance and is typically measured as a dc voltage, referred to as the transverse acoustoelectric voltage (TAV), between two electrodes deposited on the semiconductor and the substrate back surfaces (Fig. 10.8)

$$V_{\mathrm{AE}}^{\perp} = \phi^{(0)}(-d) - \phi^{(0)}(0) = \frac{\mu K^2\varepsilon_p S_0}{\varepsilon^2 v^2}f\left(\frac{\omega}{\omega_c}\right) \quad , \tag{10.1.9}$$

where $f(\omega/\omega_c)$ is, generally speaking, a complicated function of its argument [10.32]. Since TAV depends on the concentration of carriers, either free or trapped on surface and bulk states, it can provide significant information about their parameters.

(c) Longitudinal and transverse acoustoelectric nonlinearities lead to the generation of currents and voltages with frequencies harmonic to the SAW frequency or with combination frequencies, if more than one SAW propagates in a given direction. In particular, the second-order terms in (10.1.1-3) describe the nonlinear interaction of SAWs with carriers which gives rise to voltages with the doubled SAW frequency. The generation of the voltage in the direction normal to the surface, for example, results from the quadratic relation between the surface potential ϕ_s and the transverse component of the SAW field at the surface, E_{ys}:

$$\phi_s^{(2)} = \frac{\varepsilon}{6 N_d e} \left[E_{ys}^{(1)} \right]^2 \quad . \tag{10.1.10}$$

As previously mentioned, this model can be extended to include a band-bending at the semiconductor surface. This case is particularly important for the nonlinearity doubling of the frequency, since the presence of a depleted space charge layer at the surface, for instance, increases the proportionality factor in (10.1.10) by more than a factor of three compared to the case of flat bands.

SAW convolvers, for example, operate on the basis of the transverse nonlinearity in structures similar to that considered here, and numerous other SAW - semiconductor devices based on acoustoelectric interaction exploit these principles for signal processing, including frequency mixing, generation of harmonics, optical imaging and image processing, signal convolution and correlation, and data storage.

10.2 Investigation and Characterization of Materials by Surface Acoustic Waves: State of the Art and Main Results

The investigation and characterization of materials by SAWs usually make use of changes introduced either in the SAW parameters (amplitude, phase velocity, frequency, pulse duration and waveform, etc.), by internal inhomogeneities and external influences acting on wave propagation at the surface, or in the medium. As such changes are very sensitive and indicative of various physical and chemical processes, they can be effectively used for studying optical, elastic and thermal properties of materials, parameters of electronic states in semiconductors, adsorption - desorption kinetics at solid surfaces and so on. The evaluation methods based on SAWs may generally be classified into two categories depending on whether the SAW parameters are affected mechanically (or thermally, through thermoelastic deformations) or electrically - through the acoustoelectric effects.

The main parameter measured in the majority of methods based on mechanical principles is the change of the SAW phase velocity caused by perturbations in the density or elastic stiffness of the medium, by mass loading or the presence of thermal sources within the wave propagation path. Changes in SAW attenuation or frequency (in nonlinear acoustic media) also arise under mechanical influences, but they are less frequently used as far as this category of methods is concerned.

Variations in SAW velocity are usually measured either as phase shifts at the output of a SAW delay line or as frequency changes, provided the delay line is used to form an oscillator. The latter is the preferred procedure, since frequency changes are readily measurable with an accuracy of up to 10^{-6} and are the most suitable for feeding directly into digital data acquisition systems. Generally, if a properly designed SAW device is incorporated as a frequency control element of an oscillator circuit, perturbations of the order of 10 ppb in SAW phase velocity can be resolved as corresponding variations in the frequency. This principle has been widely exploited in constructing SAW gas- (or chemo-) sensors [10.33,34], but it can also be used to study adsorption-desorption kinetics and surface chemical reactions, and to measure the binding energy of molecules to various surfaces, desorption cross sections, etc. Such applications, although neither particularly popular nor developed, still look quite promising and owe their use to the high sensitivity of SAW devices to the presence of adsorbed species. Mass loading by the adsorbate is the primary mechanism of frequency change in this case. The sensitivity of the measurement can be considerably increased by using suitable chemically active overlays to enhance the adsorption or the surface reaction rate [10.33,35] and by the use of more complex SAW structures to compensate for other ambient influences [10.33,36] (for example, SAW dual-delay-line structures improve the temperature stability). In this way, the SAW structure functions as a microbalance, weighing adsorbed species with a resolution of about 50 pg/cm^2 [10.37], which corresponds to a small fraction of a monolayer coverage. Further benefits can be obtained if the device substrate is a thin solid membrane less than a few SAW wavelengths thick. The sensitivity of the structure to external mechanical influences increases, sometimes by orders of magnitude, and, furthermore, the piezoelectric coupling for the Lamb wave modes in a thin plate (Sect. 10.1.1) is significantly higher than that for Rayleigh SAWs on a thick substrate, while their sensitivity to changes in density, elastic properties or thickness of a coating is much greater.

Similar, but more pronounced effects on the SAW phase velocity and attenuation will be exerted by a thin film deposited on the substrate surface. In this case, in addition to mass loading, contributions from Love or Stoneley wave modes appear due to the changed boundary conditions, which can be used if one measures the dispersion of the SAW velocity to investigate the elastic properties of thin films. Experiments have also been conducted on SAW-induced desorption [10.38], indicating that the mechanical stresses caused by the wave can be used, in principle, to influence and where necessary modify the process of thin film growth or epitaxy on the substrate surface. These promising applications have not yet been fully investigated, but may become of practical importance in the near future.

On the other hand, changes in the material stiffness or mass density introduced by external forces and influences (thermal, electric, magnetic, optical, etc.) may be transformed and measured as corresponding changes of the SAW phase velocity. Finally, surface wave photoacoustics (Sect. 10.1.3) completes the list of methods of material investigations based on mechanical principles.

In the following some illustrative examples are presented to demonstrate the possible uses of SAWs for the analysis of various material properties.

10.2.1 Elastic Properties

The investigation of the elastic properties of materials by SAWs follows two general trends: visualization of near-surface defects and elastic inhomogeneities (SAW defectoscopy and microscopy) and determination of the elastic constants. The former application is based on the modification of SAW parameters caused by imperfections, phase boundaries, flaws, grains, dislocations, etc., which change the SAW velocity, attenuation, phase and frequency either smoothly or in a stepwise fashion. Due to the great number of affected parameters which may be combined in different ways, acoustic wave defectoscopy, and SAW defectoscopy in particular, belongs to the most universal methods of nondestructive testing of materials. Compared to BAWs, surface waves used for these purposes exhibit the advantages that they are easier to control, modify or probe along their propagation path and are more effective in nonlinear media and in interactions, see Sect. 10.1.4. This is why defectoscopy with SAWs has recently acquired a prime position among the other relevant techniques for nondestructive material evaluation, see [10.39]. This large field of well-established applications lies, however, beyond the scope of the present review.

The elastic constants of the material can be determined, most generally speaking, by measuring the velocities of acoustic waves with appropriate polarization and direction of propagation. Usually, pulse-echo [10.40] and continuous wave resonance [10.41] methods are used for this purpose, but serious additional advantages are offered by the simultaneous measurement of the velocities of optically excited shear and longitudinal bulk and Rayleigh surface waves [10.11], as considered in Sect. 10.1.3. For an isotropic material, this is sufficient to determine the complete set of elastic constants. For example, the Rayleigh wave velocity in isotropic media is approximately given by

$$v = \frac{0.87 + 1.12\nu}{1 + \nu} v_t$$

where ν is Poisson's ratio and v_t the transverse bulk wave velocity, so that ν can be calculated from the experimentally measured values of v and v_t. Other elastic constants, such as the Young's modulus and the compressibility are also related to the acoustic wave velocities by simple, well-known formulas (see e. g. [10.40]). The velocities are determined by the arrival-time delays of the corresponding pulses at a given distance along the free boundary plane and, as a whole, this approach is a variation of the traditional pulse-echo methods for measuring the elastic constants and is subject to errors of the same order. Several authors have performed such studies successfully in various metals [10.14,15,19], illustrating the following important advantages. An investigation of quite small specimens of simple geometry is possible, without the requirement for plane-parallel surfaces which exists in usual sound velocity measurements. The change in the elastic constants with temperature and pressure, or during dynamic processes of rapidly changing variables, etc. can also be studied. By varying the distance between the thermoelastic source of acoustic waves and the detecting transducers, one can measure the attenuation constants for the different wave modes easily and accurately.

10.2.2 Electrical Properties and Electronic States in Semiconductors

Electric properties of semiconductor and thin metal films and surfaces and parameters of electronic states in semiconductors can be investigated effectively and sensitively by means of the acoustoelectric effects caused by a SAW in structures such as those considered in Sect. 10.1.4. As the electrical conductivity of semiconductors changes considerably upon illumination, the majority of such studies are of a spectroscopic character, the semiconductor sample being optically excited during the interaction with a SAW. This closely relates these methods to the study of the optical properties of materials, a topic to be reviewed in the next paragraph.

The transverse acoustoelectric effect has mostly been used so far for such purposes [10.42-49]. The main advantage of the transverse geometry is that the measurement is practically contactless since no special electrodes are required on the sample (a metal foil pressed to the semiconductor back surface has usually been used as the output electrode). A typical TAV response observed when a rf pulse is applied to the input transducer of the SAW delay line is shown in Fig 10.9. It consists of fast components at the beginning and the end of the acoustic pulse (ordinary TAV) and slow relaxation features (trapping TAV). Ordinary TAV is related to free carriers in the energy bands which are moved almost instantaneously by the acoustoelectric force, while trapping TAV results from the capture of these carriers by surface states and such in the space-charge layer [10.50].

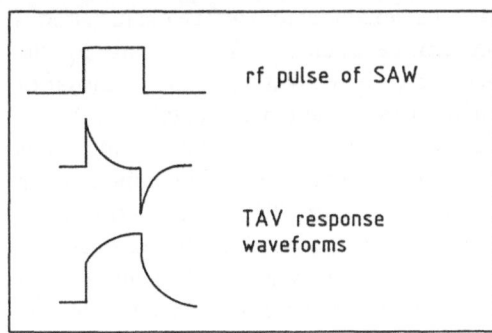

Fig. 10.9 Applied rf pulse and resulting TAV response waveforms

The main parameters in TAV experiments are the amplitude, relaxation time constant and polarity of the fast and slow components in the TAV-response waveform. The observed features can be explained qualitatively by considering the equivalent circuit of the separated-medium SAW configuration, as shown in Fig. 10.10 [10.46]. The series branch (C_{sc}, R_{sc}) represents the semiconductor space-charge layer, while the parallel branch (C_{ss}, R_{ss}) takes into account the shunting effect of the surface states which couple a part of the mobile charge processed by the acoustoelectric force. The term C_0 is the static capacitance of the sandwich air-gap-LiNbO$_3$, and R_a and C_a are the input resistance and capacitance of the measuring electronics, respectively. It can readily be shown that this equivalent circuit can transform a rectangular video input into both the waveforms presented in Fig. 10.9, depending on the ratio of C_{ss} and C_{sc}. If the space-charge layer capacitance is smaller than that of the

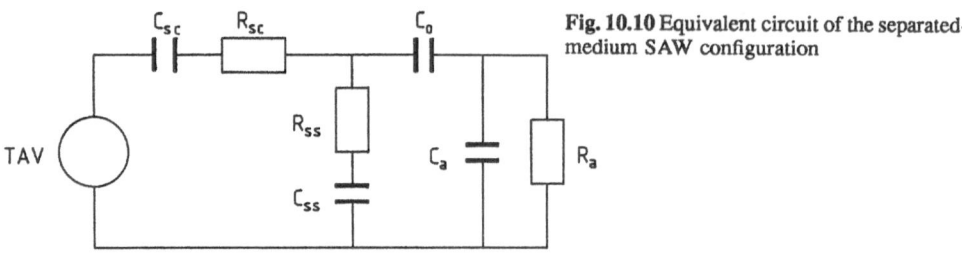

Fig. 10.10 Equivalent circuit of the separated-medium SAW configuration

surface states, the output becomes differentiated, due to the effect of the series group (C_{sc}, R_{ss}) (as shown in Fig. 10.9). In the opposite case, this exhibits the second form with the fast-to-slow components ratio determined by the ratio of the resistances R_{sc} and R_{ss}. By varying the values of elements of the equivalent circuit, one can fit the observed TAV waveforms in order to estimate the corresponding semiconductor surface effective parameters, which change substantially with the population and de-population of the surface states and the surface potential and are subject to serious variations under illumination and/or electric field application.

It is worth noting that, in practice, the distinction between the two components in the TAV waveform is not always possible or that sometimes more than two components are present, which obscures the interpretation of the obtained spectra. In addition, the recent theory of the transverse acoustoelectric effect is quite complicated (the concrete type of the function $f(\omega/\omega_c)$ (10.1.9) in many cases preventing these effective parameters being easily related to real semiconductor surface characteristics (density of states, carrier mobility, capture cross sections, etc.). This calls for the combination of TAV studies with other methods of investigation or the inclusion of model arguments in order to extract information from the experimental data.

Nevertheless, TAV spectroscopy has been widely employed to study the processes of optical generation, recombination and trapping of carriers, both in the bulk and at the surface of various semiconductor materials. Generally, TAV spectra correspond qualitatively to those of photoconductivity, and hence the width of the bandgap, the energy position of impurity and surface levels as well as relative information about their densities and kinetic parameters can be obtained therefrom. In addition, the polarity and the magnitude of the effect, usually appearing as sign-reverse features of different peak-amplitude in TAV spectra, may provide indications of the conductivity type and the character of band bending at the semiconductor surface, and the value of the conductivity frequency, etc. As an example, Fig. 10.11 shows the spectrum of the amplitude of a TAV-response waveform in n-type InAs epitaxial layers at a SAW frequency of 100 MHz [10.44]. As can be seen, TAV changes the polarity from positive to negative at a photon energy of about 1.4 eV, corresponding to an inversion of surface conductivity from n- to p-type as a result of an optically excited band-to-band electron transition. One should keep in mind, however, that such sign reverses are not always unequivocal, see, for example, [10.32] and more criteria are required in some cases to interpret them.

The transverse acoustoelectric effect, in combination with light illumination of the surface, can serve as a simple and rapid nondestructive method for the quali-tative analysis of the electronic topography of semiconductor surfaces [10.47]. The

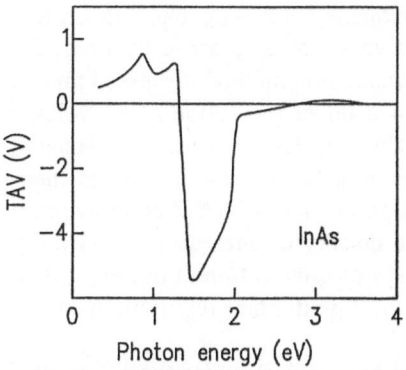

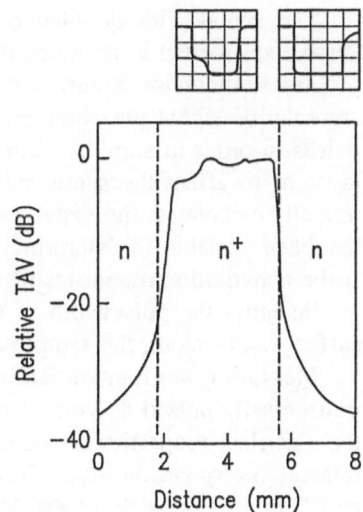

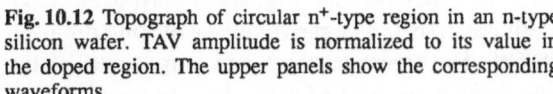

Fig. 10.11 Spectral dependence of TAV amplitude in n-type InAs epitaxial layers

Fig. 10.12 Topograph of circular n$^+$-type region in an n-type silicon wafer. TAV amplitude is normalized to its value in the doped region. The upper panels show the corresponding waveforms

method allows one to follow the spatial distribution of the surface-states' density and relaxation time along the studied surface and to check its homogeneity as well as to visualize surface electronic structures. In this case, the surface is scanned by a tightly focused light beam of suitable wavelength and the local TAV-waveform parameters are monitored. The information obtained applies only to the illuminated area because the contribution of the dark remainder is either small or can be eliminated by modulation of the light intensity and lock-in detection. As an illustration, the topograph of a circular n$^+$-type region formed in an n-type silicon wafer of specific resistance $1000\,\Omega\,\mathrm{cm}$ by local doping with phosphorus is shown in Fig. 10.12 [10.47]. This "one-line" TAV profile has been obtained by means of a 43 MHz separated-medium SAW structure, illuminated by a focused (about $100\,\mu\mathrm{m}$) He-Ne laser beam whose wavelength lies in the intrinsic absorption range of silicon. The TAV signal amplitude has been normalized to its value in the doped area, while the top figures indicate the corresponding TAV waveforms in the two regions. The horizontal scale is $100\,\mu\mathrm{s}/\mathrm{div}$ in these figures. The analysis is comparative and characteristic, with relatively high sensitivity and spatial resolution. This method is particularly effective on wide energy gap semiconductors, which are out of the scope of voltage-capacitance techniques. However, samples which are not sensitive to illumination cannot be studied effectively.

Surface electric properties can also be investigated by measuring the acoustoelectric attenuation of a SAW due to its interaction with the semiconductor. Of particularly informative value is the effect of electric pulses of varying duration, repetition rate and polarity applied to the structure of transverse electrodes, see Fig. 10.8. The pulsed electric field modulates the carrier density and the population of traps in the semiconductor space-charge layer and induces relaxation processes, similar to those arising in conventional field-induced conductivity experiments. Compared to these, however, the acoustoelectric SAW attenuation method has the advantage of being contactless and of higher sensitivity.

The signal with doubled or combination frequency resulting from the transverse acoustoelectric nonlinearity in SAW convolvers may also serve as a tool for probing semiconductor surfaces. Like the SAW-attenuation method discussed above, convolution-signal measurements can be performed under the action of dc voltage pulses in order to stimulate transient charging of energy levels in the space-charge layer or to affect the semiconductor surface potential. Thus, comparing the theoretically calculated and experimentally measured parameters of SAW convolvers, it has been possible to determine, for example, the density of surface states [10.51]. If the convolution output results from two rf pulses of quite different duration, then by changing the pulsewidth of the shorter one, the longitudinal distribution of the surface states along the sample can be scanned.

Measuring the longitudinal acoustoelectric voltage (LAV) under the effect of a transversally pulsed dc voltage applied to the structure is another way of evaluating the electrical properties of the sample. To illustrate this, Fig. 10.13 shows the LAV relaxation response in i-type Ge to both positive and negative pulses of long duration and with an amplitude of 800 V [10.52]. The time constants and amplitudes of the relaxation features appearing at the beginning and end of the pulse can be related, if certain simplifying assumptions are made, to the field-induced surface conductivity. LAV experiments in some cases make it possible to determine surface and bulk carrier mobilities [10.53].

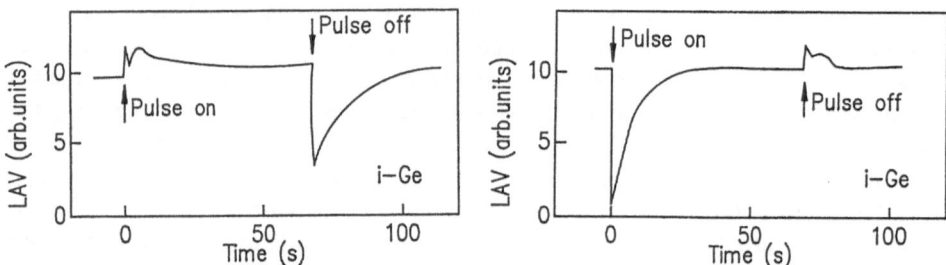

Fig. 10.13 Longitudinal acoustoelectric voltage relaxation response in i-Ge to long positive and negative pulses of 800 V amplitude

10.2.3 Optical Properties

Besides the spectroscopic applications of acoustoelectric effects, another means of studying optical properties of materials is provided by surface wave photoacoustics. Its very basic principle of converting absorbed light energy into acoustic wave energy has suggested the use of optically excited SAWs for the investigation of the optical absorption coefficient and the excitation–deexcitation kinetics. This trend is quite similar to that of traditional photoacoustics and will therefore only be considered here very briefly.

As only the energy absorbed within a few Rayleigh wavelengths below the surface contributes to the formation of SAWs [10.54] surface wave photoacoustics is especially effective in determining the optical parameters of thin films and surface layers as well as of strongly absorbing media. It is possible, in principle, to reconstruct the depth profile of the optical absorption coefficient from the SAW spectrum, but the resolution of this approach is limited by the size of the light spot, which cannot be made smaller than the light wavelength (in practice, it is often difficult to focus a laser beam to a spot smaller than the optical penetration depth).

On the other hand, photoacoustic spectroscopy using SAWs appears to be a popular tool for studying layers with submillimeter thickness of optically low absorbing materials, especially when a contactless experiment is required. The high sensitivity of the surface wave photoacoustic method in optical investigations has been particularly clearly demonstrated by measuring the optical absorption spectrum of monolayer films of rhodamine-590 dye on transparent substrates [10.13]. The experiment was performed with a pulsed dye laser of pulse energy less than $100\,\mu J$ and pulse duration of about 7 ns. The originating SAW video pulses were detected through an edge-bonded $LiNbO_3$ piezotransformer of about 50 MHz bandwidth. With this arrangement, a single shot signal-to-noise ratio of more than 100 has been achieved for a monolayer film coverage, limited mainly by the noise of the rf amplifier used. By placing an acoustic absorber on the surface between the laser spot and the detector, it has been proved that the detected pulses are due to SAWs. A decrease of SAW signals on account of long-term aging and laser-induced desorption of dye molecules has been observed, indicative of the high sensitivity of this method to the nature of coupling of the film with the surface of the substrate. A similar study of optically thin films of rare earth powders on a quartz substrate has been performed by *Tam* and *Patel* [10.55]. The acoustic wave generated in the film upon laser irradiation was found to couple effectively to the substrate and the authors have claimed a linear dependence of the photoacoustic signal on optical absorption, but it is unclear whether the measured signal results from surface waves or not.

A high resolution photoacoustic method of microscopy based on surface acoustic waves in a periodic mode of generation has been reported by *Veith* [10.30]. Like the arrangement in [10.29], the beam of a mode-locked cw laser is focused to a spot of about 5 μm diameter on the sample surface, which consists of a 10 μm-thick gold film evaporated on a $LiNbO_3$ substrate with an interdigital transducer on it. The sample can be scanned perpendicularly to the beam through an x-y stage to image the optical absorption of the film. In this experiment the lateral resolution is limited by the laser spot diameter. A problem arising in this mode of resonant SAW detection is the appearance of interference oscillations in the signal of the interdigital transducer. This can be avoided by optical detection of the SAW, but a decrease of the signal-to-noise ratio might result, on account of the broadband registration of SAWs. This method can also be used to image the distribution of elastic properties of the film or to visualize near-surface flaws.

10.2.4 Kinetic Properties

Ultrasonic properties of thin films can be conveniently studied using SAWs. The behavior of ultrasonic absorption, in particular, can reveal interesting facts about structural relaxation. As the SAW is accompanied by an elastic field, it can influence the potential energy of a defect state in a disordered network. If the network contains atoms which can undergo structural relaxation, that is, they can occupy two or more nondegenerate states in the network, the equilibrium occupation of these states will be disturbed by the elastic field of the SAW. Of course, this can only happen if the relaxation rates have the same order of magnitude as the frequency of the SAW.

From the above, it is clear that SAW absorption studies can be used to investigate structural relaxation in disordered networks. In thin films of amorphous materials they are often the only way of obtaining this information. Of course, there are also other kinetic processes which can lead to dissipation of SAW energy, e. g. phonon-assisted hopping between electronic states in semiconductors, and phonon-induced perturbation of a chemical equilibrium or desorption.

An example of such a study of the relaxation behavior of thin glassy films is given by work of *Haumeder* et al. [10.56], who studied the ultrasonic velocity and absorption of films of amorphous Si, $SiO_{0.7}$ and SiO_2 at temperatures between 0.4 and 300 K. For this purpose, they deposited 1 μm thick films of these materials on one path of a dual delay line oscillator of 300 MHz. For SiO_2, they observed acoustic properties similar to those of bulk glassy SiO_2. In SiO_2, and to a lesser extent in $SiO_{0.7}$, they found a fairly high ultrasonic absorption at temperatures below 5 K and attributed it to a two-level tunneling mechanism described above. For amorphous Si films, no such tunneling mechanism was proposed, as low-temperature absorption was very low. The amorphous Si film shows a very strong absorption around 270 K, which was interpreted as a consequence of a thermally activated structural relaxation process. *Bhatia* et al. [10.57] investigated the attenuation of SAWs of 300 MHz in sputtered films of SiH_x $(0 \leq x < 0.15)$ of 0.3 μm thickness at temperatures between 20 K and 600 K. Their results are suggestive of a thermally activated structural relaxation mechanism with a strong contribution from twofold coordinated SiH_2 centers at higher temperatures and a weaker contribution from SiH centers between 20 K and 200 K.

10.2.5 Surface Acoustic Wave Sensors

The higher density of acoustic energy in the vicinity of the surface for SAWs compared to bulk acoustic waves makes SAW devices very sensitive to all quantities acting on the near-surface elastic properties. The most important of these quantities are mass loading, temperature, force and electric field influencing the phase velocity and attenuation of the SAW.

In practice, it is often convenient to determine the change in SAW velocity by measuring the resonance frequency of a SAW delay line oscillator, because frequency changes can be measured with very high accuracy. In the feedback-oscillator method, which is used in many sensors, the output transducer of a SAW delay line is connected to its input transducer via a rf amplifier and periodic SAWs are excited. The oscillation frequency is determined by the condition that the total phase shift of the oscillator loop must be a multiple of 2π. As the phase shift between the input and output transducer is influenced by the phase velocity of the SAW, a shift of the oscillation frequency can be converted into a change in SAW velocity. The "sing-around" method measures the repetition frequency of an arrangement, where a short rf pulse is excited, detected at some distance, and the detected pulse then triggers a new pulse.

The high sensitivity of SAWs to temperature changes has been used to construct temperature sensors with a relative accuracy of 10^{-4} K [10.58]. The main advantage of a SAW temperature sensor over a quartz bulk wave resonator is that only the temperature of a near-surface layer of a thickness of an acoustic wavelength is measured. Thus, it is possible to detect temperature fluctuations of 10^{-4} K at frequencies up to hundreds of kilohertz, depending on the SAW wavelength and the thermal properties of the substrate [10.59]. The fast response of the substrate surface to small changes of the thermal conductivity of the surrounding medium can be used to construct a SAW delay line oscillator-based gas flowmeter by heating the substrate at a constant rate and measuring the oscillation frequency [10.60]. SAW devices in the shape of thin piezoelectric plates can also be used to measure high voltages, as the stiffness of the substrate changes in the presence of an electric field [10.61].

The change of the elastic properties upon absorption or adsorption of vapors at the surface of a SAW delay line affects the SAW phase velocity (usually by mass loading or by a change of the electric conductivity of the interface). This is the principle on which SAW chemosensors are based. A typical SAW chemosensor consists of a SAW dual delay line, one delay path of which is covered with a thin film of a substance which interacts selectively and reversibly with the measurand. The SAW velocity on both delay paths is measured by the feedback-oscillator method and the two resonance frequencies are compared by frequency mixing. The use of this method provides a better temperature stability of the sensor. Sensitivities in the ppm range with quite high selectivity and response times of a few minutes can be achieved with this method.

SAW chemosensors have been used with success to measure hydrogen (employing an absorptive palladium layer) [10.62], nitrogen dioxide (employing a semiconductive organic phthalocyanine layer, the electrical resistance of which varies with NO_2 absorption) [10.37] and sulfur dioxide (employing an absorbent overlay normally used in gas chromatography) [10.34].

The development of thin membrane SAW oscillators with Lamb waves propagating on both sides of the membrane may allow considerable improvements in sensor construction. The advantages of such sensors lie in the higher piezoelectric coupling factors for Lamb modes, their higher sensitivity towards coating density and the possibility of placing the delicate interdigital structure on the face opposite to the chemical interface which may be exposed to aggressive vapors.

10.3 Conclusion

Thus it can be seen that SAWs have been successfully used to study thin films and the surface properties of materials. The basis of such methods is the sensitive and complex manner in which SAW parameters are influenced by various physical and chemical processes. Compared to bulk acoustic waves, SAWs have certain practical advantages - they are easier to modify, probe and influence along their propagation path, the variety of SAW modes and types to be used in material testing applications is more numerous and SAWs are much more effective in nonlinear interactions and processes. On the other hand, the use of SAW methods of investigation is encouraged by the existing quite efficient methods for the generation and detection of SAWs in a very wide frequency range and by availability of numerous well-developed, elaborate and commercially available SAW signal processors and devices.

At the same time, surface wave photoacoustics is becoming a promising and competitive method for contactless, nondestructive evaluation of the properties of materials.

The main bottleneck in the application of SAW methods is the complex character of the processes involved, which in many cases prevent one from extracting as much information as one would like. This situation also arises in other methods, however, for example, in traditional solid-state photoacoustics, where the interconnection between optical, thermal, relaxational and acoustical effects and contributions to the experimentally measured signals can hardly be separated unequivocally except in the simplest of cases. Therefore, despite the impressive achievements of SAW methods in material investigation, a lot of effort is still required to develop them from the level of promising possibility to routine technique.

Acknowledgements: One of us (L.K.) is indebted to the Alexander von Humboldt Foundation for a fellowship. Financial support of this work by the Bundesministerium für Forschung und Technologie (BMFT) under contract No. 13N5363 8 and by the Fonds der Chemischen Industrie is gratefully acknowledged. This work was also sponsored by the European Community under Contract No. ST2J-0129-2-D.

References

10.1   G W. Farnell: In *Acoustic Surface Waves*, ed. by A. A. Oliner, Topics Appl. Phys., Vol. 24 (Springer, Berlin, Heidelberg 1978)
10.2   D. E. Oates, R. A. Becker: Appl. Phys. Lett. **38**, 761 (1981)
10.3   C. Lardat, P. Defranould: Proc. IEEE **64**, 627 (1976)
10.4   J.-P. Monchalin: IEEE Trans. UFFC-**33**, 485 (1986)
10.5   H. Sontag, A. C. Tam: IEEE Trans. UFFC-**33**, 500 (1986)
10.6   R. L. Whitman, A. Korpel: Appl. Opt. **8**, 1567 (1969)
10.7   G. I. Stegeman: IEEE Trans. SU-**23**, 33 (1976)
10.8   D. A. Hutchins, F. Hauser, T. Goetz: IEEE Trans. UFFC-**33**, 478 (1986)
10.9   O. E Mattiat (ed.): Ultrasonic Transducer Materials (Plenum, New York 1971)
10.10 A. M. Aindow, R. J. Dewhurst, S. B. Palmer: Opt. Commun. **42**, 116 (1982)
10.11 H. M. Ledbetter, J. C. Moulder: J. Acoust. Soc. Am. **65**, 840 (1979)
10.12 S. M. Avanesyan, V. E. Gusev, B. V. Zhadanov, V. I. Kuznetsov, S. A. Telenkov: Sov. Phys. Acoust. **32**, 356 (1986)
10.13 S. R. J. Brueck, T. F. Deutsch, D. E. Oates: Appl. Phys. Lett. **43**, 157 (1983)
10.14 A. C. Tam: Appl. Phys. Lett. **45**, 510 (1984)
10.15 A. C. Tam, H. Coufal: Appl. Phys. Lett. **42**, 33 (1983)
10.16 P. E. Dyer: In *Photoacoustic and Photothermal Phenomena*, ed. by P. Hess, J. Pelzl, Springer Ser. Opt. Sci., Vol. 58 (Springer, Berlin, Heidelberg 1988)
10.17 R. E. Lee, R. M. White: Appl. Phys. Lett. **12**, 12 (1968)
10.18 A. M. Aindow, R. J. Dewhurst, D. A. Hutchins, S. B. Palmer: J. Acoust. Soc. Am. **69**, 449 (1981)
10.19 A. V. Golenishev-Kutuzov, S. A. Migachev, N. R. Yafaev: Sov. Phys. Acoust. **31**, 405 (1985)
10.20 A. N. Khodinskii, L. S. Korochkin, S. A. Mikhnov: Zh. Prikl. Spectrosk. **38**, 745 (1983)
10.21 C. K. Jen, P. Cielo, J. Bussiere, F. Nadeau: Appl. Phys. Lett. **46**, 241 (1985)
10.22 D. Royer, E. Dieulesaint, X. Jia, Y. Shui: Appl. Phys. Lett. **52**, 706 (1988)
10.23 D. A. Hutchins, F. Nadeau, P. Cielo: Can. J. Phys. **64**, 1334 (1986)
10.24 J. A. Cooper, R. A. Crosbie, R. J. Dewhurst, A. D. W. McKie, S. B. Palmer: IEEE Trans. UFFC-**33**, 462 (1986)

10.25   A. Neubrand, P. Hess: Unpublished results
10.26   H. Sontag, A. C. Tam: Appl. Phys. Lett. **46**, 725 (1985)
10.27   H. Sontag, A. C. Tam: Can. J. Phys. **64**, 330 (1986)
10.28   R. E. Higashi, R. K. Mueller, W. P. Robbins: IEEE Ultrason. Symp. Proc. 357 (1983)
10.29   G. Veith, M. Kowatsch: Appl. Phys. Lett. **40**, 30 (1982)
10.30   G. Veith: Appl. Phys. Lett. **41**, 1045 (1982)
10.31   E. P. Velikhov, E. V. Dan'shchikov, V. A. Dymshakov, A. M. Dykhne, F. V. Lebedev, V. D. Pis'-mennyi, B. P. Rysev, A. V. Ryazanov: JETP Lett. **38**, 585 (1983)
10.32   J. I. Fritz: J. Appl. Phys. **52**, 6749 (1981)
10.33   A. D'Amico, A. Palma, E. Verona: Sensors Actuators **3**, 31 (1983)
10.34   A. W. Barendsz, J. C. Vis, M. S. Nieuwenhuzizen, E. Nieuwkoop, M. J. Vellekoop, W. J. Ghijsen, A. Venema: IEEE Ultrason. Symp. Proc. 586 (1985)
10.35   H. Wohltjen: Sensors Actuators **5**, 307 (1984)
10.36   A. D. Bryant, D. L. Lee, J. F. Vetelino: IEEE Ultrason. Symp. Proc. 171 (1981)
10.37   A. D. Martin, K. S. Schweizer, S. S. Schwartz, R. L. Gunshor: IEEE Ultrason. Symp. Proc. 207 (1984)
10.38   C. Krischer, D. Lichtman: Phys. Lett. A **44**, 99 (1973)
10.39   "Acoustic Microscopy", Special issue of IEEE Trans. Sonics and Ultrasonics (March 1985)
10.40   R. Truell, C. Elbaum, B. B. Chick: *Ultrasonic Methods in Solid State Physics* (Academic, New York 1969)
10.41   D. I. Bolef, M. Menes: J. Appl. Phys. **31**, 1010 (1960)
10.42   P. Das, R T. Webster, H. Estrada-Vazquez, W. C. Wang: Surf. Sci. **86**, 848 (1979)
10.43   H. Estrada-Vazquez, R. T. Webster, P. Das: J. Appl. Phys. **50**, 4942 (1979)
10.44   F. M. Mohammed Ayub, P. Das: J. Appl. Phys. **51**, 433 (1980)
10.45   P. Das, M. Tabib-Azar, B. Davary, J. H. Everson, IEEE Ultrason. Symp. Proc. 421 (1983)
10.46   L. L. Konstantinov, V. L. Strashilov, O. Ivanov: J. Phys. D **18**, L 79 (1985)
10.47   V. L. Strashilov, L. L. Konstaninov, O. Ivanov: Appl. Phys. B **43**, 17 (1987)
10.48   I. J. Fritz: J. Appl. Phys. **54**, 4457 (1983)
10.49   V. S. Bondarenko, V. L. Gromashevski, A. G. Kundzich, E. G. Miseljuk, B. V. Sobolev, A. Ph. Sharov: Fiz. Tverd. Tela **22**, 1566 (1980)
10.50   Yu. V. Gulyaev, A. Yu. Karabanov, A. M. Kmita, A. V. Medved', Sh. S. Tursunov: Fiz. Tverd. Tela **12**, 2595 (1970)
10.51   G. S. Kino: Proc. IEEE **64**, 724 (1976)
10.52   T. Shiosaki, T. Kuroda, A. Kawabata: Appl. Phys. Lett. **26**, 360 (1975)
10.53   A. Bers, J. M. Cafarella: Appl. Phys. Lett. **22**, 399 (1973)
10.54   A. A. Karabutov: Sov. Phys.-Usp. **28**, 1050 (1985)
10.55   A. C. Tam, C. K. N. Patel: Appl. Phys. Lett. **35**, 843 (1979)
10.56   M. v. Haumeder, U. Strom, S. Hunklinger: Phys. Rev. Lett. **44**, (2), 84 (1980)
10.57   K. L. Bhatia, M. v. Haumeder, S. Hunklinger: Solid State Commun. **37**, 943 (1981)
10.58   D. E. Cullen, G. K. Montress: IEEE Ultrason. Symp. Proc. 264 (1975)
10.59   R. G. Stearns, B. T. Khuri-Yakub, G. S. Kino: Appl. Phys. Lett. **43**, 748 (1983)
10.60   N. Ahmad: IEEE Ultrason. Symp. Proc. 483 (1985)
10.61   K. Toda, K. Mizutani: J. Acoust. Soc. Am. **74**, 677 (1983)
10.62   A. D'Amico, A. Palma, E. Verona: Appl. Phys. Lett. **41**, 300 (1982)

11. Heat Diffusion and Random Media

D. Fournier and A.C. Boccara

With 7 Figures

Nowadays random media are structures of paramount importance for physicists and engineers. More and more disordered structures such as glasses, sintered materials, and concrete belong to our everyday world and it is of importance to be able to describe their specific properties when we compare them to homogeneous materials. The first attempt usually consists in replacing the random media by an effective medium whose properties reflect a specific physical behavior (e.g. the thermal or the electric conductivity of a polycrystalline sample). Although very useful, the previous approach has been found to be limited in its ability to describe most of the dynamical processes, other than those occurring in the "long time" or "low frequency" range where the sample behaves like a homogeneous medium. Indeed, between this and the microscopic scale, dealing with the individual entities of the structure, one can find a full range of anomalous behaviors which are more clearly revealed by studying diffusive transport. Among all the diffusion phenomena, that of heat offers certain specific advantages: the ability to operate with various kinds of materials (insulators or conductors), the possibility of repeating an experiment, and nondestructive and noncontact testing. We intend to illustrate how, by following the heat diffusion temporal behavior, rough surfaces and random media which can be mapped on fractal structures may be characterized.

We do not claim that all the samples under study which exhibit some kind of randomness can be treated by fractal geometry. Nevertheless, we have found it helpful to analyze the geometrical aspect of heat diffusion using this concept. Moreover it has been proved, and this is of importance for heat diffusion through random structures, that a random distribution of barriers (e.g. thermal resistances between grains) can often be mapped on a fractal structure.

11.1 Diffusion Processes

The random nature of diffusion processes is related to the random displacement of carriers. These carriers can be particles (or quasiparticles) such as atoms, molecules, and ions for mass diffusion, free electrons and phonons for heat conduction, and various kinds of carriers (electrons, holes and ions) for electrical conduction.

In a homogeneous isotropic medium, the thermal conductivity of solids can be derived from the kinetic theory of gases under certain approximations:

$$K = \frac{1}{3}\varrho w \lambda C_v \tag{11.1.1}$$

where C_v is the heat capacity at constant volume, λ the mean free path of carriers between collisions, ϱ the mass density, and w the average carrier velocity.

By inserting (11.1.1) in the Fourier law, the thermal flux per unit area between two planes of temperature t_1 and t_2 separated by a distance a, can be written as

$$q = \frac{1}{3}C_v(t_1 - t_2)\frac{w}{a} . \tag{11.1.2}$$

This classical result can be interpreted in the following way: $C_v(t_1 - t_2)$ is the excess energy on one side compared to the other side, which propagates with an effective transport velocity $w\lambda/a$ [11.1].

We can check that this result (the velocity is a function of the distance a) leads to the well-known result

$$t = \frac{a}{(w\lambda/a)} = \frac{a^2}{w\lambda} , \tag{11.1.3}$$

where t is the time taken for the heat to diffuse over a path of length a. This result can easily be generalized to every kind of diffusion process in homogeneous media.

With the mean heat path in a diffusion process being proportional to $\sqrt{t}$, the time dependence of the spatio-temporal distribution of the temperature is closely related to the dimensionality of the heat source and of the space in which this source diffuses. For instance, let us recall that the spatio-temporal distribution of the temperature in a homogeneous medium (3D) heated by a unit point source located at $r = 0$ is given by [11.2]

$$T(r, t) = \frac{Q\exp(-r^2/4Dt)}{8\pi\varrho C(Dt)^{3/2}} , \quad D = K/\varrho C , \tag{11.1.4}$$

where K, ϱ, C, are the thermal conductivity, mass density and specific heat, respectively, D the thermal diffusivity, and Q the heat pulse energy. Thus the time evolution of the temperature $T(0, t)$ of the point source scales as $t^{-3/2}$. By considering a homogeneous distribution of instantaneous point sources along a line, the temperature obtained by integrating formula (11.1.4) is

$$T(r, t) = \frac{Q\exp(-r^2/4Dt)}{4\pi\varrho C(Dt)^1} \tag{11.1.5}$$

where Q is the uniform linear heat density of energy of the pulse. Thus the temperature time dependence of the line source is inversely proportional to the time: $T(0, t)$ scales as t^{-1}. Similarly, for a plane source, $T(0, t)$ scales as $t^{-1/2}$.

These results show the influence of the heat source dimensionality on the time dependence of its own temperature. More generally, one finds

$$T(0, t) \approx t^{-(d-d_s)/2} , \tag{11.1.6}$$

for the time evolution temperature of a source temperature of dimension d_s, diffusing in a Euclidean space of dimension d.

11.2 Introduction to Fractal Geometry

Since its introduction by *Mandelbrot* [11.3], the concept of fractal structure has been found to be very useful for describing geometrical shapes of either deterministic or random "selfsimilar" structures. For these structures, geometrical properties are indistinguishable as a function of length scale. It is obvious that in real physical systems, the scale range in which this self-similarity occurs is limited both in the small dimension limit (e.g. grains or atoms) and in the large dimension limit, where these objects behave like Euclidean entities (homogeneous and continuous).

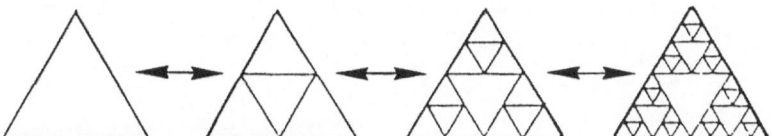

Fig. 11.1. Sierpinski gasket construction

Figure 11.1 shows an example of a self-similar geometrical construction called the Sierpinski gasket. We see that if we change the unit length by $1/2$ we have 3 equal pieces. Let us recall that for a straight line interval, a square or a cube, if we change the scale by $1/2$ we have $(1/2)^{-d}$ equal pieces of the initial object, d being equal to 1, 2 or 3 respectively. This is a definition of the usual Euclidean dimension. We can use this approach to obtain the dimension of the Sierpinski gasket:

$$(1/2)^{-\bar{d}} = 3 , \quad \text{thus} \quad \bar{d} = 1.58 ,$$

$\bar{d}$ is called the fractal dimension.

Regarding its application to heat diffusion in random media, it is worth focusing our attention on random self-similar structures. For such structures, the above properties must be considered as an average. As an example, Fig. 11.2a shows a so-called Brownian motion curve whose fractal dimension is 1.5.

Among the random structures which have been extensively used to describe the static and dynamical processes of random media (glasses, gels, polymers) one finds the so-called percolation network [11.4]. To create a site percolation network, each intersection of a d-dimensional grid is randomly occupied with a probability p. The sites are assumed to be connected if they are close neighbors along a principal direction (e.g. along lines parallel to the sides of a square grid). A critical probability p_c exists such that for $p > p_c$ a connected cluster (referred to as an "infinite cluster" for an infinite grid) will cross the grid from one site to the other. The lower cutoff for the percolation network is the elementary cell of the grid, the upper one being called the percolation correlation length. Roughly speaking, this correlation length is of the order of the biggest "holes" in the structures, and for distances larger than this length, the medium appears to be homogeneous and its mass density ϱ is constant (Fig. 11.3).

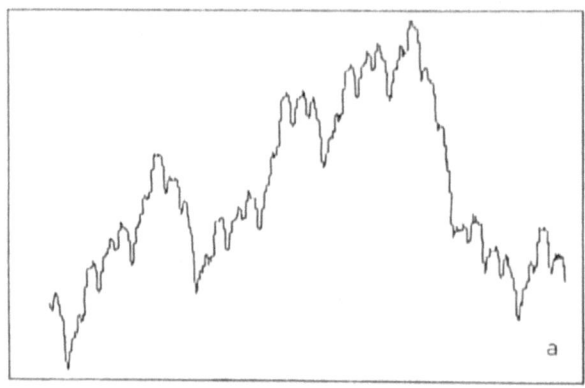

a

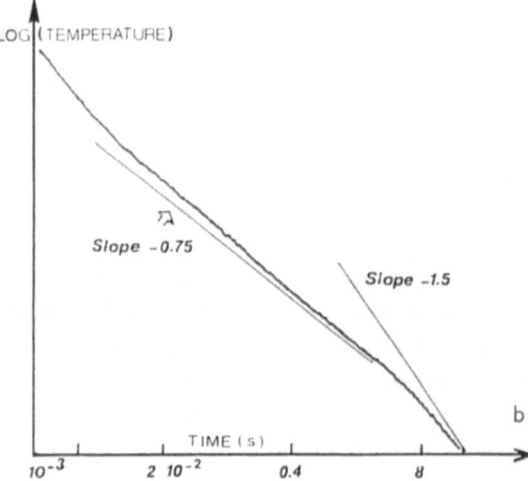

LOG (TEMPERATURE)

Slope −0.75

Slope −1.5

TIME (s)

10^{-3} $2 \; 10^{-2}$ 0.4   8

b

Fig. 11.2. (a) 2D "normal Brownian motion" (d_s = 1.5). (b) Temperature evolution of a central point of the 2D "Brownian motion curve" considered as a unit source for a 3D space (computer simulation)

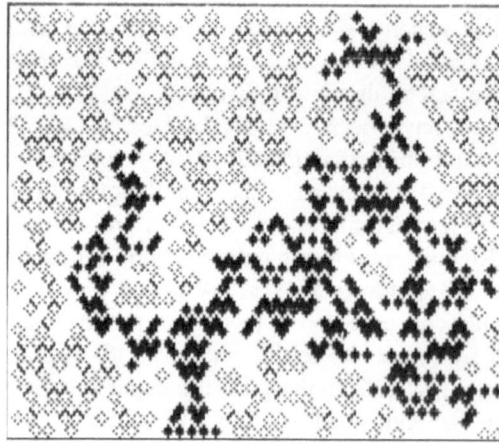

Fig. 11.3. Example of a 2D percolation network close to the threshold (the infinite cluster has been blackened)

306

Quite a few computer simulations of percolation networks have demonstrated their self-similar nature and the fractal dimensionalities $\bar{d} = 1.9$ for $d = 2$, and $\bar{d} \approx 2.6$ for $d = 3$ have been found. Moreover, predictions of dynamical processes which are of importance for transport properties have been carried out [11.5–8].

As a direct consequence of this definition of fractal dimension d, we find the concept of "space-varying density". Indeed, for fractal structures there is no proportionality between mass and volume when the length scale is modified. The amount of mass inside a sphere of radius r in a homogeneous continuous medium is given by

$$M(r) \approx \varrho r^d ,$$

where ϱ is the mass density and d the dimensionality of the space. For instance,

ϱ is length density and $d = 1$ for a one-dimensional object;
ϱ is area density and $d = 2$;
ϱ is volume density and $d = 3$.

For a fractal structure, the amount of mass inside a sphere of radius r scales as $r^{\bar{d}}$ where $\bar{d}$ is the previously defined fractal dimension of the structure.

Usually, $\bar{d}$ is smaller than the Euclidean dimension d in which the structure is immersed and the mass density $\varrho(r)$ scales as $r^{\bar{d}}/r^d = r^{\bar{d}-d}$. Thus the larger the fractal object, the smaller its density, at least within the two scales (lower and higher) where the structure can be mapped on a fractal structure.

11.3 Diffusion from Fractal Sources: A Possible Model for the Behavior of Rough Surfaces?

The problem of understanding heat diffusion in a homogeneous solid bounded by a rough surface is of importance for many practical applications. A first attempt consists in replacing the rough layer by a homogeneous one, with a thickness of the order of the perturbed depth and physical parameter (k, ϱ, c) which are different from the bulk ones. This is the usual effective medium approach [11.9].

Let us point out that among the various kinds of roughness which have been described in the literature, some of them exhibit a fractal geometry [11.10]. Indeed, some of the fractal "landscapes" which can be found in Mandelbrot's book [11.3] are very similar to some of the surfaces as they appear through a microscope.

If rough sourfaces can be mapped on fractal structures, it is of interest to ask how a fractal distribution of heat sources behaves in Euclidean spaces. The main difference, when integrating (11.1.4) is that the heat source density is no longer uniform but scales as $\bar{d}_s - d$, where $\bar{d}_s$ is the fractal dimensionality of the source and d the Euclidean dimension of the space in which the fractal structure is immersed. By integrating over the fractal distribution of heat sources one finds

$$T(0, r) \approx t^{-(d-\bar{d}_s)/2} . \tag{11.3.1}$$

Thus (11.1.6) and (11.3.1) are identical, the fractal dimension of the heat source replacing the Euclidean one.

In order to check this result, we ran a computer simulation in which the fractal heat source was built up from a set of points constituting a 2D "normal Brownian motion" ($\bar{d}_s = 1.5$) [11.4] allowed to diffuse in a 3D space. Figure 11.2b exhibits, on a log-log scale, the temperature dependence of one of the central points among the 1000 points which were used to build the curve, the temperature being averaged over 20 curves. The curve was limited in space by a $10 \times 10 \, cm^2$ square, and the thermal diffusivity of the surrounding medium is $1 \, cm^2 s^{-1}$. The temporal dependence exhibits three slopes: At short times, because the curve is constructed from discrete points, the point is isolated and the behavior is Euclidean. We then obtained a slope close to the expected -0.75 value, corresponding to $(\bar{d}_s - d)/2$. At times longer than a few tens of seconds, the heat has diffused over a distance larger than 10 cm and the whole heat source behaves like a point source.

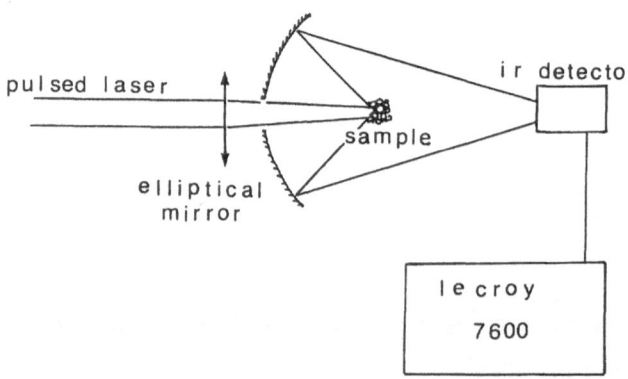

Fig. 11.4. Experimental setup for monitoring the surface temperature of opaque samples heated by a laser pulse

Figure 11.4 shows the experimental setup we used to monitor the temporal evolution of opaque sample surfaces after short (15 ns) irradiation with uniform visible light. The infrared detector is coupled with a digital oscilloscope, which averages the signal. Despite the simplicity of the experiment, a few points have to be carefully checked before such an experiment is run.

– The sample must be opaque for both visible and infrared radiation.

– The light level must be reduced to a level which allows the IR signal to be proportional to the temperature rise dT ($dT \ll T \approx 300 \, K$).

– The aperture of the heating beam must be large enough to avoid shade problems.

In Fig. 11.5 we can observe both the Euclidean and the fractal behavior of the surface temperature of a carbon sample (opaque for both the visible and infrared radiation). When the surface is polished, we obtain the $-1/2$ slope expected for a plane source diffusing in a 3D space, whereas at short times the rough surface sample curve exhibits a slope of $-1/3$, from which a fractal dimensionality $\bar{d}_s = 2.33$ can be deduced. From the time associated with the crossover region, one can get an estimation of the roughness, which is of the order of $\sqrt{4Dt}$. We point out that, if we try to account for roughness effects by using an "equivalent layer" model,

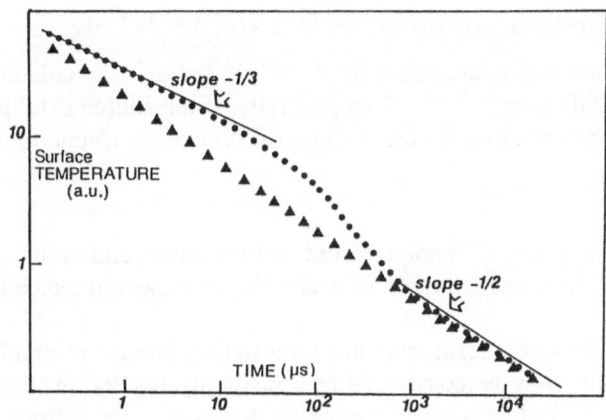

Fig. 11.5. Experimental time-dependent surface temperature for a polished (triangles) and rough surface (points) of a carbon sample

the experimental behavior cannot be explained. Indeed, for an equivalent layer of smaller thermal effusivity ($\sqrt{K\varrho C}$) than the bulk, one would expect a $-1/2$ slope at short times, when the heat has not reached the bulk-layer interface, and then a faster decrease towards another $-1/2$ slope section when heat was diffused over a path much larger than the layer thickness (bulk diffusion) [11.11,12].

11.4 Euclidean and Fractal Sources in Random Media: A Possible Model for Heat Diffusion in Random Media?

We have recalled in Sect. 11.1 that the heat diffusion length from a source is related to the mean square displacement $\langle r^2 \rangle$ of the diffusers. For instance, due to phonon–phonon or phonon–electron collisions, the heat diffusers (phonons or electrons) exhibit a "random walk" and the mean square displacement is usually $\langle r^2 \rangle \approx t$ in a Euclidean space, thus the diffusion coefficient $D = d(\langle r^2 \rangle)/dt$ is a constant. If the diffusing medium is not homogeneous but random (De Gennes' image of the drunk man in a labyrinth [11.8]), the mean square displacement exhibits a different power law: $\langle r^2 \rangle \approx t^{2/d_w}$ with $d_w > 1$ being the dimension of the random walk; the diffusion coefficient is then no longer constant. In the case of a percolating network, it has been demonstrated that d_w is equal to $2\bar{\bar{d}}/\bar{d}$, with $\bar{d}$ being the fractal dimension and $\bar{\bar{d}}$ the spectral dimension introduced in order to account for the peculiar diffusion process. *Alexander* and *Orbach* have conjectured that $\bar{\bar{d}} = 4/3$ for percolating networks for all dimensions $d > 1$ [11.13].

In order to evaluate the surface temperature variation versus time, we have to compute the average amount of mass which has been heated after a time t following the heating pulse. This mass is a function of the mean square displacement $\langle r^2 \rangle$, and of the mass density. The temperature is inversely proportional to the heated mass.

For a plane source ($d_s = 2$) diffusing in a homogeneous (constant ϱ) medium ($d = 3$), the heated mass is proportional to $(\langle r^2 \rangle)^{1/2} \approx t^{1/2}$. Thus $T(0, t) \approx t^{-1/2}$, whereas

309

in a heterogeneous (fractal) medium, $(\langle r^2 \rangle)^{1/2} \approx t^{\bar{\bar{d}}/2\bar{d}} \approx (\langle r^2 \rangle)^{(\bar{d}-3)/2}$, the heated mass scales as $t^{(\bar{\bar{d}}/2-\bar{\bar{d}}/\bar{d})}$, and the temperature as $t^{(-\bar{\bar{d}}/2+\bar{\bar{d}}/\bar{d})}$. For a percolating network, this result leads to $T(0,t) \approx t^{-0.154}$. More generally, if the source exhibits a fractal dimension $(\bar{d}_s)$ and diffuses in a fractal medium (of dimension $\bar{d}$) one finds

$$T \approx t^{-\bar{\bar{d}}/2 + \bar{d}_s \bar{\bar{d}}/2\bar{d}} \, .$$

We have looked at a large variety of random disordered structures and on many occasions have found a power law over a large time scale with an exponent typically in the range 0.15–0.25.

As an example, Fig. 11.6 shows the result obtained for a weakly bonded assembly of copper spheres. This sample may be considered as a randomly bonded medium with a planar surface $(d_s = 2)$. Below a few hundred microseconds, the diffusion occurs within the first layer of irradiated spheres. Then the heat diffuses between the spheres through a random network of bonds (the slope being 0.2). Finally, for times longer than 0.15 s the diffusion is Euclidean (slope ≈ 0.5), meaning that the heat has covered a distance larger than the biggest hole in the network structure. Notice that the value of the slope in the intermediate region is not far from the value expected for a percolating network.

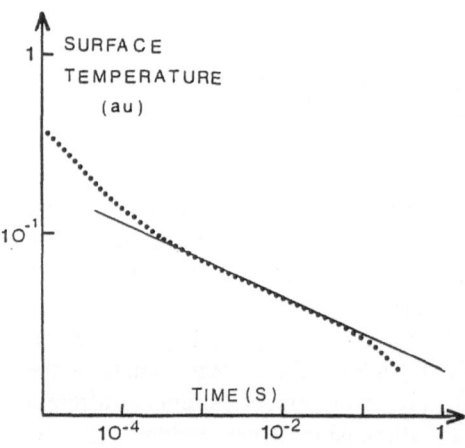

Fig. 11.6. Experimental time-dependent surface temperature of an assembly of slightly bonded spheres (100 μm)

We have also observed such "fractal" behavior (power law with a slope much smaller than the 0.5 Euclidean value) in carbon samples used for making electrical resistors in the megaohm range. Recall that for smaller values of the electrical resistance ($\approx$ a few ohms) and for a polished surface, we obtained the expected 0.5 slope (Fig. 11.7). When the carbon, which is both a good thermal and electrical conductor, is mixed with a polymer which is a bad thermal conductor and an electric insulator, the electrical conductivity can be lowered by many orders of magnitude. The thermal barriers between carbon grains introduce some kind of randomness that we have supposed to be responsible for the short time behavior reported in Fig. 11.5. At large time scales, the sample behaves like a homogeneous one. Because of the

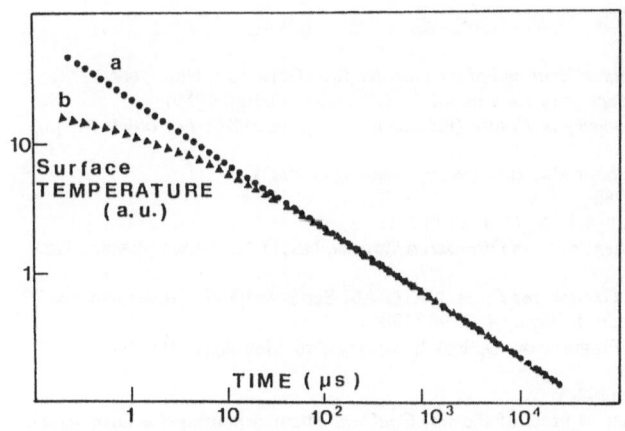

Fig. 11.7. Experimental time-dependent surface temperature for two polished surfaces: $10\,\Omega$ resistor (points) and $640\,\mathrm{k}\Omega$ resistor (triangles)

good thermal diffusivity of carbon and (or) the small size of the particles we have not been able to observe the diffusion within individual grains which occurs below the $100\,\mathrm{ns}$ range for micrometer-size particles.

In the preceding examples, we have tried to illustrate how the presence of random thermal barriers may be revealed by a time-dependent diffusive process. Such an approach can obviously be applied to other kinds of diffusion processes. As an example, let us recall that photothermal techniques are of great help in monitoring the carrier behavior in crystalline semiconductors. One can now envisage monitoring their anomalous diffusion in a polycrystalline sample, where the grain boundaries play the role of random barriers between grains.

11.5 Future Trends and Conclusion

With the few examples we have described here, we have tried to demonstrate that from the simple observation of the average surface temperature evolution of a sample, both geometrical parameters (fractal dimension of a surface, thickness of a perturbed layer) and dynamical parameters (fractal dimension and spectral dimension of a network of bonds) can be established.

More detailed analysis of the diffusion processes can be obtained by modifying the experimental scheme used here. For instance, one could separate the heating "point" source from the probed point. Or, as we did recently in collaboration with the Wayne State University group, one can use an IR camera to track the full diffusion front associated with a point (in space) and step (in time) excitation. The diffusion fronts have been shown to exhibit a fractal geometry and the crossover regions between Euclidean and fractal domains can be carefully analyzed [11.14,15].

References

11.1   E.R. Eckert, R.M. Drake: *Analysis of Heat and Mass Transfer* (MacGraw-Hill, New York 1972)

11.2   H.S. Carlsraw, J.C. Jaeger: *Conduction of Heat in Solids* (Clarendon, Oxford 1959)

11.3   B. Mandelbrot: *The Fractal Geometry of Nature* (Freeman, New York 1983); *Les objets fractals* (Flammarion, Paris 1975)

11.4   R. Zallen: *The Physics of Amorphous Materials* (Wiley, New York 1983)

11.5   R. Orbach: Science **231**, 814 (1986)

11.6   R. Rammal, G. Toulouse: J. de Phys. Lett. **44**, L-13 (1983)

11.7   R. Pynn, A. Skjeltorp: *Scaling Phenomena in Disordered Systems*, NATO ASI Series (Plenum, New York 1985)

11.8   H.E. Stanley, N. Ostrowsky: *On Growth and Form*, NATO ASI Series (Nijhoff, Amsterdam 1985)

11.9   B.K. Bein, S. Krieger, J. Pelzl: Can. J. Phys. **64**, 1208 (1986)

11.10   Int. Cong. on Opt. Sci. Eng., Program on Optical Sensing and Metrology. Hamburg 19–23 September 1988

11.11   P. Cielo: J. Appl. Phys. **56**, 230 (1984)

11.12   P. Cielo, R. Lewack, D. Balageas: In Proc. of the Int. Conf. on Thermal Sensing for Diagnostics and Control, Cambridge, MA 1985

11.13   S. Alexander, R. Orbach: J. de Phys. Lett. **43**, 1–265 (1982)

11.14   B. Sapoval, M. Rosso, J.F. Gouyet: J. Phys. Lett. **46**, 149 (1985)

11.15   J.P. Hulin, E. Clément, C. Baudet, J.F. Gouyet, M. Rosso: Phys. Rev. Lett. **61**, 333 (1988)

12. Locally Resolved Magnetic Resonance in Ferromagnetic Layers and Films

J. Pelzl and U. Netzelmann

With 41 Figures

Recent widespread interest in the research and application potential of ferromagnetic films and layers owes much to the fact that many of their magnetic properties differ from those of the bulk materials. One reason for this difference is the reduced dimensionality of the interaction space which gives rise to different anisotropic fields. Furthermore, modern processing techniques enable new materials to be produced, such as sandwiches of layers with alternating magnetic properties.

Among the experimental methods used for the investigation and characterization of magnetic films and layers, that of microwave resonance absorption has proved to be a very powerful technique providing information on internal magnetic properties such as the g-value, exchange and anisotropy fields. A major drawback of the conventional magnetic resonance technique is that it only probes the whole sample. Very recently, thermal wave detection procedures were introduced to microwave resonance spectroscopy, enabling spatially dependent resonance measurements to be made. This chapter describes different experimental procedures and their application to various ferromagnetic films and layered materials.

Section 12.1 contains a survey and comprehensive comparison of the different detection techniques of microwave resonance absorption. Some basic principles of ferromagnetic resonance absorption are recapitulated in Sect. 12.2. The subsequent sections are devoted to the three main thermal wave detection techniques. Section 12.3 deals with photoacoustically detected ferromagnetic resonance (FMR). Experimental setups are described and results are reported which were obtained from magnetic tapes and films. In Sect. 12.4, lateral magnetic inhomogeneities resolved in ferromagnetic metallic films and layers by laser beam deflection are discussed. Using the same technique, the lateral dependence of the magnetostatic surface modes in yttrium iron garnet (YIG) has been investigated. In the final section, Sect. 12.5, the novel technique of photothermally modulated ferromagnetic resonance is described. This method, which uses a laser beam to thermally modulate the magnetic resonance can improve the lateral and depth resolution of the thermal wave techniques by an order of magnitude.

12.1 Survey of Microwave Resonance Detection Techniques

The resonance absorption of microwaves provides a means of investigating the static and dynamic properties of the electron-spin system in solid or liquid matter. Usually, the sample under investigation is placed in a cavity, and the microwave frequency

ω is kept constant, while an external magnetic field B_0 is swept until the Zeeman splitting coincides with the photon energy of the microwaves. In the simplest case, tuning to resonance establishes the condition $h\omega = \gamma B_i$ where γ is the gyromagnetic ratio and B_i the local magnetic field.

The standard microwave techniques use microwave frequencies in the range between 9 and 35 GHz, corresponding to a magnetic energy of the orbital or spin magnetic moment in an internal magnetic field of a few tesla. The resonance absorption in a paramagnetic sample, commonly referred to as Electron Paramagnetic Resonance (EPR) or Electron Spin Resonance (ESR) (if there exists no orbital moment contribution), has been used successfully over the last forty years to investigate the electronic structure and hyperfine interactions of isolated paramagnetic species such as ions, radicals or lattice defects [12.1].

In magnetically ordered materials, large static and dynamic internal fields contribute to the Zeeman splitting of the electron levels. Therefore, the observation of the resonance absorption in ferromagnetic, antiferromagnetic or ferrimagnetic samples also provides information on the static internal magnetic fields [12.2].

The ferromagnetic resonance effect was first discovered in 3d transition metals [12.3]. Nowadays, modern fields of application of FMR include the inspection of ferrites and garnets used in microwave devices [12.4,5] and the investigation of magnetic properties of thin and ultrathin films of transition metal alloys [12.6].

The detection of the heat released in the course of resonance absorption as a means of studying the nuclear magnetic resonance transition was proposed by *Gorter* in 1936 [12.7]. At the magnetic resonance condition, the sample is heated by the absorbed high-frequency radiation and as a result its temperature increases. The dc temperature variation which follows the conventional resonance line is measured by a sensitive thermometer attached to the sample. In the meantime, bolometric techniques have been successfully applied to detect EPR at low temperatures [12.8] and to improve the sensitivity in broad line FMR experiments [12.9].

Photothermal methods for FMR and EPR detection have only been in use for ten years. In contrast to the bolometric technique, photothermal measurements are based on the time-dependent modulation of the thermal response. Photothermal methods rely on the generation and detection of thermal waves, providing a means of local and depth-dependent magnetic resonance spectroscopy. Different experimental schemes have been developed for the detection of thermal waves in microwave absorption experiments. The techniques already used in the domain of FMR are shown in Fig. 12.1.

The most frequently used electronic detection method is the microwave cavity measurement technique (Fig. 12.1a). The sample is placed in a resonance cavity and the variation of the loaded Q-factor of this cavity is the figure of merit that is deduced from the microwave power reflected by the loaded cavity when the magnetic field is tuned through the resonance [12.10].

In the photoacoustic method of detection (Fig. 12.1b), the sample is positioned in the cavity inside an acoustically sealed cell containing a nonabsorbing gas. By chopping the microwave power absorbed by the sample, the periodic heating of the sample causes pressure fluctuations in the surrounding gas which are measured by a microphone. The first experiments to show the possibility of photoacoustic detection

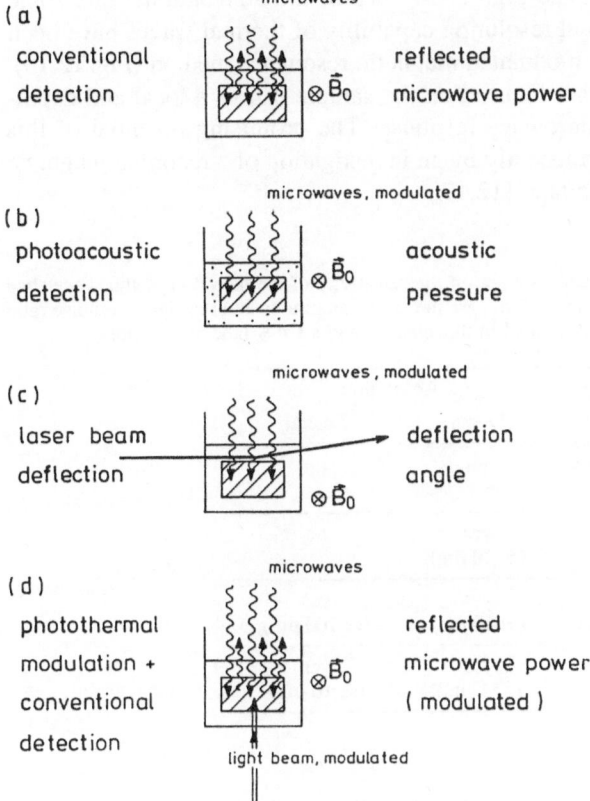

(a)
conventional
detection

microwaves

reflected
microwave power

$\otimes \vec{B}_0$

(b)
photoacoustic
detection

microwaves, modulated

acoustic
pressure

$\otimes \vec{B}_0$

(c)
laser beam
deflection

microwaves, modulated

deflection
angle

$\otimes \vec{B}_0$

(d)
photothermal
modulation +
conventional
detection

microwaves

reflected
microwave power
(modulated)

$\otimes \vec{B}_0$

light beam, modulated

Fig. 12.1a–d Schematic representation of the conventional and thermal-wave-based FMR detection techniques

of microwave resonance absorption in solids were performed by two Brasilian groups on ferromagnetic materials [12.11,12]. The photoacoustic effect due to paramagnetic resonance absorption was first observed in DPPH (α, α'-diphenyl-β-picryl-hydrazyl) [12.13]. The most promising aspect of the thermal wave technique, the capacity for magnetic resonance depth profiling, was first demonstrated on layered magnetic tapes [12.14]. Results from this latter work will be presented in Sect. 12.3.

One major drawback of microwave spectroscopy is the lack of any focusing ability like that in optical radiation. This means that local resolution in a FMR or EPR measurement has to be provided by the detection channel. A local detection method based on thermal waves is photothermal laser beam deflection, also called the mirage effect [12.15]. A probe laser beam passing in grazing incidence near the surface of the microwave heated sample is deflected in the thermal density gradient of the air at the interface (Fig. 12.1c). The deflection of the laser beam, which is oscillating at the modulation frequency of the microwave intensity, is detected by a position sensitive sensor. The mirage effect, invented and most frequently used nowadays in conjunction with optical sources, was first applied to the FMR of Ni films in order to obtain a locally resolved resonance pattern [12.16].

Decreasing sensitivity with increasing modulation frequency and a finite geometrical extension of the probe beam constitute the natural limits for the spatial resolution of the mirage effect. The high sensitivity of the conventional electronic detection technique and the spatial resolution capability of thermal waves have been combined in the photothermally modulated magnetic resonance method (Fig. 12.1d). An intensity modulated laser beam focused on the sample causes a local and depth-dependent modulation of the microwave response. The promising potential of this technique has been demonstrated recently by an investigation of a recorded magnetic signal trace on a CrO_2 magnetic tape [12.17].

Table 12.1. Comparison of the spatial resolution and of the sensitivity of the detection methods sketched in Fig. 12.1. The values given in the second column are the signal amplitude and the signal-to-noise ratio measured with the experimental setups described in this chapter and a CrO_2 tape as a sample

Method of Detection of FMR	Sensitivity	Resolution Depth	Resolution Lateral
Conventional detection	high	no	no
Photoacoustic detection	low	yes ($\approx 10\,\mu$m)	no
Detection by laser beam deflection	low	yes ($\approx 10\,\mu$m)	yes ($\approx 100\,\mu$m)
Photothermally modulated FMR	medium	yes ($\approx 1\,\mu$m)	yes ($\approx 10\,\mu$m)

In Table 12.1, some figures of merit of the detection methods shown in Fig. 12.1 are compared. The signal-to-noise ratios are those measured from a CrO_2 tape. Conventional electronic detection provides the best signal-to-noise ratio at ambient temperatures. At very low temperatures, the thermal wave method gains up to three orders of magnitude and can become as sensitive as conventional detection at about 4.2 K [12.18,19]. Whereas photoacoustic and mirage effect measurements at liquid helium temperatures face severe experimental problems, the photothermally modulated resonance can be handled as easily as conventional electronic recording.

Spatial resolution is achievable only with thermal wave techniques. The depth resolution is roughly of the order of the thermal diffusion length: $\mu = \sqrt{2\alpha/\omega_M}$, where α is the thermal diffusivity and ω_M the modulation frequency. An increase in the modulation frequency will improve the depth resolution. The upper limit of the modulation frequency is determined by the microphone response function in the case of the photoacoustic effect, and by the detection sensitivity (the signal amplitude decreases with $1/\omega_M$) in the case of the mirage effect. The limitations of the photothermally modulated technique are given by the dynamic range and sensitivity of the electronic detection channel, which can tolerate modulation frequencies up to a few megahertz for commercial spectrometers. Lateral resolutions in both the mirage effect sensing and the photothermally modulated FMR are determined by the geometrical size of the optical beam on the sample. As the mirage effect works with a laser beam in grazing incidence, whereas a focused laser beam is used for the photothermal modulation, the lateral resolutions of both techniques may differ by an order of magnitude, as indicated in Table 12.1.

The thermal wave methods listed in Table 12.1 are all noncontact techniques, but there have also been efforts in EPR experiments to detect the thermal waves with a sensor in contact with the sample. In this way, the temperature rise due to intensity modulated microwave absorption has been measured by a pyroelectric detector [12.20]. In other experiments, piezoelectric transducers have been used to measure the thermoelastic stress [12.21] or the short time ultrasonic signal [12.22] induced by the thermal wave. Although in the latter experiment a time resolution corresponding to a modulation frequency of 70 kHz could be achieved, these complementary techniques have not been developed further. For the sake of completeness, we should refer to techniques other than those using thermal waves which have been developed to perform spatially resolved microwave resonance absorption measurements. Firstly, we should mention the field gradient technique [12.23]. As in the case of NMR tomography, the resonance condition is restricted to a certain volume by a supplementary magnetic field gradient. The spatial resolution achieved is actually of the order of mm^2. A very sophisticated technique, from the experimental point of view, is represented by a combination of ferromagnetic resonance with Brillouin scattering [12.24]. This technique is best applied to samples with a good optical quality.

12.2 Basic Theory of Ferromagnetic Resonance

12.2.1 Uniform Mode Resonance

The ferromagnetic resonance (FMR) phenomenon is basically a collective excitation of the elementary magnetic moments of a ferro- or ferrimagnetic sample by a high-frequency magnetic field. The theoretical description is usually based on the equation of motion for the magnetization M of the sample inside an effective magnetic field B, given here in the Landau-Lifshitz form [12.25],

$$\dot{M} = -\gamma(M \times B) - \alpha\frac{\gamma}{M}M \times (M \times B) , \qquad (12.2.1)$$

where γ is the gyromagnetic ratio and α the damping parameter; $(\alpha\omega)^{-1}$ is the relaxation time, and ω is the microwave frequency. The first term on the right hand side describes a precession movement of the magnetization vector, while the second term phenomenologically introduces a damping force trying to re-align M into an equilibrium position. If we restrict ourselves to the linear regime, where the magnitude of M is kept constant and the precession angle is small, it is useful to express (12.2.1) in spherical coordinates (Fig. 12.2), where we obtain the following set of coupled equations for the direction (φ, ϑ) of the magnetization [12.25]:

$$\dot{\varphi}\sin\vartheta = -\gamma(B_\vartheta - \alpha B_\varphi) , \qquad (12.2.2)$$

$$\dot{\vartheta} = \gamma(B_\varphi + \alpha B_\vartheta) . \qquad (12.2.3)$$

The terms B_φ and B_ϑ are effective magnetic field components in the coordinate

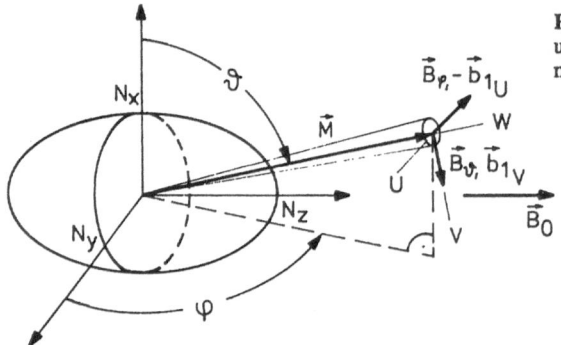

Fig. 12.2 Sketch of the coordinate system used for the derivation of the FMR resonance frequencies

axes of the magnetization system (u, v, w) defined by Fig. 12.2 and given by [12.25]

$$B_{\varphi} = \frac{-\mu_0 F_{\varphi}}{M \sin \vartheta} - b_{1u} e^{i\omega t} , \qquad (12.2.4)$$

$$B_{\vartheta} = \frac{-\mu_0 F_{\vartheta}}{M} + b_{1v} e^{i\omega t} . \qquad (12.2.5)$$

The terms F_{φ} and F_{ϑ} are the first derivatives of the free energy density of the sample with respect to φ and ϑ, respectively; b_{1u} and b_{1v} are the components of the driving high-frequency magnetic field of the microwave in the magnetization system. An equilibrium position of M is defined by the condition $F_{\vartheta} = F_{\varphi} = 0$. Expanding the free energy density around the equilibrium position (φ_0, ϑ_0) for small oscillations $\delta\varphi$ and $\delta\vartheta$,

$$F_{\varphi} = F_{\varphi\varphi}\delta\varphi + F_{\varphi\vartheta}\delta\vartheta , \qquad (12.2.6)$$

$$F_{\vartheta} = F_{\vartheta\varphi}\delta\varphi + F_{\vartheta\vartheta}\delta\vartheta , \qquad (12.2.7)$$

and using (12.2.2–5), one finally obtains [12.25] an inhomogeneous system of linear equations whose secular determinant gives the resonant frequency ω_{res},

$$\omega_{\text{res}} = \frac{\gamma(1 + \alpha^2)^{1/2}}{M \sin \vartheta_0} \mu_0 (F_{\vartheta\vartheta} F_{\varphi\varphi} - F_{\vartheta\varphi}^2)^{1/2} , \qquad (12.2.8)$$

and the linewidth $\Delta\omega$

$$\Delta\omega = \frac{\gamma\alpha}{M} \mu_0 \left(F_{\vartheta\vartheta} + \frac{F_{\varphi\varphi}}{\sin^2 \vartheta_0} \right) . \qquad (12.2.9)$$

The terms $F_{\varphi\varphi}, F_{\vartheta\vartheta}$ and $F_{\vartheta\varphi}$ are the second derivatives of the free energy density at the equilibrium point, describing the "stiffness" of the direction of magnetization. The inhomogeneous system can be solved for the oscillations $\delta\varphi$ and $\delta\vartheta$ and one obtains the relation between the transverse components of the magnetization,

$$m_u = -M\delta\varphi \sin \vartheta_0 , \qquad (12.2.10)$$

$$m_v = M\delta\vartheta , \qquad (12.2.11)$$

and the high-frequency field components b_{1u} and b_{1v}, which is expressed by the tensor of susceptibility χ in the system (u, v, w). The element χ_{uu}, for example, which is a relevant quantity for the common linear polarized high-frequency driving field b_1, is then given by

$$\chi_{uu} = \frac{\gamma^2(1+\alpha^2)\mu_0 F_{\vartheta\vartheta} + i\alpha\omega\gamma M}{\omega_{\mathrm{res}}^2 - \omega^2 + i\omega\Delta\omega} = \chi' - i\chi'' , \tag{12.2.12}$$

where

$$\chi' = \frac{(\omega_{\mathrm{res}}^2 - \omega^2)\gamma^2(1+\alpha^2)\mu_0 F_{\vartheta\vartheta} + \alpha\gamma M\omega^2\Delta\omega}{(\omega_{\mathrm{res}}^2 - \omega^2)^2 + \omega^2\Delta\omega^2} \tag{12.2.13}$$

and

$$\chi'' = \frac{\omega\Delta\omega\gamma^2(1+\alpha^2)\mu_0 F_{\vartheta\vartheta} - \alpha\gamma M\omega(\omega_{\mathrm{res}}^2 - \omega^2)}{(\omega_{\mathrm{res}}^2 - \omega^2)^2 + \omega^2\Delta\omega^2} . \tag{12.2.14}$$

The susceptibility has been separated as usual into its real and imaginary parts, where χ' is related to the dispersion signal and χ'' to the absorption signal. The typical shapes of the FMR absorption and dispersion signals are shown in Fig. 12.3.

In contrast to EPR, in the FMR experiment the energy of the magnetic dipoles themselves contributes significantly to the total free energy. Therefore, a term related to the sample shape has to be considered for all calculations of resonance frequencies. For the simple case of a homogeneously magnetized sample of ellipsoidal shape, the free energy comprises two terms: the Zeeman term and the demagnetization term [12.25]

$$F = \frac{1}{\mu_0}(-M \cdot B_0 + \tfrac{1}{2}M\underline{\underline{N}}M) . \tag{12.2.15}$$

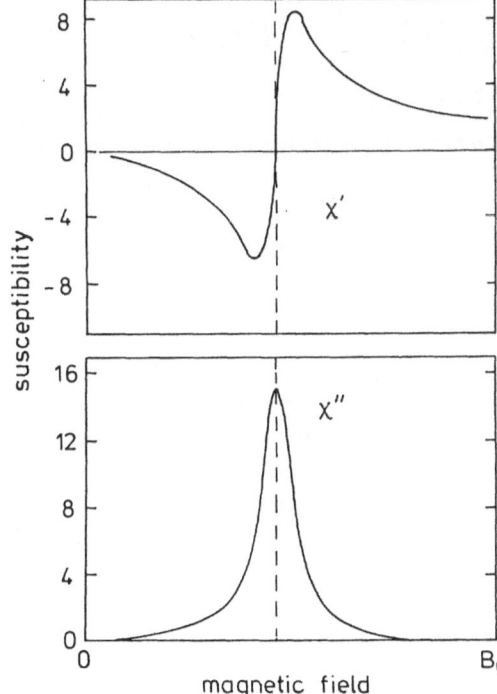

Fig. 12.3 Typical shape of the FMR dispersion (*upper*) and absorption (*lower*) signal

The demagnetization tensor $\underline{\underline{N}}$ can be represented by three diagonal elements N_x, N_y and $N_z (N_x + N_y + N_z = 1)$ for a sample oriented as shown in Fig. 12.2. If the equilibrium position of M is parallel to the external field B_0 ($\vartheta_0 = \pi/2$) and $\alpha \ll 1$, from (12.2.13) and (12.2.8) one obtains *Kittel*'s famous resonance formula [12.26]:

$$\omega_{\text{res}}^2 = \gamma^2 [B_0 + M(N_x - N_z)][B_0 + M(N_y - N_z)] . \qquad (12.2.16)$$

The FMR experiment is usually performed at a fixed microwave frequency ($\omega = \omega_{\text{res}}$) as a function of the external magnetic field B_0. The external field B_0 that meets the resonance condition (12.2.16) distinctly depends on the demagnetizing fields. The effect of sample shape on the position of the FMR line for some important sample shapes is shown schematically in Fig. 12.4.

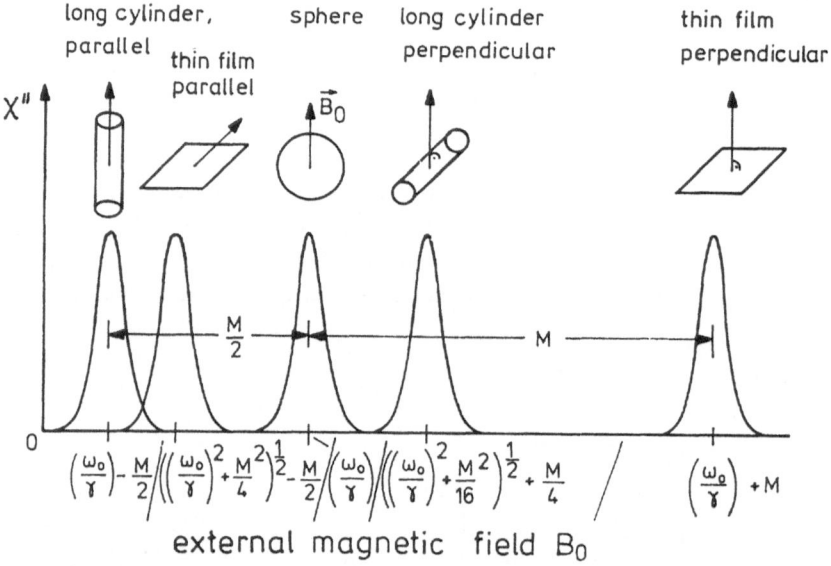

Fig. 12.4. Resonance line positions of samples with different shapes and orientations with respect to the external magnetic field

In addition to the Zeeman and the dipolar energy of (12.2.15), there are other possible contributions to the free energy. The magnetocrystalline energy, represented by the uniaxial anisotropy constant K_u or cubic anisotropy constants K_1 and K_2, influences the line positions as a function of the crystal orientation and can be measured by FMR. Surface anisotropies can be present. A magnetoelastic energy term introduces the coupling of mechanical strain to the magnetization. At lower magnetic fields or microwave frequencies, the influence of the domain structure of the sample has to be considered. If some shape or crystalline anisotropy is present to produce a suitable directional stiffness of M, FMR is possible even without any external magnetic field (natural FMR [12.26]).

12.2.2 Magnetostatic Modes and Spin Waves

In the discussion in the preceding section the magnetization of the sample has been assumed to be homogeneous over the sample. In the case of nonuniform excitations, we have to allow for a magnetization dependent on both space and time:

$$M = M(r, t) .$$ (12.2.17)

The magnetization is often expanded in the form of a Fourier series for plane waves in terms of a wave vector k:

$$M(r, t) = \sum_k M_k(t) \exp(i k \cdot r) .$$ (12.2.18)

As the magnetization is no longer parallel inside the sample, internal magnetic dipolar interactions and exchange interactions have to be introduced. This can be accomplished by using the equation of motion from (12.2.1) extended by an exchange term and by simultaneously solving Maxwell's equations for the whole space. Depending on the relation of the size of the sample to the wavelength $2\pi/k$ and the strengths of the interactions mentioned above, one can distinguish between three main regimes in a dispersion diagram, as shown for a typical experimental situation in Fig. 12.5.

Beside the uniform precession found for $k = 0$, in the range up to $k \approx 2 \times 10^2 \, \mathrm{m^{-1}}$ there is a region with two branches of electromagnetic propagation in samples larger than or comparable with the wavelength, where $\omega \sim k$ holds. The wavelength is of the same order as the wavelength in free space and interference phenomena such as body resonances [12.4] are likely to occur.

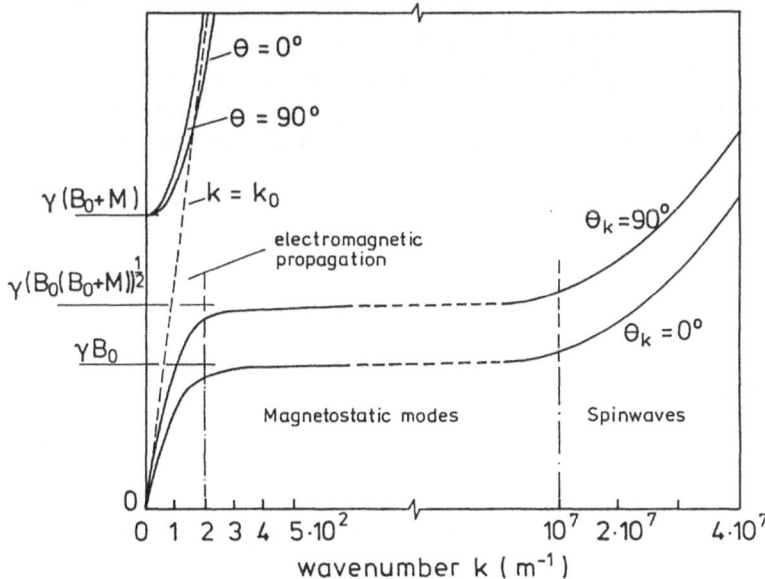

Fig. 12.5. Dispersion diagram for wave propagation in a typical infinite ferromagnetic medium in an external magnetic field. Θ_k is the angle between the external field and the direction of propagation. (After [12.4])

In the range of $k \approx 2 \times 10^2 \, \mathrm{m}^{-1}$ to $k \approx 10^7 \, \mathrm{m}^{-1}$, we are in the magnetostatic region, characterized by a flat dispersion. The retardation or propagation effects can be neglected for sufficiently small sample sizes and one uses Maxwell's equations in their magnetostatic approximation,

$$\nabla \times \boldsymbol{H} = 0 \,, \tag{12.2.19}$$

$$\nabla \cdot \boldsymbol{B} = \nabla \cdot (\boldsymbol{B_0} + \boldsymbol{M}) = 0 \,, \tag{12.2.20}$$

including the electromagnetic boundary conditions at the sample surface, as well as the equation of motion (12.2.1) to calculate characteristic oscillations of the magnetization of the sample. These magnetostatic modes or Walker modes were first calculated [12.27] for spheres and were observed experimentally as additional sharp resonance peaks beside the uniform resonance mode in disks and rods [12.28] of ferrite materials.

For thin films if infinite extension, a first theory of magnetostatic wave propagation was set up by *Damon* and *Eshbach* [12.29]. They predicted the existence of several types of propagating magnetostatic waves, namely the magnetostatic forward volume waves (MSFVW) and backward volume waves (MSBVW), and the magnetostatic surface waves (MSSW) (Fig. 12.6, left). Each type of wave requires a certain orientation of the external magnetic field and has its characteristic dispersion relation (Fig. 12.6, right). In thin films of finite size, magnetostatic modes occur according to the additional selection caused by the boundary conditions. Since the first experimental observations [12.30], magnetostatic waves in thin films have been attracting considerable interest because of applications in microwave devices such as delay lines [12.31,32].

When proceeding to high orders of magnetostatic modes and increasing values of k, the surface boundary conditions of the sample become more and more negligible and the modes acquire more and more of the character of plane standing waves. When k exceeds a value of about $10^7 \, \mathrm{m}^{-1}$, one is entering the region of spin-wave modes or magnons (Fig. 12.5). As neighboring spins become increasingly noncollinear, the

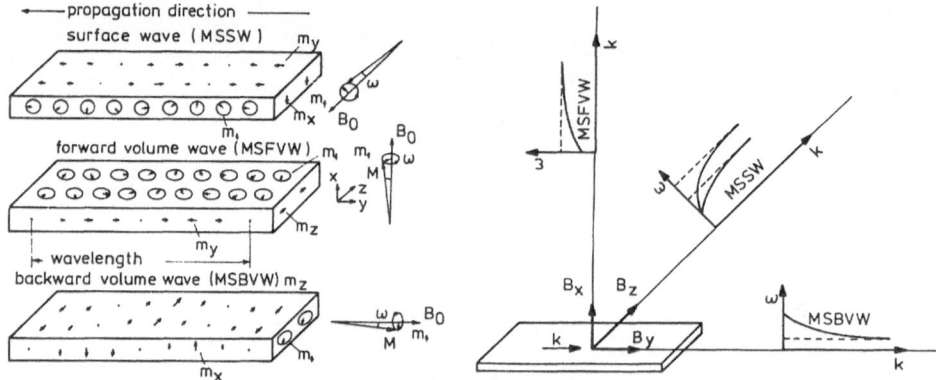

Fig. 12.6 Characteristic oscillation modes of magnetostatic forward volume waves (MSFVW), backward volume waves (MSBVW) and surface waves (MSSW) in an infinite, plane medium (*upper*) and their typical dispersion behavior (*lower*). (After [12.31,32])

effect of exchange interaction, which has been neglected up to now, has to be introduced into the basic equation [12.1] by an additional term in the form

$$\gamma D(M \times \nabla^2 M) \,, \tag{12.2.21}$$

where D is the exchange stiffness constant. The equation of motion can be separated into a linear and a nonlinear part. The linear part will give the dispersion relation of the spin waves:

$$\begin{aligned}\omega_k^2 =& \gamma^2 (B_0 - N_z M + Dk^2) \\ & \times (B_0 - N_z M + Dk^2 + M \sin^2 \Theta_k) \,.\end{aligned} \tag{12.2.22}$$

Here, Θ_k is the angle between the propagation of the spin wave and the direction of the effective magnetic field in the sample. The dispersion is characterized by a typical k^2 dependence. The nonlinear part in the equation of motion will describe multimagnon processes such as Suhl's first and second spin-wave instabilites [12.33].

Quadratic dispersion has been observed experimentally in thin films with a thickness of a few hundred nanometers, where standing spin waves occur as additional resonance peaks beside the uniform mode [12.34] (Fig. 12.7). Due to the boundary conditions, only discrete values of the wavevector k are admitted: $k = n\pi/d$, where d is the thickness of the film. The line shifts of the spin wave peaks then usually obey an n^2 law, where n is the order of the mode, providing an excellent means of obtaining the exchange constant.

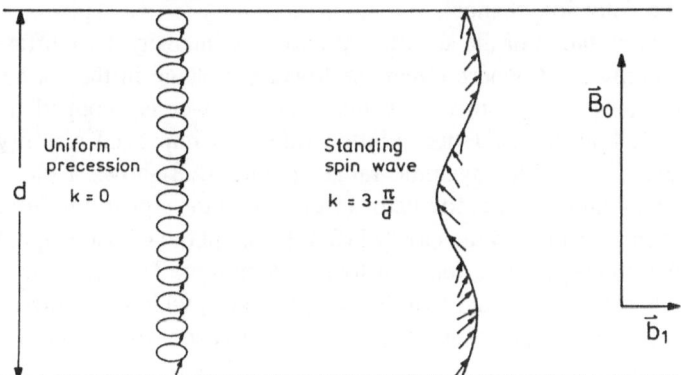

Fig. 12.7. Standing spin wave and uniform mode in a thin film. (After [12.4])

12.3 Photoacoustically Detected Ferromagnetic Resonance

12.3.1 Experimental Setups and Procedures

The photoacoustic (PA) effect relies on the measurement of the modulated heat flow from the sample into the surrounding gas. In the gas, the heat is converted to pressure changes, which are detected by means of a microphone. The photoacoustic effect has been explored and exploited predominantly in connection with light absorption

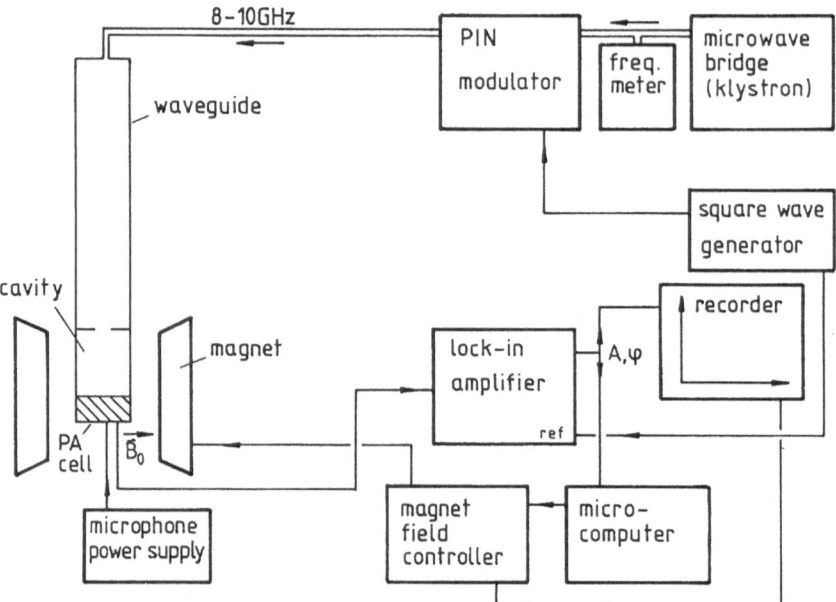

Fig. 12.8. Block diagram of an EPR spectrometer modified for photoacoustic detection

as a heating source [12.35,36]. The utilization of the PA technique in FMR or EPR requires some modifications of a conventional resonance spectrometer along the excitation path and at the detection channel.

In order to provide a modulation of the absorbed microwave intensity, two different procedures are commonly used, where either the klystron voltage in the source is modulated [12.37,38,39,40] or the outcoming microwave power is chopped by a *pin*-diode modulator [12.41,42]. The latter solution offers an improved stability of the microwave system and a wider dynamic range. Figure 12.8 shows a block diagram of an arrangement using the *pin* modulator inserted into the signal line from the klystron of the microwave bridge to the cavity [12.43]. The photoacoustic signal detected by a gas-coupled microphone is fed to a lock-in amplifier. The phase sensitive amplifier obtains its reference signal from a sine generator, which also drives the modulator. The amplitude and the phase angle of the photoacoustic signal are measured either as a function of the external magnetic field at a fixed modulation frequency or as a function of the modulation frequency at a fixed magnetic field.

The photoacoustic cell has to meet competing requirements in order to provide a maximum signal. The microwave intensity at the sample has to be large and the gas volume with the microphone in contact has to be as small as possible, moreover, the whole arrangement has to fit commercial spectrometers. Two types of PA cells have been designed which are either compatible with the standard TE_{102} cavity or which can be used as the shorted end of a waveguide. Figure 12.9 shows two designs which have already been used in microwave resonance experiments [12.37,41,43].

The quartz tube cell (Fig. 12.9, left), originally developed for powdered samples, is used in the standard TE_{102} resonator and consists of an ordinary quartz tube, which contains the sample, with an electret microphone in a Teflon housing on

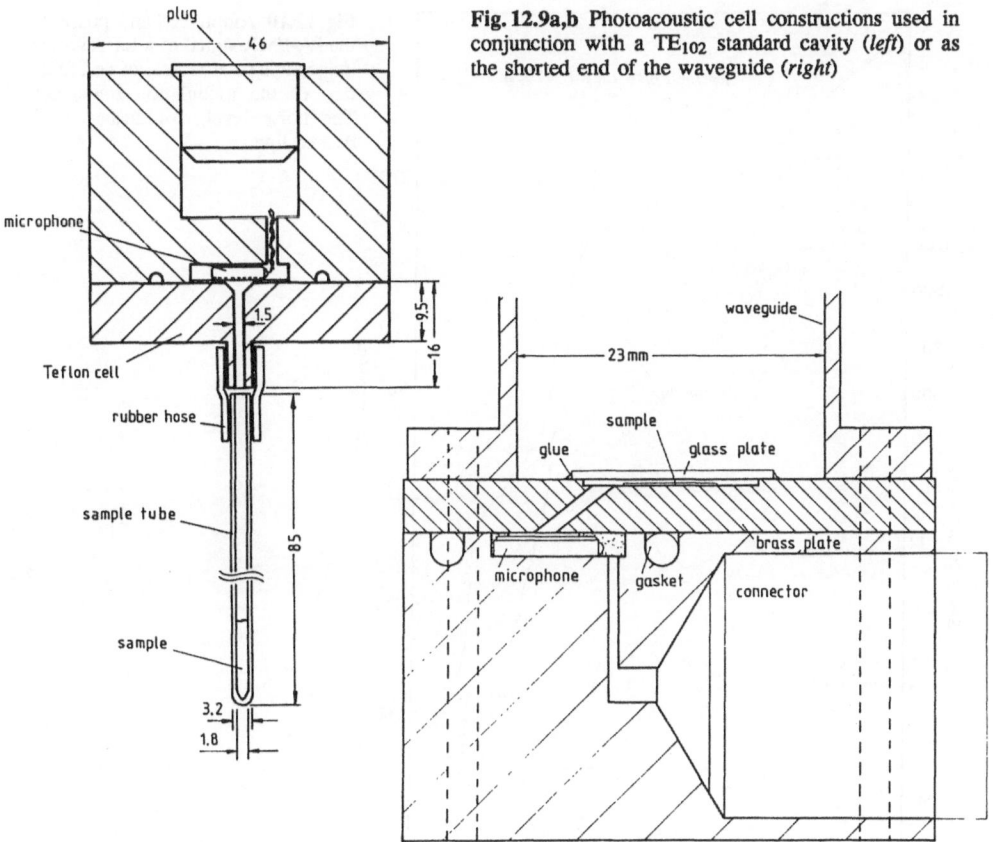

Fig. 12.9a,b Photoacoustic cell constructions used in conjunction with a TE$_{102}$ standard cavity (*left*) or as the shorted end of the waveguide (*right*)

top. This cell takes advantage of the high Q-value of the standard cavities and requires minimal modification of the microwave circuit of the spectrometer. Due to the distance between the sample and the microphone outside the resonator, the cell volume is fairly high, lowering the first acoustic Helmholtz resonance down to about 1100 Hz and limiting the range of usable modulation frequencies.

The waveguide cell (Fig. 12.9, right) has been optimized for a minimum gas volume and for the study of thin flat specimens. The cell has been machined as a small cavity of 0.5 mm depth in a brass plate, which forms one wall of a TE$_{10n}$-type rectangular resonator constructed from X-band waveguides and a coupling iris. The sample is positioned at a point of maximum intensity of the microwave magnetic field. The PA cell wall dividing it from the interior of the microwave resonator is formed by a plate made out of glass or mica. The sample chamber is acoustically connected by a duct to the microphone at the other side of the brass plate. The residual volume has been reduced by a factor of 4.5 compared with the quartz tube cell. Thus the PA yield of the cell is increased and the range of modulation frequencies is extended. The orientation of the sample, with respect to static and high-frequency magnetic fields, is well defined. The Q-value of the microwave resonator, however, is usually a factor 2 or 3 below that of a commercial resonator. For very small PA signals, the FMR signal contribution has to compete with background

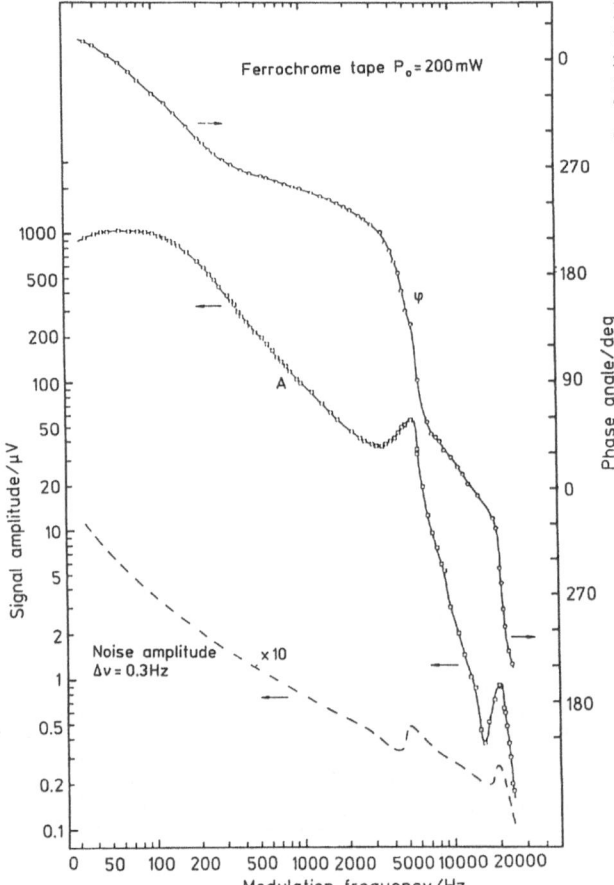

Fig. 12.10 Amplitude and phase of the PA-FMR signal of a ferrochrome magnetic tape at resonance as a function of the modulation frequency. The noise level is indicated as a dashed line

signals due to the skin effect, microwave absorption in the brass, and dielectric absorption in the cell cover plate. These background signals are independent of the magnetic field and can be subtracted vectorially to obtain the pure PA-FMR signal. In both cells, an electret microphone (Knowles BT 1759) with integrated amplifier ist used.

For the case of the waveguide cell, we briefly discuss the variation of the PA-FMR signal with the modulation frequency. Figure 12.10 shows a typical frequency dependence of the PA-FMR signal at the absorption maximum. The sample was a ferrochrome magnetic tape of size 3.8 mm × 3.8 mm, and the microwave power was 200 mW.

The frequency dependence of the thermal diffusion processes in the sample and the gas is obscured by the frequency response of the microphone below 100 Hz and by acoustic resonances in the PA cell above 2000 Hz. Due to the small cell volume, the first Helmholtz resonance [12.44] appears at a high frequency of 5300 Hz. Although in the range 100–2000 Hz the signal decreases as $\omega_M^{-1.2}$, due to thermal wave attenuation in the solid and in the gas, the signal-to-noise ratio is quite constant in the whole frequency range up to the first Helmholtz resonance, as can be deduced by the noise level shown as a dashed line in Fig. 12.10. Amplifier noise below 1 kHz and background noise and vibrations from the laboratory are the predominant contributions to the noise signal, in addition to the ultimate limitation caused by the Brownian motion statistical fluctuations in the gas.

Another important experimental parameter is the linearity of the PA-FMR response with respect to the incident microwave radiation. In contrast to conventional FMR detection techniques, where the FMR signal varies as $P_0^{1/2}$, with the incident microwave power P_0, the thermal response should be proportional to the incident power. This behavior has been verified at the fixed modulation frequency of 320 Hz for the ferrochrome magnetic tape sample, which was the same as that used in the first experiment (Fig. 12.11). For more than three orders of magnitude of the incident power, the PA response turns out to be proportional to P_0. The signal phase is constant. This linear behavior in connection with the utilization of a heterodyne lock-in amplifier, which is not sensitive to higher harmonics of the modulation frequency, justifies the application of a square-wave modulated microwave input instead of the sinusoidal one assumed in most theoretical calculations [12.45].

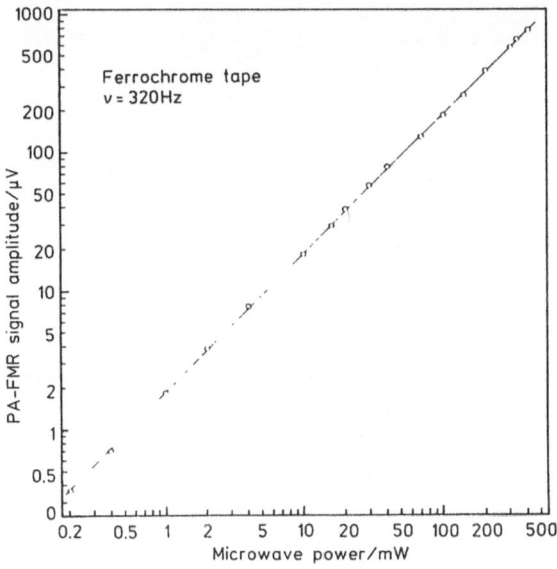

Fig. 12.11 PA-FMR signal amplitude of a ferrochrome magnetic tape at resonance as a function of the incident microwave power

The correlation of conventionally and thermally detected FMR when using high-Q microwave resonators is not obvious, but has been studied theoretically and experimentally [12.46]. We can define the FMR signal as the absorbed microwave power in the sample normalized with respect to the incident microwave intensity at the sample position. In the conventional resonator perturbation method, the FMR signal defined above is proportional to a change of the inverse unloaded Q-value of the resonator, $\Delta(1/Q_u)$, which is measured exactly by a determination of the change of the voltage standing wave ratio (VSWR) [12.14]. This measurement corrects implicitly for the effect of a varying microwave intensity in the resonator due to varying loading of the resonator when passing the FMR resonance line.

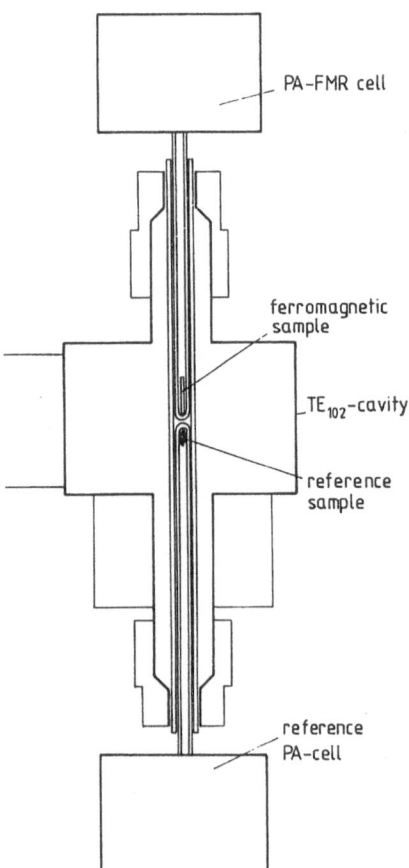

Fig. 12.12 Experimental arrangement for measuring simultaneously the PA-FMR signal and a reference PA signal proportional to the microwave intensity in a standard TE_{102} cavity

The thermally detected FMR, however, only measures the absorbed microwave power in the sample. A normalization with respect to the microwave intensity has to be performed in addition. *Cesar* [12.47] obtained this normalization signal by using a second reference PA cell in the resonator, measuring wall current losses in a thin aluminum foil in the resonator wall. A similar approach used in connection with the quartz tube cell described above is shown in Fig. 12.12. A reference quartz tube cell with a diamagnetic reference sample, e.g. a powdered resin, is placed in the resonator simultaneously with the PA-FMR cell containing the ferromagnetic sample. Figure 12.13 shows the PA-FMR signal of a strongly absorbing metallic sample and the reference PA signal. The varying microwave intensity in the resonator is clearly visible as a dip in the reference signal at the resonance position. The normalized signal (dashed curve) can be shown to correlate with the conventionally measured signal [12.46].

The normalization procedure in thermally detected FMR in cavities can very often be omitted if the condition $\Delta(1/Q_u) \ll 1/Q_{u_0}$ is fulfilled, where Q_{u_0} is the unloaded Q-value of the resonator without FMR absorption. This can be accomplished by using resonators with moderate Q-values and small filling factors, at the expense of a higher microwave input.

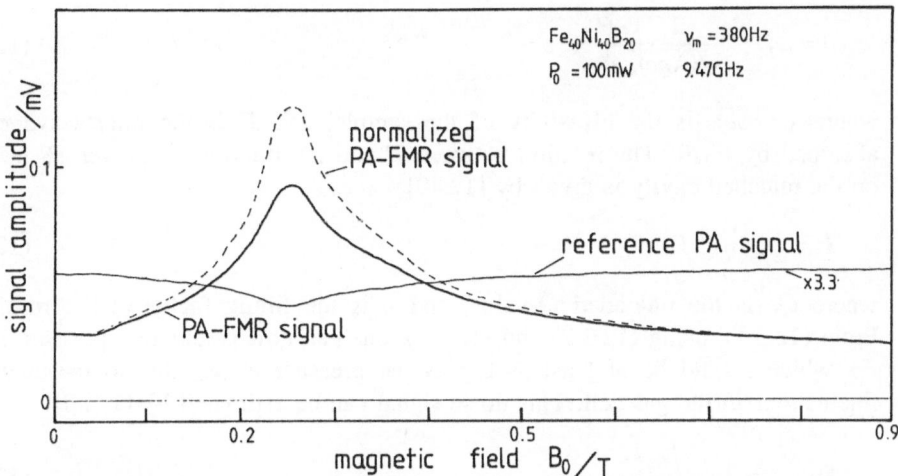

Fig. 12.13 PA-FMR spectrum of a metallic $Fe_{40}Ni_{40}B_{20}$ amorphous ribbon, including the reference signal measured simultaneously in the arrangement shown in Fig. 12.12. The normalized PA-FMR spectrum is shown as a dashed line

At the end of this section, we will briefly discuss the detection sensitivity of PA-FMR compared with conventional detection [12.41]. The minimal temperature oscillation detectable experimentally may be calculated for the case of the PA-FMR signal of the ferrochrome magnetic tape measured with the waveguide cell at a modulation frequency ν_M of 300 Hz. The noise level at this frequency can be taken from Fig. 12.10 and is 0.15 μV for a bandwidth of 1 Hz. Using a microphone sensitivity of 10 mV/Pa, the pressure amplitude in the gas cell can be calculated. For a one-dimensional heat flow in a thermally thick gas cell (thickness of the gas layer greater than the thermal diffusion length) with a homogeneous pressure distribution, the relation between the surface temperature oscillation ϑ and pressure oscillation δp is given by [12.48]

$$\vartheta = \frac{1}{(\kappa - 1)} \frac{V_g}{A_s} \sqrt{\omega_M} \frac{1}{\sqrt{(\lambda \varrho c)_g}} \delta p \,, \tag{12.3.1}$$

where κ is the adiabatic exponent of the gas, V_g the PA cell volume, A_s the area of the surface of the sample, $\sqrt{(\lambda \varrho c)_g}$ the effusivity of the gas, λ the thermal conductivity, ϱ the density, and c the specific heat capacity. Using $\kappa = 1.4$, $A_s = (3.8 \text{ mm})^2$, $V_g = 63 \text{ mm}^3$ and $(\lambda \varrho c)_g^{1/2} = 5.24 \text{ Ws}^{1/2}\text{K}^{-1}\text{m}^{-2}$, a minimum detectable temperature oscillation of $\vartheta = 4 \times 10^{-7}$ K is obtained.

To compare the ultimate sensitivities of the PA-FMR and conventional FMR, we will calculate the minimum detectable susceptibility χ''_{min}, for a given filling factor of the microwave resonator for both detection schemes. We will consider a surface microwave absorption in a thermally thick sample, assuming one-dimensional heat flow. This will apply for the case of a thin ferromagnetic film on a thick substrate or for a thick ferromagnetic metal sample. If the effusivity of the gas is much smaller than the effusivity of the sample, the temperature oscillation ϑ at the gas–sample interface is given by [12.48]

$$\vartheta = \omega_M^{-1/2} \frac{1}{\sqrt{(\lambda \varrho c)_s}} \frac{P}{A_s} \, , \tag{12.3.2}$$

where $\sqrt{(\lambda \varrho c)_s}$ is the effusivity of the sample, and P is the microwave power absorbed by FMR. The relation between P and the microwave power P_0 incident on the matched cavity is given by [12.49]

$$P = \tfrac{1}{2} Q_u \eta \chi'' P_0 \, , \tag{12.3.3}$$

where Q_u is the unloaded Q-value, and η is the filling factor of the resonator. From (12.3.1), using (12.3.2) and (12.3.3), one can now calculate a pressure signal δp, which should be at least as big as the pressure δp_{noise} due to the statistical fluctuations in the gas cell. This noise signal can be estimated by [12.50]

$$\delta p_{noise} = 2 v_g \sqrt{\frac{\varrho_g k_B T \Delta \nu}{\omega_{ac} Q_{ac} V_g}} \, , \tag{12.3.4}$$

where v_g is the velocity of sound, k_B Boltzmann's constant, $\Delta \nu$ the noise bandwidth, and ω_{ac}, Q_{ac} the resonant frequency and quality factor of the principal acoustic resonance in the gas cell. For the waveguide PA cell, a value of $Q_{ac} \approx 1$ and $\omega_{ac} \approx 2\pi \times 5300 \, \text{s}^{-1}$ will be used to estimate a noise level of $1 \, \mu\text{Pa}/\text{Hz}^{-1/2}$.

By equating δp_{noise} with δp from (12.3.1), one obtains the minimum susceptibility detectable by PA, $\chi''_{min,PA}$:

$$\chi''_{min,PA} = \frac{\omega_M}{\kappa - 1} \sqrt{V_g} \sqrt{\frac{(\lambda \varrho c)_s}{(\lambda \varrho c)_g}} \frac{4 v_g}{Q_u \eta P_0} \sqrt{\frac{\varrho_g k_B T \Delta \nu}{\omega_{ac} Q_{ac}}} \, . \tag{12.3.5}$$

As mentioned above, the conventional FMR signal is determined as a change $\Delta(1/Q_u)$ of the cavity Q-factor by diode microwave reflection measurement. The change ΔR in the reflection factor for a resonator matched to the microwave circuits is [12.49]

$$\Delta R = \sqrt{\Delta P / P_0} = \tfrac{1}{2} Q_u \eta \chi'' \, . \tag{12.3.6}$$

A microwave diode matched to the impedance of the waveguide Z_0 will convert the change of the reflected microwave power $\Delta P = \Delta U^2 / Z_0$ into a change of the output voltage:

$$\Delta U = \tfrac{1}{2} Q_u \eta \chi'' \sqrt{P_0 Z_0} \, . \tag{12.3.7}$$

If the diode is connected to an ideal amplifier, the smallest detectable signal is equal to the detector noise

$$\Delta U_{noise} = \sqrt{4 k_B T Z_0 \Delta \nu} \, . \tag{12.3.8}$$

Thus, from (12.3.7) and (12.3.8), the minimum detectable susceptibility by conventional FMR will be

$$\chi''_{min,conv} = \frac{2}{Q_u \eta} \sqrt{\frac{4 k_B T \Delta \nu}{P_0}} \, . \tag{12.3.9}$$

Now we can calculate the ratio of the detection limits of PA (12.3.5) and conventionally detected FMR (12.3.9):

$$\frac{\chi''_{\text{min,PA}}}{\chi''_{\text{min,conv}}} = \frac{\omega_M}{\kappa - 1} \sqrt{\frac{(\lambda \varrho c)_s}{(\lambda \varrho c)_g}} \sqrt{\frac{V_g}{P_0}} \sqrt{\frac{\varrho_g v_g^2}{\omega_{ac} Q_{ac}}} . \tag{12.3.10}$$

For the case of a ferrochrome tape with an effusivity of $\sqrt{(\lambda \varrho c)_s} \approx 1000 \, \text{Ws}^{1/2}$ $\text{K}^{-1}\text{m}^{-2}$ measured in the waveguide cell at a microwave power of $P_0 = 200 \, \text{mW}$ and a modulation frequency of 300 Hz, the ratio is 1020, which means that PA-detected FMR is three orders of magnitude less sensitive than conventional detection. Roughly, this relation is also found experimentally from the signal-to-noise ratios measured.

From (12.3.10), we can deduce that thermally detected FMR can become much more sensitive at low temperatures, where the specific heat capacity decreases rapidly. Such an increase of sensitivity has also been observed experimentally [12.18].

12.3.2 Depth-Dependent FMR from Layered Magnetic Tapes

One of the most promising applications of thermal wave techniques is provided by the depth profiling capability. In this section, we describe an application of the PA-FMR technique to detect the layered structure of a magnetic tape [12.14,41].

Particulate magnetic recording tapes, as used for audio or video recording, consist of small, acicular particles of γ-Fe_2O_3, CrO_2 or pure metals embedded in an organic binder matrix. The volumetric filling factor ranges from 25 % to 60 %. To obtain a high coercivity and remanent magnetization of the recording medium, the particles are produced in the shape of needles with a length of about 300 nm and a thickness of about 50 nm (γ-Fe_2O_3), which are oriented preferably in the recording direction of the medium. Such small particles are usually single domain with dominant shape anisotropy and contributions of magnetocrystalline anisotropy [12.51].

The ferrochrome magnetic tape used for the depth resolved experiments is a commercial cassette tape for audio recording (BASF ferrochrom III). It consists of a surface layer with a thickness of 2.2 μm containing ferromagnetic CrO_2 particles and a 3 μm thick layer containing ferrimagnetic γ-Fe_2O_3 particles in the depth (Fig. 12.14). Both layers are backed by a 7.9 μm thick polyester foil. A sample of

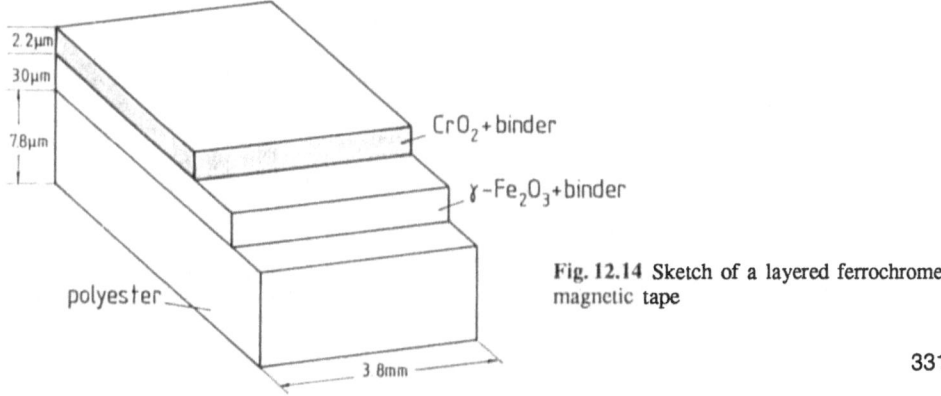

Fig. 12.14 Sketch of a layered ferrochrome magnetic tape

1.5 mm × 20 mm was cut out of the tape, glued to a piece of cardboard and placed in the PA quartz tube cell described in the preceding section. For comparison, samples with single layers of γ-Fe$_2$O$_3$ and CrO$_2$ particles were also prepared in the same way. According to the results of Sect. 12.2.1, the FMR signals obtained with these samples are strongly dependent on the orientation of the magnetic tape sample with respect to the external magnetic field. In the results presented here, the external field was applied perpendicular to the tape and the long particle axis, the microwave magnetic field was in the plane of the tape and parallel to the particle axis.

For a fixed orientation of the tape, the FMR spectrum is very characteristic of the magnetic layer studied, due to different magnetizations ($M = 0.47$ T for γ-Fe$_2$O$_3$ and $M = 0.61$ T for CrO$_2$ at room temperature), relaxation times and magnetocrystalline anisotropy of the materials and due to orientation and geometrical properties of the particles and the filling factor of the layer. In Fig. 12.15 (lower part), the characteristic

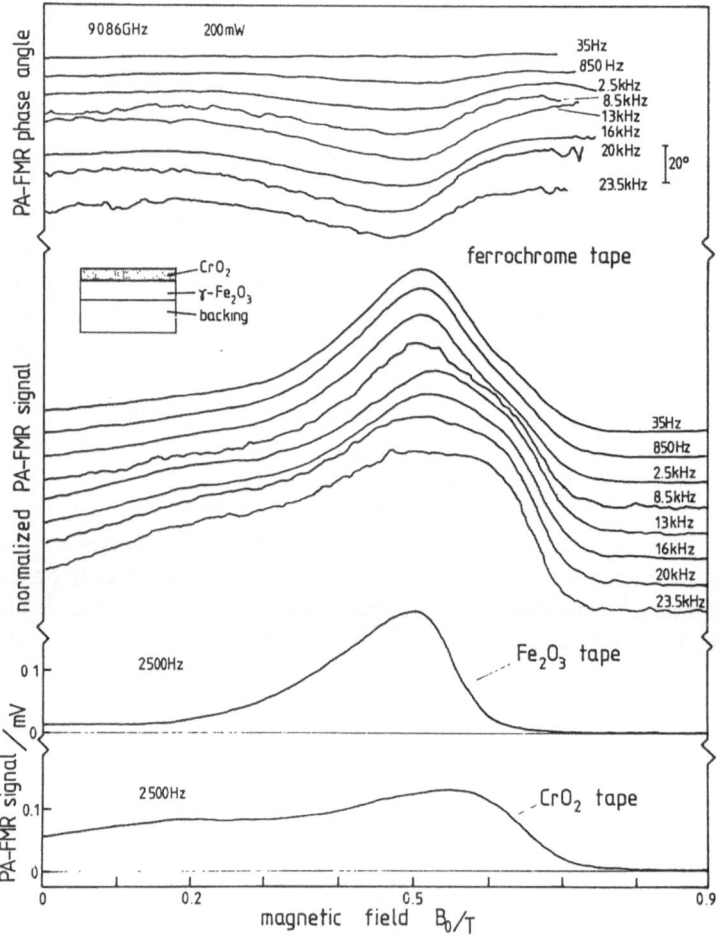

Fig. 12.15. PA-FMR spectra of a ferrochrome magnetic tape at different modulation frequencies (*upper part*). PA-FMR spectra of single-layer γ-Fe$_2$O$_3$ and CrO$_2$ tapes at a modulation frequency of 2500 Hz (*lower part*)

332

FMR spectra for the single layer γ-Fe_2O_3 and CrO_2 tapes are shown. The CrO_2 resonance maximum is found at a higher field than for the γ-Fe_2O_3 (due to a higher magnetization). As the total anisotropy field in the CrO_2 particles favors natural FMR at X-band frequencies, the line is very broad.

In Fig. 12.15 (upper part), we have shown PA-detected FMR spectra of the ferrochrome tape at different modulation frequencies. At a low modulation frequency of 35 Hz the magnetic layers are thermally thin (the total thickness is small compared with the thermal diffusion length) and the thermal response of both layers contributes equally to the total signal. The FMR signal is therefore a simple superposition of the single layer signals shown in the lower part of Fig. 12.15. At higher modulation frequencies, the CrO_2 resonance peak emerges more and more from the spectrum, until the signal has nearly changed to the signal of the CrO_2 single layer at 23.5 kHz. The double layer structure has become thermally thick, and only the surface layer contributes to the signal. In the PA-FMR signal phase, a phase minimum with increasing depth occurs at the resonance field of the γ-Fe_2O_3 layer. The phase delay at this field roughly represents the traveling time which the thermal wave launched in the γ-Fe_2O_3 layer needs to pass the CrO_2 layer. Similar results are obtained from an inverted ferrochrome tape, where the sequence of the layers is reversed [12.41,42].

We will now compare the experimental results presented above with the results of a theoretical model of thermal diffusion in a double layer system [12.41]. A one-dimensional model will be used, as the thickness of the layers is very small compared with the lateral dimensions of the sample. Figure 12.16 shows a sketch of the two-layer system studied, consisting of a thermally thick gas space (g), two absorbing layers (s_1) and (s_2) with thickness x_1 and ($x_2 - x_1$), respectively, and a thermally thick, nonabsorbing backing material (b). The magnetic particles are small compared with any length scale relevant to the thermal problem, thus homogeneous thermal properties can be assumed in the layers. As the magnetic layers are weak microwave absorbers, a homogeneous absorbed microwave power density P_i is assumed for each region (s_i). The time-dependent absorption of microwave power is given by

$$P_i(t) = \tfrac{1}{2} P_i[1 + \exp(i\omega_M t)] . \qquad (12.3.11)$$

We will calculate the temperature oscillation at the solid-gas interface, which is representative of the PA signal. After transients have settled, a steady-state temperature distribution

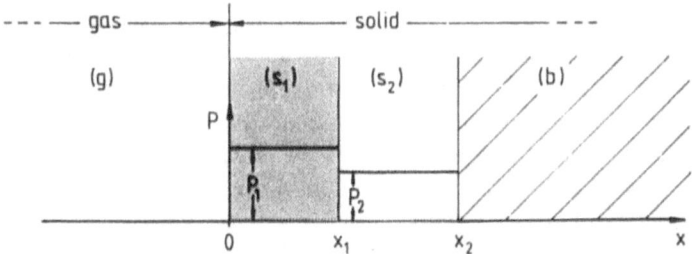

Fig. 12.16. The two-layer model used for the calculation of the temperature distribution in the ferrochrome magnetic tape

$$T(x,t) = T_{dc}(x) + \vartheta(x)\exp(i\omega_M t) \tag{12.3.12}$$

will develop, where the temperature oscillation $\vartheta(x)$ is the interesting quantity. To calculate $\vartheta(x)$, the thermal diffusion equation has to be solved for the four regions of the model, considering the boundary conditions for steadiness of temperature and heat flow at the three interfaces between the regions [12.52]. Here, we will only present the results of this lengthy, but straightforward calculation [12.41]. Using the abbreviations

$$\sigma = (1+i)/\mu , \tag{12.3.13}$$

where μ is the thermal diffusion length,

$$g = \sqrt{\frac{(\lambda\varrho c)_g}{(\lambda\varrho c)_{s_1}}} ,$$

$$s = \sqrt{\frac{(\lambda\varrho c)_{s_1}}{(\lambda\varrho c)_{s_2}}} ,$$

$$b = \sqrt{\frac{(\lambda\varrho c)_{s_2}}{(\lambda\varrho c)_b}} , \tag{12.3.14}$$

and

$$
\begin{aligned}
U &= \exp(\sigma_{s_1} x_1)\{(1-b)(s-1)\exp[-\sigma_{s_2}(x_2-x_1)] \\
&\quad - (1+b)(s+1)\exp[\sigma_{s_2}(x_2-x_1)]\} , \\
V &= \exp(-\sigma_{s_1} x_1)\{(1+b)(s-1)\exp[\sigma_{s_2}(x_2-x_1)] \\
&\quad - (1-b)(s+1)\exp[-\sigma_{s_2}(x_2-x_1)]\} , \\
W &= (1-b)\exp[-\sigma_{s_2}(x_2-x_1)] + (1+b)\exp[\sigma_{s_2}(x_2-x_1)] ,
\end{aligned}
\tag{12.3.15}
$$

we find that the temperature oscillation at the gas-solid interface is given by

$$
\begin{aligned}
\vartheta_{s_1}(0) =& \frac{-i}{\omega_M}\frac{\alpha_{s_1}}{2\lambda_{s_1}}\left(\frac{V+U+2W}{V(1-g)+U(g+1)}\right)P_1 + \frac{-i\alpha_{s_2}}{\omega_M 2\lambda_{s_2}} \\
&\times \left(\frac{2(2-W)}{V(1-g)+U(g+1)}\right)P_2 .
\end{aligned}
\tag{12.3.16}
$$

In (12.3.16), the resulting temperature oscillation has been expressed by the contribution from the surface layer, which is proportional to P_1 and the contribution from the buried layer, which is proportional to P_2. A numerical analysis of (12.3.16) using estimated values for the thermal parameters of the magnetic tape and the known thicknesses shows that, in accordance with the qualitative discussion given above, at modulation frequencies over 2 kHz the amplitude contribution of the depth layer decreases and the phase delay increases rapidly.

Moreover, the measured frequency-dependent PA-FMR spectra could be fitted by spectra calculated theoretically from (12.3.16). The single CrO_2- and γ-Fe_2O_3-

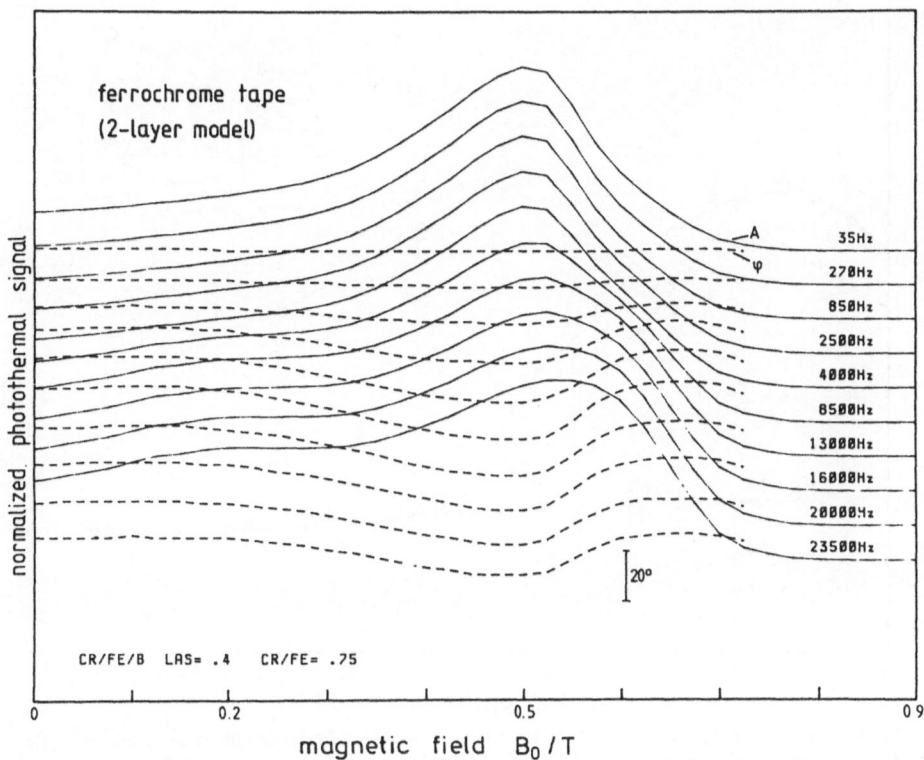

Fig. 12.17. Calculated PA-FMR spectra of the ferrochrome magnetic tape using the single layer spectra shown in the lower part of Fig. 12.15

layer FMR signals of Fig. 12.15 (lower part) were used as magnetic field dependent heat source functions $P_i(B_0)$. It was assumed that these functions are the same in the ferrochrome tape. One has to allow for a first free parameter to take account of the unknown relative absorption strength of the single layers. All geometrical parameters and the thermal parameters of gas and backing material are known. The average density of the magnetic layers could be measured and a fixed value of $1000\,\mathrm{Jkg^{-1}K^{-1}}$ for the average heat capacity was used, mainly determined by the thermal properties of the binder material. The remaining second free parameter was the effective thermal conductivity of the magnetic layers, which was assumed to be equal for both layers. In Fig. 12.17, the result of the fitting procedure is shown. The amplitude and phase signals agree very well with the experimental result in Fig. 12.15. A value of 0.75 for the relative absorption strength at the resonance maximum between CrO_2 and γ-Fe_2O_3, and an average thermal conductivity of $0.4\,\mathrm{Wm^{-1}K^{-1}}$ for the layers has been extracted from the fit.

12.3.3 PA-FMR from Metallic Foils and Films

The depth profiling analysis is generally carried out by studying the PA signal as a function of the modulation frequency (Sect. 12.3.2). An alternative procedure is to measure the PA signal at a constant frequency but change the phase angle offsets.

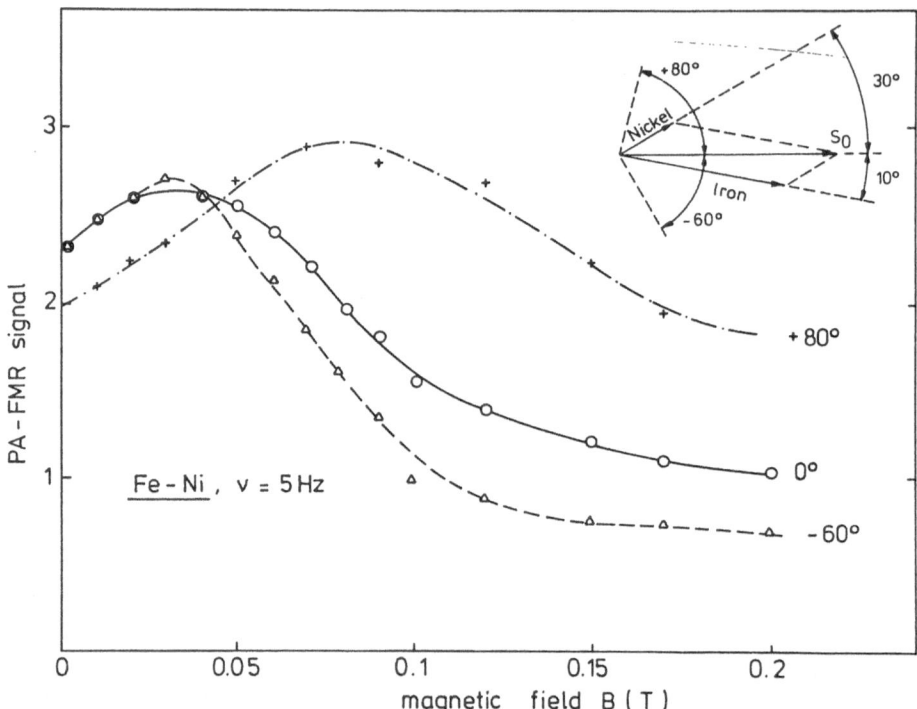

Fig. 12.18. Magnetic field dependence of the PA signal of a layered Fe-Ni sample at three different phase offsets φ. The modulation frequency is 50 Hz. The signals at $\varphi = -60°$ and at $\varphi = 80°$ correspond to the FMR of pure iron and nickel, respectively. (After [12.53])

The usefulness of the latter technique when applied to FMR has been demonstrated for a layered sample made of a nickel film bonded to an iron film, both of 50 μm thickness [12.53,54]. The sample was placed in a shorted waveguide-type PA cell and the PA signal arising from the thermal wave traveling through the sandwich was detected. A second reference cell was mounted in a cavity wall using an Al foil as a reference sample, see Sect. 12.3.1. Figure 12.18 shows the magnetic field dependence of normalized PA signals of the Fe-Ni layer due to the resonance absorption at 9.2 GHz. Measurements were performed at a fixed modulation frequency with different phase offsets φ. For a given phase φ, the PA signal amplitude can be written in terms of the in-phase (A_0) and out-of-phase (A_{90}) components [12.53]: $A(B_0) = A_0 \cos \varphi + A_{90} \sin \varphi$. As depicted in the diagram in Fig. 12.18, the signal at $\varphi = -60°$ corresponds to the pure iron resonance and that at $\varphi = 80°$ to the pure nickel resonance. The authors showed that this considerable phase difference between the pure signals is due to the glue interface layer between the Fe and the Ni foil. On the basis of the phase lag data, they were able to deduce the reduced thermal thickness of this layer.

Photoacoustic detection only responds to the energy dissipated in the resonance process and is therefore insensitive to the spurious electronic effects occurring in conventional detection techniques. This advantage of PA detection has been exploited in the investigation of spin wave resonance in thin films of Permalloy [12.38].

The photoacoustic FMR response of the films evaporated on a glass substrate were measured at room and liquid nitrogen temperature using a PA detection arrangement similar to that described above. With the magnetic field perpendicular to the film plane, several spin wave modes could be observed, but the mode intensities did not show the behavior expected from the Kittel mode. This discrepancy has not been resolved. In the following section, we present results of locally resolved measurement of standing spin waves using the mirage effect. The results of these experiments point towards the strong influence of lateral inhomogeneities on the intensity behavior of the spin wave modes.

12.4 FMR Detection by Photothermal Laser Beam Deflection

The use of laser beam deflection or the mirage effect [12.15] as a sensor for the heat dissipated in the sample enables measurements to be made of the local magnetic resonance absorption. Periodic heating of the sample due to the absorption of the chopped microwave radiation gives rise to an oscillating refractive index in the gas layer near the sample surface. The refractive index gradient can be probed by the deflection of a laser beam passing at grazing incidence over a heated solid. The amplitude of the time modulated deflection angle is proportional to the oscillating local surface temperature [12.55–57]. Therefore, measurement of the mirage effect as a function of the modulation frequency also yields depth-dependent information.

Compared to the photoacoustic detection method discussed previously, the laser beam deflection technique has to overcome two additional experimental problems. Firstly, the probe beam has to pass through the microwave cavity and to reach every point on the sample. Secondly, the microwave power density has to be increased by at least one order of magnitude, as only the small portion absorbed at the actual position of the probe beam contributes to the signal.

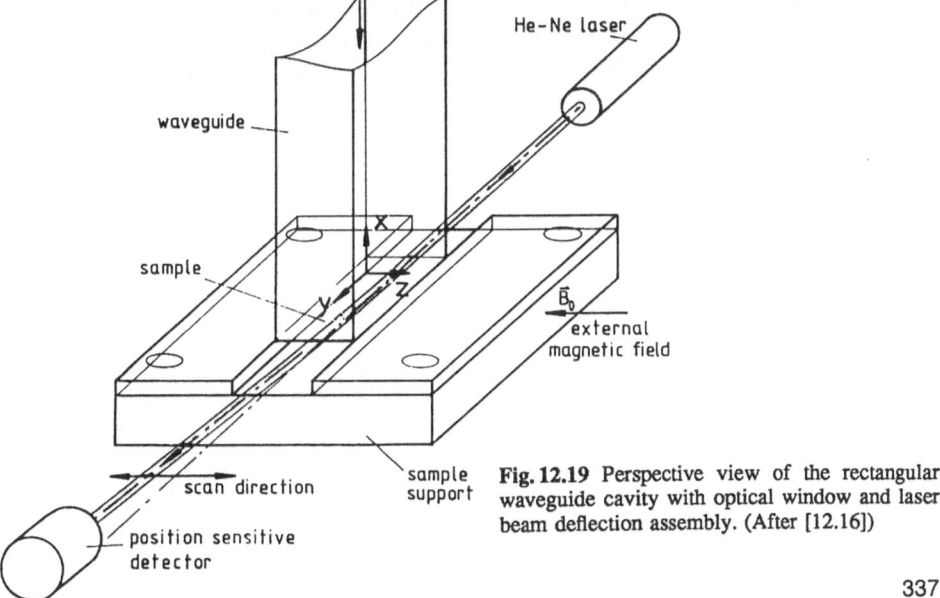

Fig. 12.19 Perspective view of the rectangular waveguide cavity with optical window and laser beam deflection assembly. (After [12.16])

In order to meet the first requirement, a shorted waveguide with two narrow slots at the bottom can be used [12.16,42]. The slotted rectangular waveguide constructed for the beam deflection experiments discussed in the following section is shown in Fig. 12.19. It is designed to work in the transverse electric TE_{105} mode and has a loaded Q-value of about 1200. The unfocused probe beam of a He-Ne laser passes in grazing incidence near the sample at the bottom of the cavity, parallel to the x direction. The deflection of the beam is detected by a position sensor. The laser and the sensor are moved together relative to the waveguide in the y direction. The detection unit forms part of an electron paramagnetic resonance spectrometer working at X-band frequencies of about 9.2 GHz (Fig. 12.20). Modifications are also made to allow intensity modulation and to raise the microwave power level at the sample. Using a traveling wave tube amplifier (TWTA), the microwave power supplied by the klystron (400 mW max) can be boosted up to 20 W. The microwave intensity is chopped by a pin-diode modulator inserted into the line between the klystron and the TWTA.

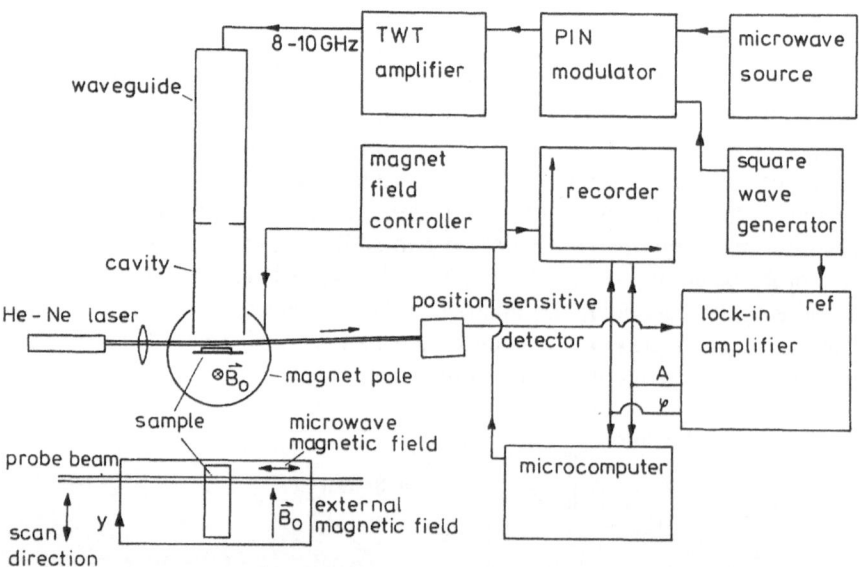

Fig. 12.20. Block diagram of the experimental setup for the detection of FMR by the laser beam deflection technique

12.4.1 Specifications of the Detection Unit

The amplitude and phase angle of the vertical deflection of the laser beam experience a much stronger frequency dependence than the corresponding quantities of the photoacoustic signal (Sect. 12.3.1). This is due to the additional attenuation of the thermal wave in the gas layer between the probe beam and the sample. For a beam passing at the distance z_0 above a homogeneously heated surface with an oscillating temperature amplitude ϑ_0, the complex normal deflection ϕ_n is given by [12.55]

$$\phi_n \propto \frac{\vartheta_0}{\mu} \exp\left[-\frac{z_0}{\mu} + i(\omega_M t - \frac{z_0}{\mu} + \frac{\pi}{4})\right] . \tag{12.4.1}$$

As the thermal diffusion length μ in the gas varies as $1/\sqrt{\omega_M}$, the amplitude suffers a strong exponential decrease in addition to the frequency dependence of the oscillating surface temperature.

According to (12.4.1), the phase angle of the normal deflection signal varies linearly with the distance z_0:

$$\varphi_n = -\frac{z_0}{\mu} + \frac{\pi}{4} . \tag{12.4.2}$$

Figure 12.21 shows experimental results from a ferrochrome tape [12.41] which had also been investigated with photoacoustically detected FMR (Sect. 12.3.2). The phase angle values of the normal deflection signal have been corrected for the frequency dependence of the oscillating surface temperature and of the electronic filters and amplifiers. As a function of the square root of the modulation frequency, the corrected value shows the expected linear dependence. Inserting the thermal diffusivity value of air of $\alpha = 0.22\,\mathrm{cm}^2/\mathrm{s}$, the slope yields an effective distance of the laser beam from the surface of $z_0 = 0.23\,\mathrm{mm}$. The deviations observed at the most elevated frequencies are due to the finite diameter of the probe beam. A thorough investigation of the influence of the probe beam size has recently been published by *Legal Lasalle* et al. [12.58].

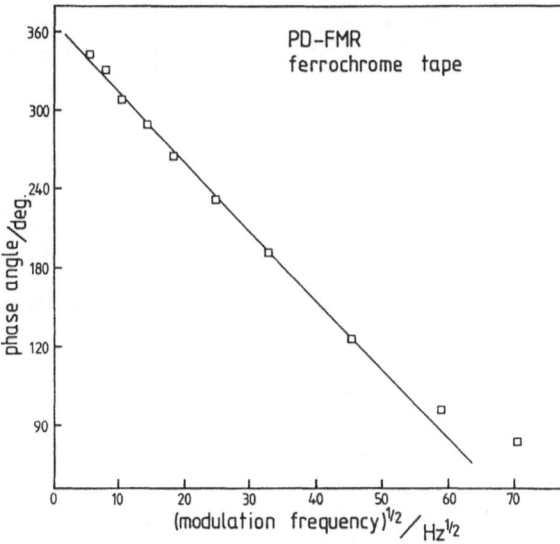

Fig. 12.21 Corrected phase angle of the vertical deflection signal from FMR of a ferrochrome tape as a function of the modulation frequency

The detection limits are determined at the low frequency side by irregular motions of the air and by fluctuations of the probe beam direction. At high modulation frequencies from 1 kHz up, the electronic noise of the diode and of the amplifier yield the dominant, whilst nearly frequency-independent, contribution. This constant noise level, in conjunction with the rapidly decreasing signal amplitude with increasing modulation frequency (12.4.1) depresses the signal-to-noise ratio more violently for the beam deflection signal than it does for the photoacoustic detection technique (Sect. 12.3) or for photothermally modulated FMR (Sect. 12.5).

The main aim of using the mirage effect as a detection method for FMR is to make laterally resolved measurements. The observed FMR absorption profile as a function of the probe beam position (which provides the local information) results from the convolution of the spatial variation of χ'' and the local distribution of the microwave energy density at the sample. At the bottom ($z = 0$) of the resonator shown in Fig. 12.19, the microwave magnetic field does not depend on the y direction whereas along the x direction it should vary as $b_x = b_1 \sin(\pi x/a)$. One of the first applications of PD-FMR was devoted to examining the predicted behavior [12.41]. A strip of the ferrochrome tape already investigated by PA-FMR was fixed in the cavity of Fig. 12.19 along the diagonal of the rectangular base plate, as no slots allowing optical access are permitted in the long side (y direction) of the cavity.

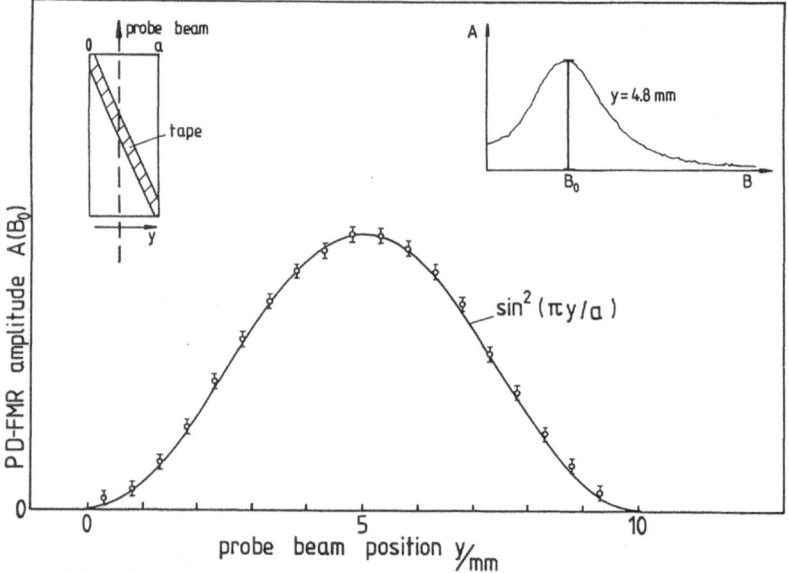

Fig. 12.22. Local dependence of the vertical deflection signal corresponding to the FMR maximum at B_0 (*upper right inset*) of a ferrochrome tape which was fixed at the bottom of the rectangular cavity (Fig. 12.19) along the diagonal (*upper left inset*)

The vertical deflection signal was measured as a function of the external magnetic field at different positions y of the probe laser beam. In the lower part of Fig. 12.22, the peak heights of the FMR signals are plotted as a function of the probe beam position. The experimental values follow the $\sin^2$ behavior very precisely, as the energy density is proportional to b_x^2. This result may be considered as a test of the reliability of PD-FMR as a quantitative technique. It also shows that the slots cut into the cavity to provide optical access, do not markedly perturb the energy density distribution inside the cavity.

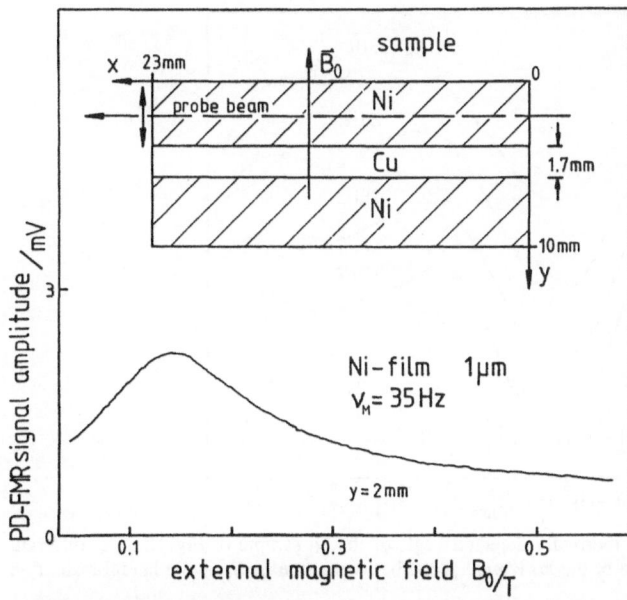

Fig. 12.23. (*Upper part*) Orientation of the Ni sample at the bottom of the shorted waveguide with respect to the external magnetic field and the probe beam direction. (*Lower part*) Magnetic field dependence of the photothermally detected FMR signal measured with the probe beam at $y = 2\,\mathrm{mm}$ and a modulation frequency of 35 Hz. (After [12.16])

12.4.2 PD-FMR Imaging of a Ni Film

Laterally resolved FMR signals detected by photothermal beam deflection (the mirage effect) were first reported from a Ni film with artificial inhomogeneities [12.16]. In this experiment, a nonconducting substrate was covered with a 1 μm thick copper film. Finally, the sample, a 1 μm thick Ni film, was evaporated on the Cu surface, leaving a path of 1.7 mm width (Fig. 12.23, upper part). Pure nickel is a ferromagnet with a Curie point at 631 K [12.59]. The sample was positioned at the bottom of the slotted waveguide with the orientation as indicated in Fig. 12.23. At a fixed probe beam position, the resonance pattern was measured at different modulation frequencies. In Fig. 12.24, the peak height of the resonance pattern is traced as a function of the probe beam position for different modulation frequencies. The nickel free path is clearly visible at all modulation frequencies. The origins of the structure at the location of the Ni film are accounted for by variations of the beam height z_0, due to buckling of the metallic film, particular near the borders. This interpretation is supported by the results of the transverse deflection measurements, which enable us to differentiate contributions to the PM-FMR from inhomogeneities of the magnetic properties and from variations of the probe beam distance [12.60].

The behavior at the Cu-Ni border has been used to investigate the lateral resolution of the PD-FMR technique at the edge of a homogeneously heated surface. The experimental curves of Fig. 12.24 demonstrate the increase of the lateral resolution with increasing modulation frequency. In addition, the steepness at the boundaries strongly depends on the distance z_0 of the probe beam from the surface.

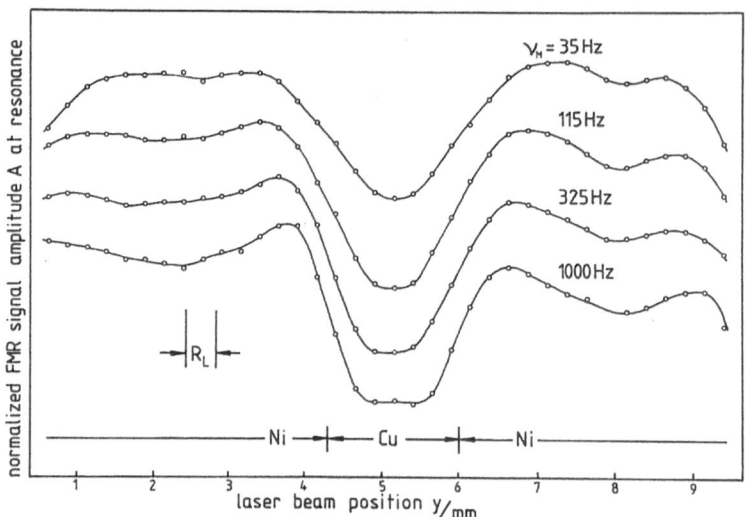

Fig. 12.24. Peak height of the photothermally detected FMR of the Ni sample of Fig. 12.23 at different modulation frequencies as a function of the horizontal probe beam position y. R_L is the lateral resolution limit. (After [12.16])

In order to quantify the lateral resolution, the "spatial step resolution" R has been defined. R represents the distance in the y direction where the amplitude of the vertical deflection signal changes from 0.1 to 0.9 of its maximum value and vice versa (Fig. 12.26, inset). With the assumption of a semi-infinite sample with a homogeneous but time-modulated plane heat source for $y \geq 0$, the oscillating temperature gradient has been calculated and plotted in Fig. 12.25 as a function of the reduced vertical z/μ and horizontal y/μ probe beam position, where μ is the thermal diffusion length of the air above the sample [12.60]. The constant gradient curves of Fig. 12.25 indicate that the oscillating temperature gradient extends into

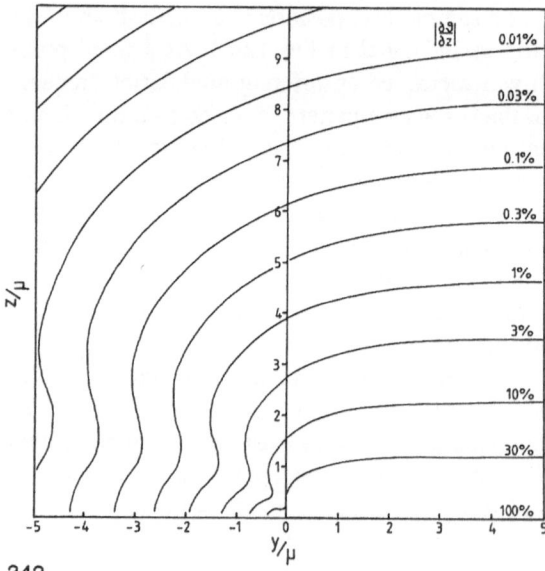

Fig. 12.25 Curves of constant (normal) temperature gradient in the air above a semi-infinite plane heat source ($y \geq 0$, $-\infty \leq x \leq +\infty$) plotted as a function of the coordinates normalized with respect to the thermal diffusion length μ of air

the nonheated area at $y \leq 0$. From the intersections of the constant z/μ line with the constant gradient curves, the reduced step resolution R/μ is determined as a function of z/μ. As a result of this analysis, a linear dependence has been found for beam heights $z/\mu > 3$:

$$R = 2.6\mu + 0.44z . \tag{12.4.3}$$

Experimental values of the step resolution R have been determined from Fig. 12.24. In Fig. 12.26, these data are plotted as a function of the inverse square root of the modulation frequency together with results of the exact theory (full curve) and the linearized approximation (dashed line). From the intersection $R_L = 0.4\,\text{mm}$, in the limit of very high modulation frequencies, representing the maximum achievable resolution, and from the slope of the straight line in Fig. 12.26 the diffusivity of the air $\alpha = 2.0 \times 10^{-5}\,\text{m}^2\text{s}^{-1}$ and the beam distance $z = 0.91\,\text{mm}$ are obtained. These fit parameters were then used to calculate the exact behavior represented by the full curve.

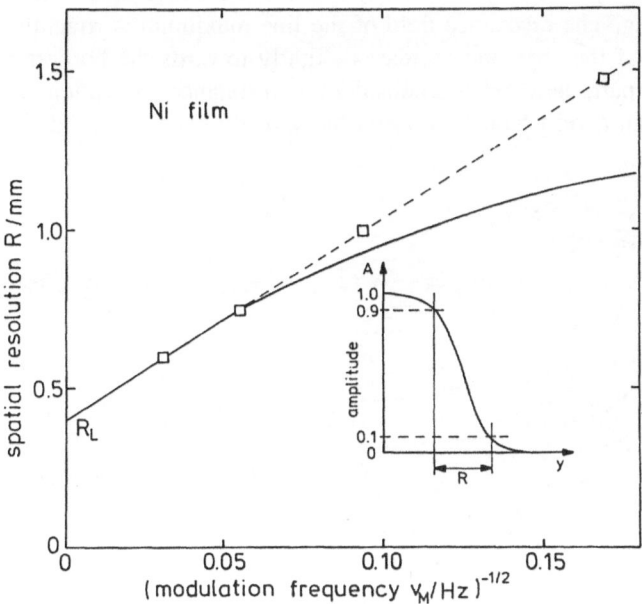

Fig. 12.26. Spatial step resolution R as a function of the inverse square root modulation frequency. Experimental points and theoretical results (*full and dashed curves*). The inset illustrates the definition of the step resolution R

12.4.3 PD-FMR Imaging of Amorphous Metallic Foils

Metglass ribbons produced by melt-spinning techniques [12.61] are materials which possess very strong inhomogeneities of the in-plane magnetic properties. The photothermal beam deflection method has been applied very successfully to investigate the lateral variation of the FMR on commercially available metglass ribbons of composition $Fe_{40}Ni_{40}B_{20}$ [12.41,42,62]. The amorphous alloy, $Fe_{40}Ni_{40}B_{20}$ is ferromagnetic below $T_c = 662\,K$ and has a saturation magnetization of 1.03 T at room temperature [12.63]. In the PD-FMR experiments, strips of 0.91 mm × 9.2 mm were cut from the metglass band, which was about 38 μm thick, and these strips were attached at the bottom of the slotted rectangular microwave cavity in Fig. 12.19. The vertical beam deflection signal was measured as a function of the magnetic field at different positions of the laser beam in the y direction. The distance of the probe beam from the surface was kept constant at $z_0 = 0.2\,mm$ and the modulation frequency was fixed at 430 Hz. At 9.2 GHz, the skin depth out of resonance is 5.8 μm and at resonance is only a few tenths of a micrometer, which is always less than the smallest thermal diffusion length achievable in these experiments. Therefore, the modulation frequency was chosen to form a compromise between an optimum signal-to-noise ratio (Sect. 12.4.1) and the maximum lateral resolution (Sect. 12.4.2). Figure 12.27 shows PD-FMR signals which were recorded every 0.5 mm along the long edge of the strip. The external magnetic field was aligned parallel to the long edge. In the lower part of the same figure, the conventionally recorded signal is shown together with the trace formed by the average of the PD-FMR lines. Two observations are worth noting. The resonance field of the line maximum is roughly constant in the central part of the strip and increases slightly towards the borders. And secondly, in the central part, there exists a considerable resonance absorption at zero magnetic field. Although a zero field absorption has already been observed in

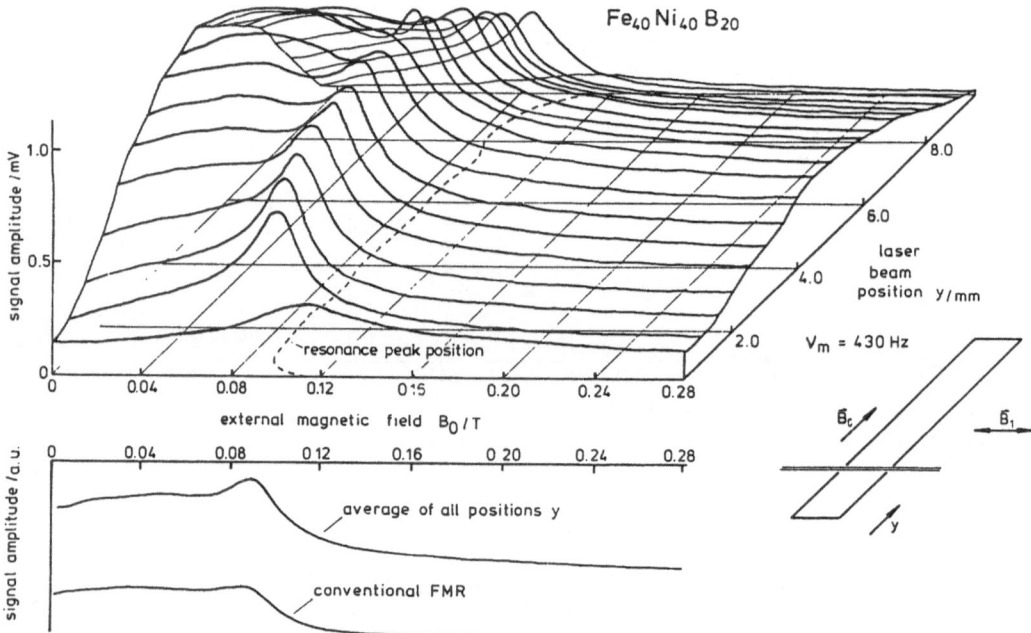

Fig. 12.27. Vertical deflection signal of the photothermally detected FMR from a strip of a $Fe_{40}Ni_{40}B_{20}$ metglass ribbon at different probe beam positions. The external magnetic field is oriented along the long edge of the strip and perpendicular to the probe beam direction. The dashed line indicates the field position of the resonance maximum. The lower part shows the conventionally recorded FMR signal and the trace obtained by averaging the PD-FMR lines. (After [12.42])

polycrystalline and multidomain samples of other materials with magnetic anisotropy [12.2], its origins in the metglass must be of a different nature. For the FeNiB amorphous alloys, the saturation fields and the uniaxial surface anisotropy fields are both of the order of 0.01 T [12.61,64,65,66] which is far below the low-lying resonance absorption which extends up to fields of about 0.1 T. A probable source of the observed zero field resonance in the metglass might be strain-induced anisotropy fields. The preferential orientation should then be parallel to the long edge of the ribbon as the resonance peak at low fields is strongly suppressed when the strip is rotated by 90° in the cavity [12.42]. The occurrence of strain effects is also supported by the results from recent photothermal modulated FMR measurements on the same samples (Sect. 12.5.3).

The lateral variation of the resonance field of the main resonance at about 0.1 T has been satisfactorily explained by *Netzelmann* [12.41] on the basis of a spatially dependent demagnetizing field [12.67]. Figure 12.28 shows the results of his cal-

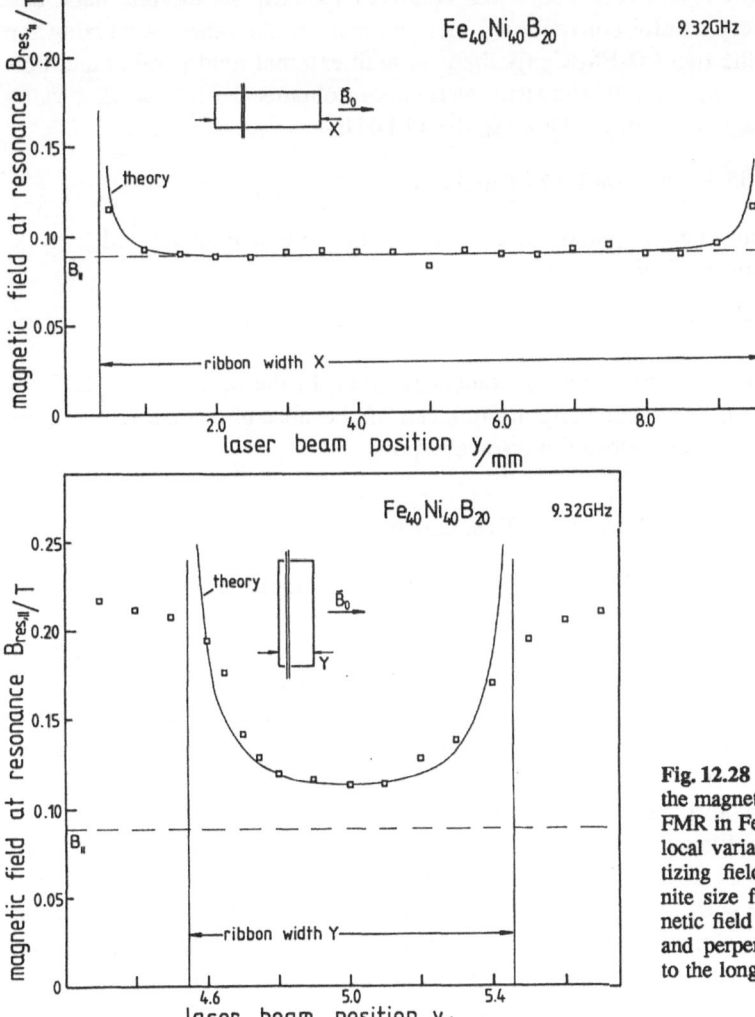

Fig. 12.28 Local dependence of the magnetic resonance field for FMR in $Fe_{40}Ni_{40}B_{20}$ due to the local variation of the demagnetizing field in a strip with finite size for the external magnetic field parallel (*upper part*) and perpendicular (*lower part*) to the long edge

culations for the metglass strip compared with the experimental data. The most pronounced effects are seen when the external magnetic field is aligned perpendicular to the long edge of the strip (lower part of Fig. 12.28). The deviations of the experimental points from the theoretical curve at the border of the strip are due to the finite resolution imposed by the beam size and the lateral resolution R (Sect. 12.4.2). But altogether there is good quantitative agreement between the experiment and the theory.

Finally, one has to compare the PD-FMR results with the data obtained with the conventional microwave detection technique. For this purpose, the FMR of the metglass ribbon was also measured conventionally and compared with the average of the PD-FMR signals recorded at the different positions on the sample [12.41,42]. The two curves resemble each other, but at second glance the line shapes, particularly in the tails, are slightly different. These deviations are presumably due to different relative weights of the contributions to the total sign from the outer perturbed regions.

A reliable comparison of the quantitative numbers may be obtained if one considers the PD-FMR signal recorded at the centre of the strip on the one hand and the results of a very careful conventional measurement on the other. Analyzing the central traces of the two PD-FMR experiments, with external field parallel and perpendicular to the long edge of the strip, *Netzelmann* obtains the following g-value and saturation magnetization for $Fe_{40}Ni_{40}B_{20}$ [12.41]:

$$g = 2.13 \pm 0.05 \quad \text{and} \quad M_0 = (1.01 \pm 0.02)\text{T} \ .$$

These values should be compared with the results of the conventional FMR investigation carried out by *Frait* [12.66]:

$$g = 2.08 \quad \text{and} \quad M_0 = 1.033\text{T} \ .$$

Within the experimental error, the agreement is good. In the case of the PD-FMR experiments, one might expect a slightly reduced M_0 because of a small temperature rise due to heating by the boosted microwave power.

12.4.4 Locally Resolved Spin Wave Resonance

In thin films of magnetic alloys, a microwave field can excite standing spin waves (Sect. 12.2.2), which may be detected conventionally or by thermal wave techniques as is done in the case of the uniform mode of FMR. Compared to conventional recording, photoacoustic detection provides a direct measure of the dissipative part of the susceptibility (Sect. 12.3.3). The use of photothermal deflection for the detection of spin wave resonance (SWR), also enables laterally resolved measurements to be made. The first study of SWR by laser beam deflection was reported by *Kordecki* et al. [12.68,69]. They investigated evaporated films of Permalloy ($Ni_{80}Fe_{20}$) in the normal configuration with the external magnetic field perpendicular to the film plane. Films of different thicknesses and different quality were prepared. The experiments were performed using the arrangement shown in Fig. 12.20, but instead of the rectangular slotted waveguide resonator of Fig. 12.19, a wall-less cylindrical microwave cavity composed of equally spaced annular plates was used (Fig. 12.34).

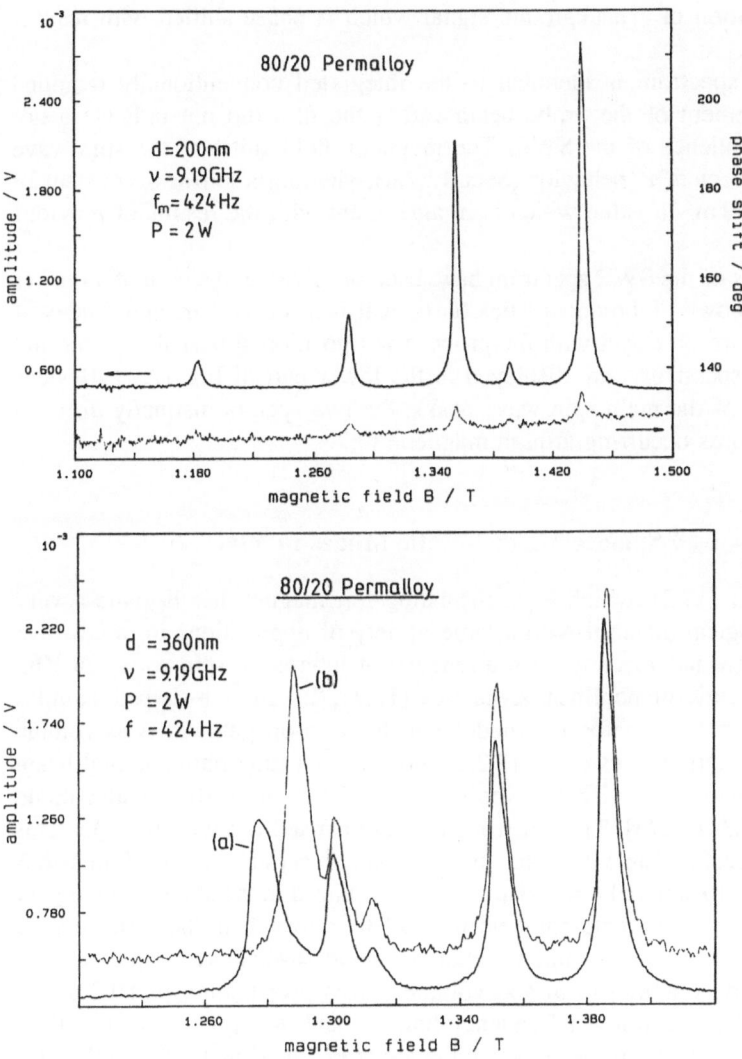

Fig. 12.29. Spin wave resonance of Permalloy films detected by laser beam deflection with the magnetic field perpendicular to the film plane. (*Upper part*) Magnetic field dependence of the amplitude and the phase angle of the vertical deflection signal recorded from a 200 nm thick good-quality film with the probe beam positioned at the center of the sample. (*Lower part*) Deflection signal from the center (a) and the boundary area (b) of an inhomogeneous Permalloy film of 360 nm thickness

This cavity, which allows optical access for the probe beam as well as the magnetic field aligned perpendicular to the film, was also used for the photothermally modulated FMR measurements. The design will be described in detail in connection with these later experiments in Sect. 12.5.1. Results of PD-SWR on Permalloy films are presented in Fig. 12.29. The two upper traces show the amplitude and phase angle of the vertical beam deflection signal detected in the center of a good quality $Ni_{80}Fe_{20}$ film that was 200 nm thick. The microwave intensity of about 2 W was modulated with 424 Hz. The phase angle change at resonance maximum most probably results from the superposition of a background signal which is phase shifted with respect to the FMR signal.

The PD-SWR spectrum is identical to the integrated conventionally recorded spectrum. Displacement of the probe beam across the film did not indicate a noticeable local dependence of the SWR. The magnetic field shifts of the spin wave peaks obey the expected n^2 behavior (Sect. 12.2.4), yielding a stiffness constant of $D\hbar\gamma = 4.5 \times 10^{-40} \, \text{J m}^2$, a value which is in agreement with the results of previous studies [12.70].

Local variations of the SWR spectrum have been observed in the case of a thicker film that already showed inhomogeneities on optical inspection. The amplitudes of two PD-SWR spectra observed with the probe beam positioned near the center and near the border, respectively, are displayed in the lower part of Fig. 12.29. Besides a measurable shift of the main spin wave peaks, the two spectra distinctly differ in the subsidiary features occurring at high magnetic fields.

12.4.5 Locally Resolved Surface Magnetostatic Modes in YIG

Yttrium iron garnet (YIG), which is an insulating ferrimagnet, has become a very important technological material with a large variety of applications in microwave devices [12.5]. There has recently been a renewal of interest in spin waves in YIG from the point of view of nonlinear dynamics [12.71,72]. In slab-shaped samples microwaves can excite magnetostatic modes which can propagate either as volume or surface magnetostatic waves (Sect. 12.2.2). For the first time photothermal beam deflection has been used to visualize the spatial variation of surface and volume modes in a YIG slab [12.73]. The experiments were carried out with the apparatus described in Sect. 12.4.2. The YIG slab, cut from a single crystal, had the dimensions $a = 0.98 \, \text{mm}$, $b = 8.40 \, \text{mm}$ and $c = 4.40 \, \text{mm}$ and was placed in the slotted rectangular cavity (Fig. 12.19) with the external field parallel to c, which is the [100] crystal direction. In this case, the slab supports magnetostatic surface waves propagating along the y direction (parallel to b) and volume waves propagating parallel to the external magnetic field. The lower frequency limit ω_1 and the upper frequency limit ω_2 of the volume modes are marked out by the angle Θ_k at $0°$ and $90°$ of (12.2.22) of Sect. 12.2.2 [12.74]:

$$\omega_1 = \gamma(B_0 - N_z M) , \tag{12.4.4}$$

$$\omega_2 = \gamma\sqrt{(B_0 - N_z M)(B_0 - N_z M + M)} , \tag{12.4.5}$$

where B_0 is the external field in the z direction, M and N_z are the saturation magnetization and the z component of the demagnetization tensor.

The surface modes have the lower frequency limit ω_2 and the upper limit ω_3 [12.74,29]:

$$\omega_3 = \gamma(B_0 - N_z M + M/2) . \tag{12.4.6}$$

In the relations (12.4.4–6) the anisotropy fields have been omitted. Due to the boundary conditions, the surface and volume modes form standing waves which give rise

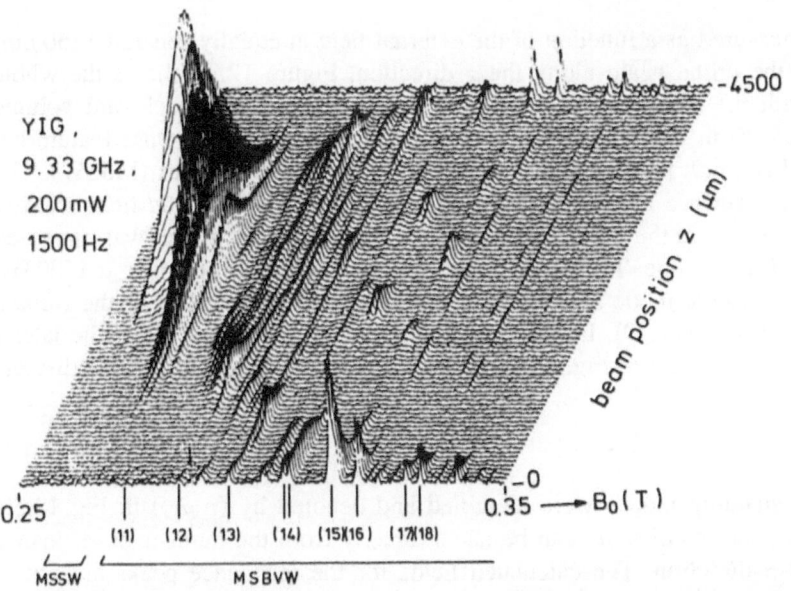

Fig. 12.30 PD-FMR image of the magnetostatic surface waves (MSSW) and of the magnetostatic backward volume waves (MSBVW) in a YIG slab. The probe laser beam was aligned parallel to the z direction (see inset of Fig. 12.31). The modulation frequency of the microwave power (200 mW) was 1500 Hz

to discrete resonance peaks. At the constant microwave frequency, which is 9.2 GHz, the peaks become visible at a distinct external field, but also display a spatial distribution corresponding to the nodal points and loops of the standing waves. The lateral images obtained with the photothermally detected magnetic resonance clearly show the spatial behavior of the surface and volume magnetostatic modes of the YIG slab (Figs. 12.30 and 12.31). The geometrical conditions are indicated in the inset of Fig. 12.31. The probe beam passes along the y direction perpendicular to the external magnetic field, which is aligned in the z direction. The vertical deflection

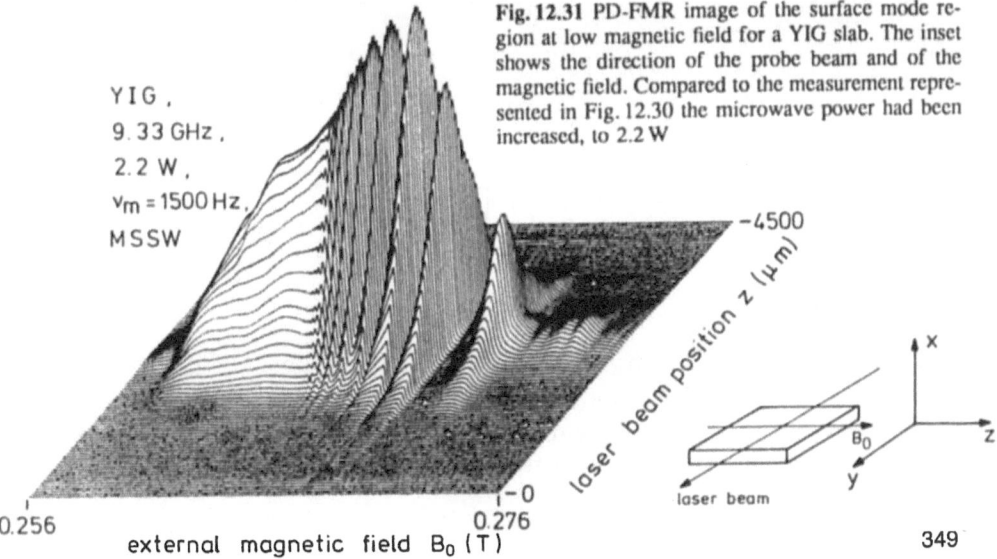

Fig. 12.31 PD-FMR image of the surface mode region at low magnetic field for a YIG slab. The inset shows the direction of the probe beam and of the magnetic field. Compared to the measurement represented in Fig. 12.30 the microwave power had been increased, to 2.2 W

signal was measured as a function of the external field at equally separated (50 µm) positions of the probe beam along the z direction. Figure 12.30 shows the whole spectrum, with the resonance peaks arising from magnetostatic backward volume waves (MSBVW) in the field range above 0.276 T and the very intense feature below this field strength that is due to the magnetostatic surface waves (MSSW). The region with the surface modes is displayed on an enlarged magnetic field scale in Fig. 12.31. To improve the signal-to-noise ratio, the microwave power was increased by an order of magnitude. The modulation frequency was kept constant at 1500 Hz.

The magnetostatic mode spectra have been analyzed on the basis of the *Damon* and *Eshbach* theory [12.29]. In the finite slab, with the thickness d and the lateral dimensions b and c (inset of Fig. 13.31), the wave vectors k_y and k_z assume discrete values:

$$k_y = \frac{n_y \pi}{b} \quad \text{and} \quad k_z = \frac{n_z \pi}{c} .$$

(12.4.7)

The volume standing modes were identified and denoted by $(n_y n_z)$ in Fig. 12.30. The standing wave number n_z can be taken directly from the number of resonance peaks in the z direction. The calculated fields for the resonance peaks are not in exact agreement with observations. The discrepancy is attributed to a gradient field across the sample [12.73].

Volume magnetostatic modes and surface modes have been distinguished by the dependence of the PD-FMR signal on the modulation frequency. The vertical beam deflection due to the microwave resonance could be observed in a wide frequency range, extending from a few hertz up to 100 kHz. The evolution of the resonance pattern with increasing modulation frequency in the lower field region from 0.256 T to 0.276 T is shown in Fig. 12.32. These measurements were conducted with the

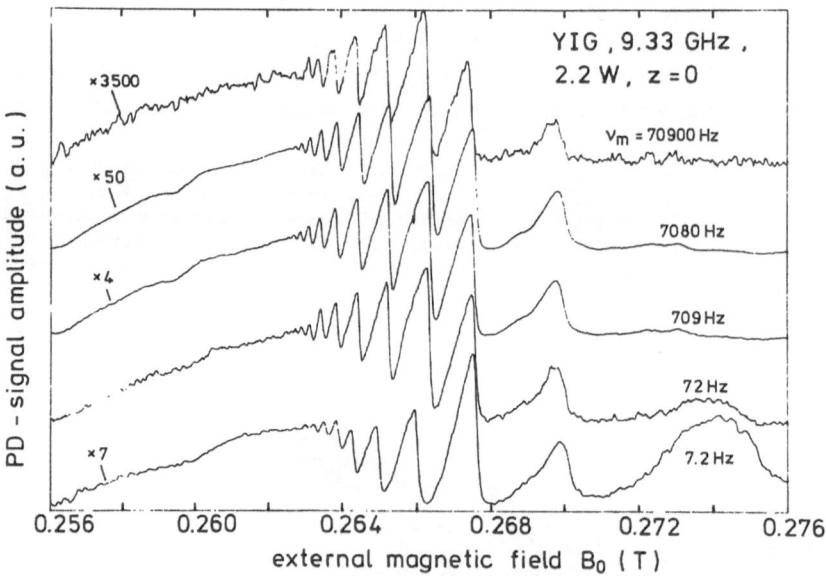

Fig. 12.32. PD-FMR recordings from a YIG slab at different modulation frequencies in the low field region. The measurements were made with the probe beam positioned in the middle of the slab at $z = 0$ (see Fig. 12.31 inset)

probe beam passing through the center of the slab at $z = 0$. Whereas the resonance peaks above 0.270 T have already disappeared at about a few kilohertz, the peaks at the lower fields are still visible at the highest frequencies, indicating that these excitations are confined to the surface region. Two further remarkable properties of the surface modes are made apparent by the PD-FMR studies. The surface magnetostatic modes are confined to the center of the slab and, apart from the sharp resonance peak, there appears a broad feature at very low magnetic fields. This broad absorption is attributed to a continuum of high k surface magnetoelastic waves which are excited in the nonuniform field near the edges in the y direction. This interpretation was confirmed by an accidental observation. The sample had been broken in the middle and after putting the two pieces together again, the broad feature had disappeared while the sharp resonance peaks could be reproduced. This result supports the former assumptions that the sharp resonances must arise from long wavelength dipolar coupled excitations, whereas the excitations responsible for the broad feature must be coupled by short wavelength exchange or elastic interactions. The confinement of the surface modes to the middle of the slab most probably results from the nonuniformity of the internal field [12.75].

12.5 Photothermally Modulated Ferromagnetic Resonance

A major drawback of photoacoustic detection (Sect. 12.3) or of photothermal beam deflection sensing (Sect. 12.4) is their low sensitivity as compared to the conventional diode detection of microwaves, see Table 12.1. Photothermally modulated ferromagnetic resonance (PM-FMR) combines the spatial resolution inherent to thermal wave techniques with the high sensitivity and large bandwidth of direct microwave detection [12.17,76,77]. PM-FMR relies on the temperature dependence of the resonance absorption. The microwave absorption of a sample in the cavity is thermally modulated by an intensity modulated light beam which hits the sample. The heat penetration depth can be varied by changing the modulation frequency and local variation can be achieved by scanning the beam focus over the sample. The periodic heating of the sample modulates the microwave power reflected at the cavity that can be detected conventionally. Altogether, PM-FMR provides a real three-dimensional method for the spatial evaluation of magnetic properties. The first PM-FMR measurements were performed on the magnetic layer of a CrO_2 recording tape [12.17]. In this section, we also report on very recent studies of the uniform mode in amorphous metallic bands [12.76,77] and on spin waves in metallic films [12.77,78].

12.5.1 Experimental Arrangement

The experimental arrangement for a photothermally modulated FMR measurement is very similar to that of an optically excited EPR study [12.79]. The apparatus used in the experiments discribed in [12.17,76-78] was basically a Varian E-Line EPR spectrometer equipped with an X-band microwave bridge. Modifications were made to the cavity and the detection unit. Figure 12.33 shows the block diagram of the

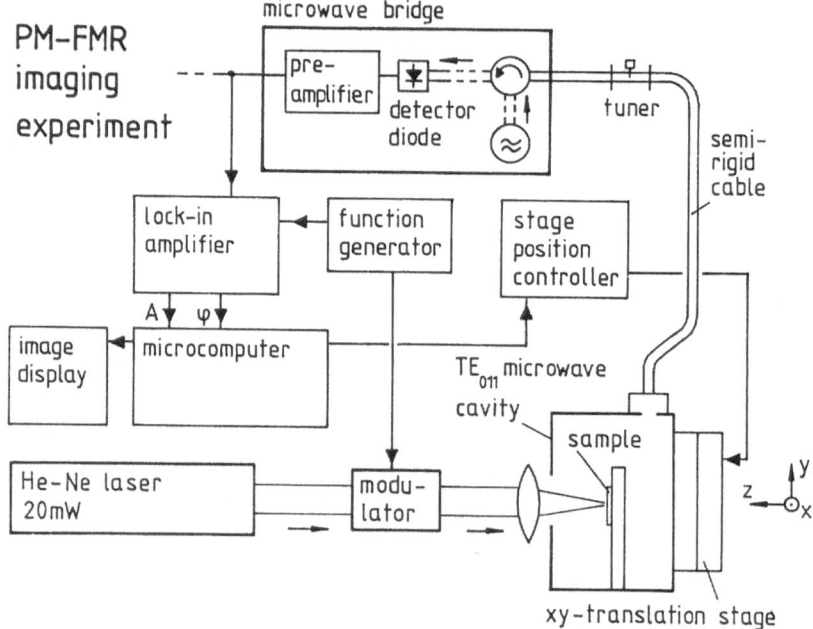

Fig. 12.33. Block diagram of the experimental setup used for photothermally modulated FMR imaging. The system fits into the EPR spectrometer of Fig. 12.20

modified section of the spectrometer. The sample, which is mounted in a wall-less microwave cavity, is illuminated with a focused laser beam from a 20 mW He-Ne laser. The intensity of the beam is modulated by an acousto-optical modulator. To measure the temperature derivatives of the FMR or EPR only, the laser and modulator could be replaced by a halogen lamp and a mechanical chopper. The cavity was mounted on a translation stage allowing a lateral variation of the laser spot on the sample with a resolution of 2.5 μm. In order to detect the photothermally modulated microwave response, the conventional microwave signal was branched off from the microwave bridge after the detector diode and the preamplifier. The ac component of the microwave signal at the frequency of the laser beam modulation was detected synchronously by a lock-in amplifier, which received the reference signal from the function generator of the acousto-optical modulator. The photothermally modulated microwave absorption was usually detected in the frequency range 10 Hz to 100 kHz. A personal computer was used to control the translational stage and to perform the acquisition of the signals available at the output of the lock-in amplifier. The microwave cavity was a wall-less cylindrical TE_{011} resonator (Fig. 12.34) which had been constructed following the design of *Chamel* et al. [12.80]. The body of this cavity is composed of five equally spaced concentric annular plates which are connected to each other by Teflon bolts. A tuning screw above the coupling hole is used to match the resonator to the waveguide. Residual microwave radiation losses are reduced by shielding parts of the cavity with an aluminum foil. The cavity has a loaded quality factor of about 3100 and a resonance frequency of 9.2 GHz.

12.5.2 Signal Generation Process

The photothermally modulated response is proportional to the local temperature derivative of the microwave resonance absorption signal. The electrical signal detected by the diode is proportional to the square root of the microwave power reflected at the impedance matched cavity, ΔP, with [12.49]

$$\sqrt{\frac{\Delta P}{P_0}} = \frac{\eta Q_u \chi''(T)}{2} ,$$ (12.5.1)

where $\chi''(T)$ is the dissipative part of the high frequency susceptibility of the sample, and η and Q_u are the filling factor and the unloaded Q-factor of the cavity, respectively. The susceptibility oscillates about its steady-state value due to periodic heating by the absorbed light beam:

$$\chi''(T) = \chi''(T_0) + \left(\frac{\partial \chi''}{\partial T}\right)_{T_0} \vartheta \exp(i\omega_M t + \varphi) ,$$ (12.5.2)

where ϑ is the amplitude of the ac temperature at the modulation frequency ω_M, averaged over the characteristic volume, e.g. determined by the thermal diffusion length (see below). For simplicity, we assume an instantaneous heat production after the photon has been absorbed and also a relaxation rate of the energy in the spin system or from the spin system to the lattice that is much larger than the modulation frequency. The temperature derivative of the imaginary part of the susceptibility has several contributions, which have to be considered individually:

$$\chi''(T) = \chi''(M_s(T), K(T), \lambda(T), \tau(T), \ldots) ,$$ (12.5.3)

with the spontaneous magnetization $M_s(T)$, the anisotropy constant K, the magnetostriction λ and the relaxation time τ. The first PM-FMR investigations were

353

performed on the commercially available magnetic tapes of CrO_2 and γ-Fe_2O_3 and on tapes supporting a double layer of chromium and iron oxide (ferrochrome). These are the same tapes investigated with PA-FMR (Sect. 12.3.2). They are composed of a 13 μm polyester backing and a few micrometer thick magnetic layer consisting of CrO_2 or γ-Fe_2O_3 needle-shaped particles which are immersed in a polyester matrix (Fig. 12.14). The needles are aligned in the recording direction. For the measurements, pieces of about $4 \times 4\,mm^2$ were glued on a glass support and then mounted in the open cavity (Fig. 12.34). The magnetic field was always directed parallel to the sample surface, and either parallel or perpendicular to the recording direction of the tape.

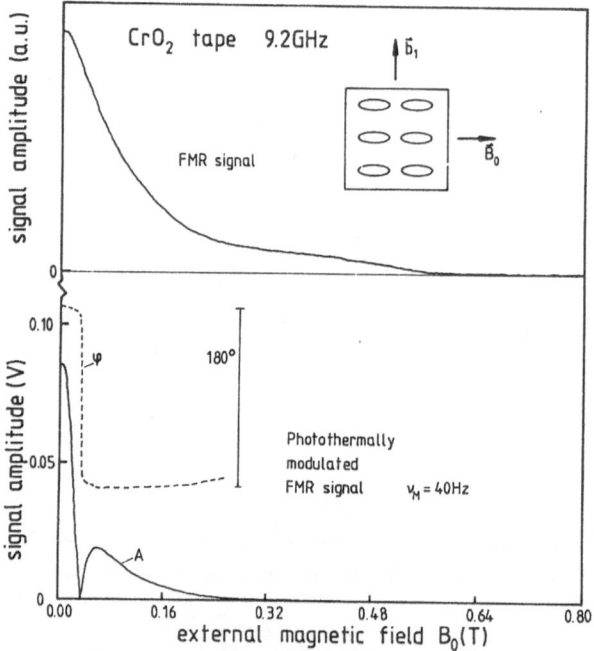

Fig. 12.35. FMR absorption signal (*upper trace*) and the amplitude A and the phase angle φ of the photothermally modulated FMR signal (*lower traces*) of a CrO_2 recording tape. The external magnetic field was aligned parallel to the recording direction

Figure 12.35 shows the FMR signal (upper part) and the photothermally modulated FMR signal from a CrO_2 tape with the external magnetic field lying in the recording direction [12.17]. Photothermal modulation was performed with a halogen lamp as the light source and a chopper. The modulation frequency was 40 Hz. Two features of the PM-FMR signal are worth noting. There exists a finite signal amplitude at zero external field and the temperature derivative changes its sign at about 0.09 T. This reversal of sign is reproduced by the phase-sensitive lock-in detection as a phase change of 180°, while simultaneously the signal amplitude becomes zero. The PM-FMR spectrum of the Fe_2O_3 tape qualitatively displays the same magnetic field dependence (Fig. 12.36).

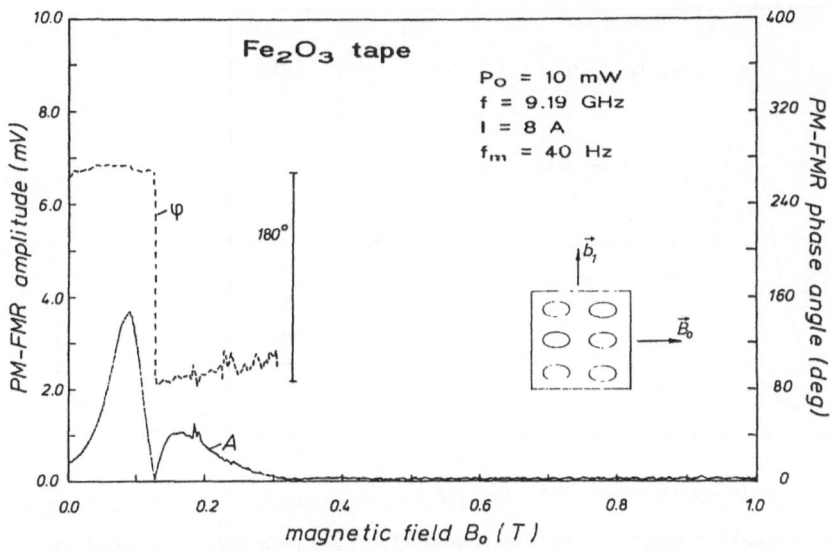

Fig. 12.36. Magnetic field dependence of the PM-FMR amplitude and phase angle measured for a Fe_2O_3 type at 40 Hz. The external field was aligned parallel to the recording direction of the tape

For both tapes, the PM-FMR amplitude, which is proportional to the temperature derivative of the susceptibility (12.5.1) is dominated by the behavior of the spontaneous magnetization $M_s(T)$. There are two competing effects of $M_s(T)$ on the FMR line. First, the intensity of the microwave absorption is proportional to M_s, which implies a negative value of the temperature derivative of χ'' in the whole magnetic field range. Second, the line position of the resonance line is a function of the dynamic demagnetization fields in the sample which, in the case where the external field is parallel to the surface, are proportional to M_s. Therefore, with an increase in temperature according to (12.2.18), the resonance shifts to higher fields, which produces a negative $\partial\chi''/\partial T$ in the low field wing of the resonance line and a positive one in the high field wing (Fig. 12.4). Due to the latter effect, the PM-FMR signal of the chrome and iron oxide tapes change sign with an increasing magnetic field. This interpretation is supported by a theoretical investigation of the FMR resonance line shapes, including the effects resulting from the temperature dependence of the spontaneous magnetization [12.76,81]. The magnetic needles are approximated by ellipsoidal shaped single domain particles with their long axis aligned in the recording direction. The magnetic field dependence of the resonance absorption of a single particle is treated within the scope of a simple isolated particle model and calculated with the procedure described in Sect. 12.2.1 and making use of the measured $M_s(T)$ behavior of γ-Fe_2O_3 [12.82] and of CrO_2 [12.83]. Figure 12.37 shows the resonance curves of γ-Fe_2O_3 for two temperatures obtained with this procedure [12.76]. The calculations have been carried out assuming a ratio of the long axis to the small axis of the ellipsoid of 7.

The two temperature effects mentioned above become clearly visible in Fig. 12.37. With an increase in temperature, the line is shifted to higher fields and simultaneously the intensity is diminished. The temperature derivative deduced from these calcu-

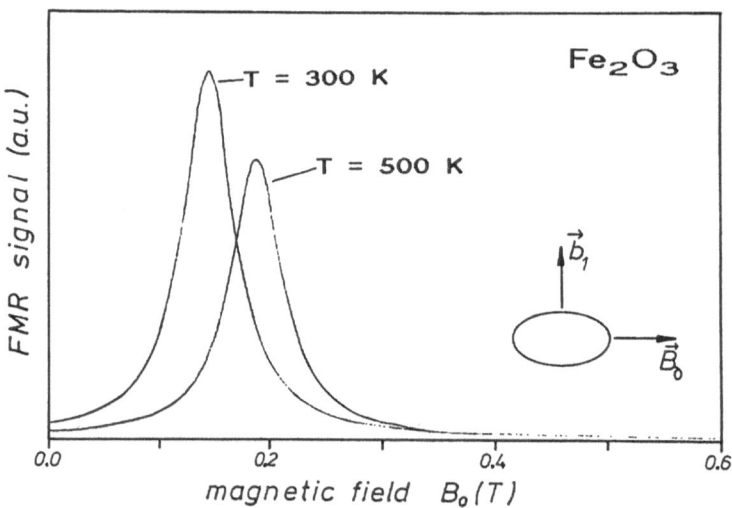

Fig. 12.37. Theoretical FMR absorption line of an ellipsioidal γ-Fe$_2$O$_3$ particle at two temperatures

lations reproduces all the observed features. The quantitative numbers are slightly different between experiment and theory, which is due to the fact that contributions other than that from M_s in (12.5.2,3), particularly the influence of the anisotropy energy and the particle interaction in the tape, have been neglected.

The photothermally modulated FMR signal represents the temperature derivative of the FMR signal which originates from that region of the sample which is reached by the thermal wave. In the case of a short optical absorption length or surface absorption, the signal depth corresponds to the thermal diffusion length μ, which is a function of the modulation frequency. The oscillating part of the susceptibility has to be integrated over a layer of thickness $2\pi\mu$. As the dc temperature varies only slightly across this layer, the PM-FMR signal will become proportional to the spatially averaged ac temperature amplitude, which is itself proportional to the surface temperature (12.5.2). For the simple situation where the light beam is absorbed at the surface and sample is thermally thick, the PM-FMR signal should decrease with the inverse power of the modulation frequency ω_M:

$$\text{PM-FMR} \propto \bar{\vartheta} 2\pi\mu \propto \omega_M^{-1} . \tag{12.5.4}$$

In order to verify the thermal wave nature of the PM-FMR, the variation of the amplitude and phase angle of the signals from the magnetic tapes were investigated [12.76]. Figure 12.38 shows experimental and theoretical results obtained from the CrO$_2$ tape illuminated from the top. The PM-FMR signal was measured without an external field, where the PM-FMR amplitude has its highest value. The theoretical curves were calculated assuming a one-dimensional heat flow according to the Rosencwaig-Gersho model [12.52]. The diagrams give a satisfying confirmation of the agreement between the experiment and the theory for both the amplitude and the phase angle.

As the reported frequency-dependent measurements were conducted with a halogen lamp as the light source, the upper limit of the modulation frequency of about

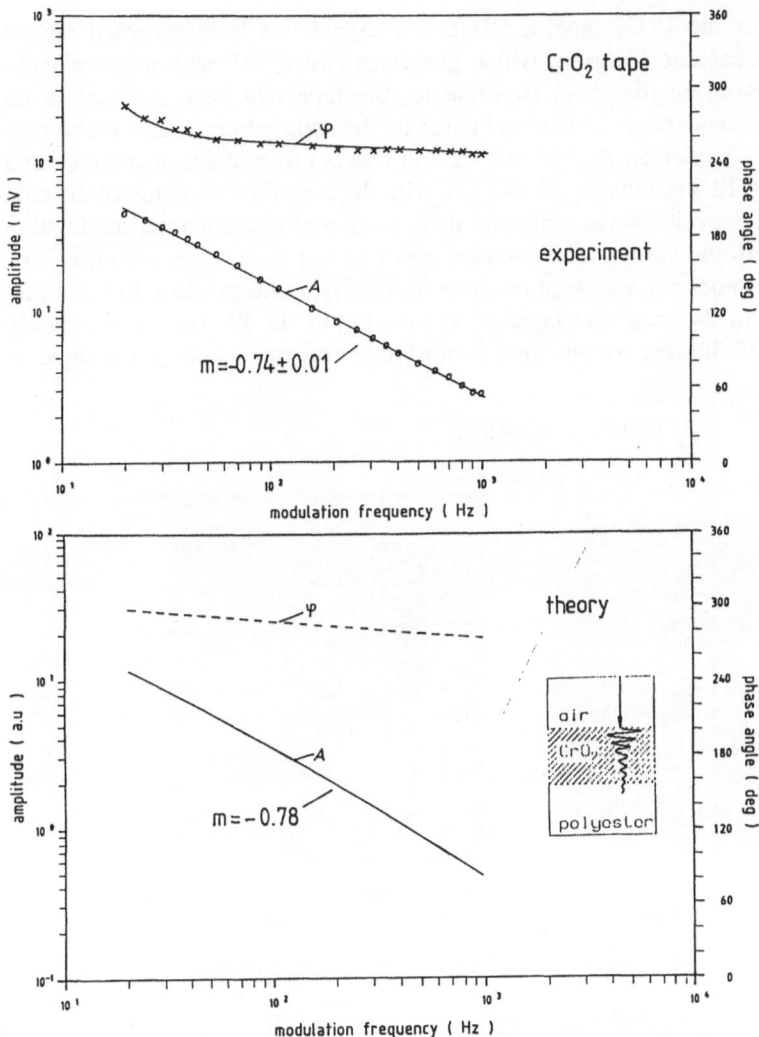

Fig. 12.38. Phase angle and amplitude of the PM-FMR signal from the CrO_2 tape as a function of the modulation frequency. Experimental results (*top*) and theoretical predictions (*bottom*)

1 kHz was set by the mechanical chopper. In more recent experiments, which used a laser in conjunction with an acousto-optical modulator, the PM-FMR signal could be easily observed up to 100 kHz [12.77,84].

12.5.3 PM-FMR Imaging of Magnetic Tapes and Foils

Lateral resolved PM-FMR images can be obtained by recording the PM-FMR signal at constant external field and at constant frequency as a function of the light beam position on the sample. In this way, PM-FMR images were obtained from a CrO_2 band and from metglass ribbons [12.17,76,77].

In the case of the CrO_2 tape, a 50 Hz sine signal had been recorded on the tape by a stereo cassette recorder, which generates two tracks with a periodically changing magnetization direction. Beforehand, the tape had been exposed to an intense external magnetic field perpendicular to the tape to erase any previously recorded signals completely. A quadratic section was cut from the tape and mounted in the cavity (Fig. 12.34), outside the magnet, with the recording direction of the tape perpendicular to the microwave magnetic field, b_1. Local photothermal modulation was achieved with the He-Ne laser focused down to a spot of about 90 μm radius. The modulation frequency was kept constant at 415 Hz, corresponding to a thermal diffusion length in the magnetic layer of 11 μm. Figure 12.39 shows the laterally resolved PM-FMR images for the signal amplitude and the signal phase angle in

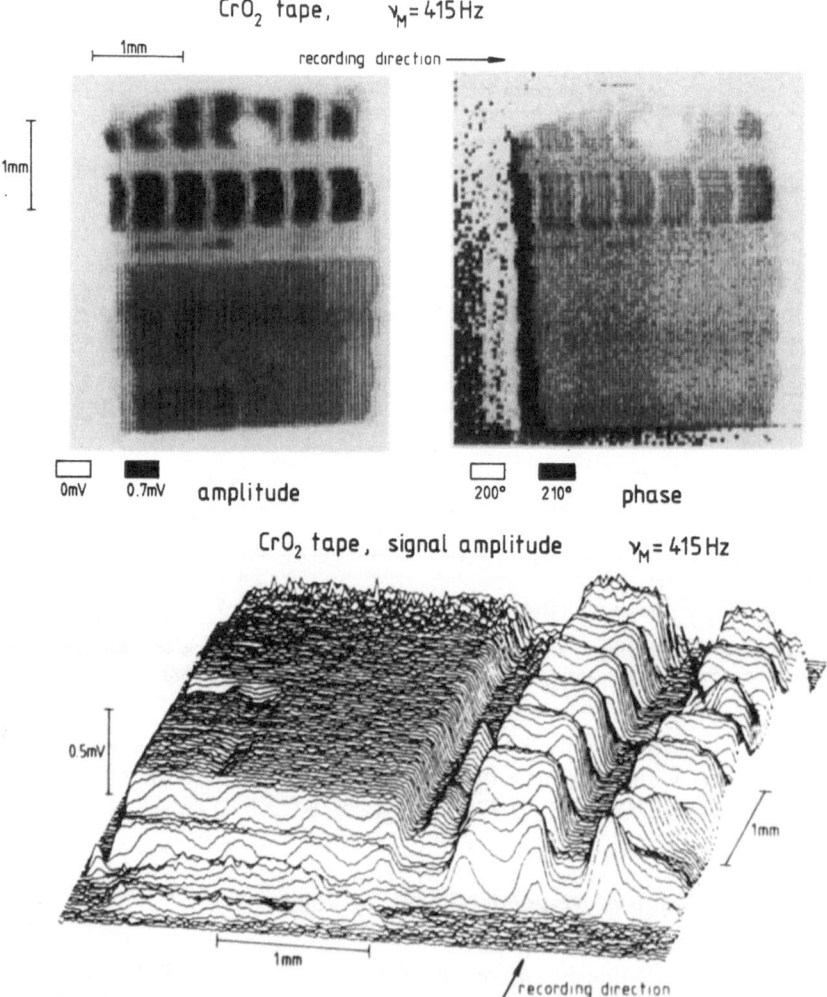

Fig. 12.39. Amplitude (*upper left part*) and phase (*upper right part*) images and perspective view of the amplitude image (*bottom*) of the photothermally modulated FMR signal from a CrO_2 recording tape where a 50 Hz signal had been recorded on two tracks by a stereo recorder. (After [12.17])

the grey density representation and the amplitude a second time in perspective, at the bottom of the same figure. The magnetic patterns produced by the erasing and recording head are most distinctly reproduced in the amplitude image. The signal height is proportional to the effective local magnetization, which has a preferential orientation in the recording direction. The main features of the amplitude image seem to be governed by the orientation of the local magnetization. The phase angle is more sensitive to the thermal properties [12.85]. As the bright spot in the upper section of the images shows up more strongly in the phase image, one would ascribe it to a thermal defect. On subsequent optical inspection of the sample, this defect was identified as an air bubble in the glue at the interface of the tape and the glass support.

Whereas in the magnetic tape the spatial variation of the magnetization direction was induced artificially, in the metglass ribbons of $Fe_{40}Ni_{40}B_{20}$, lateral magnetic inhomogeneities already exist as a characteristic property of the material (Sect. 12.4.4). Photothermal beam deflection provided a first insight into the lateral inhomogeneities but suffered primarily from a low spatial resolution because of the averaging in the direction of the grazing incidence of the probe laser beam and of the limitations imposed on the modulation frequency. The photothermally modulated FMR technique can profit from a focused laser beam and, due to electronic detection, the upper limit of the modulation frequencies can be shifted to higher values. This considerable increase of the spatial resolution enables one to investigate the PM-FMR within the domain structure. In the amorphous ribbons, one can distinguish between two types of magnetic domains [12.61,64]. First, large wave-like domains ranging in size from about 25 μm to 150 μm, which disappear in a small magnetic field applied parallel to the long edge of the ribbon. Second, narrow channel domains of size 3 μm to 5 μm. In these domains, the magnetization near the surface lies in the plane, whereas in the bulk the magnetization is directed perpendicular to the surface.

For the PM-FMR investigations, a small strip was cut from the band and a thin graphite film sprayed onto it in order to enhance the optical absorption. In these experiments, the laser was focused down to a spot of about 19 μm diameter and the PM-FMR was recorded in an external magnetic field. The modulation frequency was set to 1 kHz, corresponding to a thermal diffusion length of about 25 μm, which has to be compared with the skin depth of about 5.8 μm at a microwave frequency of 9.2 GHz. Figure 12.40 shows the amplitude (left hand side) and the phase image of the PM-FMR, recorded from the front side of the FeNiB metglass strip at three magnetic field settings. The scanning comprises 115 steps in the horizontal direction and 109 steps in the vertical direction. Neighboring points are separated by 10 μm. In order to eliminate contributions which may result from thermal or optical inhomogeneities, the pure photoacoustic amplitude and phase images [12.84] have also been recorded. These measurements did not reveal any noticeable structure across the sample. Therefore, the pattern observed in Fig. 12.40 should be of magnetic origin. At very low magnetic field values, the phase image shows some structure which disappears as the magnetic field is increased from 6 mT to 12 mT. The most remarkable observation is the appearance of a strip-like structure at the resonance maximum at 89 mT.

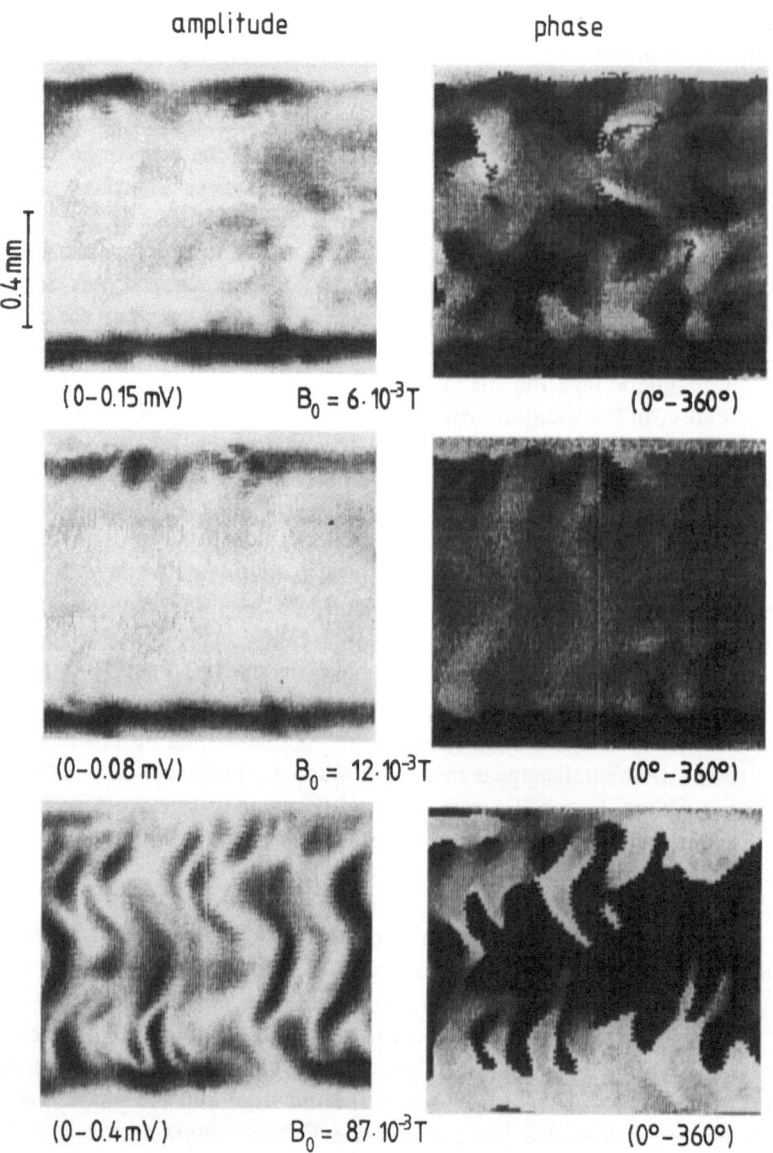

amplitude   phase

0.4 mm

(0 – 0.15 mV) $B_0 = 6 \cdot 10^{-3}$ T   (0° – 360°)

(0 – 0.08 mV) $B_0 = 12 \cdot 10^{-3}$ T   (0° – 360°)

(0 – 0.4 mV) $B_0 = 87 \cdot 10^{-3}$ T   (0° – 360°)

Fig. 12.40. Amplitude and phase PM-FMR images from a $Fe_{40}Ni_{40}B_{20}$ metglass band at three different magnetic fields recorded at a modulation frequency of 1 kHz. The size of the strip is about $1 \times 1\,mm^2$ and the lateral scanning resolution 10 μm

Even at the lowest values of 6 mT, the external fields are too large to allow the observation to be explained in terms of the domain types discussed above. At the present stage, any interpretations of these results are still hypothetical. As the structure in Fig. 12.40 is observed at the rear as well as at the front surface of the metglass bands, one may rule out the different surface properties which are a consequence of the preparation process. During the spinning process, the tarnished rear surface was in contact with the wheel, whereas the glossy front surface was

360

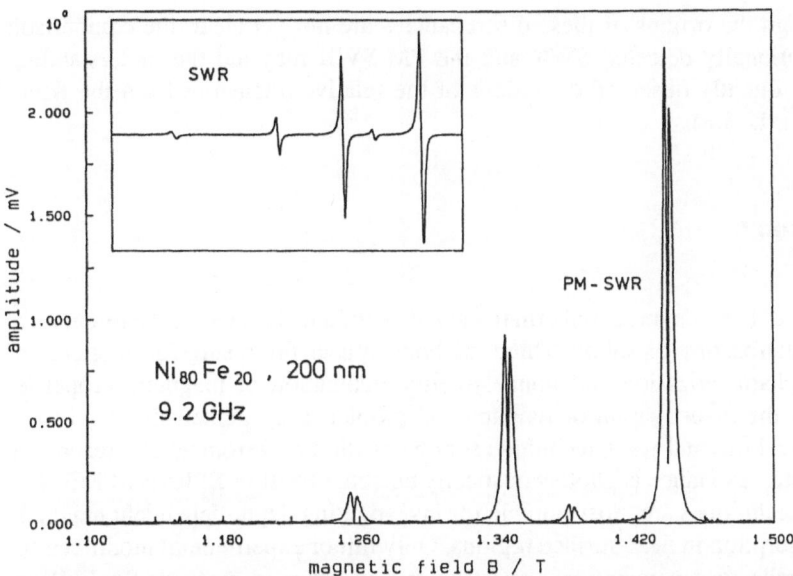

Fig. 12.41 Photothermally modulated spin wave resonance from a $Ni_{80}fe_{20}$ Permalloy film of about 200 nm thickness. The inset shows the conventionally recorded spin wave spectrum

turned to the air. The most probable origins of the strip-like structure seen by PM-FMR are strain-induced local anisotropies which cause the local magnetization at different positions of the sample to respond with a different orientational change as the thermal perturbation is applied.

12.5.4 Photothermally Modulated Spin Wave Resonance

Very recently, the photothermal modulation technique was also applied to the microwave resonance of spin waves in metallic films [12.77,86]. Figure 12.41 shows photothermally modulated spin wave resonance (PM-SWR) from a 200 nm thick Permalloy on glass, which was measured in the perpendicular geometry with the magnetic field and the laser beam perpendicular to the surface. The inset shows the conventionally recorded spin wave spectrum of the same sample. The PM-SWR lines have roughly the same type of shapes as the PM-FMR signals of chromium and iron in the magnetic recording tapes (Sect. 12.5.2). At about the center of the line, the temperature derivative changes its sign. At this field strength, the amplitude becomes zero while the phase angle (not shown in the figure) changes by 180°. Apart from this effect, the phase corrected PM-SWR signals are likely to be the same as the spin wave lines conventionally detected by field modulation. This implies that the main contribution to the PM-SWR comes from the thermal modulation of the spin wave position due to the temperature dependence of the spontaneous magnetization. Therefore, PM-SWR may be considered to be a resonance technique with an internal magnetization field modulation. Although the line positions are the same for the external and internal modulated resonance absorption processes, the two techniques yield different results with regard to the relative intensities of the spin wave peaks. Although the origins of these discrepancies are not yet clear, the comparison of the conventionally detected SWR and the PM-SWR may aid the understanding of the very frequently observed deviations of the relative intensities from the Kittel behavior (Sect. 12.3.3).

12.6 Summary

This chapter has been devoted to thermal wave detection techniques in the microwave resonance spectroscopy of solids. Three methods appear promising for applications in materials characterization and nondestructive evaluation of magnetic properties as well as for the investigation of fundamental problems of magnetism.

The classical thermal wave technique for observation of ferromagnetic resonance or paramagnetic resonance is photoacoustically detected FMR or EPR (PA-FMR, PA-EPR). These techniques are most suitable for investigating depth-dependent magnetic resonance absorption in near-surface regions. Only minor experimental modifications need to be made to a conventional spectrometer in order to measure PA-FMR or PA-EPR in addition to the conventional signal. Such a hybrid detection system may be used to discriminate between surface absorption, which is the most prominant contribution to the PA detected signal, and the bulk absorption, as these two cannot be differentiated by means of the conventional detection method.

The second technique, the photothermal deflection sensing of FMR (PD-FMR) provides a means for a laterally resolved resonance spectroscopy. Much more sophisticated technical equipment is required than for the PA-FMR, which may be a handicap for the industrial application of PD-FMR. Both photoacoustic detection and sensing with the laser beam deflection rely on the measurement of the heat released during the resonance absorption. Therefore PA-FMR and PD-FMR constitute pure absorption spectroscopies in the microwave region. On the other hand, monitoring the temperature changes in the gas layer that is in contact with the sample surface suffers from great signal losses, particular when the heat is generated in the inner part of the sample. Photothermally modulated FMR (PM-FMR) seems to overcome this problem as it combines the sensitivity of the conventional microwave detection with the spatial resolution inherent to thermal wave techniques. In addition, PM-FMR does not require any gas or solid contact with a detector. Thus PM-FMR can be used to investigate the spatial variation of FMR under extreme conditions, such as high or low temperatures or under pressure. Although the first results obtained by photothermally modulated FMR are certainly encouraging, more detailed experimental and theoretical investigations of the signal generation process are needed.

Finally, one has to compare the photothermal detection techniques of FMR with other experimental methods which can provide information on local magnetic properties. The Bitter technique [12.87] is a very common experimental procedure for determining the magnetic field distribution on the surface of opaque, magnetically ordered samples by optical inspection of the orientation of small iron-oxide particles dispersed on the specimen. The magneto-optical Kerr effect [12.88] relies on the rotation of the polarization direction of light by the local magnetization of the material. As the light beam has to traverse the sample or the reflected beam has to be analyzed, the sample has to be of good optical quality.

Spatial resolutions superior to those of thermal wave techniques are achieved with microscopes. Lorentz microscopy [12.89], based on the deviation of an electron beam in the magnetic field of a magnetized region, attains a resolution of 0.1 μm. Resolution in the range of a few nanometers is now available with the development of the Atomic Force Microscope [12.90] or the Scanning Electron Microscope with Polarization Analysis [12.91]. But the sensing depths of these microscopic techniques are restricted to the outermost atomic layers.

Among these different experimental methods, photothermal detection may be considered as a complementary technique for depth resolutions in the micrometer range and for materials which have poor surface and bulk properties and so cannot be investigated by the other techniques. But there is also a fundamental difference. Whereas the previously mentioned techniques are sensitive only to the static magnetization, the photothermally based detection of FMR (PA-FMR, PD-FMR and PM-FMR) constitutes spectroscopic methods which provide information about both the static magnetization and dynamical processes.

Acknowledgement: The authors are indebted to S.M. Rezende, Th. Orth, O. von Geisau and R. Kordecki for providing unpublished data and for stimulating discussions. Parts of this research work have been supported by the Bundesministerium für Forschung und Technologie (Project No. 13N379/1) and by the Deutsche Forschungsgemeinschaft (SFB 166). Their help is gratefully acknowledged.

References

12.1 A. Abragam, B. Bleaney: *Electron Paramagnetic Resonance of Transition Ions* (Clarendon, Oxford 1970)

12.2 A.G. Gurevich: *Ferrites at Microwave Frequencies* (Heywood, London 1963)

12.3 J. Griffiths: Nature **158**, 670 (1946)

12.4 B. Lax, K.J. Button: *Microwave Ferrites and Ferrimagnets* (McGraw-Hill, New York 1962)

12.5 J. Helszajn: *YIG Resonators and Filters* (Wiley, Chichester 1985)

12.6 Z. Frait, D. Fraitova: "Spin Wave Resonance in Metals", in *Spin Waves and Magnetic Excitations* 2, ed. by A.S. Borovik-Romanov, S.K. Sinha. Modern Problems in Condensed Matter Sciences, Vol. 22. (North-Holland, Amsterdam 1988) pp. 1–70

12.7 C. Gorter: Physica **3**, 995 (1936)

12.8 J. Schmidt, I. Solomon: J. Appl. Phys. **37**, 3719 (1967)

12.9 J. Rudd, K. Myrtle, J. Cochran, B.J. Heinrich: J. Appl. Phys. **57**, 3693 (1985)

12.10 J.W. Orton: *Electron Paramagnetic Resonance* (Iliffe Books, London 1968)

12.11 O. Cleves Nunes, A. Nonteiro, K. Skeff Neto: Appl. Phys. Lett. **35**, 656 (1979)

12.12 C. Evora, R. Landers, H. Vargas: Appl. Phys. Lett. **36**, 864 (1980)

12.13 R.L. Melcher: Appl. Phys. Lett. **37**, 895 (1980)

12.14 U. Netzelmann, J. Pelzl: Appl. Phys. Lett. **44**, 854 (1984)

12.15 A. Boccara, D. Fournier, J. Badoz: Appl. Phys. Lett. **36**, 136 (1980)

12.16 U. Netzelmann, U. Krebs, J. Pelzl: Appl. Phys. Lett. **44**, 1161 (1984)

12.17 Th. Orth, U. Netzelmann, J. Pelzl: Appl. Phys. Lett. **53**, 1979 (1988)

12.18 A. Vasson, A.M. Vasson: J. Phys. D **14**, L39 (1981)

12.19 J. Pelzl, K. Klein, O. Nordhaus: Appl. Opt. **21**, 94 (1982)

12.20 R. Melcher, G. Arbach: Appl. Phys. Lett. **40**, 910 (1982)

12.21 R.C. DuVarney, A.K. Garrison, G. Busse: Appl. Phys. Lett. **38**, 675 (1981)

12.22 U. Netzelmann, H. Lerchner, J. Pelzl, M.W. Sigrist: J. de Phys. **44**, C6–221 (1983)

12.23 M. Maltempo, S. Eaton, G. Eaton: J. Magn. Reson. **77**, 75 (1988)

12.24 W. Wettling, W. Jantz: J. Magn. Magn. Mater. **45**, 364 (1984)

12.25 S. V. Vonsovskii (ed.): *Ferromagnetic Resonance* (Pergamon, Oxford 1966)

12.26 C. Kittel: Phys. Rev. **73**, 155 (1948)

12.27 L. R. Walker: Phys. Rev. **105**, 390 (1957)

12.28 J. F. Dillon: J. Appl. Phys. **112**, 59 (1958)

12.29 R. Damon, J. Eshbach: J. Phys. Chem. Solids **19**, 308 (1961)

12.30 L. K. Brundle, N. J. Freedman: Electron. Lett. **4**, 132 (1968)

12.31 J. C. Sethares: J. Appl. Phys. **53**, 2646 (1982)

12.32 J.-P. Castera: J. Appl. Phys. **55**, 2506 (1984)

12.33 H. Suhl: Proc. Inst. Radio Eng. **44**, 1270 (1956)

12.34 M. H. Seavey, P. E. Tannenwald: Phys. Rev. Lett. **1**, 168 (1958)

12.35 A. Tam: Rev. Mod. Phys. **58**, 381 (1986)

12.36 P. Hess, J. Pelzl (eds.): *Photoacoustic and Photothermal Phenomena*, Springer Ser. Opt. Sci., Vol. 58 (Springer, Berlin, Heidelberg 1988)

12.37 U. Netzelmann, E. v. Goldammer, J. Pelzl, H. Vargas: Appl. Opt. **21**, 32 (1982)

12.38 M. Davies, M. Heath: J. Magn. Magn. Mater. **31–34**, 661 (1983)

12.39 W. Wettling, W. Jantz, L. Engelhardt: Appl. Phys. A **26**, 19 (1981)

12.40 H. Vargas: In *Photoacoustic Effect: Principles and Applications*, ed. by E. Lüscher, P. Korpiun, H. Coufal, R. Tilgner (Vieweg, Braunschweig 1984) p. 347

12.41 U. Netzelmann: Ph. D. Thesis, Ruhr Universität Bochum (1986)

12.42 U. Netzelmann, J. Pelzl, D. Fournier, A. C. Boccara: Can. J. Phys. **64**, 1307 (1985)

12.43 O. v. Geisau, U. Netzelmann, J. Pelzl: J. Appl. Phys. **63**, 3347 (1988)

12.44 O. Nordhaus, J. Pelzl: Appl. Phys. **25**, 221 (1981)

12.45 M. Davies, M. Heath: J. Phys. D **18**, 1655 (1985)

12.46 C. Cesar, H. Vargas, U. Netzelmann, J. Pelzl: J. Magn. Magn. Mater. **54–57**, 1185 (1986)

12.47 C. Cesar: Ph. D. Thesis, Universidade Estadual de Campinas (SP), Brazil (1985)

12.48 B. K. Bein, J. Pelzl: J. de Phys. **44**, C6–27 (1983)

12.49 C. Poole: *Electron Spin Resonance* (Interscience, New York 1967)

12.50 L. Kreuzer: In *Optoacoustic Spectroscopy and Detection*, ed. by Y.-H. Pao (Academic, New York 1977)

12.51 E. Köster, T. C. Arnoldussen: "Recording Materials" in *Magnetic Recording I*, ed. by D. Mee, E. D. Daniel (McGraw-Hill, New York 1987)

12.52 A. Rosencwaig, A. Gersho: J. Appl. Phys. **47**, 64 (1976)

12.53 C. L. Cesar, H. Vargas, J. Pelzl, L. C. M. Miranda: J. Appl. Phys. **55**, 3460 (1984)

12.54 U. Netzelmann, J. Pelzl, H. Vargas, C. L. Cesar, L. C. M. Miranda: IEEE Trans. MAG-20, 1252 (1984)

12.55 W. D. Jackson, N. M. Amer, A. C. Boccara, D. Fournier: Appl. Opt. **20**, 1333 (1981)

12.56 J. C. Murphy, L. C. Aamodt: J. Appl. Phys. **51**, 4580 (1980); ibid. **52**, 4903 (1981)

12.57 D. Fournier, A. C. Boccara: in *Photoacoustic Effect: Principles and Applications*, ed. by E. Lüscher, P. Korpiun, H. Coufal, R. Tilgner (Vieweg, Braunschweig 1984) p. 80

12.58 E. Legal Lasalle, F. Lepoutre, J. P. Roger: J. Appl. Phys. **64**, 1 (1988)

12.59 D. L. Connelly, J. S. Loomis, D. E. Mapother: Phys. Rev. B **3**, 924 (1971)

12.60 U. Krebs: Diplomarbeit, Fakultät für Physik und Astronomie, Ruhr Universität Bochum (1986)

12.61 H. Kronmüller, R. Schäfer, G. Schröder: J. Magn. Magn. Mater. **6**, 61 (1977)

12.62 C. L. Cesar, H. Vargas, U. Netzelmann, J. Pelzl: J. Magn. Magn. Mater. **54–57**, 1185 (1986)

12.63 F. Luborsky: *Amorphous Ferromagnets*, Technical Information Series (General Electric Company, Schenectady, NY 1978)

12.64 H. Kronmüller: J. de Phys. **41**, C8–618 (1980)

12.65 J. Gonzales, J. Vicent: J. Appl. Phys. **57**, 5400 (1985)

12.66 Z. Frait: J. Magn. Magn. Mater. **35**, 37 (1983)

12.67 R. I. Joseph, E. Schlömann: J. Appl. Phys. **36**, 1579 (1965)

12.68 R. Kordecki, T. Kochmann, G. Sievers, J. Pelzl, G. Dumpich: Verh. Deutsch. Phys. Ges. **4**, 45 (1988)

12.69 R. Kordecki: Ph. D. Thesis, Ruhr Universität Bochum (1989)

12.70 R. F. Soohoo: *Magnetic Films* (Harper and Row, New York 1965)

12.71 F. M. de Aguiar, S. M. Rezende: Phys. Rev. Lett. **56**, 1070 (1986)

12.72 Y. T. Zhang, C. E. Patton, G. Srinivasan: J. Appl. Phys. **63**, 5433 (1988)

12.73 O. von Geisau, U. Netzelmann, S. M. Rezende, J. Pelzl: In *Microwave and Optronics '89*, Conf. Proc. (Network, Hagenburg, FRG 1989)

12.74 J. P. Castera: J. Appl. Phys. **55**, 2506 (1984)

12.75 O. von Geisau: Private communcation

12.76 T. Orth: Diplomarbeit, Fakultät für Physik und Astronomie, Ruhr Universität Bochum (1988)

12.77 J. Pelzl, U. Netzelmann, T. Orth, R. Kordecki: In *Photoacoustic and Photothermal Phenomena VI*, ed. by J.C. Murphy, Springer Ser. Opt. Sci. (Springer, Berlin, Heidelberg) in press

12.78 R. Kordecki, T. Kochmann, H. Volz, J. Pelzl, E. Becker, G. Dumpich, Z. Frait: Physica, in press

12.79 S. Geschwind: "Optical Techniques in EPR in Solids", in *Electron Paramagnetic Resonance*, ed. by S. Geschwind (Plenum, New York 1972) p. 353

12.80 M. Chamel, R. Chicault, Y. Merle d Aubigne: J. Phys. E **9**, 87 (1976)

12.81 U. Netzelmann: J. Appl. Phys., in press

12.82 W. Fuller Brown, C.E. Johnson: J. Appl. Phys. **33**, 2752 (1962)

12.83 Ch. Guillaud: C. R. Acad. Sci. **223**, 1110 (1946)

12.84 Th. Orth: Private communication

12.85 G. Busse, A. Rosencwaig: Appl. Phys. Lett. **36**, 815 (1980)

12.86 R. Kordecki: Private communication

12.87 F. Bitter: Phys. Rev. **38**, 1903 (1931)

12.88 J. Kranz, A. Huber: Z. Angew. Phys. **15**, 220 (1963)

12.89 L. Reimer: *Transmission Electron Microscopy*, Springer Ser. Opt. Sci., Vol. 36 (Springer, Berlin, Heidelberg 1984)

12.90 Y. Martin, D. Rugar, H.K. Wickramasinghe: Appl. Phys. Lett. **52**, 244 (1988)

12.91 R.J. Celotta, D.T. Pierce: Science **234**, 333 (1986)

Subject Index

Topics in Current Physics

Founded by Helmut K. V. Lotsch